The Classes of Higher Dimensional Polytopes in Chemical, Physical, and Biological Systems

Gennadiy Vladimirovich Zhizhin
Russian Academy of Natural Sciences, Russia

IGI Global
PUBLISHER of TIMELY KNOWLEDGE

A volume in the Advances in Chemical and Materials Engineering (ACME) Book Series

Published in the United States of America by
IGI Global
Engineering Science Reference (an imprint of IGI Global)
701 E. Chocolate Avenue
Hershey PA, USA 17033
Tel: 717-533-8845
Fax: 717-533-8661
E-mail: cust@igi-global.com
Web site: http://www.igi-global.com

Library of Congress Cataloging-in-Publication Data

Names: Zhizhin, G. V. (Gennadiĭ Vladimirovich), author.
Title: The classes of higher dimensional polytopes in chemical, physical, and biological systems / Gennadiy Zhizhin.
Description: Hershey, PA : Engineering Science Reference (an imprint of IGI Global), [2022] | Includes bibliographical references and index. | Summary: "This monograph explains the meaning of higher dimensions of classical geometry and systematically generalizes the results of geometric research in various fields of knowledge leading to the development of branches of science related to human activities"-- Provided by publisher.
Identifiers: LCCN 2022015788 (print) | LCCN 2022015789 (ebook) | ISBN 9781799883746 (hardcover) | ISBN 9781799883753 (paperback) | ISBN 9781799883760 (ebook)
Subjects: LCSH: Polytopes. | Polygons. | Molecules--Models.
Classification: LCC QA691 .Z44 2022 (print) | LCC QA691 (ebook) | DDC 516/.158--dc23/eng20220517
LC record available at https://lccn.loc.gov/2022015788
LC ebook record available at https://lccn.loc.gov/2022015789

This book is published in the IGI Global book series Advances in Chemical and Materials Engineering (ACME) (ISSN: 2327-5448; eISSN: 2327-5456)

British Cataloguing in Publication Data
A Cataloguing in Publication record for this book is available from the British Library.

Advances in Chemical and Materials Engineering (ACME) Book Series

J. Paulo Davim
University of Aveiro, Portugal

ISSN:2327-5448
EISSN:2327-5456

Mission

The cross disciplinary approach of chemical and materials engineering is rapidly growing as it applies to the study of educational, scientific and industrial research activities by solving complex chemical problems using computational techniques and statistical methods.

The **Advances in Chemical and Materials Engineering (ACME) Book Series** provides research on the recent advances throughout computational and statistical methods of analysis and modeling. This series brings together collaboration between chemists, engineers, statisticians, and computer scientists and offers a wealth of knowledge and useful tools to academics, practitioners, and professionals through high quality publications.

Coverage

- Polymers Engineering
- Metallic Alloys
- Fatigue and Creep
- Wear of Materials
- Electrochemical and Corrosion
- Fracture Mechanics
- Heat Treatments
- Artificial Intelligence Methods
- Ductility and Crack-Resistance
- Computational methods

IGI Global is currently accepting manuscripts for publication within this series. To submit a proposal for a volume in this series, please contact our Acquisition Editors at acquisitions@igi-global.com or visit: https://www.igi-global.com/publish/.

The Advances in Chemical and Materials Engineering (ACME) Book Series (ISSN 2327-5448) is published by IGI Global, 701 E. Chocolate Avenue, Hershey, PA 17033-1240, USA, www.igi-global.com. This series is composed of titles available for purchase individually; each title is edited to be contextually exclusive from any other title within the series. For pricing and ordering information please visit https://www.igi-global.com/book-series/advances-chemical-materials-engineering/73687. Postmaster: Send all address changes to above address.

Titles in this Series

For a list of additional titles in this series, please visit: https://www.igi-global.com/book-series/advances-chemical-materials-engineering/73687

Advances in the Modelling of Thermodynamic Systems
Elhoucine Essefi (Higher Institute of Applied Sciences and Technology of Gabes, Tunisia) and Ibtissem Jendoubi (Faculty of Sciences of Bizerte,Tunisia)
Engineering Science Reference • © 2022 • 315pp • H/C (ISBN: 9781799888017) • US $245.00

Handbook of Research on Tribology in Coatings and Surface Treatment
Amirhossein Pakseresht (University of Trencin, Slovakia) and Omid Sharifahmadian (University of Trencin, Slovakia)
Engineering Science Reference • © 2022 • 430pp • H/C (ISBN: 9781799896838) • US $285.00

Advanced Manufacturing Techniques for Engineering and Engineered Materials
R. Thanigaivelan (Muthayammal Engineering College, India) N. Rajan (Vinayaga Mission Kripananda Variyar Engineering College, India) and T.G. Argul (St. Mary's Engineering College, India)
Engineering Science Reference • © 2022 • 351pp • H/C (ISBN: 9781799895749) • US $245.00

Handbook of Research on Green Synthesis and Applications of Nanomaterials
Rajni Garg (Rayat Bahra University, India) Rishav Garg (Galgotias College of Engineering and Technology, India) and Nnabuk Okon Eddy (Department of Pure and Industrial Chemistry, University of Nigeria, Nsukka, Nigeria)
Engineering Science Reference • © 2022 • 569pp • H/C (ISBN: 9781799889366) • US $295.00

Handbook of Research on Nanoemulsion Applications in Agriculture, Food, Health, and Biomedical Sciences
Karthikeyan Ramalingam (B.S. Abdur Rahman Crescent Institute of Science and Technology, India)
Engineering Science Reference • © 2022 • 671pp • H/C (ISBN: 9781799883784) • US $295.00

Handbook of Research on Advancements in the Processing, Characterization, and Application of Lightweight Materials
Kaushik Kumar (Birla Institute of Technology, India) B. Sridhar Babu (CMR Institute of Technology, India) and J. Paulo Davim (University of Aveiro, Portugal)
Engineering Science Reference • © 2022 • 458pp • H/C (ISBN: 9781799878643) • US $285.00

Emerging Applications and Implementations of Metal-Organic Frameworks
Shimaa Mohamed Elsaeed (Egyptian Petroleum Research Institute, Egypt) Elsayed Zaki (Egyptian Petroleum Research Institute, Egypt) and Abdel-Azim Abdel-Azim (Egyptian Petroleum Research Institute, Egypt)
Engineering Science Reference • © 2021 • 254pp • H/C (ISBN: 9781799847601) • US $225.00

IGI Global
PUBLISHER of TIMELY KNOWLEDGE

701 East Chocolate Avenue, Hershey, PA 17033, USA
Tel: 717-533-8845 x100 • Fax: 717-533-8661
E-Mail: cust@igi-global.com • www.igi-global.com

Table of Contents

Preface vii

Chapter 1
Regular and Semi-Regular Three-Dimensional Polytopes 1

Chapter 2
Regularly Faceted Three-Dimensional Polytopes 20

Chapter 3
Necessary and Sufficient Conditions for Proving the Existence of Polytopes of Higher Dimension ... 64

Chapter 4
Regular and Semi-Regular Polytopes of Higher Dimension 91

Chapter 5
Irregular Polytopes of Higher Dimension 123

Chapter 6
Higher Dimensional Polytopes That Are Products of Lower Dimensional Polytopes 146

Chapter 7
Higher-Dimensional Polytopes With Regular-Faceted Polyhedrons Are Different From Archimedes Bodies 184

Chapter 8
Polytopes Dual to Higher-Dimensional Polytopes 221

Chapter 9
The Polytopes of the Higher Dimension in Physics and Chemistry and Construction of Spaces of the Higher Dimension 254

Chapter 10
Polytopes of the Highest Dimension in Biology 293

Chapter 11
Polytopes of the Highest Dimension in the Theory of Heredity 335

Afterword 363

About the Author 364

Index 365

Preface

In the Euclidean geometry familiar to us, everyone knows the correct three-dimensional polyhedrons: tetrahedron, octahedron, cube, dodecahedron, icosahedron (Vaschenko-Zakharchenko, 1880). These ideal polyhedrons (Plato's bodies) are easily present in our minds, like an idea. But in real nature they are rarely present. Deviations from ideality (correctness) can be different. All deviations not related to metric characteristics fit into one wording. Elements of different dimensions in three-dimensional polyhedrons can have, for a given dimension (in this case 0, 1, 2), different incidence values with respect to elements of a different dimension. Zero-dimensional elements (i.e. vertices) can differ in the same figure by the number of edges emanating from the vertex (i.e., by the incidence with respect to the edges), as well as by the number of two-dimensional faces in which they are vertices (i.e. incidence with respect to two-dimensional faces). One-dimensional elements (i.e. edges) can have different incidence values for the same shape in relation to two-dimensional faces (i.e. belong to a different number of two-dimensional faces). Two-dimensional elements of the same shape can have a different number of vertices and edges. This means that the elements of a given dimension can have different incidence values in relation to both elements of a higher dimension and in relation to elements of a lower dimension. Metric deviations from the correctness of three-dimensional polyhedrons are that in the same figure there can be different lengths of edges, different areas and shapes of flat faces, as well as the values of the angles between edges and flat faces of the same figure can have different values. There is a class of convex regular-faced three-dimensional polyhedrons (Johnson, 1960; Zalgaller, 1966; Zhizhin, 2009), i.e. convex polyhedrons, whose two-dimensional faces have regular polygons, however, in this case, the same polyhedron may have different regular polygons in the set of faces. Obviously, semiregular polyhedrons are a subset of the set of regular polyhedrons. However, there are convex regular three-dimensional polyhedrons that are not semiregular polyhedrons.
With all these differences, any three-dimensional convex polyhedron must satisfy the Euler equation

$$f_0 - f_1 + f_2 = 2.$$

Here f_i is the number of elements of a given dimension *i* $(i = 0, 1, 2)$.

Analysis of the forms of molecules and clusters of chemical compounds in inorganic and organic chemistry, physics, geochemistry, biology, and medicine, carried out in the works of the author (Zhizhin, 2014a, b; 2015; 2016 a, b, c; 2018 a, b; 2019 a, b, c; 2020 a, b, c, c, d, e; Zhizhin, & Diudea, 2016; Zhizhin, Khalai, & Diudea, 2016) showed that most chemical compounds in nature have the highest dimension, i.e. 4D, 5D, etc. Here the difference from ideal geometric shapes is even more varied and significant. In this case, the concept of the incidence of a given element in relation to elements of higher and lower dimensions in a given polytope plays an important role. In the general case, for geometric figures of any dimension n (polytope of dimension n) with convex faces of various dimensions from 0 to n - 1, the Euler - Poincaré equation should hold (Poincaré, 1895)

$$\sum_{i=0}^{i=n-1} (-1)^i f_i(n) = 1 + (-1)^{n-1}, \tag{1}$$

$f_i(n)$ is the number of elements of a dimension i (i=1÷n–1) in polytope of dimension n.

The Euler equation is obviously a special case of the Euler-Poincaré equation.

As shown in the works of the author (Zhizhin, 2013, 2019), the fulfillment of the Euler - Poincaré equation is not the only condition for the existence of a polytope of higher dimension. Another condition is related to the incidence values of faces of different dimensions to other faces. For example, the incidence value of a two-dimensional face in relation to three-dimensional faces should increase linearly with increasing polytope dimension. If this condition is not met, and the Euler-Poincaré equation is satisfied, then the polytope of this dimension does not exist. Since the considered polytopes of the highest dimension are models of molecules, the vertices of the polytopes are the places in which atoms or functional groups of chemical compounds are located. In this case, the incidence of any element of the polytope to another element of the polytope means that some atoms or functional groups belong to, for example, larger formations in the polytope. And this, in turn, can be interpreted as the transfer of information from some atoms or functional groups to larger material formations. Of interest is the question of information flow from elements of one dimension to elements of another dimension in a high-dimensional polytope. In the works of the author (Zhizhin, 2019, 2021), the law of conservation of incidents in polytopes of the highest dimension is established. According to this law, the integral incidence flow from low-dimensional elements to high-dimensional elements is equal to the integral incidence flux from high-dimensional elements to low-dimensional elements. Moreover, the value of this integral flow of incidents increases sharply with an increase in the dimension of the polytope. This turned out to be very important in explaining the patterns of transmission of hereditary information in the processes of interaction of nucleic acids. Since the transfer of hereditary information from one nucleic acid to another nucleic acid (Zhizhin, 2019, 2020) occurs in a high-dimensional polytope (polytope of hereditary information).

The variety of higher-dimensional polytopes discovered during specific studies of chemical compounds turned out to be so significant and the properties of individual representatives of this diversity are so common that it is time to describe and list the types of possible higher-dimensional polytopes based on these properties. It should be emphasized that the classification of polytopes of the highest dimension, to which this monograph is devoted, is based on the analysis of specific geometric forms in nature and is not abstract. It should also be noted that the description of geometric shapes based on the study of symmetry transformation groups, which is so widespread in the study of three-dimensional polytopes (Fedorov, 1885, 1889, 1891; Coxeter,1963), becomes very time consuming when analyzing polytopes of higher dimensions. Such an analysis was carried out in the study of regular polytopes of the highest dimension (Zhizhin, 2020). However, when analyzing various kinds of irregular polytopes of the highest dimension (and this is the main subject of the monograph), it becomes practically impossible due to mathematical difficulties and loss of clarity. Here one should refrain from mechanical transfer of the laws of three-dimensional geometry to the figures of space of the highest dimension. Such mechanical transfer, as shown in the monograph (Zhizhin, 2021), leads to incorrect results.

In connection with what is stated in the monograph, the consideration of classes of polytopes begins with regular (Plato's body) and semiregular three-dimensional polytopes (Chapter 1). Semi-regular polytopes are convex polytopes with faces of regular polygons of two or more types if all polyhedral angles (gonohedrons) have the same composition and order of arrangement of faces. It is proved that all of the listed semiregular polytopes can be obtained from regular convex polytopes using the operations of truncation of vertices and deformations of polyhedral angles to keep the figure closed. It is also proved that from each convex polyhedron from the set of regular and semiregular polytopes one can pass using a finite number of transformations to any other polyhedron from the indicated set of polytopes. This is important to know, since in the processes in the processes of mechanical, thermal and chemical materials, the cells of the structure of materials can change their shape, while maintaining their closedness and convexity. The previously considered abstract transformations of three-dimensional polytopes could not solve the problem (McMullen, & Schulte, 2002; Diudea, & Nagy, 2007; Ashrafi, Cataldo, Iraumanesh, & Ori, 2013; Diudea, 2018).

Further in the monograph, regularly faceted three-dimensional polytopes are considered (Chapter 2), i.e., convex polytopes with faces with regular polygons of two or more types, if among the polyhedral angles (gonohedrons) there can be angles with different composition or order of arrangement of faces.

Semiregular polytopes form a subset of the set of regularly faceted polytopes. The list of regularly faceted polytopes proposed earlier (Johnson, 1960; Zalgaller, 1966) was refined, and the existence of regularly faceted polytopes of a new type was found (Zhizhin, 2009).

The conditions for the existence of high-dimensional convex polytopes (Grünboum, 1967) taking into account the Euler-Poincaré equation and the incidence values of the faces are considered in Chapter 3 (Zhizhin, 2019a).

In the fourth chapter, regular polytopes of higher dimensions are considered, in particular, four-dimensional regular polytopes, the form of which was first proposed in the work of Stringham (Stringham, 1880). It is proved that the notation and formulas introduced by Schläfli (1901) are not a proof of the existence of a polytope of higher dimension but are only a possible algebraic balance of the number of faces of different dimensions. The proof of the existence of a polytope of higher dimension, as follows from Chapter 3 of this monograph, can only be a proof of the fulfillment of the Euler-Poincaré equation under the necessary conditions for the incidence of faces. When deriving Schläfli' formulas, the Euler-Poincaré equation is not used, but only the so-called vertex polytopes of the polytope of the highest dimension are considered. In order to prove the existence of, for example, a four-dimensional polytope, it is necessary to construct it, i. e. list all its components of lower dimension and indicate how they touch each other. In the monograph, for the first time, this is how a four-dimensional polytope was constructed, consisting of 24 octahedrons (24-cell), and the fulfillment of the Euler-Poincaré equation for this polytope was proved. In the mathematical literature, the concepts of a 600-cell (600 tetrahedrons) and 120-cell (120 dodecahedrons) have become widespread. However, no convincing evidence of their existence has been presented. This monograph shows that attempts to construct these polytopes inevitably lead to a contradiction with the necessary condition of incidence for four-dimensional polytopes. This can be seen as proof of the impossibility of their existence.

In the fifth chapter of the book, irregular polytopes of the highest dimension are considered. Here we should note the proof of the existence of higher-dimensional polytopes with an external boundary in the form of a closed surface made up of two-dimensional flat faces. The existence of such polytopes is consistent with the principles of Riemann's geometry (Riemann, 1948), which provides for the boundedness of a higher-dimensional space.

In the sixth chapter of the monograph, polytopes of higher dimension are considered, which are the product of polytopes of low dimension. These polytopes are important for the solution of Hilbert's 18th problem on the normal partition of n-dimensional space (Hilbert, 1901; Alexandrov, 1969). It is these polytopes that are the necessary parallelotopes of higher dimension that provide a solution to the 18th Hilbert problem (Zhizhin, 2021). The parallelohedrons and the stereohedrons previously proposed by Alexandrov (Alexandrov, 1934, 1954) and by Delaunay (Delaunay, 1961; Delanunay, & Sandakova, 1961) cannot provide a solution to this problem.

The ninth chapter of the book is devoted to the application of high-dimensional polytopes in the field of inorganic and organic chemistry when describing the molecules of chemical compounds (Zhizhin, 2014; 2014 a; 2016 a, b; 2017; 2018). The tenth chapter of the book is devoted to the application of high-dimensional polytopes in the field of physics and mathematics when describing clusters of chemical compounds and constructing higher-dimensional spaces from congruent figures (Zhizhin, 2015; 2016; 2016 c; 2019 c, f, h; 2020 e; 2021; Shevchenko, Zhizhin, & Mackay, 2013; Zhizhin, & Diudea, 2016). The eleventh chapter of the book is devoted to the application of high-dimensional polytopes in biology (Zhizhin, 2016 d; 2019 b; 2020 a, b, c, d; 2018 a). The last 12th chapter is devoted to higher-dimensional polytopes used in the theory of heredity (Zhizhin, 2019 d, e, g; 2020; 2021 a, b).

Gennadiy Vladimirovich Zhizhin
Russian Academy of Natural Sciences, Russia

REFERENCES

Alexandrov, A. D. (1934). The derivation of four - dimensional abnormal parallelohedrons. *Proceedings of the U.S.S.R. Academy of Sciences. Atmospheric and Oceanic Physics*, *6*, 803–817.

Alexandrov, A. D. (1954). On filling space with polyhedrons. *Bulletin of the Leningrad University. Mathematics Series.*, *1*, 33–43.

Alexandrov, P. S. (Ed.). (1969). *Hilbert's problems*. Science.

Ashrafi, A. R., Cataldo, F., Iranmanesh, A., & Ori, O. (Eds.). (2013). *Topological Modelling of Nanostructures and Extended Systems*. Springer. doi:10.1007/978-94-007-6413-2

Coxeter, Y. S. M. (1963). Regular Polytopes. Academic Press.

Delaunay, B. N. (1961). Proof of the main theorem of the theory of stereohedrons. *Reports. Academy of Sciences of the USSR*, *138*(6), 1270–1272.

Delaunay, B. N., & Sandakova, N. N. (1961). Theory of stereohedrons. *Proceeding Mathematical In. name V.A. Steklov of Academy Sciences of the USSR*, *64*, 28–51.

Diudea, M. V. (2018). *Multi-shell Polyhedral Clusters*. Springer. doi:10.1007/978-3-319-64123-2

Diudea, M. V., & Nagy, C. L. (2007). *Periodic Nanostructures*. Springer. doi:10.1007/978-1-4020-6020-5

Fedorov, E.S. (1885). Beginning of the teachings on the figures. *Notes of the St. Petersburg Mineralogical Society, 21.*

Fedorov, E.S. (1889). Symmetry of finite figures. *Notes of the St. Petersburg Mineralogical Society, 25.*

Fedorov, E.S. (1891). Symmetry of the correct systems of figures. *Notes of the St. Petersburg Mineralogical Society, 28.*

Grünboum, B. (1967). *Convex Polytopes.* Springer.

Hilbert, D. (1901). Gesammelte Abhandlangen. *Archiv f. Mathematics and Physics, 3*(1), 44–63, 213–237.

Johnson, W. W. (1960). Convex polyhedrons with regular faces. *Notices of the American Mathematical Society, 7*, 952.

McMullen, P., & Schulte, P. (2002). *Abstract Regular Polytopes.* Cambridge Univ. Press. doi:10.1017/CBO9780511546686

Poincaré A. (1895). Analysis situs. *J. de é Ecole Polytechnique, 1.*

Rimann, B. (1948). *Compositions.* OGIZ.

Schläfli, L. (1901). *Theorie der vielfachen Kontinuität Zürcher und Furrer.* doi:10.1007/978-3-0348-5118-3

Shevchenko, V. Ya., Zhizhin, G. V., & Mackay, A. L. (2013). On the structure of the quasicrystals in the space of higher dimensions. In Diamonds and related nanostructures. Springer.

Stringham, W. I. (1880). Regular figures in *n* - dimensional space. *American Journal of Mathematics, 3*(1), 1–14. doi:10.2307/2369441

Vaschenko-Zakharchenko, M. E. (1880). *Beginnings of Euclid with an explanatory introduction and interpretations.* Imperial University of St. Vladimir.

Zalgaller, V. A. (1966). *Convex polyhedrons with correct sides.* Science.

Zalgaller, V.A. (1966). List of all convex polyhedrons with correct sides. *The International Congress of Mathematicians*, 28 – 29.

Zhizhin, G. V. (2014). *4D world.* St. Petersburg: Polytechnica-Service.

Zhizhin, G. V. (2014a). On higher dimension in nature. *Biosphere., 4*, 1–10.

Zhizhin, G. V. (2015). *Polytopic prismahedrons–fundamental regions of the n-dimension nanostructures.* Report on the International conference "Nanoscience in Chemistry, Physics, Biolodgy and Mathematics", Cluj-Napoca, Romania.

Zhizhin, G. V. (2016a). Dimensions of compounds in supramolecular chemistry. *Int. J. Chem. Modeling., 7*(2), 125–135.

Zhizhin, G. V. (2016b). N-prismahedrons and their dual polytopes. *Int. J. Chem. Modeling., 7*(2), 136–139.

Zhizhin, G. V. (2016c). From the multidimensional phase space to a multidimensional space in nature. *Biosphere., 8*(3), 258–267.

Zhizhin, G. V. (2016d). The structure, topological and functional dimension of biomolecules. *Journal of Chemoinformatics and Chemical Engineering*, *5*(2), 44–58. doi:10.4018/IJCCE.2016070104

Zhizhin, G. V. (2017). Systematics of Anomalies in the Filling of Electron Orbitals of Chemical Elements. *Journal of Chemoinformatics and Chemical Engineering*, *6*(1), 44–58. doi:10.4018/IJCCE.2017010104

Zhizhin, G. V. (2018). *Chemical Compound Structures and the higher Dimension of Molecules: Emerging research and Opportunities*. IGI Global. doi:10.4018/978-1-5225-4108-0

Zhizhin, G. V. (2018a). The Metric Structure of Nucleic Acids and Higher Dimension of their Constituents. *Journal of Chemoinformatics and Chemical Engineering*, *7*(2), 1–15. doi:10.4018/IJCCE.2018070101

Zhizhin, G. V. (2019a). *The Geometry of Higher - Dimensional polytopes*. IGI Global. doi:10.4018/978-1-5225-6968-8

Zhizhin, G. V. (2019b). *Attractors and Higher Dimensions in Population and Molecular Biology: Emerging Research and Opportunities*. IGI Global. doi:10.4018/978-1-5225-9651-6

Zhizhin, G. V. (2019c). Higher Dimensions of Clusters of Intermetallic Compounds: Dimensions of Metallic Nanoclusters. *Int. Journal of Applied Nanotechnology Research*, *4*(1), 8–25. doi:10.4018/IJANR.2019010102

Zhizhin, G. V. (2019d). The Law of Conservation of Incidents. *International Journal of Chemoinformatics and Chemical Engineering*, *8*(1), 1–15. doi:10.4018/IJCCE.2019010103

Zhizhin, G. V. (2019e). The polytope of hereditary information: The structure, location, signification. *Biochemistry and Modern Applications*, *2*, 56–62. doi:10.33805/2638-7735.124

Zhizhin, G. V. (2019f). On the structure of parallelohedrons of higher dimension. Hilbert's 18 the problem. *International Journal of Chemoinformatics and Chemical Engineering*, *8*(2), 69–96. doi:10.4018/IJCCE.20190701.oa2

Zhizhin, G. V. (2019g). The structure of the polytope of hereditary information. *International Journal of Chemoinformatics and Chemical Engineering*, *8*(2), 7–22. doi:10.4018/IJCCE.2019070102

Zhizhin, G. V. (2019h). The Geometry of Higher-Dimensional Multi-Shell Clusters with Common Center and Different Centers: The Geometry of Metal Clusters with Ligands. *Int. Journal of Applied Nanotechnology Research*, *4*(2), 45–65. doi:10.4018/IJANR.2019070103

Zhizhin, G. V. (2020). Hidden nucleic acids bond order? *Acta Scientific Biotechnology*, *1*(4), 34–36.

Zhizhin, G. V. (2020a). Spirals of Polysaccharides Simultaneously Rotating in Opposite Directions. *Acta Scientific Biotechnology*, *1*(7), 13–16.

Zhizhin, G. V. (2020b). Enantiomorphism of Tartaric Acid as a Property of Five-Dimensional Polytopes. *Acta Scientific Biotechnology*, *1*(12), 12–16.

Zhizhin, G. V. (2020c). The Dimension and Life. *Acta Scientific Biotechnology.*, *1*(11), 36–40.

Zhizhin, G. V. (2020d). Higher Dimensions in Linear, Helicoidal and Folded Structures of Protein Molecules. *Acta Scientific Biotechnology*, *1*(12), 3–9.

Zhizhin, G. V. (2020e). *Nanotechnologies and Clusters in the Spaces of Higher Dimension: Emerging Research and Opportunities*. IGI Global.

Zhizhin, G. V. (2021). *Normal Partitions and Hierarchical Fillings of N-Dimensional Spaces*. IGI Global. doi:10.4018/978-1-7998-6768-5

Zhizhin, G. V. (2021a). The Structure of DNA Taking into Account the Higher Dimension of Its Components. In M. Khosrow-Pour (Ed.), *Encyclopedia of Information Science and Technology* (5th ed., pp. 730–747). doi:10.4018/978-1-7998-3479-3.ch050

Zhizhin, G. V. (2021b). Methylated Polytope of Hereditary Information. *Int. Journal of Applied Research on Public Health Management.*, *6*(2).

Zhizhin, G. V., & Diudea, M. V. (2016). Spaces of nanoword. In Sustainable nanosystems, development, properties and applications. IGI Global.

Zhizhin, G. V., Khalaj, Z., & Diudea, M. V. (2016). Geometrical and topological dimensions of the diamond. In Distance, symmetry and topology in carbon nanomaterials. Springer. doi:10.1007/978-3-319-31584-3_12

Chapter 1
Regular and Semi-Regular Three-Dimensional Polytopes

ABSTRACT

The convex three-dimensional regular and semiregular polyhedrons were investigated using mental mechanical operations on polyhedrons. They include cutting polyhedrons (cutting off vertices), the necessary deformations of the section sections to shape the sections into regular polygons, and rotating parts of the polyhedrons relative to each other. There is proved the existence of 16 semiregular polyhedrons, that is, three more polyhedrons than in the study of "operations on maps." It is shown that any regular or semiregular convex three-dimensional polytope can be passed to any other regular or semiregular polyhedron in a finite number of steps.

INTRODUCTION

Before considering polytopes of higher dimensions, it is advisable to recall convex three-dimensional polyhedrons, both of an ideal shape and with certain deviations from ideality, which will allow us to further determine the influence of an increase in the dimension of a figure on its properties and features. An important question here is the method of studying both three-dimensional figures (polyhedrons) and polytopes of higher dimensions. Traditionally, in the study of polyhedrons, the method of operations on maps is used (Johnson, 1966; Pesanski, & Randie, 2000; Diudea, 2005, 2018). In this monograph, when describing three-dimensional polyhedrons, a mechanical research method will be used (Zhizhin, 2005, 2006, 2009, 2014, 2019) since it was established that using this method it is possible to prove the existence of a larger number of both semiregular and regular-faceted polyhedrons.

FIVE PLATO BODIES

Even in Ancient Greece, the existence of five regular convex polyhedrons - Platonic solids was established (Aleksandrov, 1950): tetrahedron, octahedron, cube, dodecahedron, icosahedron (Euler, 1736). In each

DOI: 10.4018/978-1-7998-8374-6.ch001

of these polyhedrons, all the vertices are the vertices of polyhedral angles - gonohedrons, in Fedorov's terminology (Fedorov, 1885, 1889, 1891). Since the faces of the gonohedron are regular polygons, the plane angles at the vertices of the gonohedron satisfy equality

$$\alpha = (n - 2)\pi/n, \tag{1.1}$$

n is the number of sides at face of gonohedron.

In addition, planar angles around the vertex of the gonohedron must satisfy three natural conditions necessary for the existence of a gonohedron

1. Around the top of the gonohedron, there must be at least three flat corners.
2. The sum of the planar angles located around the vertex of the gonohedron must be less than 2π.
3. The sum of the plane angles located around the vertex of the gonohedron, minus one angle, must be greater than this angle.

Taking into account these conditions and equality (1.1), only five different gonohedrons and the corresponding previously indicated regular convex polyhedrons can exist (see Table 1).

There cannot be other regular three-dimensional polyhedrons since there cannot be other gonohedrons with faces of regular polygons. Images of regular convex three-dimensional polyhedrons in projection onto a two-dimensional plane with infinitely distant center of the projection (while the projection retains the parallelism of the edges of the polyhedrons) are shown in Figure 1.

Figure 1. Regular convex polyhedrons
a) tetrahedron, b) octahedron, c) cube, d) dodecahedron, e) icosahedron

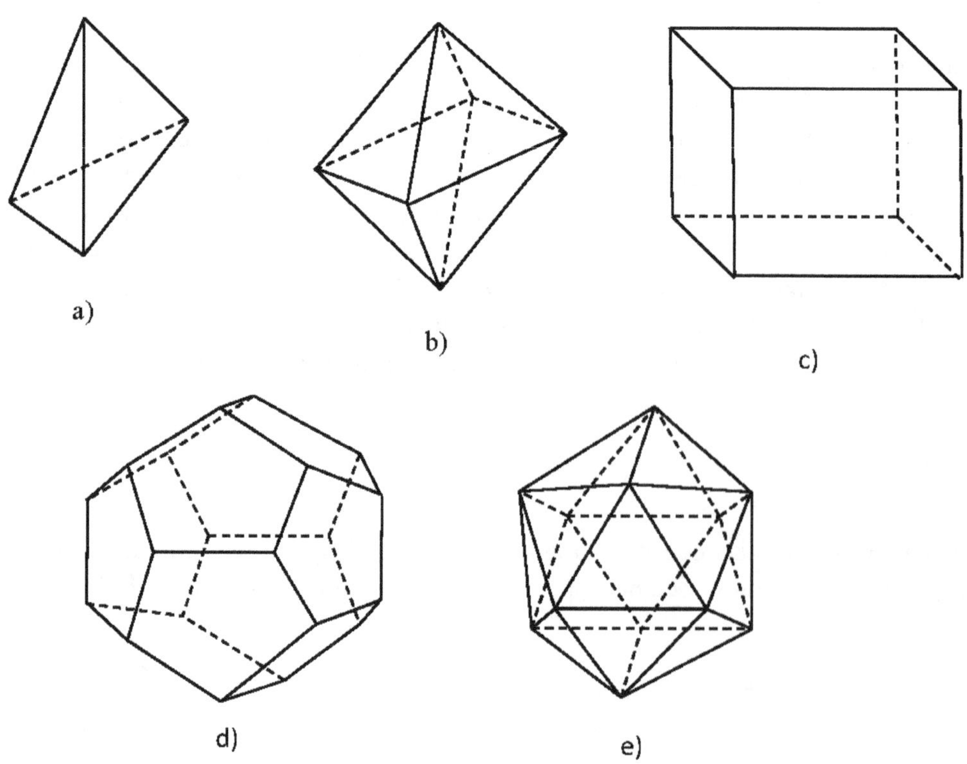

Table 1. The regular convex polyhedrons

The name of polyhedron	N	K	M	l	r	{n, k}
tetrahedron	3	3	4	4	6	{3,3}
octahedron	3	4	8	6	12	{3,4}
cube	4	3	6	8	12	{4,3}
dodecahedron	5	3	12	20	30	{5,3}
icosahedron	3	5	20	12	30	{3,5}

In Table 1, for each regular polyhedron, in addition to the number of sides at the face of the gonohedron n, the number of faces of the gonohedron k, the number of faces of the polyhedron m, the number of vertices of the polyhedron l, the number of edges of the polyhedron r, and the Schläfli symbol {n, k} are indicated (Schläfli, 1901).

All regular convex polytopes can be inscribed in a ball, and each of the polytopes leads to a discrete group of motions of the ball, for which the vertices of the polytope coincide, i. e. the vertices of the polyhedron form systems of equivalent points. If we draw through all the vertices of the polyhedron the planes tangent to the ball, then by virtue of the equivalence of the vertices, we again get a regular convex polyhedron describing the ball. In this case, a tetrahedron gives rise to a larger tetrahedron, an octahedron gives rise to a larger cube, a cube gives rise to a larger octahedron, a dodecahedron gives rise to a larger icosahedron, an icosahedron gives rise to a larger dodecahedron. This is a consequence of the duality of regular three-dimensional convex polyhedrons: the tetrahedron is dual to itself, the octahedron and the cube are dual to each other, the dodecahedron and icosahedron are dual to each other. For dual polytopes, the numbers of edges coincide, and the numbers of faces and vertices are mutually interchanged (see Table 1).

THE CONVEX SEMIREGULAR POLYHEDRONS

We will call convex polyhedron with faces from regular polygons of two or more types semi-regular polyhedron if all polyhedral angles (gonohedron) of the polyhedron have the same composition and order of arrangement of faces. All gonohedrons of a regular polyhedron are congruent by this definition, i. e. compatible with a motion not necessarily associated with the motion of the entire polyhedron as a whole. Semiregular polyhedrons are obtained from regular polyhedrons using a sufficiently large number of so-called "map operations" using representations of graph theory. A detailed description of "card operations" and a list of semiregular polyhedrons obtained by this method can be found in the literature cited. However, in this monograph we will not use this method. Let us show that all convex semiregular polyhedrons can be obtained from regular convex polyhedrons using the concepts of three mechanical operations on convex regular polyhedrons: 1) cutting the polyhedron along some selected section (we denote this operation by t); 2) deformation of the cut off part of the polyhedron so that a regular polygon is formed in the cut section with the length of the edges equal to the length of the edges of the original polyhedron (we denote this operation by δ); 3) rotation of the cut parts relative to each other by some angle (we denote this operation ω). It is important that the use of these three operations makes it possible to obtain all the known semiregular polytopes, while some of the semiregular polytopes were not obtained when using "map operations". The repetition of the operation will be marked by raising it to a power.

Theorem 1.1. There are 16 convex semiregular three-dimensional polytopes listed in Table 2, and only them.

Evidence

We will consider convex regular polyhedrons (Table 1 and Figure 1) as the basis for obtaining convex semiregular polyhedrons, i. e. polyhedrons, the surface of which is made up of different regular polygons, but all polyhedral angles of each polyhedron are the same and can be combined by movement. Let us show that in this case 16 polyhedrons, shown in Table 2, and only they can be obtained.

The method for obtaining semiregular polyhedrons from regular polyhedrons includes: the operation of cutting off polyhedral angles from regular polyhedron with their subsequent replacement by regular polygons with the same side lengths as the edges of the remaining parts of the polyhedron; subsequent operations of cutting off polyhedral angles from the obtained semi-regular polyhedrons with the necessary deformation of the dihedral angles and side lengths of the polygons in order to obtain closed bodies with the same edge lengths throughout the body; movement of one part of the polyhedron relative to the other. All these transformations must satisfy the condition of the identity of the polyhedral angles on the entire surface of the polyhedron and the sum of the plane angles that make up the polyhedral angles, naturally, should not exceed 360 degrees.

Let us determine which semiregular polyhedron can be obtained from the tetrahedron.

Cutting off the triangular corners of the tetrahedron with a plane (operation t_3, the subscript indicates the number of sides in the section) so that regular triangles with sides equal to the edges of the rest of the tetrahedron are formed in the sections, we get a polyhedron with triangular and hexagonal faces (hexagonal faces are formed from triangular faces of the tetrahedron). Dihedral angles between hexagonal faces coincide with dihedral angles in a tetrahedron, and dihedral angles between hexagonal and triangular faces are determined by the formulas of spherical geometry (next section of the chapter). The result is (see Figure 2) a semiregular polyhedron - a truncated tetrahedron (polyhedron 1 in Table 2).

Figure 2. Truncated tetrahedron

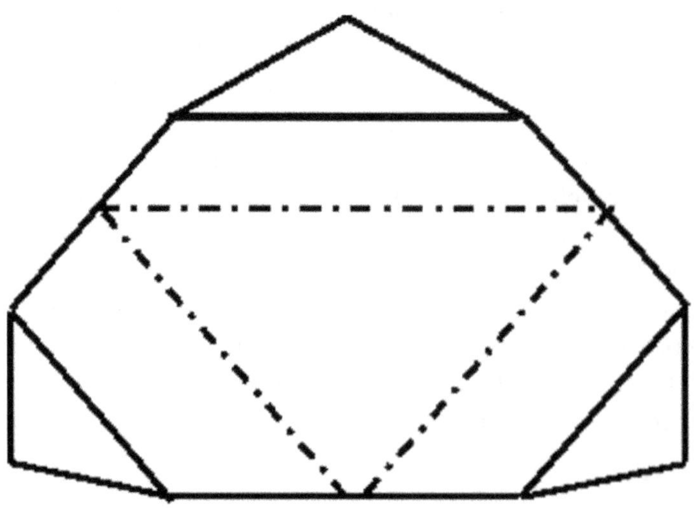

Trimming the trihedral vertices of a truncated tetrahedron converts hexagons to dodecagons, triangles to hexagons, and triangles should form at the cut.

In this case, polyhedral angles with different sets of faces are formed, that is, a semiregular polyhedron cannot be obtained. However, cutting off the truncated tetrahedron by planes parallel to the triangular faces (operation t_3) passing through the midpoints of the sides of the hexagonal faces (dash-dotted lines in Figure 2), we get a regular polyhedron - an octahedron!

Let us determine which semiregular polyhedron can be obtained from the octahedron.

Cutting off the tetrahedral corners of the octahedron (operation t_4), we obtain (see Figure 3) a truncated octahedron with hexagonal and square faces (polyhedron 5, table 2).

Table 2. Semiregular polyhedrons

N°	The name	The number of vertices	The number of faces	The composition of faces	The type of vertices
1	Truncated tetrahedron	12	8	4 hexagons 4 triangles	Trihedral corners 6 – 6 – 3
2	Truncated Cube	24	14	6 octagons 8 triangles	Trihedral corners 8 – 8 – 3
3	Truncated Dodecahedron	60	32	12 decagons 20 triangles	Trihedral corners 10 – 10 – 3
4	Truncated Icosahedron	60	32	20 hexagons 12 pentagons	Trihedral corners 6 – 6 – 5
5	Truncated Octahedron	24	14	8 hexagons 6 squares	Trihedral corners 6 – 6 – 4
6	Truncated Cuboctahedron	48	26	6 octagons, 8 hexagons, 12 squares	Trihedral corners 6 – 8 – 4
7	Truncated Icosidodecahedron	120	62	30 squares, 20 hexagons, 12 decagons	Trihedral corners 10 – 6 – 4
8	Cuboctahedron	12	14	6 squares 8 triangles	Tetrahedral corners 4 – 3 – 4 – 3
9	Rombocuboctahedron	24	26	18 squares 8 triangles	Tetrahedral corners 4 – 4 – 4 – 3
10	Ashkinuze Polyhedron	24	26	18 squares 8 triangles	Tetrahedral corners 4 – 4 – 4 – 3
11	Icosadodecahedron	60	32	12 pentagons 20 triangles	Tetrahedral corners 5 – 3 – 5 – 3
12	Romboicosadodecahedron	60	62	30 squares, 20 triangles, 12 pentagons	Tetrahedral corners 5 – 4 – 3 – 4
13	Flat-nosed cube (left)	24	38	6 squares 32 triangles	Pentahedral corners 4 – 3 – 3 – 3 – 3
14	Flat-nosed cube (right)	24	38	6 squares 32 triangles	Pentahedral corners 4 – 3 – 3 – 3 – 3
15	Flat-nosed dodecahedral (left)	60	92	80 triangles 12 pentagons	Pentahedral corners 5 – 3 – 3 – 3 – 3
16	Flat-nosed dodecahedral (right)	60	92	80 triangles 12 pentagons	Pentahedral corners 5 – 3 – 3 – 3 – 3

Figure 3. Truncated octahedron

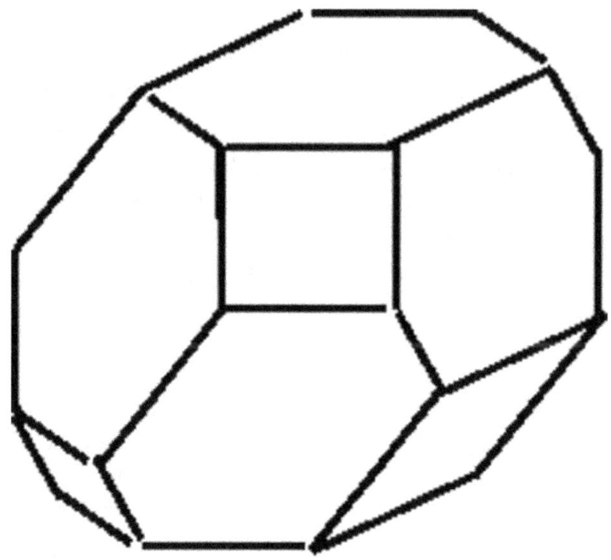

Cutting off the truncated octahedron by planes parallel to the squares and passing through the midpoints of the sides of the hexagons (operation t_4), we obtain (see Figure 4) a cuboctahedron (polyhedron 8 in Table 2).

Figure 4. Cuboctahedron

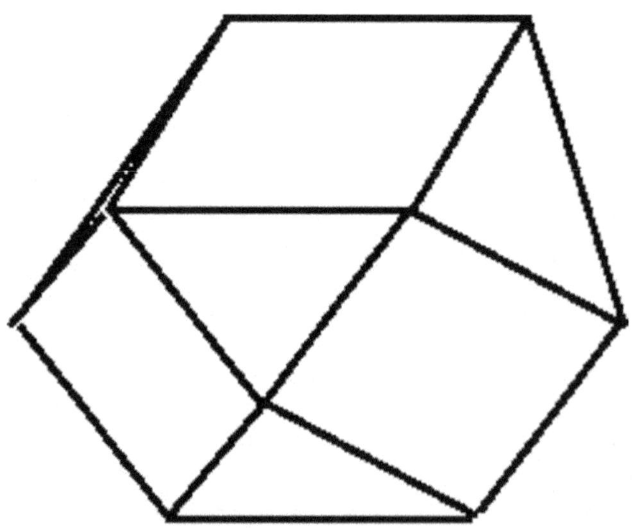

We cut off the quadrangular corners of the cuboctahedron to form squares in the sections, so that the squares turn into octagons, and the triangles into hexagons. In this case, to form trihedral angles with faces from an octagon, hexagon and square, it is necessary to slightly increase the dihedral angles of the cuboctahedron and compress the squares of the cuboctahedron along the perpendiculars to the sides so that the lengths of the sides formed from the section of triangles and squares are the same (operation $t_4\delta$). This gives (see Figure 5) a truncated cuboctahedron (polyhedron 6 in table 2).

Figure 5. Truncated cuboctahedron

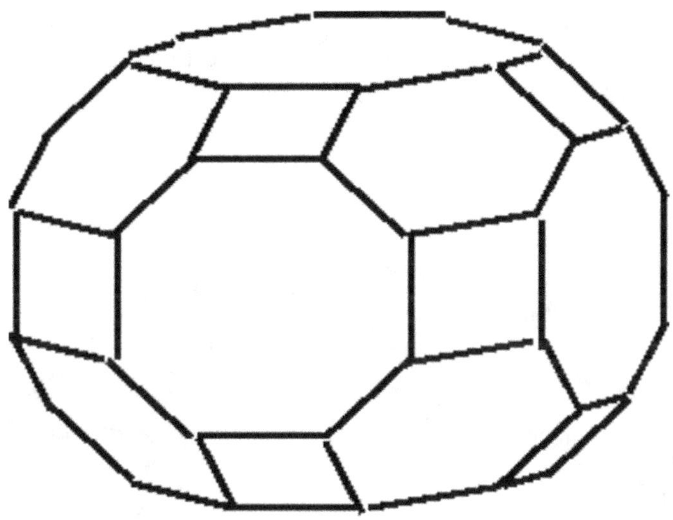

The quadrangular corners of a cuboctahedron can be cut deeper to form smaller squares and smaller triangles into triangles, also increasing dihedral angles and contracting the squares (operation). This will give (see Figure 6) a rhombocubooctahedron (polyhedron 9 in Table 2).

Figure 6. Rombocuboctahedron

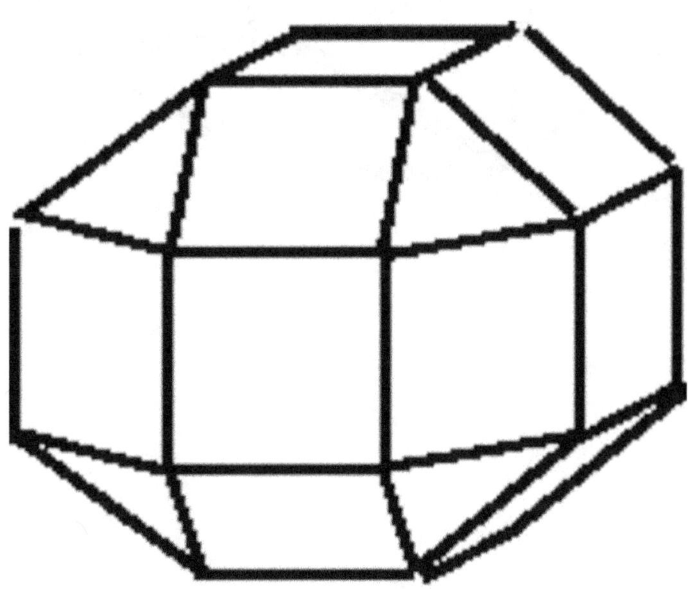

The quadrangular corners of a cuboctahedron can be cut deeper to form smaller squares and smaller triangles into triangles, also increasing dihedral angles and contracting the squares (operation $t_4\delta$). This will give (see Figure 6) a rhombocubooctahedron (polyhedron 9 in Table 2). The truncated cuboctahedron allows further transformations. Cutting off the triangular corners (operation $t3_)$ of the truncated cuboctahedron so that the hexagons turn into triangles, the octagons - into squares, and the squares of the truncated cuboctahedron are cut in half, with the formation of triangles in the sections, we get a flat-nosed cube. At the same time, since squares can be inscribed in an octagon in two ways, as well as triangles in a hexagon, a flat-nosed cube exists in two modifications (operation ω) - left and right (polyhedrons 13 and 14 in table 2, Figure 7 and Figure 8).

Figure 7. Flat-nosed cube (left)

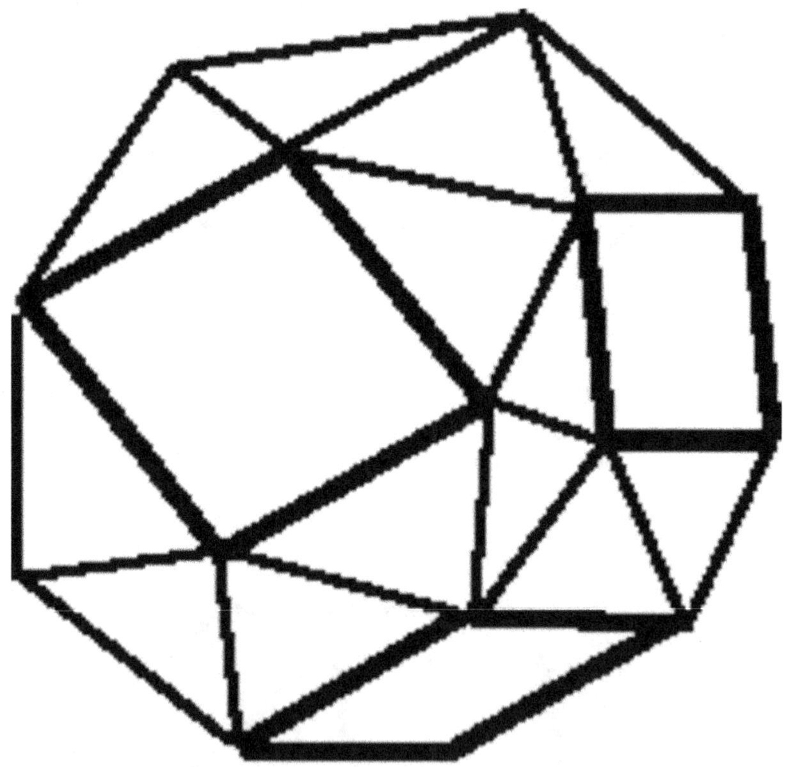

Figure 8. Flat-nosed cube (right)

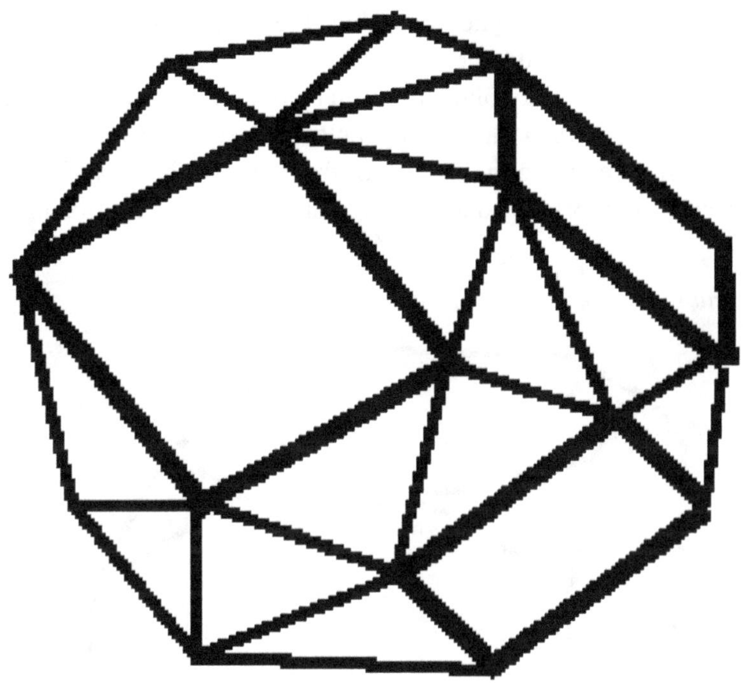

The rhombocuboctahedron also allows a modification associated with the possibility of rotating the part of the polyhedron lying above the belt of squares by one face without breaking the tetrahedral angles (operation ω). This operation forms (see Figure 9) the Ashkinuze polyhedron (polyhedron 10 in Table 2).

Figure 9. Ashkinuze polyhedron

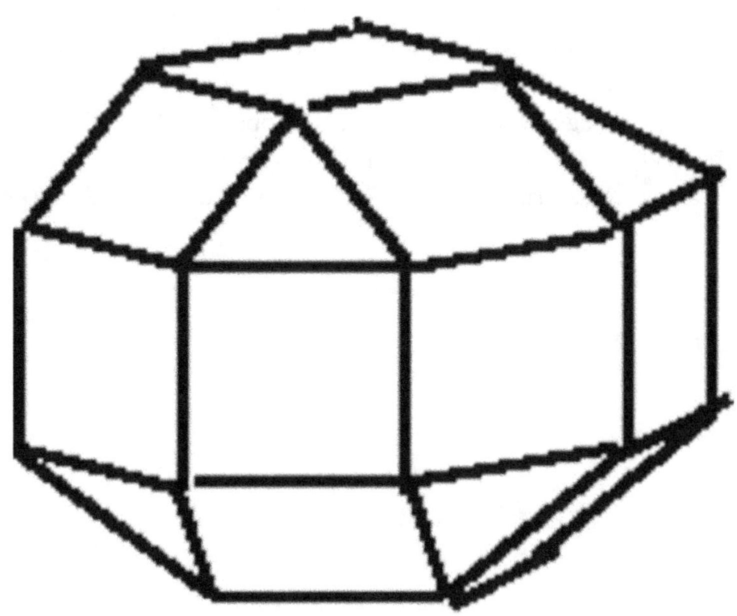

Further cutoffs of the rhombocubooctahedron, the Ashkinuze polyhedron and flat-nosed cubes cannot lead to the formation of semi-regular polyhedrons, since the sum of the flat angles of the faces formed at the cross-sections is more than 360 degrees.

Let us determine which semiregular polyhedron can be obtained from the cube.

Cutting off the triangular corners of the cube by a plane (operation t_3) with the formation of regular triangles in sections with sides equal to the lengths of the edges of the rest of the cube, we obtain (see Figure 10) a truncated cube with octagonal and triangular faces (polyhedron 2 in Table 2).

Figure 10. Truncated cube

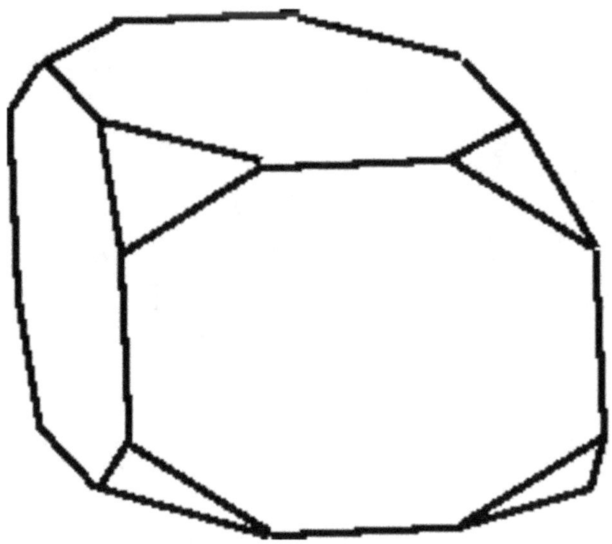

The truncated cube can be cropped even further so that the octagons become squares and the triangles are replaced with larger triangles (operation t_3). This will give us a cuboctahedron again. Further transformations of the cuboctahedron and the resulting figures are repeated as in the already considered transformations of the octahedron.

Let us determine which semiregular polyhedrons can be obtained from the dodecahedron.

Cutting off the trihedral corners of the dodecahedron (operation t_3), we turn the pentagons into decagons, and in the sections we get triangles. This is a truncated dodecahedron (polyhedron 3 in Table 2, Figure 11).

Figure 11. Truncated dodecahedron

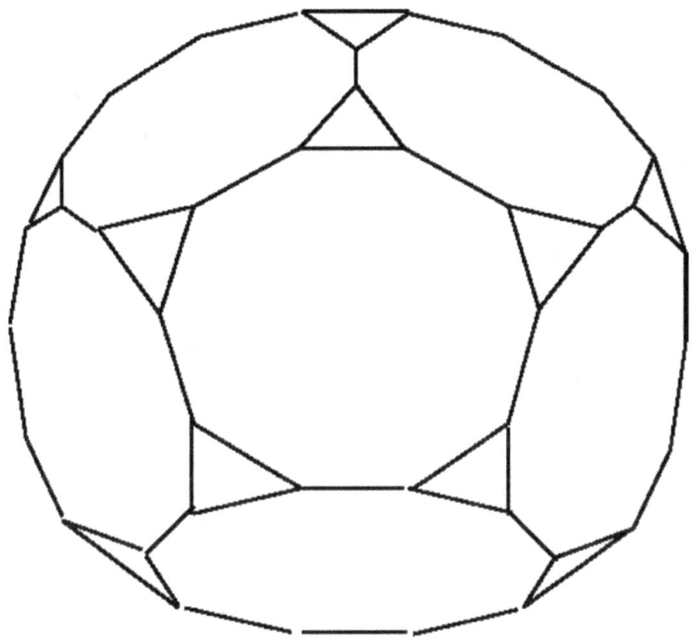

Cutting off the deeper triangular corners of the dodecahedron (operation t_3), the pentagons will decrease in size, and triangles will form at the cross-sections. This is an icosadodecahedron (polyhedron 11 of Table 2, Figure 12).

Figure 12. Icosadodecahedron

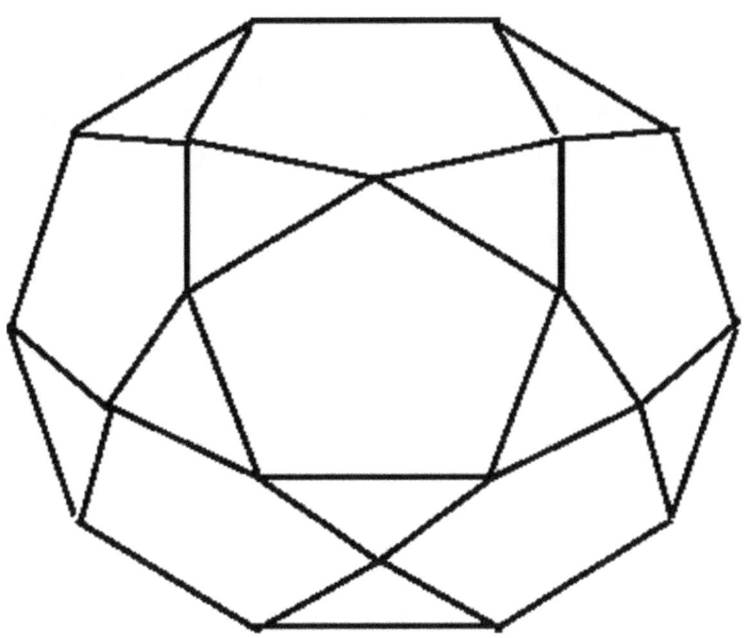

Cutting off the trihedral angles of a truncated dodecahedron does not lead to semiregular polyhedrons due to the excess of the sum of plane angles for possible polyhedral angles of 360 degrees.

Cutting off the quadrangular corners of the icosadodecahedron, we get squares at the section, pentagons turn into decagons, triangles into hexagons. At the same time, we increase the dihedral angles by squeezing the pentagons along the perpendiculars to the sides (operation $t_4\delta$). As a result, we get a truncated icosadodecahedron (polyhedron 7 in table 2, Figure 13).

Figure 13. Truncated icosadodecahedron

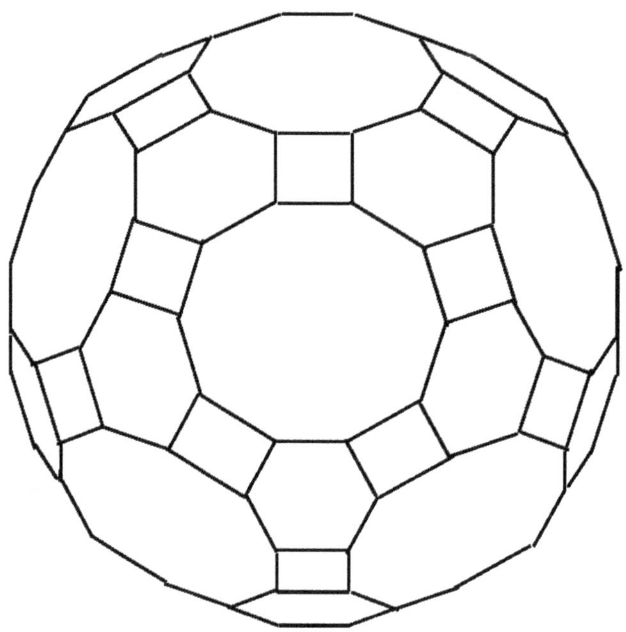

Cutting off the quadrangular corners of the icosadodecahedron more deeply, we turn pentagons into smaller pentagons, triangles into smaller triangles, and squares are formed at the sections of the section. Still applying the required deformation (operation $t_4\delta$), we obtain a rhomboicosadodecahedron (polyhedron 12 in Table 2, Figure 14).

Figure 14. Rhomboicosadodecahedron

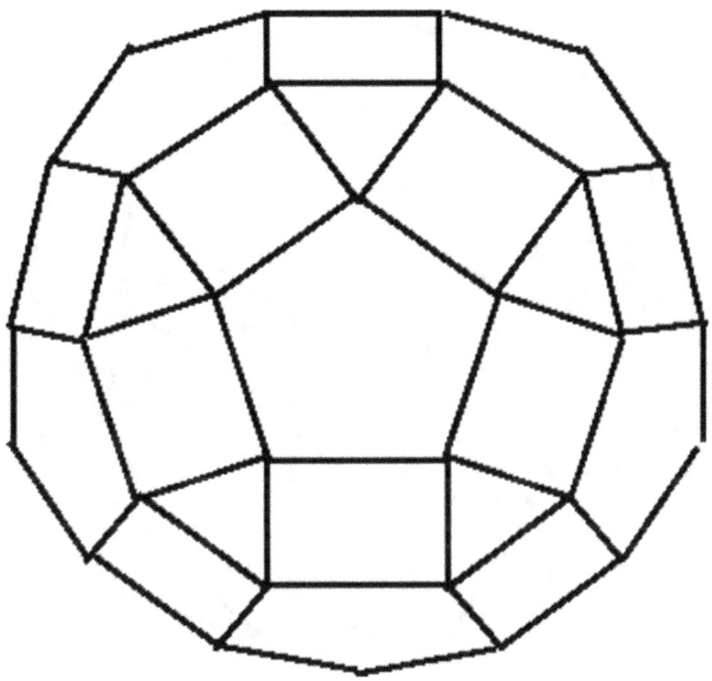

Deeply cutting off the trihedral vertices of the truncated icosadodecahedron, we turn decagons into pentagons, hexagons into triangles (squares are cut in half), triangles are formed in the sections. As a result, applying deformation (operation $t_3\delta$), we get a flat-nosed dodecahedron. Since a triangle can be inscribed in a hexagon in two ways, a flat-nosed dodecahedron exists (operation ω) in two versions (polyhedrons 15 and 16 in Table 2, Figure 15, and Figure 16).

Figure 15. Flat-nosed dodecahedron (left)

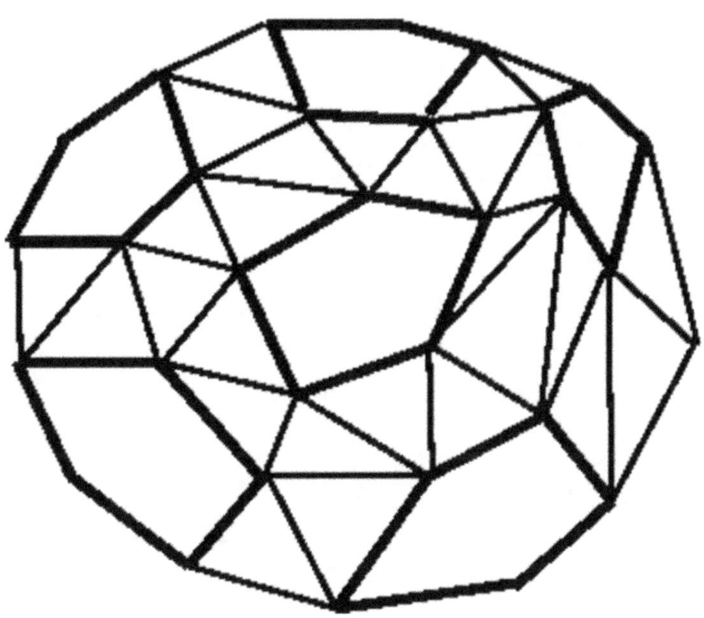

Figure 16. Flat-nosed dodecahedron (right)

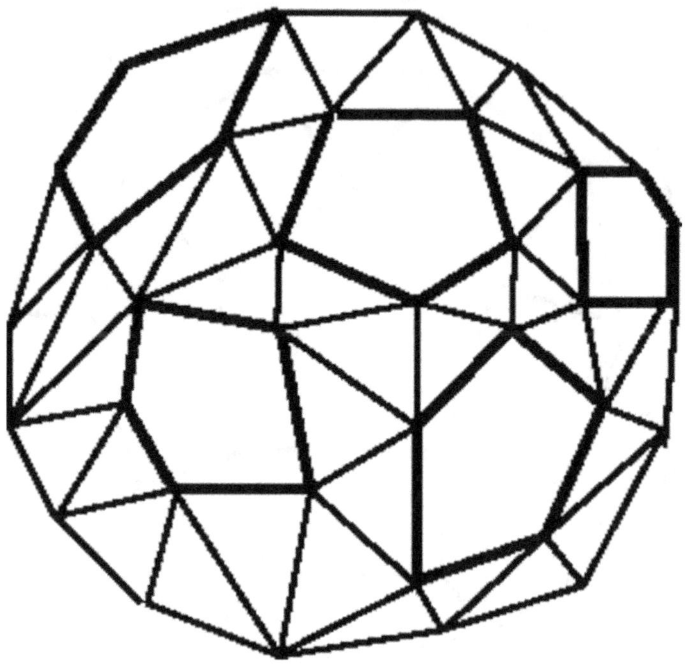

Further cutting off the polyhedral angles of flat-nosed cubes and a rhomboicosadodecahedron does not lead to semi-regular polyhedrons due to the formation of various polyhedral angles in the same polyhedron.

Let us determine which semiregular polyhedrons can be obtained from the icosahedron.

Cutting off the pentahedral corners of the icosahedron (operation t_5), we get pentagons at the sections of the section, and the triangles turn into hexagons. This is a truncated icosahedron (polyhedron 4 in Table 1, 2, Figure 17).

Figure 17. Truncated icosahedron

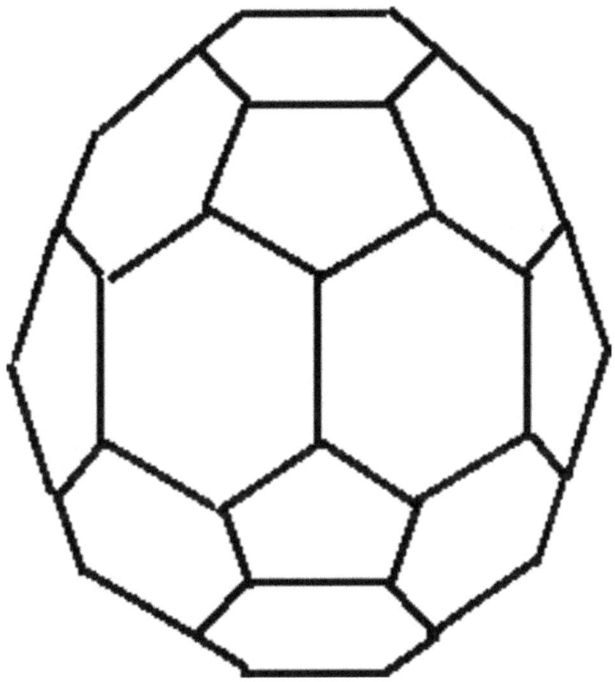

Further cutting off the trihedral angles of the truncated icosahedron does not lead to the formation of semiregular polyhedrons due to the formation of various polyhedral angles. However, from an icosahedron, you can get a regular polyhedron - a dodecahedron. To do this, it is enough to cut off all pentahedral corners of the icosahedron (operation t_5), until the sides of the formed adjacent pentagons coincide.

Since all the polytopes listed in Table 2 are obtained in the listed transformations, and the impossibility of obtaining other semiregular polytopes is shown, Theorem 1.1 is proved.

Comment. As follows from Table 2, there are three pairs of semiregular polyhedrons in which the parameters reflected in the table inside the pair coincide. These are pairs of polyhedrons 9 and 10, 13 and 14, 15 and 16. However, as you can see, the polyhedrons inside the pair differ from each other by the arrangement of regular polygons on the surface of the polyhedrons. Therefore, they are different. This difference could not be noticed when using the "operations on cards" method. In this regard, the set of semiregular polyhedrons found when using the method of "operations on maps" is three units less than the set of semiregular polyhedrons given in Table 2.

Consequence. In the proof of Theorem 1.1, it was found that all regular and semiregular polytopes can be divided into two families. The first family includes a tetrahedron, octahedron, cube, truncated tetrahedron, truncated octahedron, cuboctahedron, truncated cube, truncated cuboctahedron, rhombicuboctahedron, Ashkinuse polyhedron, flat-nosed cube (right and left). The second family includes the icosahedron, dodecahedron, truncated icosahedron, truncated dodecahedron, icosadodecahedron, truncated icosadodecahedron, rhomboicosadodecahedron, flat-nosed dodecahedron (right and left). Within each family of polyhedrons (if we introduce into consideration for each previously introduced transformation operation its inverse operation), it is possible, in a finite number of transformations, to pass from any of the polytopes of this family to any other polytope from this family. Moreover, with the help of the above transformations, it is impossible to pass from one family of polyhedrons to another family of polyhedrons. The first family of polyhedrons can be called the octahedron family. The second family of polyhedrons can be called the icosahedron family.

The diagram of transitions between polyhedrons in the octahedron family is shown in Figure 18.

Figure 18. Transitions between polyhedrons in the octahedron family

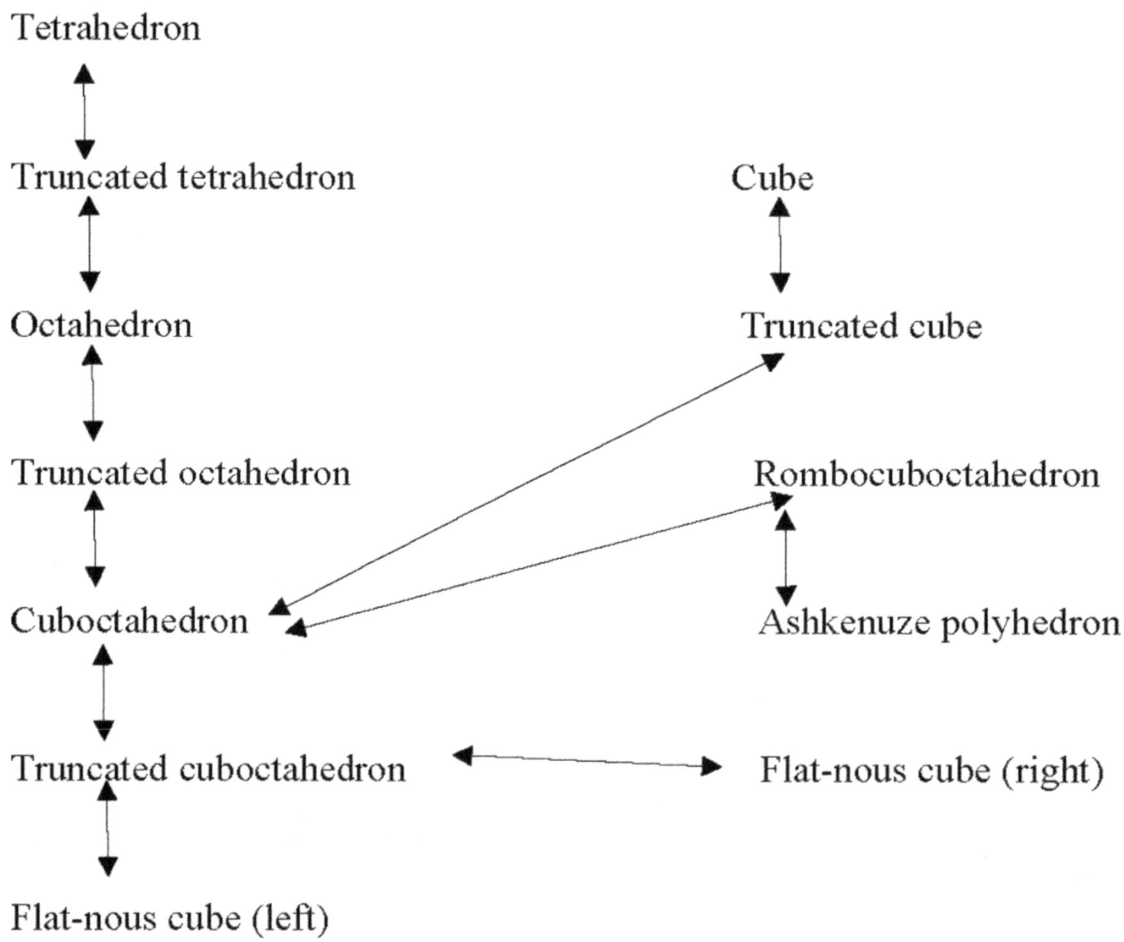

The diagram of transitions between polyhedrons in the icosahedron family is shown in Figure 19.

Figure 19. Transitions between polyhedrons in the icosahedron family

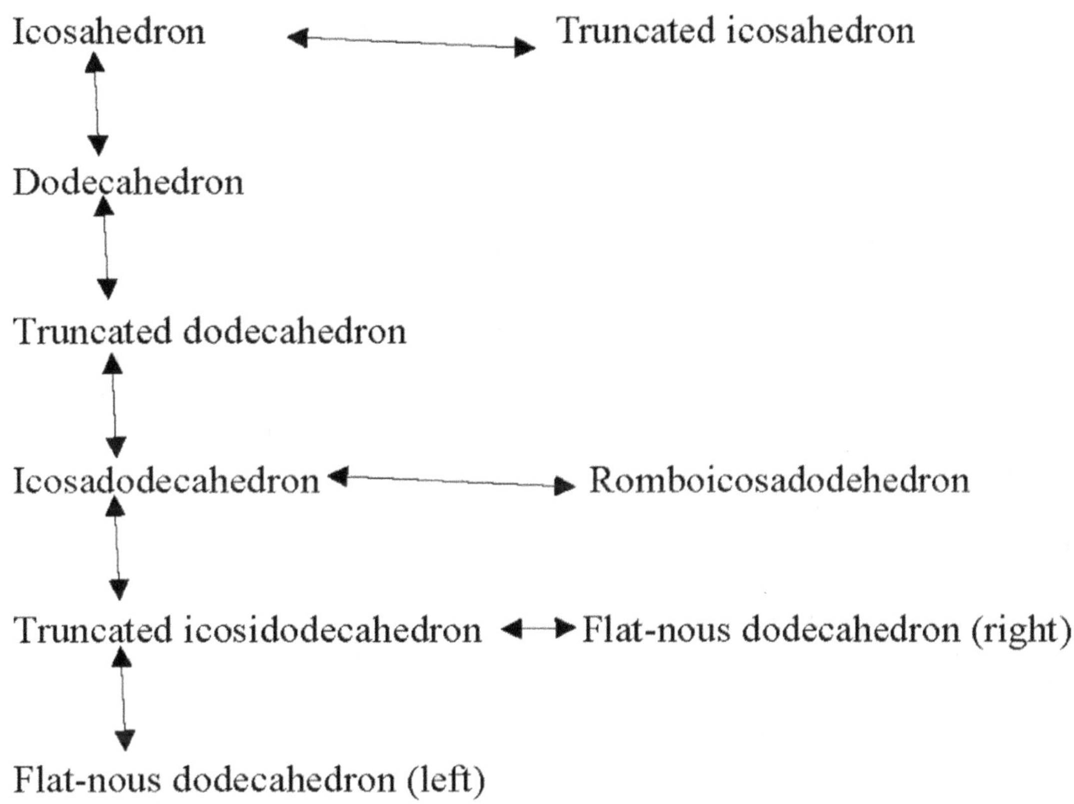

Theorem 1.2. From each convex polyhedron from the set of regular and semiregular polyhedrons, one can pass using a finite number of transformations to any other polytope from the specified set of polyhedrons.

Evidence

The proof of Theorem 1.2 is based on Theorem 1.1 and its corollary. In addition to the transformations of polyhedrons listed in the proof of Theorem 1.1, there is one more, one might say, special transformation that transforms an octahedron into an icosahedron. It consists (Coxeter, 1963; Grunbaum, 1967) in dividing each edge of the octahedron by one third and two thirds of the edge so that the division points become the vertices of a regular triangle inscribed in the triangular face of the octahedron. Cutting off the parts in each triangular face of the octahedron external to the inscribed triangle and placing pairs of triangles equal to the inscribed ones in the resulting holes, we obtain an icosahedron. The transition from octahedron to icosahedron unites both families of polyhedrons (octahedron and icosahedron) into one common family of polyhedrons. This provides (taking into account possible inverse transformations) a transition from any regular or semiregular polytope to any other polyhedron from the specified family of polyhedrons in a finite number of transformation steps. Q.E.D.

Comment. The transition from one type of polyhedron to another type of polyhedron occurs in nature in the process of evolution. For example, the shape of tissue cells changes in the course of tissue cancers.

CONCLUSION

In contrast to the widespread method of analysis of "operations on maps" based on the concepts of graph theory, for the first time three-dimensional regular and semiregular polyhedrons were investigated using mental mechanical operations on polyhedrons. They include cutting polyhedrons (cutting off vertices), the necessary deformations of the section sections to shape the sections into regular polygons, and rotating parts of the polyhedrons relative to each other. With the help of these three operations, in contrast to numerous operations on maps, it was possible to prove the existence of 16 semiregular polyhedrons, i. e. three more polyhedrons than in the study of " operations on maps". In addition, it is shown that the whole family of regular and semiregular convex polytopes falls into two classes: the class of octahedron polyhedrons and the class of icosahedral polyhedrons. With the help of these operations, within each class, you can transition from any polyhedron in the class to any other polyhedron in the class. Transition from one class to another class using the specified operations is not possible. If we introduce into consideration the special operation of transition from an octahedron to an icosahedron, introduced by Coxeter (Coxeter, 1963), then any regular or semiregular convex three-dimensional polytope can be passed to any other regular or semiregular polyhedron in a finite number of steps using the listed operations.

The question of the transition from one three-dimensional polyhedron to another is of interest in connection with the observed possibility of changing the shape of polyhedrons in various dissipative systems during the evolution of systems under the influence of external influences and internal reactions to these influences.

REFERENCES

Alexandrov, A. D. (1950). *The convex polyhedrons*. Gostechizdat.

Coxeter, Y. S. M. (1963). *Regular Polytopes*. Academic Press.

Diudea, M. V. (2005). Nanoporous carbon allotropes by septupling map operations. *Journal of Chemical Information and Modeling*, *45*(4), 1002–1009. doi:10.1021/ci050054y PMID:16045294

Diudea, M. V. (2018). *Multi – shell Polyhedral Clusters*. Springer. doi:10.1007/978-3-319-64123-2

Euler, L. (1736). Solutio problematisad geometriam situs pertinentis. *Comment Acad Sci I Petropolitanae*, *8*, 128–140.

Fedorov, E.S. (1885). Beginning of the teachings on the figures. *Notes of the St. Petersburg Mineralogical Society*, *21*.

Fedorov, E.S. (1889). Symmetry of finite figures. *Notes of the St. Petersburg Mineralogical Society*, *25*, 1 - 52.

Fedorov, E.S. (1891). Symmetry of the correct systems of figures. *Notes of the St. Petersburg Mineralogical Society, 28.*

Grünbaum, B. (1967). *Convex Polytopes*. Springer.

Johnson, W. W. (1960). Convex polyhedrons with regular faces. *Notices of the American Mathematical Society*, *7*, 952.

Pisanski, T., & Randić, M. (2000). Bridges between geometry and graph theory, geometry at work. *MMA Notes*, *53*, 174–194.

Schläfli, L. (1901). *Theorie der vielfachen Kontinuität Zürcher und Furrer.* doi:10.1007/978-3-0348-5118-3

Zalgaller, V. A. (1966). *Convex polyhedrons with correct sides*. Science.

Zalgaller, V.A. (1966). List of all convex polyhedrons with correct sides. *The International Congress of Mathematicians*, 28 – 29.

Zhihin, G.V. (2005). Correct both semiregular polyhedrons and filling of space with them. *Life and Safety*, *1-2*, 551-559.

Zhizhin, G. V. (2005). *Dissipative of structures in chemical, geological and ecological systems*. Science.

Zhizhin, G. V. (2014). *4D world.* St. Petersburg: Polytechnica - Service.

Zhizhin, G. V. (2019). *The Geometry of Higher - Dimensional polytopes*. IGI Global. doi:10.4018/978-1-5225-6968-8

KEY TERMS AND DEFINITIONS

Gonohedron: A polyhedral corner whose faces are regular polygons.

Graph Theory: A graph *G (V, E)* is a pair of two sets, V and *E, V=V(G)* being a finite nonempty set and *E=E(G)* a binary relation defined on V.

Mechanical Operations on a Polyhedron: Conceivable transformations of a polyhedron, including cutting a polyhedron (cutting off vertices), deforming a section of a cut to obtain a regular polygon in a section, rotating parts of a polyhedron after cutting relative to each other.

Operation on Maps: Operation on maps are topological modifications of a parent graph.

Regular Polyhedron: Convex polyhedron with faces of regular polygons of one type.

Semiregular Polyhedrons: Convex polyhedrons with faces of regular polygons of two or more types, if all gonohedrons of the polyhedron have the same composition and arrangement of faces.

Chapter 2
Regularly Faceted Three-Dimensional Polytopes

ABSTRACT

It is shown that known classification of convex polyhedrons with sides from correct polygons offered by V.A. Zallgaller as full transfer of all possible convex figures of this type is far from completeness. There exist nine new convex polyhedrons with sides from correct polygons and the impossibility of existence of two figures from Zallgeller classification. Thus, the general number of regularly faceted polyhedrons except for the regular polyhedrons, known semiregular polyhedrons, prisms, and antiprisms make 100 figures instead of 92 as Johnson earlier assumed. The existence of a convex polyhedron with faces of regular polygons was discovered, which differs as much as possible from the bodies of Archimedes and Plato by the presence of six different types of gonohedrons in it (polytypichedron-max).

INTRODUCTION

Regularly faceted three-dimensional polytopes are polytopes whose faces are regular polygons. Moreover, in the same polyhedron there can be faces with a different number of sides. The question of determining all possible convex regularly faceted polyhedrons and their classification (in what follows we will only talk about convex polyhedrons and therefore there is no need to emphasize this each time) has a long history. A definite generalization of these works can be considered the monograph by V.A. Zalgaller (Zalgaller, 1966). This monograph contains a bibliography of works in which, to one degree or another, the question of the existence of convex regularly faceted polyhedrons is considered. It, in essence, is devoted to proving one theorem: apart from prisms and antiprisms, there are only 28 simple convex regularly faceted polyhedrons. (A regularly faceted polyhedron is called simple if it does not allow cutting by a plane into two other regularly faceted polyhedrons.) Schematic representations of these polyhedrons are given. In addition, it is proved that the assertion, expressed earlier (Johnson, 1960), that (apart from prisms and antiprisms) there are only 92 regularly faceted polytopes, which have 28 simple regularly faceted polytopes in their composition (Zalgaller, 1966 a).

DOI: 10.4018/978-1-7998-8374-6.ch002

Currently, interest in convex regularly faceted polyhedrons is associated with the analysis of various dissipative structures that arise in various natural phenomena and in the processes of human activity (Zhizhin, 2005 a, b; 2006). In many cases, a dissipative structure is a spatial network, the unit cells of which are regularly faceted polyhedrons. Examples are biological tissues, the cell shapes of which are often close to regularly faceted polyhedrons (Lewis, 1923; D`Arsy, 1942; Petrov, 2007), and as it was recently established based on the analysis of photographs of the Universe, the large-scale structure of the Universe can be considered as a spatial network with a unit cell in the form of a truncated octahedron (Zhizhin, 2008). The shapes of regular polyhedrons are of interest in construction as elements playing the role of space fillers, as well as shapes of clusters in nanotechnology (Zhizhin, 2006, 2021).

This chapter shows that the statement that there are only 28 simple convex regularly faceted polyhedrons (Zalgaller, 1966) is wrong. The existence of a much larger number of simple regularly faceted polyhedrons was discovered and proved (Zhizhin, 2009), while it was proved that some of the polyhedrons declared by V.A. Zalgaller do not actually exist. Accordingly, N. Johnson's assumption on the existence of only 92 total number regularly faceted polyhedrons is wrong too. In the proof Johnson's assumption Zalgaller used Cauchy's lemma (Cauchy, 1944) and the theorem of A.D. Aleksandrov (Aleksandrov, 1950), indicating the conditions for increasing or decreasing plane angles during deformation of polyhedral angles formed by plane angles. This made it possible in many cases to find the maxima and minima of plane angles under deformation of polyhedral angles, preserving the convexity of a closed polyhedral surface and the hardness of the faces. However, these conditions alone are not enough for the construction of regularly faceted polyhedrons. This was also pointed of Zalgaller, but at the same time Zalgaller no presentation of methods for finding plane angles with deformation of polyhedral angles compatible with the closedness of the polyhedral surface and the hardness of the faces, in those cases when the application of the Cauchy lemma and Alexandrov theorem does not lead to results. As a result, in addition to the well-known regularly faceted polyhedrons, which are regular and semiregular polytopes and their parts, Zalgaller discovered 9 simple regularly faceted polyhedrons. In this paper, it is proved that of these 9 polyhedrons (i.e., regularly faceted polyhedrons that are not regular and semiregular polytopes and their parts), there are only 8, and in addition there are 9 more new simple regularly faceted polyhedrons. For the proof, this chapter develops the approach proposed of author in 2005 (Zhizhin, 2005). According to this approach, the construction of all polyhedral closed and non-closed surfaces is carried out not mentally, but naturally (materially), i. e. after cutting a sufficient number of different regular polygons from hard paper, they are glued to each other with adhesive tape, forming a multifaceted surface. In this case, the possibility of continuation of the figure to the formation of a closed surface, i. e. regularly faceted polyhedron. This reveals many forms that can be regularly faceted polyhedrons (forms suspicious of regularly faceted polyhedrons). To exclude from this set of cases of the formation of conditional edges and bends of faces, a detailed analysis of all possible polyhedral angles with faces from regular polygons is preliminary carried out and for the first times functional relationships between various planar and dihedral angles of these polyhedral angles are established during their deformation. The possibility of extending each polyhedral angle to the formation of a closed figure is investigated taking into account these functional dependencies. For each figure suspicious of a regularly faceted polyhedron, taking into account the dependencies between the angles established using the methods of descriptive geometry (projections onto three coordinate planes of three-dimensional space), a closed system of algebraic equations is drawn up for the characteristic angles of the polyhedron. The existence of a regularly faceted polyhedron is considered proven only when the resulting system of algebraic equations has a solution. When obtaining such solutions, all characteristic angles and dimensions of the polyhedron are automatically determined. Such an analysis was completely absent in the work of Zalgaller. With such constructions, both Cauchy's lemma and A.D. Aleksandrov, and other necessary conditions, for example, homeomorphism to the sphere of unfolding of the obtained polytopes. In this work, to denote a polyhedral angle with faces of regular polygons, the term gonohedron is used, introduced by N.E. Fedorov (Fedorov, 1885, 1889, 1891), since not every gonohedron is a vertex of a regularly faceted polyhedron. In this chapter, a new principle of searching and classifying regularly faceted polyhedrons is applied. After compiling a table of all possible gonohedrons with faces, from regular polygons, a certain motive is selected, consisting of a regular polygon with the same or other regular polygons attached to its sides. Examples of motives are shown in Figure 1.

Figure 1. Motives from regular polygons

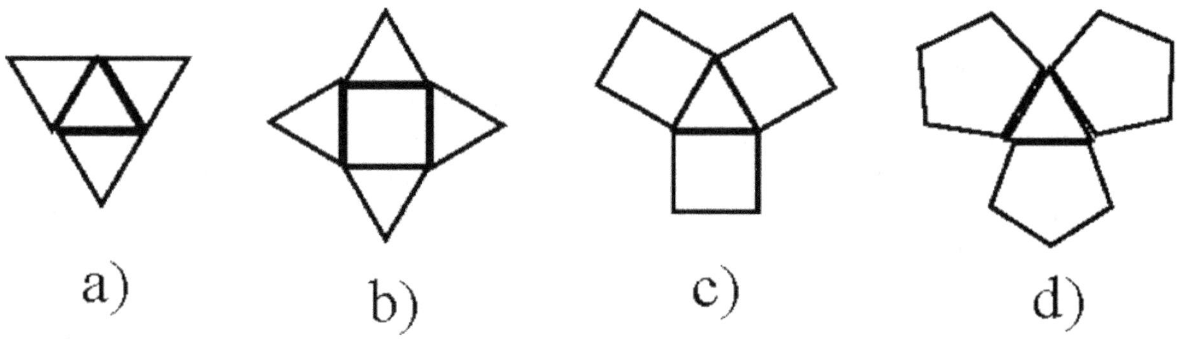

Then, various possibilities of forming various gonohedrons from tables of gonohedrons at the common vertices of regular polygons of a given motive are considered, including by inserting other regular polygons between the polygons of the motive. Further, we consider the possibility of extending the obtained non-closed polyhedral surfaces to obtain closed figures. Consistently considering various motives and ways of their continuation with natural construction of spatial closed and non-closed figures.

GONOHEDRAS WITH FACES FROM REGULAR POLYGONS

Systematics of Gonohedrons with Faces of Regular Polygons

Theorem 2.1. There are 17 different variants of gonohedrons formed by regular polygons. They are listed in Tables 1 – 3. Table 1 corresponds to gonohedrons with identical faces. Table 2 corresponds to gonohedrons with two types of faces. Table 3 corresponds to gonohedrons with three types of faces. Indices n, i, j in Tables 1 – 3 correspond to the number of sides of regular polygons, and indices k, l, m correspond, respectively, to the number of regular n-, i-, j-gons forming a gonohedron.

Evidence

Plane angles located around the vertex of the gonohedron must satisfy the following conditions necessary for the existence of the gonohedron:

1) Around the top of the gonohedron, there must be at least 3 flat corners;
2) the sum of flat angles located around the vertex of the gonohedron must be less than 2 π;
3) The sum of all plane angles around the vertex of the gonohedron, minus any one of these angles, must be greater than the angle to be subtracted.

Let us consider the case when all plane angles of the gonohedron are equal, i.e. they are the angles of a regular n-gon (face Γn) . Then these angles are determined by the equality

$$\alpha = (n - 2)\pi/n, \tag{2.1}$$

n is the number of sides at face of gonohedron.
In accordance with condition 2) we have

Table 1. The gonohedrons with equal faces

The original number of gonohedron, N°	The number of sides of a regular polygon, n	The number faces of gonohedron, k	Face symbol, Γ_n
1	3	3	Γ_3
2	3	4	Γ_3
3	3	5	Γ_3
4	4	3	Γ_4
5	5	3	Γ_4

Table 2. The gonohedrons with faces of two views (n < i)

N°	The gonohedron symbol $\begin{bmatrix} k,l \\ n,i \end{bmatrix}$	The regular triangle is part of the faces of the gonohedron, n = 3	The regular triangle is not part of the faces of the gonohedron, n = 4	The regular triangle and square is part of the faces of the gonohedron, n = 5
		I	I	I
6	$\begin{bmatrix} 1,2 \\ n,i \end{bmatrix}$	4÷11	5÷7	6
7	$\begin{bmatrix} 1,3 \\ n,i \end{bmatrix}$	4	-	-

Continued on following page

Table 2. Continued

Nº	The gonohedron symbol $\begin{bmatrix} k,l \\ n,i \end{bmatrix}$	The regular triangle is part of the faces of the gonohedron, n = 3	The regular triangle is not part of the faces of the gonohedron, n = 4	The regular triangle and square is part of the faces of the gonohedron, n = 5
		I	I	I
8	$\begin{bmatrix} 2,1 \\ n,i \end{bmatrix}$	4÷7	> 4	6-9
9	$\begin{bmatrix} 2,2 \\ n,i \end{bmatrix}$	4, 5	-	-
10	$\begin{bmatrix} 2,2 \\ n,i \end{bmatrix}_d$	4, 5	-	-
11	$\begin{bmatrix} 3,1 \\ n,i \end{bmatrix}$	> 3	-	-
12	$\begin{bmatrix} 4,1 \\ n,i \end{bmatrix}$	4, 5	-	-

$$k(n-2)/n < 2 \tag{2.2}$$

Taking into account condition 3) from (2.2) we have

$$3 \le k < 2n/(n-2)\,. \tag{2.3}$$

Condition (2.3) for equal plane angles of the gonohedron, it is satisfied identically.

Let $n = 3$. Then it follows from (2.3) that $3 \le k < 6$, that is, if the faces of the gonohedron are regular triangles, then the number of faces of the gonohedron can be equal to 3, 4, 5. Let $n = 4$. Then it follows from (2.3) that $3 \le k < 4$, i. e. if the faces of the gonohedron are squares, then the number of faces of the gonohedron is 3. Let $n = 5$. Then it follows from (2.3) that $3 \le k < 3{,}3333$, that is, if the faces of the gonohedron are correct pentagons, then the number of faces of the gonohedron is also 3. For $n > 5$, as follows from (2.3), there are no gonohedron with faces in the form of regular n-gons. Possible variants of gonohedrons formed by equal regular polygons are presented in Table 1.

Let us consider the case when polygons of different types participate in the formation of gonohedrons. Then condition 1) takes the form

Table 3. The gonohedrons with faces three views (n < i < j)

N°	The gonohedron symbol $\begin{bmatrix} k,l,m \\ n,i,j \end{bmatrix}$	n = 3								n = 4			n=5	
		i=4	i=5	i=6	i=7	i=8	i=9	i=10	i=11	i=5	i=6	i=7	i=6	
		j	j	j	J	J	j	j	j	J	j	j	j	
13	$\begin{bmatrix} 1,1,1 \\ n,i,j \end{bmatrix}$	5÷11	6÷29	>6	8÷41	9÷23	10÷17	11÷	12	6÷19	7÷	8	7	
14	$\begin{bmatrix} 1,2,1 \\ n,i,j \end{bmatrix}$	5	-	-	-	-	-	14	13	-	11	9	-	
15	$\begin{bmatrix} 1,2,1 \\ n,i,j \end{bmatrix}_d$	5	-	-	-	-	-	-	-	-	-	-	-	
16	$\begin{bmatrix} 2,1,1 \\ n,i,j \end{bmatrix}$	5÷11	6,7	-	-	-	-	-	-	-	-	-	-	
17	$\begin{bmatrix} 2,1,1 \\ n,i,j \end{bmatrix}_d$	5÷11	6,7	-	-	-	-	-	-	-	-	-	-	

$$k + l \geq 3, \tag{2.4}$$

where k is the number of polygons of the first type ($k > 0$), 1 is the number of polygons of the second type ($l > 0$).

Let us denote the number of sides for a regular polygon of the first type - n, for the second type - i. Let $n < i$. Then condition 2) is written as

$$k(n-2)/n + l(i-2)/i < 2. \tag{2.5}$$

Condition 3) is written in this case by two inequalities:

$$k(n-2)/n + l(i-2)/i < 2(i-2)/i, \tag{2.6}$$

$$k(n-2)/n + l(i-2)/i > 2(n-2)/n. \tag{2.7}$$

Since $n < l$, inequality (2.7) holds identically if inequality (2.6) holds. Thus, to find gonohedrons formed by polygons of two different types, it is necessary to find positive integer solutions to the system of inequalities (2.4) - (2.6) with variables $k, l, n, i,$ as in nonlinear integer programming problems.

Let the plane angles of the gonohedron include the angle $\pi/3$, i. e. $n = 3$ and an equilateral triangle is part of the faces of the gonohedron. Then conditions (2.5), (2.6) take the form:

$$6(i-2)/i < k + 3l(i-2)/i < 6. \tag{2.8}$$

Let $i = 4$, i. e. the second kind of polygon is a square. Then inequalities (2.4), (2.8) imply the existence of the following distributions of regular triangles and squares around the vertex of the gonohedron k=1,l=2; k=1,l=3; k=2,l=1; k=2,l=2; k=3,l=1. Moreover, for $k = 2$, $l = 2$, there are two modifications of the gonohedron, differing in the arrangement of triangles and squares: triangles and squares alternate with each other, or two triangles are adjacent to each other (then two squares are adjacent to each other).

Let $i = 5$, i. e. the second kind of polygon is a regular pentagon. Then inequalities (2.4), (2.8) imply the existence of the following distributions of triangles and pentagons around the vertex of the gonohedron – k=1,l=2; k=2,l=1; k=2,l=2; k=3,l=1. Moreover, for $k = 2$, $l = 2$, as in the previous case, two modifications of the gonohedron are possible: triangles and pentagons are arranged in pairs, or they alternate with each other. For $i = 6$ and $i = 7$, from (2.4), (2.8) it follows that there are three types of gonohedron - k=1,l=2; k=2,l=1; k=3,l=1. For $i = 8 \div 11$ there are two types of gonohedron – k=1,l=2; k=3,l=1. For $i > 11$ there is only one type of gonohedron - k=3,l=1. If the faces of the gonohedron do not include a regular triangle, but a square is included, i.e. $n = 4$, $i > 4$, then conditions (2.6), (2.5) take the form

$$4(i-2)/i < k + 2l(i-2)/i < 4. \tag{2.9}$$

From (2.4), (2.9) it follows that $i = 5 \div 7$ there are two types of gonohedron with the participation of a square – k=1,l=2; k=2,l=1. For $i > 7$ there is one type of gonohedron -k=3,l=1. If the faces of the gonohedron do not include a regular triangle and a square ($n = 5$), then, as follows from (2.4) - (2.6), a gonohedron with two types of faces exists only when $k = 1$, $l = 2$, $i = 6$; $k = 2$, $l = 1$, $i = 6 \div 9$.

The gonohedron formed by the faces in the form of two different regular polygons are shown in Table 2. In this table, the numbering of the gonohedron starts from 6, since the 5 gonohedron are listed in Table 1. Each gonohedron in Table 2 is denoted as a 2 × 2 matrix, where the numbers in the first row of the matrix corresponds to the number of polygons forming a gonohedron. The first place of the first line shows the number of polygons with fewer sides. The second row of the matrix indicates the number of sides of the polygons, the number of which is indicated in the first row, respectively. Gonohedrons 9 and 10, with the same composition of plane angles around the vertex, differ in the arrangement of plane angles, namely, in gonohedron 10, the faces Γ_i are combined into pairs.

Let us consider the case when regular polygons of three different types participate in the formation of a gonohedron. Then condition 1) is satisfied identically, and conditions 2) and 3) are written as

$$2(j-2)/j < k(n-2)/n + l(i-2)/i + m(j-2)/j < 2, n < i < j. \tag{2.10}$$

From inequalities (2.10), as in the previous cases, there are variants of the arrangement of polygons around the vertex of the gonohedron, which are given in Table 3. The numbering of the gonohedrons in Table 3 is a continuation of the numbering of the gonohedrons in Table 2. Each gonohedron in Table 3 is denoted as a matrix 3 × 2, where the first row of the matrix corresponds to the number of polygons of various types forming a gonohedron, where the number of polygons with fewer sides is indicated in the first place, and the number of polygons with the greatest number of sides is indicated in third place. The second row of the matrix indicates the number of sides of the polygons, the number of which is given in the first row of the matrix, respectively. Note that gonohedrons 14 and 15, as well as gonohedrons 16 and 17, differ in the arrangement of flat angles around the top of the gonohedron. For gonohedrons 15 and 17, the faces Γ_i and, respectively, Γ_n are paired.

Writing down condition 2) for a gonohedron with four different regular polygons, it is easy to verify that it does not hold, i. e. there are no gonohedron with four distinct faces.

The proof of Theorem 2.1 is complete.

It should be noted that the above classification of gonohedrons with faces of regular polygons was first obtained author in 2005 (Zhizhin, 2005). A list of gonohedrons is also available in the work of Zalgaller (Zalgaller, 1966). However, in the work of Zalgaller is no proof of the existence of gonohedrons. Moreover, in it is no gonohedrons 8 for $n = 3$, $i = 6$ and $n = 3$, $i = 7$, whose existence is proved in Theorem 2.1. The resulting classification of gonohedrons, taking into account the composition of flat faces and the sequence of their arrangement includes 17 classes of gonohedrons, the numbering of which will be used in the future when constructing regularly faceted polyhedrons.

Dependences Between Planar and Dihedral Angles of Gonohedrons during their Deformation

Triangular Gonohedrons

Gonohedrons 1, 4, 5, 6, 8, 13 of the resulting taxonomy are trihedral angles. Between the angles of their flat faces and the dihedral angles (the angles between the faces) there is a relationship determined by the formula of spherical geometry (Zalgaller, 1966). Let *a, b, c* be flat angles of a trihedral angle. If we denote by *A* the dihedral angle lying opposite the angle *a*, then between the plane angles and the dihedral angle there is a relation:

$$\cos a = \cos b \cos c + \sin b \sin c \cos A \tag{2.11}$$

By relation (2.11), knowing the three values of the parameters from the four *a, b, c, A*, you can find the value of the fourth (Zalgaller, 1966).

Using relation (2.11), one can find the relationship between the dihedral angle formed by two given regular polygons adjacent on a common side and the angle between the sides of these polygons having a common vertex. For example, let the square *OADM* lie in the horizontal plane, and the (regular) triangle is attached to the side of the square *AD* (see Figure 2) so that this triangle *ADC* lies above the plane, resting on the common side with the square.

Let the dihedral angle between the square and the triangle be α. The question is, what is the angle *OAC* equal to? Consider a trihedral angle with vertex *A* and sides Δ *OAC*, Δ *DAC*, and square *OADM*. Dihedral angle $DA = \alpha$. The angle $OAC = \gamma$ lies opposite it. Other angles that make up the dihedral are $\pi / 2 = OAD$ and $\pi / 3 = DAC$. According to formula (2.11), we have in this case:

Figure 2. To calculate the angles between a square and a triangle

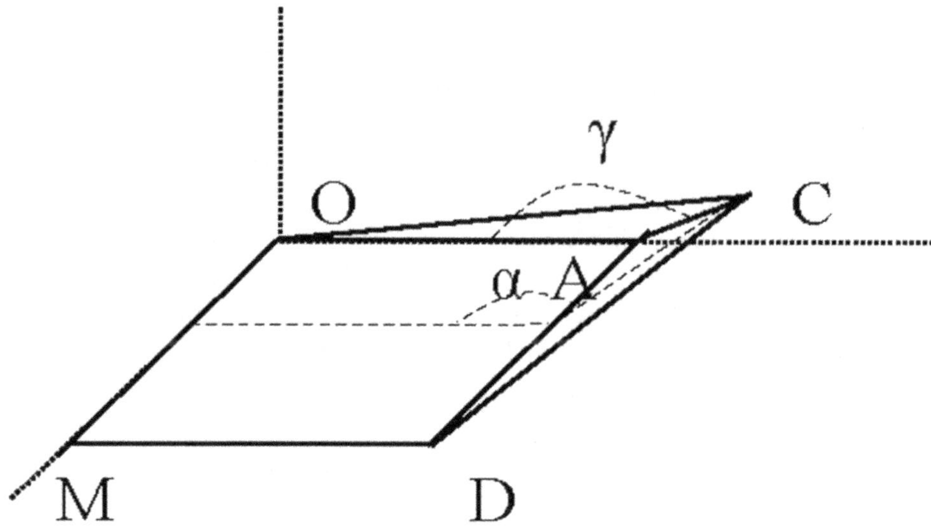

$$\cos\gamma = (\sqrt{3}/2)\cos\alpha \,. \tag{2.12}$$

Formula (2.11) can also be used when considering polyhedral angles, highlighting trihedral angles in them. In any trihedral angle, the relationship between the angles is determined by formula (2.11), and for given plane angles, the trihedral angle cannot be deformed. Polyhedral corners can be deformed. The polyhedral angles in the resulting systematics of gonohedrons include the tetrahedral gonohedrons 2, 7, 9, 10, 11, 14, 15, 16, 17 and the pentahedral gonohedrons 3 and 12.

Tetrahedral Gonohedrons

Each polyhedral corner has deformation limits associated with the conditions of the rigidity of the edges and the continuity of the edges along the edges. Many of the tetrahedral gonohedrons can be either symmetric or asymmetrical, depending on the location of the faces and the value of the dihedral angles on its edges. Let us consider sequentially the possibilities of deformation of tetrahedral gonohedrons.

Gonohedron 2

Since all faces of this gonohedron are the same, it has two limiting states in the process of deformation. One limiting state is characterized by the equality of all dihedral angles between triangles adjacent along the edges (109^0 28'16 ") and the equality of all dihedral angles between triangles and the square at the base (54^0 44'08"). In this state, the triangular faces and the square at the base form a four-sided pyramid. The gonohedron in this state has two mutually perpendicular planes of a symmetry and a symmetry axis of the 4th order. The second limiting state is associated with the degeneration of a tetrahedral angle into a flat angle of 120 °. It arises when gonohedron 2 is completely compressed along any of the two planes of symmetry. When the gonohedron is compressed along any of the planes of symmetry, the dihedral angles, the edges of which lie in this plane, increase, and the two dihedral angles, the edges of which lie in the other plane, increase. If an increase in a dihedral edge is denoted by a plus sign, and a decrease in a dihedral angle is denoted by a minus sign, then, in accordance with Cauchy's lemma (Hadamard, 1938; Cauchy, 1944), signs alternate on four edges of a gonohedron under compression. For the purposes of further construction of regularly faceted polyhedrons, of interest are those deformations and states to which they lead, which can preserve the convexity of figures and the possibility of continuing figures with edges of the same length. Therefore, in the process of compression of gonohedron 2, of interest is the state in which the former diagonal of the square at the base takes on the length of the edge. In this case, as a result of the diagonal fracture of the square, it takes the form of two regular triangles, forming a dihedral angle along the former diagonal of the square. This is how a body is formed, consisting of two tetrahedrons.

Gonohedron 7

This gonohedron is composed of three squares and one triangle. It has two limiting states. One of them is symmetrical about the plane that cuts the square, lying on the horizontal plane and the triangle along the diagonals (see Figure 3).

In this state, the dihedral angle between the squares is 135 °. Indeed, it follows from Figure 3 that the angle γ is the dihedral angle between the squares. Taking into account the symmetry of the gonohedron with respect to the plane perpendicular to the horizontal plane and passing through the diagonals of the square *AEMB* and the triangle *MGK*, assuming the length of the edge equal to *a*, we obtain *BD* = - *a* cos γ = *MN* = *HN* = *MO*, i. e. *OHMN* is a square and its diagonal is $ON = -a\sqrt{2}\cos\gamma$. But *ON* =

Figure 3. The gonohedron 7 in symmetry state

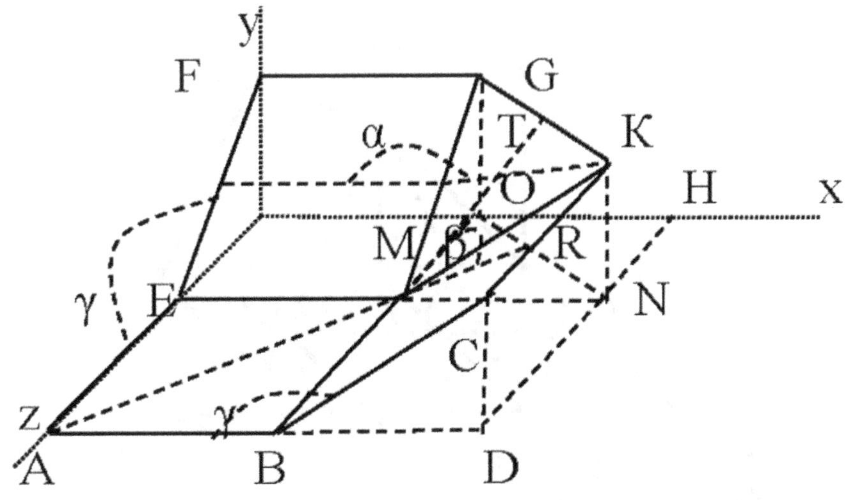

GK = *a*, i.e. $\cos\gamma = -1/\sqrt{2}$, therefore cos γ=1350· You can also find the dihedral angle between the squares and the triangle α. Using (2.12), we have $\cos\alpha = 2/\sqrt{3}\cos\gamma = -\sqrt{2/3}$, i. e. α = 144.73⁰. From *ΔMTR*, determining the coordinates *x, y, z* of points *T, R*, we find the angle β between the bisector of the angle *GMK* and the segment *MR*, i. e. the angle of inclination of the triangle to the horizontal plane $\beta = 55^0$.

The second limiting state of gonohedron 7 is associated with the transformation of a tetrahedral angle into a trihedral angle, i. e. the triangle and one of the squares are located in the same plane, and the other two squares form a dihedral angle of 150 °. Here, also, during the deformation of the gonohedron, there is an alternation of increase and decrease in dihedral angles on four edges and the loss of symmetry elements of the gonohedron. For example, if we decrease the dihedral angle *MB* (in the limit it is equal to 90 °), then the dihedral angle *EM* increases, and the dihedral angle *MG* also decreases (in the limit it is also equal to 90 °) and the dihedral angle *MK* increases (in the limit it is equal to 180 °).

Gonohedron 9

Gonohedron 9 exists in two versions - with alternating triangles and squares (denote it 9_4) and with alternating triangles and pentagons (9_5). Let us first consider the first variant of the gonohedron. In a symmetrical state, it is shown in Figure 4.

From the symmetry of the gonohedron (see Figure 4) it follows that Ð*MOF* = Ð*MAC* = α, Ð*BDF* = Ð*CDE* = γ. The diagonal of the upper square $CF = a\sqrt{2}$, *a* is the length of the edge (throughout the text of chapter).

Figure 4. Gonohedron 9 with the participation of squares in a symmetric state

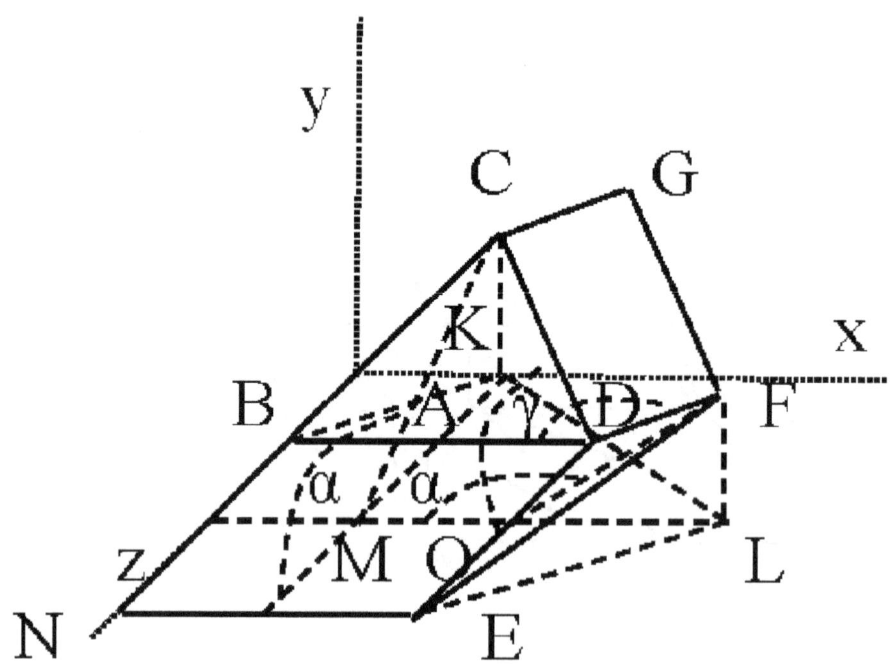

On the other hand, we have $CF = KL = \sqrt{MK^2 + ML^2} = MK\sqrt{2}$. But *MK*=*ML*=*MA*+*AK*= $a/2-(a/2)\sqrt{3}\cos\alpha$. Thus, $a\sqrt{2} = a/2-(a/2)\sqrt{3}\cos\alpha$. Hence it follows that $\cos\alpha = -1/\sqrt{3}$,

i. e. α = 125,3⁰. Taking into account (2.12), we have cos γ = - 1/2, that is, γ = 120⁰. Drawing the bisectors of the angles with the vertex *D* in the triangles *BCD, DEF*, determining the coordinates *x, y, z* of the midpoints of the sides *BC, EF*, we find the angle ψ between these bisectors, i. e. the angle between the triangles of the gonohedron 9_4 $\sin\psi / 2 = \sqrt{2/3}$, i. e. ψ = 110⁰. From the coordinates of the points

$$G(x_G = a, y_G = a\sqrt{2}, z_G = a/2), N(x_N = 0, y_N = 0, z_N = a/2),$$

we find that the distance *GN* = *2a*.

We emphasize that Zalgaller (1966), when considered tetrahedral angles, an incorrect conclusion obtained, from which it follows that for a given tetrahedral angle the angle γ = 135⁰. However, the analysis carried out using the methods of descriptive geometry showed that this angle is 120 °.

Deformation of the gonohedron violates the equality of the angles Ð*MOF* and Ð*MAC*, as well as the angles Ð*BDF* and Ð*CDE*, and the gonohedron loses its symmetry. In accordance with the Cauchy lemma, these angles are divided into pairs, some of which increase, and others decrease. From the point of view of the continuation of the gonohedron, of interest is its state, in which the decreasing angles become equal to π / 2 (see Figure 5).

In Figure 5, the angles are $\alpha 1_{=}\pi\backslash 2$. Then the angles $\gamma 1 = \pi_{\backslash} 2$. From Δ OF*L it* follows that

Figure 5. Gonohedron 9 with squares and rectangular dihedral angles

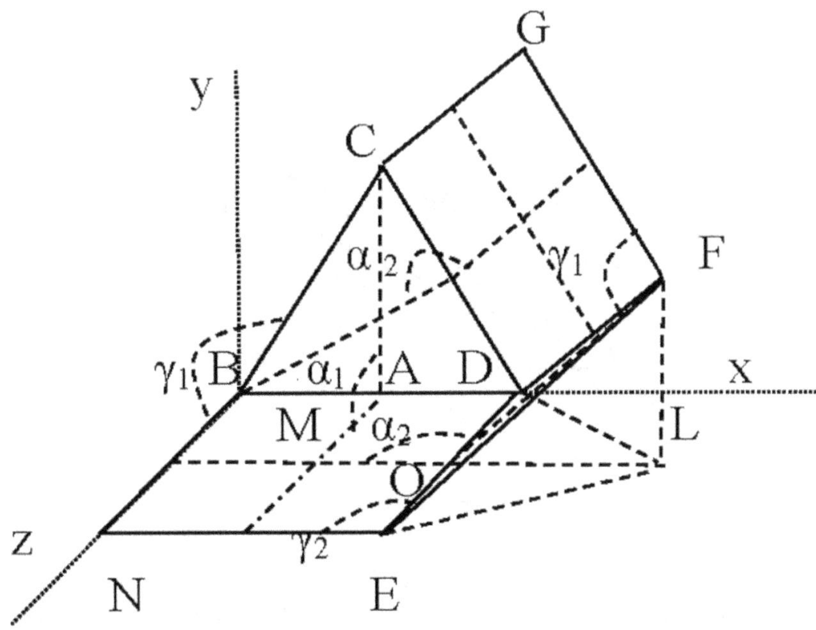

$$FL = (\sqrt{3}/2)a\sin\alpha_2, OL = (\sqrt{3}/2)a\cos\alpha_2.$$

The distance between points C and F can be determined by the coordinates of these points

$$! F=\sqrt{(a/2)^2+(\sqrt{3}/2a-\sqrt{3}/2a\sin\alpha_2)^2+(a/2-a+\sqrt{3}/2a\cos\alpha_2)^2}$$

Equating this value of $a\sqrt{2}$, as the diagonals of the square *CGFD*, we obtain that $tg\alpha_2=-1/\sqrt{3}$, hence $\alpha_2 = 150^0$. Taking into account relation (2.12), we find that $a\sqrt{2}\cos\gamma_2=-0.75, \gamma_2=138.6^0$.

With further deformation of the gonohedron, all its faces are superimposed on each other. In the works of Zalgaller, when considering tetrahedral angles with the participation of squares and triangles, this asymmetric state of the angle was not considered, although the angle in this state participates in the formation of regularly faceted polyhedrons.

Consider now a gonohedron 9, whose faces include pentagons. In a symmetric state, this gonohedron is shown in Figure 6.

In a symmetric state dihedral angles *OG, OH, OC, OF* are equal to α, and angles Ð*HOA,* Ð*FOC* are equal γ. It follows from the triangle ABG that

Figure 6. Gonohedron 9 with the participation of pentagons in a symmetric state

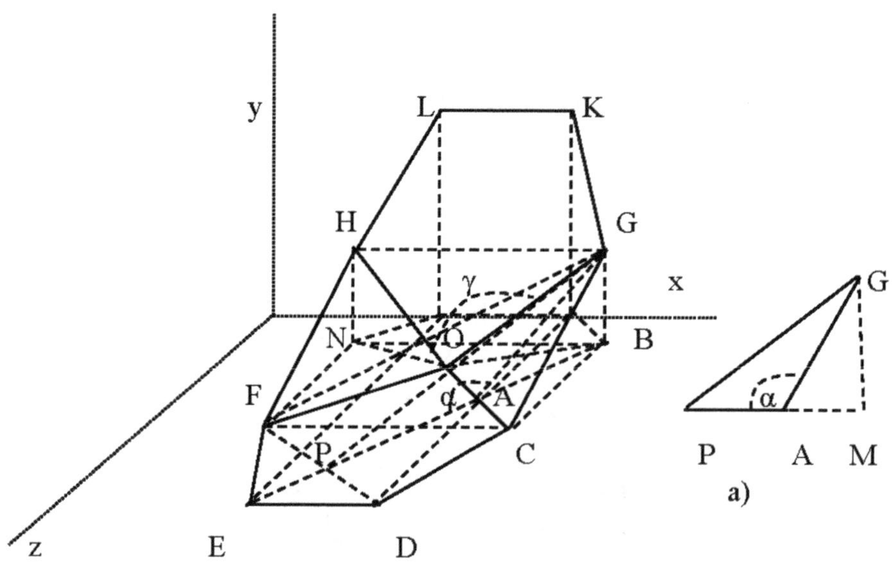

$$BG=\sqrt{3}/2a\sin\alpha, AB=-\sqrt{3}/2a\cos\alpha, OB+CB=FN,$$

$$OB=\sqrt{a^2/4+AB^2}=a/2\sqrt{1+3\cos^2\alpha}, FB=\sqrt{FC^2+CB^2}=a\sqrt{4\sin^2 54+0,25(1+3\cos^2\alpha}$$

From triangle *FGB* we find $FG=a\sqrt{4\sin^2 54+1}$, as from triangle *FOG* we received that

$$\sin(\gamma / 2) = FG / (2a) = 0,5\sqrt{1 + 4\sin^2 54}, \gamma = 144^0.$$

From triangle *PAG* we find angle *α* complementing a triangle to a right triangle P*AG* (see Figure 6 a). We received, that

$$AM = AG\cos(\pi - \alpha) = (\sqrt{3} / 2)a\cos\alpha, GM = (\sqrt{3} / 2)a\sin\alpha, DG = FG, PM = PA + AM,$$

$PG^2=PM^2+MG^2$. From triangle *PDG* we find that $PG^2=a^2(1+3\sin^2 54)$. Comparing the two found expressions for PG^2, we obtain:

$$\cos\alpha = (\sin^2 72 - 0,25 - 3\sin^2 54) / (\sqrt{3}\sin^2 72), \alpha = 142,8^0.$$

Having drawn the bisectors of the angles of triangles *FHO, OGC* with vertex *O*, taking into account that the segments *FN, CB* are parallel to the *z* axis and *FC* = *NB* = *a2sin54*, we find the angle between these bisectors, i. e. the angle ψ between the triangles of the gonohedron 9 with the participation of pentagons $\sin\psi / 2 = (2 / \sqrt{3})\sin 54, \psi = 138,18^0$. In the same way, having drawn the bisectors of the angles of the pentagons with the vertex O, determining the distance between the midpoints of the segments by their coordinates *(x, y, z),* we find the angle between the pentagons $\varphi = 116, 565^0$.

When the gonohedron is deformed, the elements of its symmetry disappear.

Gonohedron 10

This gonohedron also exists in two versions: with squares (10_4) and pentagons (10_5). Let the gonohedron consist of doubled triangles and doubled squares. For any deformations compatible with the existence of a gonohedron, it has a plane of symmetry passing through an edge common to two squares and an edge common to two triangles. In Figure 7 these are the edges *OA* and *AC*, respectively.

The maximum dihedral angle *AC* (denote it by ψ) is π. Then the dihedral angle *OA* (we denote it by φ) is equal to 120^0. The minimum values of the angles ψ and φ are equal to 0 (the tetrahedral angle degener-

Figure 7. Gonohedron 10 with squares

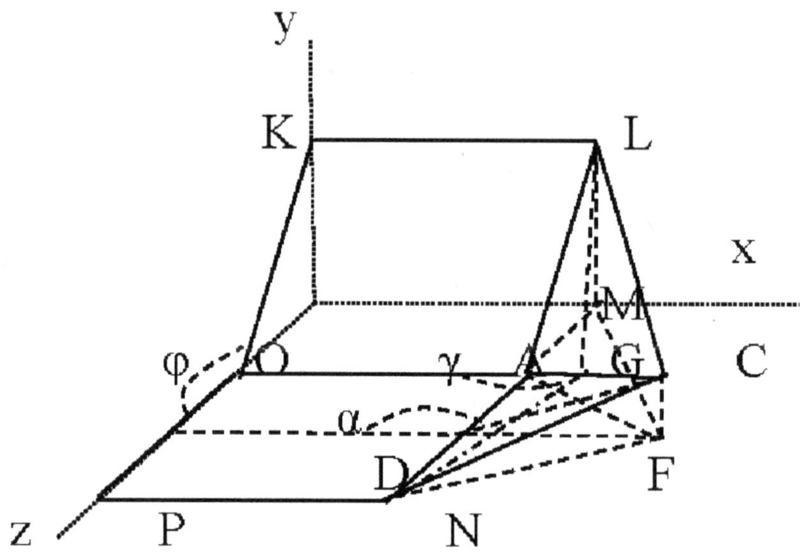

ates into a flat one). In the general case, the relationship between the angles φ and α (the dihedral angle between triangles) can be found from the expressions for the distance between the points L and C. On the one hand, $LC = a$, on the other hand. Comparing the two expressions for LC, taking into account that

$$LM = a \sin \varphi,\ AM = -a \cos \varphi,\ CF = (\sqrt{3}/2)a\sin\alpha, DF = -(\sqrt{3}/2)a\cos\alpha\ ,$$

we obtain

$$\sin\varphi = (\sqrt{3}\sin\alpha)^{-1}. \tag{2.13}$$

Therefore, if $\varphi = \pi / 2$, then $\alpha = 145^0$. Then the angle γ between the edges OA and AC, according to (2.12), is 135.2^0. From triangles NGL, NAL it is possible to establish the relationship between the angles φ and ψ

$$\sin\varphi / 2 = (\sqrt{3}/2)\sin\psi / 2. \tag{2.14}$$

Consider a gonohedron 10 with pentagons. This gonohedron (10_5) also, for all deformations compatible with the existence of a gonohedron, has a plane of symmetry passing through an edge common to two pentagons and an edge common to two triangles. The maximum opening angle of the gonohedron is associated with reaching a state in which two triangles lie in the same plane. In this case, the dihedral angle between the pentagons is $\varphi = 120^0$. The sequence for determining the functional dependencies between the characteristic angles of the gonohedron in this case is the same as in the previous case. However, the form of these dependencies is much more complicated. When depicting a gonohedron (see Figure 8), let us introduce the notation of Cartesian coordinates.

From triangle *TPR* it follows that $PR = TP\sin\varphi$, $TR = -TP\cos\varphi$, $TP = a\sin 72$. Since $PK = a\sin 54$, $RO = -PL\cos\varphi$, $PL = a\cos 54$, so coordinates of point *K* are: $x_K = 2a\sin 54$, $y_K = a\sin 72\sin\varphi$, z_K

Figure 8. Gonohedron 10 with pentagons.

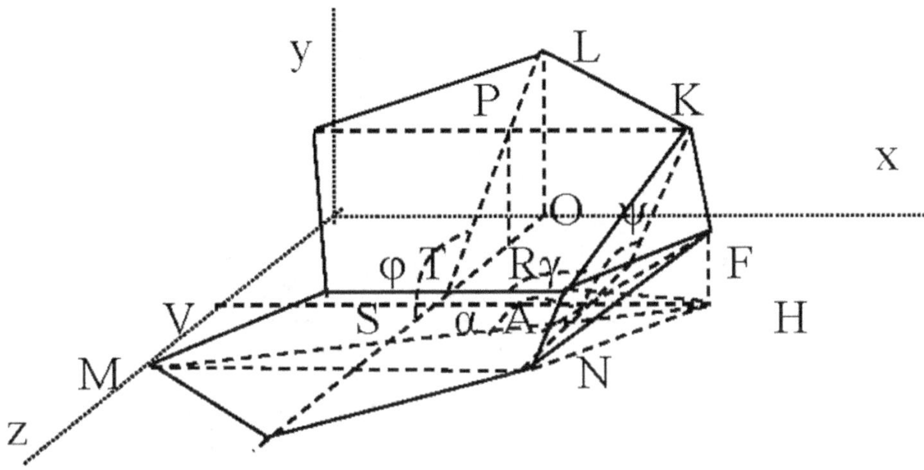

$= -a\cos 54\cos\varphi$. We find coordinates of point *F*. From triangle *SFH* we have $SH = -FS\cos\alpha$, $FH = FS\sin\alpha$, $FS = (\sqrt{3}/2)a$. Since $MH = TL + SH = a(\cos 54 + \sin 72) - a(\sqrt{3}/2)\cos 72$, so $VH = MH\sin 72$, $MV = MH\cos 72$. Thus, coordinates of point *F*: $x_F = a[\cos 54 + \sin 72 - (\sqrt{3}/2)\cos\alpha]\sin 72$, $y_F = (\sqrt{3}/2)a\sin\alpha$, $z_F = TO + TP - MV = -a(\cos 54 + \sin 72)\cos\varphi + a\sin 72 - a[\cos 54 + \sin 72 - (\sqrt{3}/2)\cos\alpha]\cos 72$. Since

$$\mathrm{KF} = a = \sqrt{(x_K - x_F)^2 + (y_K - y_F)^2 + (z_k - z_F)^2}, \tag{2.15}$$

then the last equation sets the functional dependence of the angles φ and α. It is solved numerically if necessary. The dependences of other angles on each other are found in a similar way.

Gonohedron 11

Gonohedron 11 has three triangular faces and one face in the form of an n-gon ($n > 3$). Gonohedron 11 with the participation of the square (11_4) is shown in Figure 9 in a symmetric state, when the dihedral angles *AN, NE* are equal to each other (we denote them by α).

Any deformation breaks the symmetry of the gonohedron with respect to the plane passing through the diagonals of the square and the central triangle. In the symmetric state $BL = CR$ and the angle α can

Figure 9. Gonohedron 11 with square.

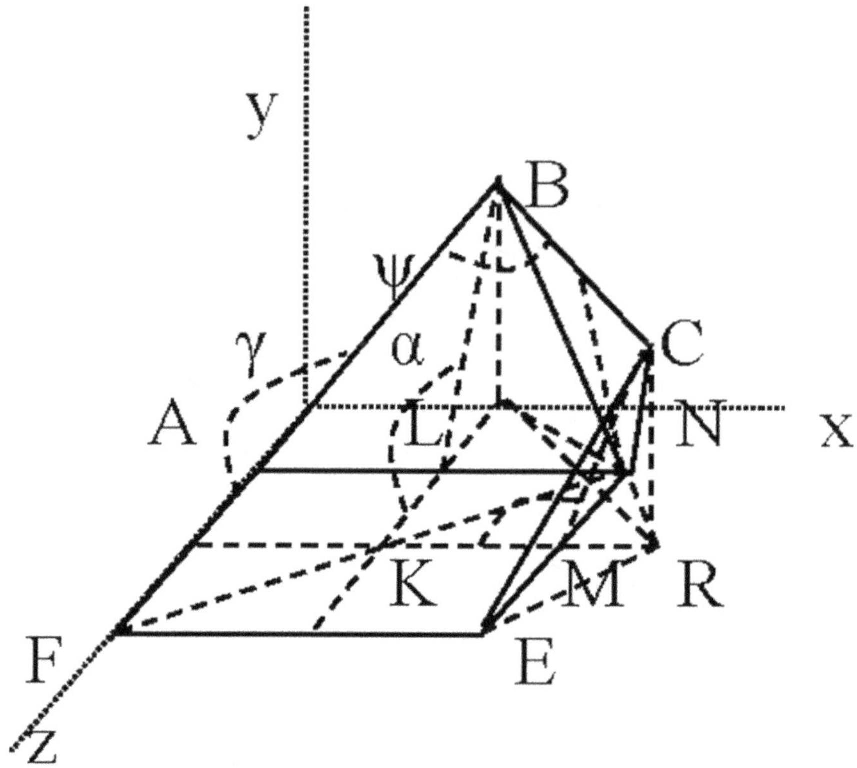

be found from the equation $a^2 = 2\,(KM + MR)^2$, where $KM = a / 2$, $MR = -\ (\sqrt{3}/2)\,a\cos\alpha$, i. e. $\cos\alpha = (1 - \sqrt{2})/\sqrt{3}$, $\alpha = 103.83^0$. From (2.12) we find that $\gamma = 102^0$. Find the distance AC. Point A coordinates: x_A=0, y_A=0, $z_A = -(\sqrt{3}/2)a\cos\alpha$. The coordinates of the point C:

$$x_C = a - (\sqrt{3}/2)a\cos\alpha, y_C = (\sqrt{3}/2)a\sin\alpha, z_C = a/2 - (\sqrt{3}/2)a\cos\alpha.$$

Then, $(AC/a)^2 = (1-\sqrt{3}/2a\cos\alpha)^2 + 3/4\sin^2\alpha + 0{,}25$, i. e. $AC = a\ 1.55377$. From triangle ABC we find $\sin(\psi/2) = 0.5\,(AC/2) = 0.776886983$, i. e. $\psi = \gamma = 102^0$.

Upon deformation, the symmetry of the gonohedron is violated and the limiting state occurs when two triangles are in the same plane, i. e. $\gamma = 120^0$. Then, according to (2.12), $\cos\alpha = -0.5735$, i. e. $\alpha = 125.3^0$.

Let us now consider the case when gonohedron 11 has a face with more than four sides (gonohedron 11_n). From triangle ARB (see Figure 10) we have $AB = -(\sqrt{3}/2)A\cos\alpha, BR = (\sqrt{3}/2)a\sin\alpha$.

From triangle OAK we get $OA = a/2\ \mathrm{ctg}\,(\pi/n)$, $OB = OA + AB = a/2\ \mathrm{ctg}\,(\pi/n) - (\sqrt{3}/2)\,a\cos\alpha$. From triangle OLB we find $LB = OB\sin(\pi/n) = a/2$, i.e.

Figure 10. Gonohedron with n-gon (n > 4).

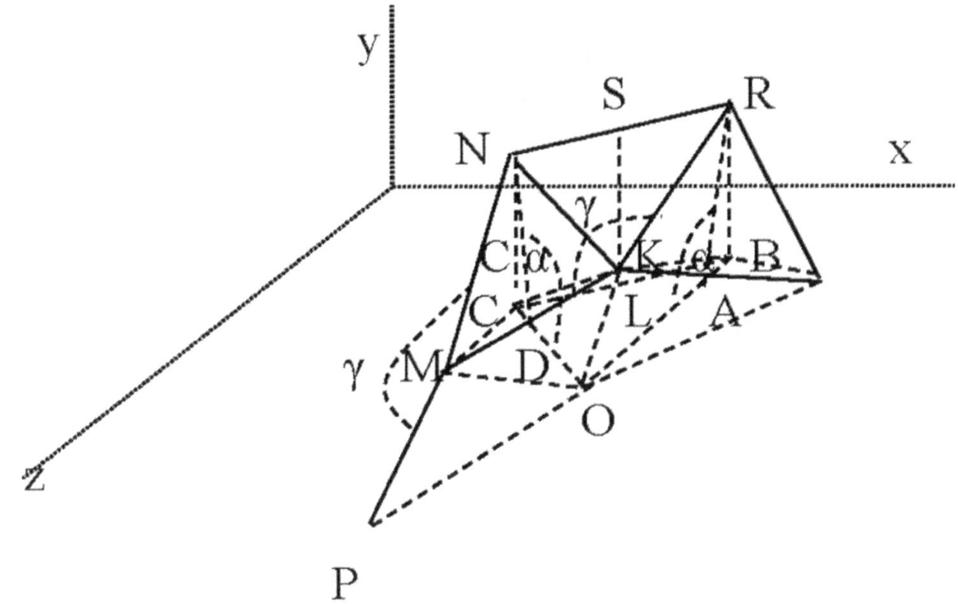

$$\cos\alpha = \frac{\cos(\pi / n) - 1}{\sqrt{3}\sin(\pi / n)}. \qquad (2.16)$$

We find an angle $\gamma =$ Ð$PMN =$ ÐMLR. We designate of coordinate point B: $x_B, y_B=0, z_B$. Thana point R has coordinates $x_R = x_B, y_R = (\sqrt{3}/2)a\sin\alpha, z_R = z_B$. We find coordinates of point M. Evident that

$$y_M = 0, z_A = z_B - (\sqrt{3}/2)a\cos\alpha, x_K = x_B - a/2, z_K = z_A,$$

$$z_M - z_K = (x_K - x_M)tg(\pi - 2\pi / n), (z_M - z_K)^2 + (x_K - x_M)^2 = a^2.$$

From two last equality (with a count coordinates of point K) we have

$$z_M = z_B - (\sqrt{3}/2)a\cos\alpha + a\sin(\pi - 2\pi / n), x_M = x_B - a/2 - a\cos(\pi - 2\pi / n).$$

Than from triangle MKR we received $MR^2 = (z_M - z_R)^2 + (y_m - y_R)^2 + (x_M - x_R)^2 = 4a^2\sin^2(\gamma/2)$, т.е.

$$\sin(\gamma / 2) = 0,5\sqrt{2 + \sqrt{3}\cos\alpha\sin(\pi - 2\pi / n) + \cos(\pi - 2\pi / n)}. \qquad (2.17)$$

When the gonohedron is deformed, the angles α and γ are again divided into increasing and decreasing. The limiting state occurs when two triangles are located in the same plane.

Gonohedron 14

Gonohedron 14 has one symmetric state (see Figure 11).

We find a dihedral angle α between pentagon and squares. From triangle *ADC* are $DC = a\sin\alpha$, $AC = -a\cos\alpha$. So Ð$LKM = 54^o$, $MK = a/2$, $LK = AC$, $MK/LK = \cos 54$, hence $\cos\alpha = -0{,}5/\cos54$, and $\alpha = 148{,}3^o$.

Figure 11. Genohedron 14.

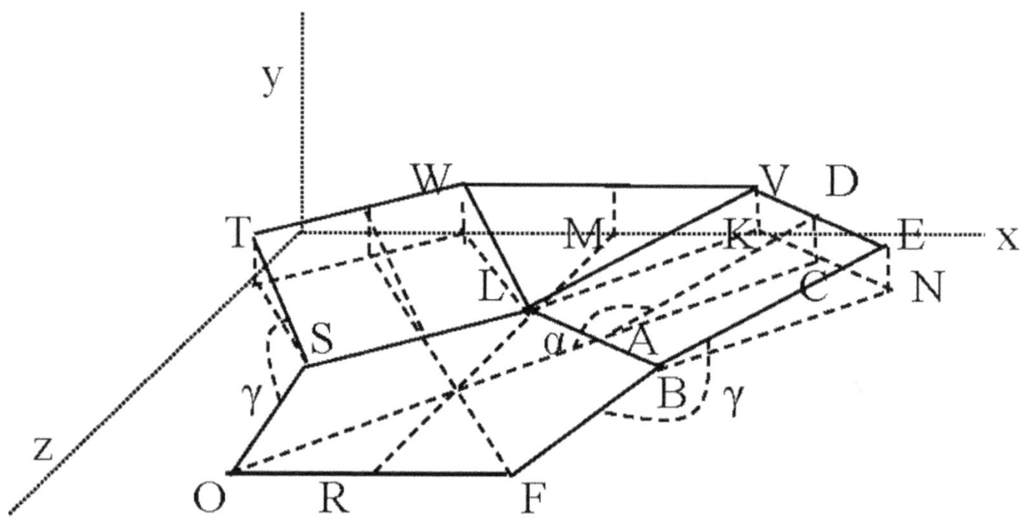

We find the angle between the sides of the pentagon and squares that have a common vertex from triangle *FEB*. Point *E* coordinates: $x_E = a\,(2\cos36 + 1)$, $y_E = a\sin\alpha$, $z_E = NP = a\sin36$. Point *F* coordinates: $x_F = a\,(\cos36 + 1)$, $y_F = 0$, $z_F = ML + LR = a\,(\cos54 + \sin72 + 0.5\mathrm{tg}54)$. Then

$$EF = 2a\sin(\gamma/2) = \sqrt{(x_F - x_E)^2 + (y_F - y_E)^2 + (z_F - z_E)^2} = a\sqrt{7/4 + \sin^2 72 - \sin^2 36 + \sin 72\, tg54}\ .$$

From the last equality it follows that $\gamma = 144^0$. In addition, ÐTWV =Ð WVE =Ð $MKN = 144^{)}$. By formula (2.12) we find the dihedral angle between squares and triangles 159.1^0. At the slightest deformation, the gonohedron loses its symmetry and rather quickly approaches its limiting state, when the triangle and one of the squares lie in the same plane and the angle $\gamma = 150^0$.

Gonohedron 16

Gonohedron 16 has a finite set of modifications (see Table 3). However, it is unlikely that gonohedron 16 with polygons with a large number of sides can be used to construct regularly faced polyhedrons. Therefore, we will consider the gonohedron 16 in two variants most convenient for such constructions. Let the gonohedron 16 have two triangles, one square, and one pentagon as faces. Its symmetric (relative to the plane passing through the diagonal of the square, perpendicular to the plane of the square) is shown in Figure 12.

Let us calculate the characteristic angles of the gonohedron in this state. From triangle *PRC* we have $PR = -(\sqrt{3}/2)a\cos\alpha_1, CR = (\sqrt{3}/2)a\sin\alpha_1$. From triangle *ONR* we get *OR* = *OP* + *PR*, *OP* = *a* /

Figure 12. Gonohedron 16 with square and pentagon

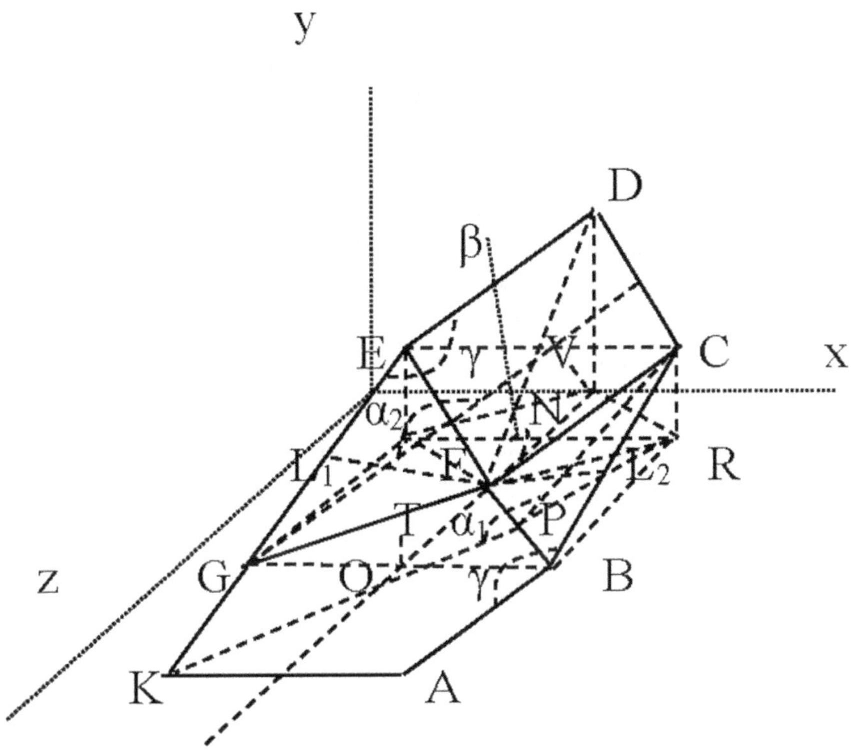

2tg54, *NR* = *OR*sin36 = $[a / 2\text{tg}54 - (\sqrt{3}/2)a\cos\alpha_1]\sin 36$. In addition, *NR* = $a / \sqrt{2}$, as half the diagonal of the *FEDC* square, whose diagonal *EC* is parallel to the horizontal coordinate plane. Equating the two expressions for *NR*, we get that $\cos\alpha_1 = \frac{1}{\sqrt{3}}(tg54 - \frac{\sqrt{2}}{\sin 36})$, i.e. the dihedral angle between the pentagon and the triangle is $\alpha 1 \cong 126.50$·

Let us calculate the angles unknown so far between the edges of the gonahedron on its outer contour. From the symmetry conditions *ABC* = *KGE* = γ. From triangle *ABC* we find value this angle. Let coordinates of point *V* are $x_V=x_0$, $y_V=0$, $z_V=0$. Then coordinates of point *C* are $x_C=x_0+NR$, $y_C=CR$, $z_C=VN=FN=ON$

Coordinates of point *A*: $x_A=x_0+a/2$, $y_A=0$, $z_A=OF+OP+2FN$. Hence $CA = 2a\sin(\gamma/2) = \sqrt{(x_C-x_A)^2+(y_C-y_A)^2+(z_C-z_A)^2}$. Из последнего равенства с учётом значений координат точек находим $\gamma = 130°$.

From condition of symmetry ÐGED = ÐDCD. Find this angle from triangle *BCD*. The coordinates of point *D*: $x_D=x_0$, $y_D=2CR$, $z_D=0$. The coordinates of point B: $x_B=x_0+OF\sin72$, yB=0, $z_B=2FN+OF-TO$,

$TO=OF\cos72$. Hence $DB = 2a \sin(\text{ÐBCD}/2) = \sqrt{(x_B - x_D)^2 + (y_B - x_D)^2 - (z_B - z_D)^2}$. From later equality we find that ÐBCD = ÐDED = 130° = γ. By (2.12) we find dihedral angle α_2 between square and triangles 137,95°. We find an angle TFD. From triangle FND we have $\cos\beta = -\cos(\text{Ð}TFD) = \sqrt{2}$ VN/a. But $VN = OR\cos36 - OF = (a/2)\cos36(\text{tg}54 - \sqrt{3}\cos\alpha_1) - (a/2)(2\sin72 + 2\cos54 - \text{ctg}36)$. Hence $\cos(\text{Ð}TFD) = \sqrt{2}(\sin72 + \cos54) - (1+\sqrt{2})\text{ctg}36$, т.е. Ð$TFD$ = 99,98°, β = 80,02°. Having drawn the bisectors of the angles of the triangles with the vertex F, by the coordinates x, y, z of the points L_1, L_2 we find the angle between the triangles in the gonohedron $16_{4,5}$, i. e. ÐL_1FL_2 = 138,18°.

During deformation, the gonohedron loses symmetry and the equality of the four angles between the edges at its boundary, up to the limiting state, when the square and one of the triangles are located in the same plane.

Gonohedron 16 with the participation of a square and a hexagon ($16_{4,6}$) qualitatively retains the features of a gonohedron with the participation of a square and a pentagon, only the values of the characteristic angles change. The gonohedron $16_{4,6}$ in the symmetric state is shown in Figure 13.

Performing transformations similar to the transformations carried out in the previous case, we calculate the characteristic angles of this gonohedron. In this case, the dihedral angles between the triangles and

Figure 13. Genohedron 16 with hexagon and square in symmetric state

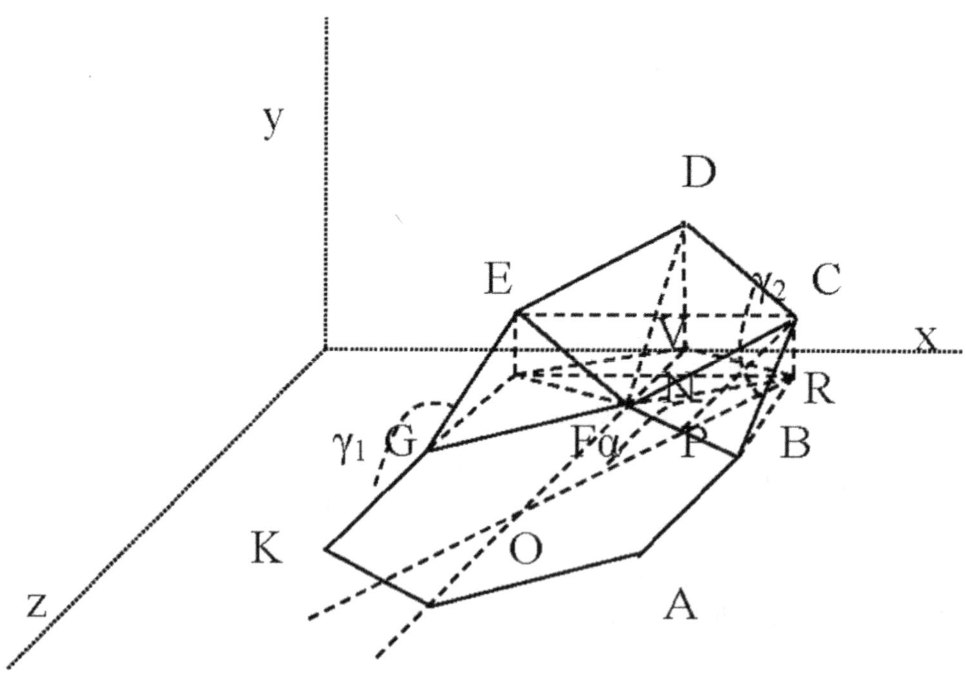

the hexagon decreased slightly, α = 129, 2°. The angles between the sides of the hexagon and triangles having a common vertex on the outer boundary of the gonohedron ÐABC = ÐKGE = $\gamma 1$_1640· The angles between the sides of the square and triangles having a common vertex on the outer boundary of the gonohedron are ÐBCD = ÐGDE = $\gamma 2$=,46.80. Using formula (2.12), by the value of $\gamma 2$, we find the dihedral angle between triangles and the square 146,80. 't is essential that in this case. When the gonohedron is deformed, it becomes asymmetrical.

In the gonohedron 16 with the participation of the pentagon and the hexagon, all characteristic angles in the symmetric state of the gonohedron are slightly larger than the corresponding angles of the gonohedron 16 with the participation of polygons with fewer sides.

Gonohedron 17

The last family of tetrahedral angles represents gonohedron 17. However, of the variants of gonohedron 17 listed in Table 3, the most suitable for use in constructing regularly faceted polyhedrons is gonohedron 17 with the participation of a square and a pentagon. All 17 gonohedrons, including the selected one, are asymmetrical in any state. But it can become symmetrical when glued to its outer contour between the square and the pentagon of two connected triangles. Moreover, the resulting figure remains symmetrical for any deformation compatible with the convexity of this figure. This gives the gonohedron 17 the possibility of wide use in the construction of convex figures. The gonohedron 17 with the participation of a square and a pentagon ($17_{4,5}$) in an arbitrary state other than the limiting ones is shown in Figure 14.

Let us determine the functional relationships between the characteristic angles of the gonohedron and consider its limiting states.

Figure 14. Gonohedron 17 with square and pentagon

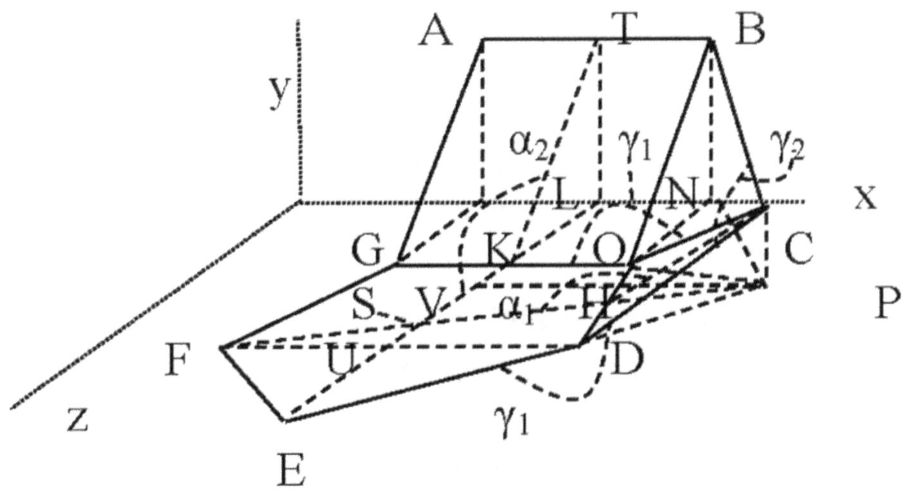

From the triangles *HCP, KLT* we find $HP =-(\sqrt{3}/2)a\cos\alpha_1, CP = (\sqrt{3}/2)a\sin\alpha_1$, $TL = NB = a \sin\alpha 2$ $KL = ON = -a\cos\alpha 2$. From the pentagon FG*ODE* and the triangle OD*P* it follows that SH = *SK* = $a/2\text{ctg}36$, SP = $a/2\text{ctg}36 - (\sqrt{3}/2)a\cos\alpha_1, VP = SP\sin 72, SV = SP\cos 72$. Therefore, it is possible to determine the coordinates of points B and C:

$$x_B = x_L + a/2, y_B = a\sin\alpha_2, z_B = 0; x_C = x_L + 0,5a(1+\cos 72) - \sqrt{3}/2\cos\alpha_1 \sin 72,$$

$$y_C = \sqrt{3}/2a\sin\alpha_1, z_C = a(0,5\sin 72 - \sqrt{3}/2\cos 72\cos\alpha_1 - \cos\alpha_2).$$

Then $BC^2=a^2=(x_B-x_C)^2+(y_B-y_C)^2+(z_B-z_C)^2$. From the last equality, taking into account the obtained coordinate values after transformations, we find the equation of the functional connection between the dihedral angles between the pentagon and the triangle and between the square and the triangle:

$$\cos\alpha_1 = \frac{\cos 72\cos\alpha_2\sqrt{1-\sin 72\cos\alpha_2} \pm \sqrt{2}\sin\alpha_2\sqrt{1+2\sin 72\cos\alpha_2}}{\sqrt{3}(1+\sin 72\cos\alpha_2)\sqrt{1-\sin 72\cos\alpha_2}}, \quad (2.18)$$

where the sign in front of the root in the numerator is chosen so that cos α_1 is negative and is less than 1 in absolute value.

Let us establish a functional connection between the angles of the outer contour of the gonohedron and the dihedral angles. Determine the coordinates of point *D*: $x_D=x_L+a\sin 54$, $y_D=0$, $z_D=a(\sin 72 - \cos \alpha 2_)$. Then

$$DB^2 = a^2 = (x_B - x_D)^2 + (y_B - y_D)^2 + (z_B - z_D)^2, \sin(\gamma_2/2) = DB/(2a).$$

From the last two equalities we find

$$4\sin^2\frac{\gamma_2}{2} = 5/4 + \sin^2 54 - \sin 54 + \sin^2 72 - 2\sin 72\cos\alpha_2. \quad (2.19)$$

Determine the coordinates of the point *E*: $x_E=x_L$, $y_E=0$, $z_E=a(\cos 54+\sin 72 - \cos \alpha 2_)$. Then, after calculating the distance EC by the coordinates of points E and C from the equality sin (γ1 / 2) = EC / (2a) we find

$$4\sin^2(\gamma_1/2) = 5/4 + 0,5\cos 72 + \cos 54(\sin 72 + \cos 54) \\ -\sqrt{3}/2\cos\alpha_1(\sin 72 + 2\cos 72\sin 72 + 2\cos 54\cos 72) \quad (2.20)$$

The gonohedron $17_{4,5}$ has two limiting states. One of them corresponds to the case when two triangles are located in the same plane, i. e. angle $\gamma_2 = 120^0$. The obtained functional dependences can be used to calculate the values of other characteristic angles $\alpha_1 = 100.8^0$, $\gamma_1 = 108^0$, $\alpha_2 = 121.7^0$. The second limiting state corresponds to the case when the square and the nearest triangle are located in the same plane, i. e. $\gamma_1 = 150^0$. The values of other characteristic angles calculated from the functional dependences are as follows: $\alpha_1 = 149.7^0$, $\alpha_2 = 80^0$, $\gamma_2 = 81^0$. It is unlikely that gonohedron with states close to the second limiting state can be used to construct convex polyhedrons, since in this case the dihedral angle α_2 is acute.

In the gonohedron 17 with the participation of a square and a hexagon, all characteristic angles are slightly larger than the corresponding angles of the gonohedron 17 with the participation of a square and a pentagon.

Pentahedral Gonohedrons

Due to the large number of edges, pentahedral gonohedrons have the greatest potential for deformation while maintaining flat faces. Therefore, with a small number of variants of such gonohedra, they are most suitable for creating regularly faceted convex polyhedrons. The pentahedral gonohedrons include gonohedron 3 and gonohedron 12 with a square, as well as gonohedron 12 with a pentagon. Each of these gonohedrons has an infinite set of states symmetric with respect to a plane passing through some edge, one of the end points of which coincides with the vertex of the gonohedron. Figures 15, 16, 17 show images in similar states of gonohedron 3, gonohedron 12 with the participation of a square, and gonohedron 12 with the participation of a pentagon, respectively.

Due to the presence in these gonohedrons of the same part, consisting of four triangles in contact with each other along the edges, it is possible to introduce a universal consideration of these gonohedrons with the derivation of general functional dependencies between the characteristic angles. To do this, it is sufficient to introduce the angle δ_i $(i = 3, 4, 5)$ between the line along which the plane of symmetry inter-

Figure 15. Gonohedron 3

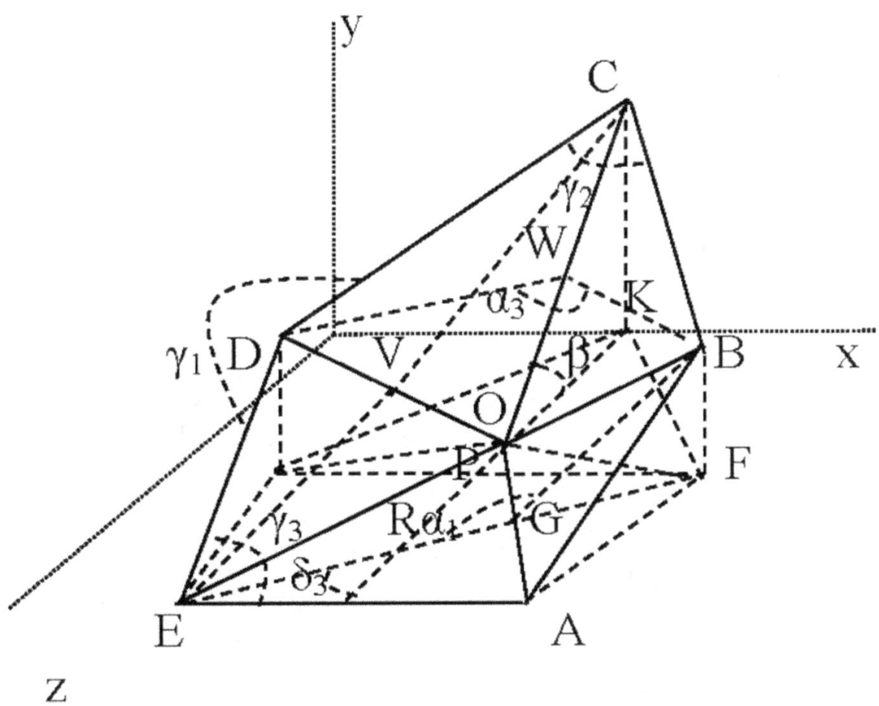

Figure 17. Gonohedron 12 with pentagon

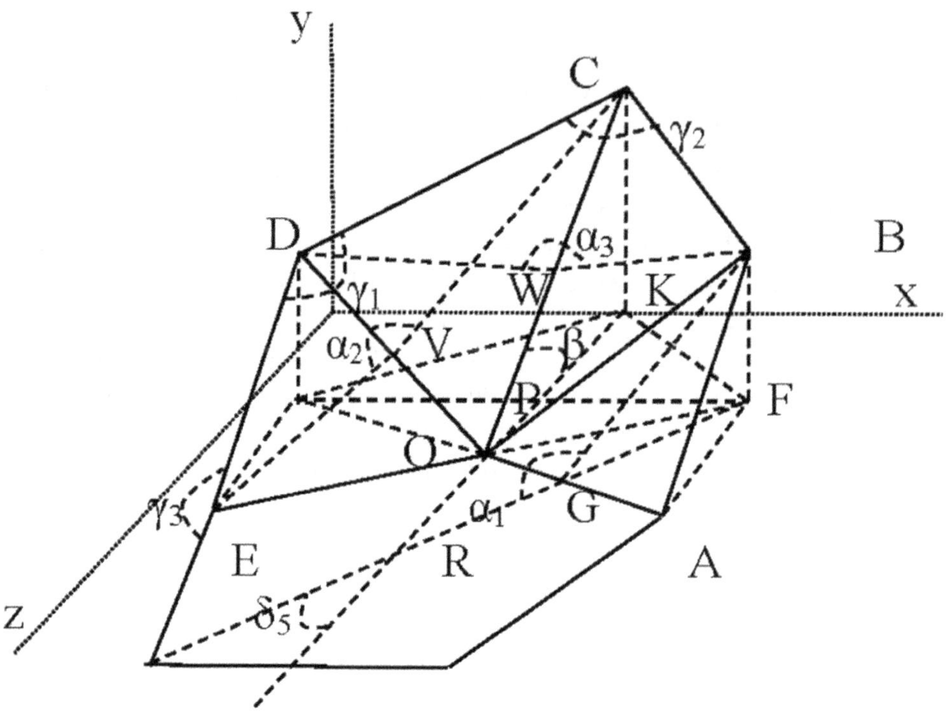

Figure 16. Gonohedron 12 with square

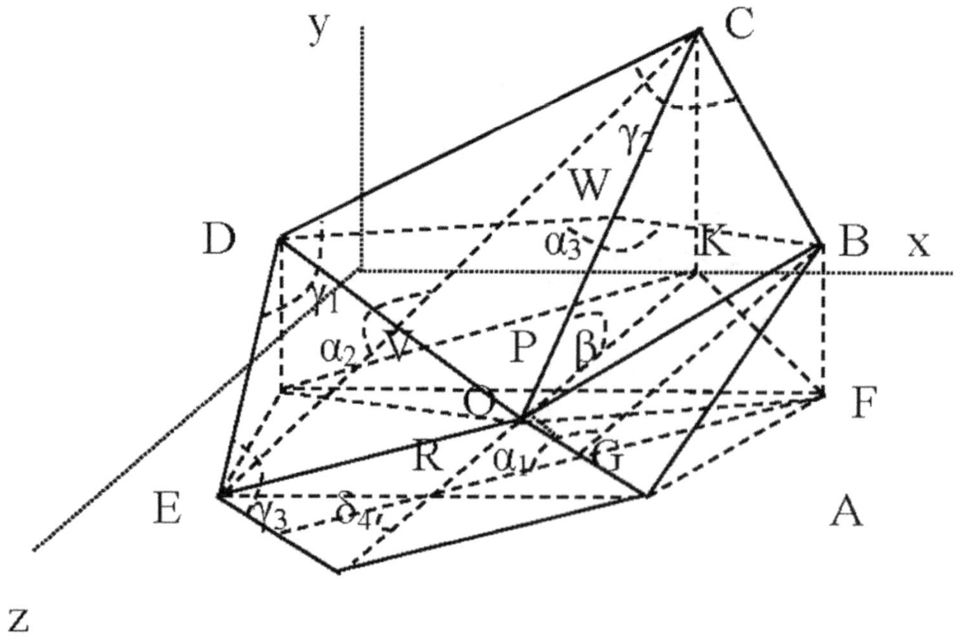

sects the base of the gonohedron (a polygon lying in the horizontal coordinate plane) and perpendicular to the middle of one of the sides of the base passing through the vertex of the gonohedron (see Figure 15 - Figure 17). Moreover, $\delta_3 = 60^0$ corresponds to gonohedron 3 (at its base there is a triangle), $\delta_4 = 45^0$ corresponds to gonohedron 12 with the participation of a square (base is a square), $\delta_5 = 36^0$ corresponds to gonohedron 12 with the participation of a pentagon (base is a pentagon). Then the coordinates of the characteristic points of the gonohedrons can be written in general form. Point *B* coordinates:

$$x_B = x_K + (a/2)\sin\delta_i(ctg\delta_i - \sqrt{3}\cos\alpha_1),\ y_B = (\sqrt{3}/2)a\sin\alpha_1,$$

$$z_B = KO + OR - RP = a\cos\beta + a/(2\sin\delta_i) - (a/2)\cos\delta_i(ctg\delta_i - \sqrt{3}\cos\alpha_1).$$

Point *C* coordinates: $x_C = x_K$, $y_C = a\sin\beta$, zC_0. Point A coordinates:

$$x_A = x_K + a\cos\delta_i, y_A = 0, z_A = a(\cos\beta + \sin\delta_i).$$

Point *D* coordinates:

$$x_D = x_K - (a/2)\sin\delta_i(ctg\delta_i - \sqrt{3}\cos\alpha_1),\ y_D = (\sqrt{3}/2)a\sin\alpha_1,$$

$$z_D = a\cos\beta + a/(2\sin\delta_i) - (a/2)\cos\delta_i(ctg\delta_i - \sqrt{3}\cos\alpha_1).$$

From the values of the coordinates of these points, one can find the necessary functional dependencies for calculating all the angles of the indicated gonohedrons.

From the condition $CB^2 = a^2 = (x_B - x_C)^2 + (y_B - y_C)^2 + (z_B - z_C)^2$, taking into account the obtained coordinate values after transformations, we find the relationship between the angles β and $\alpha 1$

$$\cos\beta = \frac{-\phi \pm \sqrt{3}\sin\alpha_1\sqrt{3\sin^2\alpha_1 - 1 + \phi^2}}{3\sin^2\alpha_1 + \phi^2},\quad \phi = \sin\delta_i + \sqrt{3}\cos\delta_i\cos\alpha_1. \tag{2.21}$$

Feedback is often needed in calculations:

$$\cos\alpha_1 = \frac{-\zeta \pm \sqrt{\zeta^2 - \left[(1+\cos\beta\sin\delta_i)^2 - 3\sin^2\beta\right]3(\sin^2\beta + \cos^2\delta_i\cos^2\beta)}}{3(\sin^2\beta + \cos^2\delta_i\cos^2\beta)}, \tag{2.22}$$

$$\zeta = (1+\cos\beta\sin\delta_i)\sqrt{3}\cos\delta_i\cos\beta.$$

From triangle *CVA* we find $CA = a\sqrt{3}\sin(\alpha_2/2)$, but $CA^2 = (x_A - x_C)^2 + (y_A - y_C)^2 + (z_A - z_C)^2$. Hence

$$\sin(\alpha_2 / 2) = \sqrt{2/3}\sqrt{1+\sin\delta_i \cos\beta}\,. \tag{2.23}$$

From triangle *COA* we find *CA*=2*a*sin(γ1$_{/}$2), hence

$$\sin(\gamma_1 / 2) = \sqrt{1/2}\sqrt{1+\sin\delta_i \cos\beta}\,, \tag{2.24}$$

$$\sin(\alpha_2 / 2) = (2\sqrt{3})\sin(\gamma_1 / 2). \tag{2.25}$$

Analogous can received two similar equality:

$$\sin(\alpha_3 / 2) = (2\sqrt{3})\sin(\gamma_2 / 2), \tag{2.26}$$

$$\sin(\alpha_1 / 2) = (2\sqrt{3})\sin(\gamma_3 / 2). \tag{2.27}$$

From triangle *DWB* we find *DB*=$a\sqrt{3}\sin(\alpha_3 / 2)$. But

$$DB = 2PF = a(\cos\delta_i - \sqrt{3}\sin\delta_i \cos\alpha_1),$$

hence we have:

$$\sin(\alpha_3 / 2) = (1/\sqrt{3})\cos\delta_i - \sin\delta_i \cos\alpha_1. \tag{2.28}$$

The obtained equalities accurately determine the values of all angles of the gonohedrons by the value of one of the angles.
Consider two limiting cases.

1. 1. Let α3$_{=}\pi$, γ2=1$_2$00. Then:

$$\cos\alpha_1 = -\frac{1-(1/\sqrt{3})\cos\delta_i}{\sin\delta_i}, \quad \cos\beta = \frac{\sqrt{3}\cos\delta_i - 1}{\sin\delta_i}, \tag{2.29}$$

$$\sin(\alpha_2/2) = \sqrt{(2/\sqrt{3})\cos\delta_i}, \quad \sin(\gamma_1 / 2) = \sqrt{(\sqrt{3}/2)\cos\delta_i}\,.$$

2. 2. Let α2$_{=}\pi$, γ1=1$_2$00. Then:

$$\cos\beta = 1/(2\sin\delta i),\ \cos\alpha 1 = \frac{2\sqrt{2}-3}{\sqrt{3}}\mathrm{ctg}\delta i. \tag{2.30}$$

$$\sin(\gamma_3/2) = 0,5\sqrt{3/2}\sqrt{1-\frac{2\sqrt{2}-3}{\sqrt{3}}ctg\delta_i},\ \sin(\alpha_3/2) = 2\frac{2-\sqrt{2}}{\sqrt{3}}\cos\delta_i,$$

$$\sin(\gamma_2/2) = (2 - \sqrt{2})\cos\delta_1 .$$

Using the obtained analytical expressions, setting various values of δ_i, one can calculate the values of the characteristic angles of various gonohedrons. Table 4 shows the results of calculations of the characteristic angles of pentahedral gonohedrons in limiting states.

Pentahedral gonohedrons can be used to construct regularly faceted polyhedrons even when their states are asymmetric. In these cases, the asymmetry of some gonohedron compensates for the asymmetry of other gonohedrons.

Table 4. Values of characteristic angles of pentahedral gonohedrons in limiting symmetric states

Gonohedron	State	β	α_1	α_2	α_3	γ_1	γ_2	γ_3
3 $\delta_3 = 60°$	1	98.9	145.22	98.9	180	82.3	120	111.5
	2	54.75	93.28	180	17.8	120	34	78
12 with square $\delta_4=45°$	1	71.5	146.8	129.36	180	103	120	112.14
	2	45	95.685	180	57.16	120	49	79.8
12 with pentagon $\delta_5 = 36°$	1	46.95	155	150.2	180	113.62	120	115.46
	2	31.4	97.85	180	66.4	120	56.6	81.5

CONVEX SIMPLE REGULARLY FACETED POLYTOPES

Theorem 2.2. The set of convex simple regularly faceted polyhedrons consists of 36 polyhedrons. It includes 2 regular and 34 irregular polyhedrons. Among the irregular polyhedrons there are 2 pyramids, 9 known semiregular polyhedrons, 7 polyhedrons that are parts of known semiregular polyhedrons, 7 out of 8 polyhedrons declared for the first time by V.A. Zalgaller, and 9 new polytopes established in this paper, two of which are new semiregular polytopes.

Evidence

By the method described in the introduction, there is a set of polyhedrons suspicious of regularly faceted polyhedrons (except for regularly faceted polyhedrons, which are known regular and semiregular polyhedrons and their parts). Then each polyhedron from this set (with the exception of rare obvious cases) is investigated analytically and numerically for the consistency of the figure with the functional dependencies obtained in the previous chapter between the angles of the gonohedrons, the conditions for the figure to be closed, the absence of false edges and fracture of the edges. All characteristic angles and dimensions of the polyhedron are determined simultaneously. The search results for regularly faceted polyhedrons are given in Table 5, which includes the known regular and semiregular polyhedrons and their parts. It indicates from which motives and by pasting which polygons between the sides of the motives, each polyhedron is obtained. Polyhedrons are denoted by the symbol PG with a subscript indicating the number of the polyhedron that appears sequentially during the study (first column of the table). The second column of the table schematically depicts the motive and the polygons glued between the sides of the motive. In this case, the central polygon remains explicitly from the motive. The sides of this polygon with a dotted line continue outside the polygon. In the obtained areas, adjacent to the sides of the central polygon of the motive (outside it), numbers are put that coincide with the number of sides of the polygons of the motive surrounding the central polygon of the motive. And in the areas bounded by the dotted extensions of the sides of the central polygon of the motif, numbers are placed corresponding to the designations of the pasted polygons between the sides of the motif. The designations of the pasted polygons are shown in Figure 18.

Figure 18. Numeric designations for glued polygons

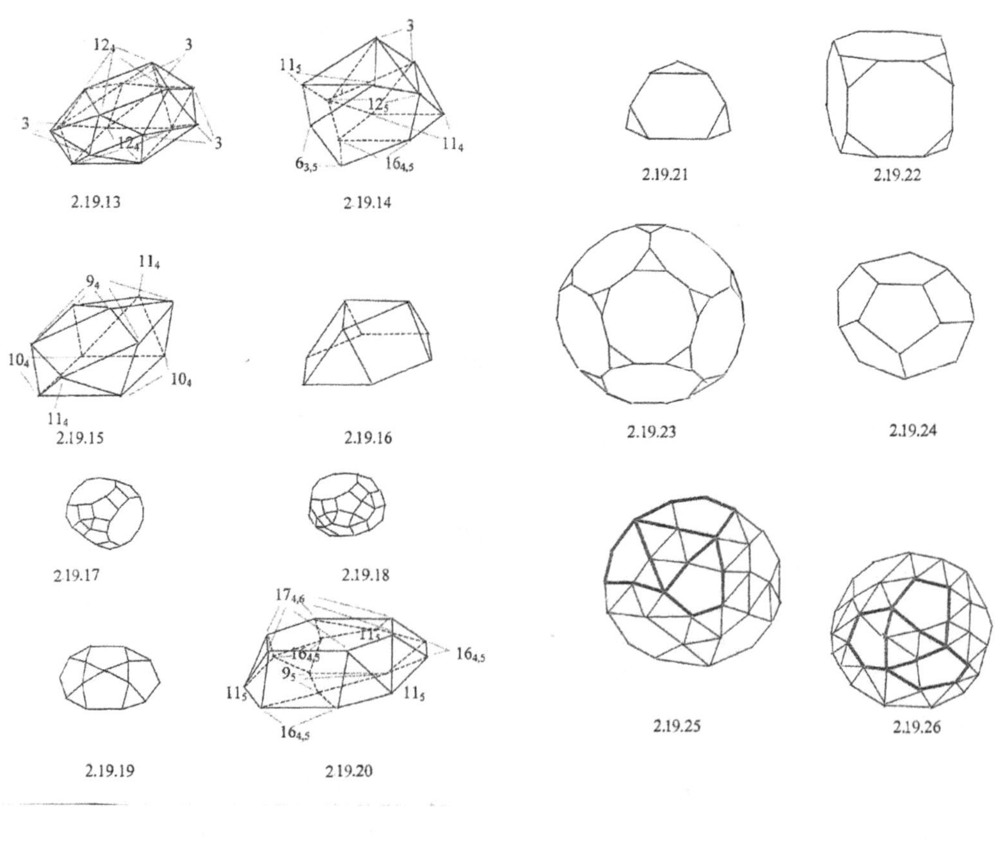

If the sides of the motive are connected to each other, i. e. an empty set is pasted in, then in the area between the dotted extensions of the sides of the central polygon of the motive, 0 is put on the schematic image. If a triangle is pasted in, then in the area between the dotted continuation of the sides of the central polygon of the motif on the image, 3 is put, if a square is pasted, then 4 is put, etc. Separately, images of inserted complexes from a square and a triangle are entered when joining a triangle to the left (3 + 4) and right (4 + 3) triangles, since such insertions often lead to different shapes. Similarly, images of inserted complexes from a pentagon and a triangle are entered when a triangle is attached to the pentagon on the left (3 + 5) and on the right (5 + 3). Listed in Figure 18 ten insert options exhaust all possible insert options for any motif from regular polygons. The third column of the table lists the established names of known polyhedrons. In Figure 19 shows the polyhedrons shown in Table 5.

Figure 19. Regularly faceted polyhedrons

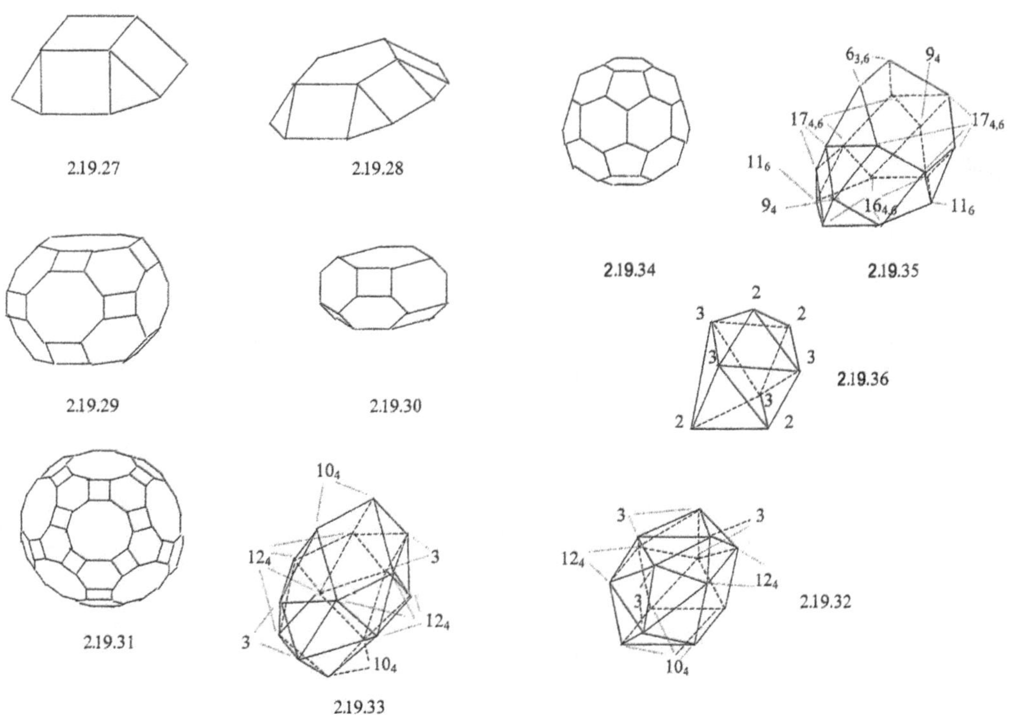

Since the existence of a figure is considered proven only when all its characteristic angles and dimensions are determined, the proof of the existence is carried out not only for the figures first discovered in this work, but also for the figures declared by Zalgaller, as regularly faceted polyhedrons, since there were no such proofs of existence in works of Zalgaller. For one from the figures declared by Zalgaller, as regularly faceted polyhedron that do not satisfy the listed conditions of existence, provide proofs of the impossibility of their existence.

Table 5. Regularly simple faceted polyhedrons

Polyhedron	Motive diagram and inserts	Name
1	2	3
PG1	3 0 3 0 3 0	tetrahedron
PG2	3+4 3 3 4+3 3 4+3	flat-nosed cube (left)
PG3*	4+3 3 3 3+4 3 3+4	flat-nosed cube (right)
PG4	3+3 3 3 3+3 3 3+3	pentagonal pyramid
PG5	5 3 3 5 3 5	triple-cut icosahedron
PG6	3 3 3 4 3 4	quadrangular pyramid
PG7	4 3 3 3+3 3 3+3	wedgecrown
PG8*	4+3 3 3 3+3 3 3+3	....
PG9*	3+3 3 3 4+3 3 3+4	....
PG10*	5 3 3 4+3 3 3+4	....
PG11*	5 3 3 3+4 3 3+4	....
PG12	5 3 5 3 3 5 3 5	double serporotonda

Polyhedron	Motive diagram and inserts	Name
1	2	3
PG15*	4 3 4 3 3 4 3 3	....
PG16	3 4 4 3 4 3	gable dome
PG17	5 4 4 5 4 5	rhomboicosi-dodecahedron thrice-truncated
PG18	5 4 4 5 4 5	rhomboicosi-dodecahedron twice-truncated
PG19	3 5 5 3 5 3	five-slope rotunda
PG20	3 5 5 3 5 3	flattened triangular clinorotonda
PG21	0 6 6 0 6 0	truncated tetrahedron
PG22	0 8 8 0 8 0	truncated cube
PG23	0 10 10 0 10 0	truncated dodecahedron
PG24	0 5 0 5 0 5 5 0 5 0	dodecahedron
PG25	3+3 3 3+3 3 3+3 3 3 3+3 3 3+3	flat-nosed dodecahedron (left)
PG26*	3+3 3 3+3 3 3+3 3 3 3+3 3 3+3	flat-nosed dodecahedron (right)

Continued on following page

Table 5. Continued

1	2	3	1	2	3
PG_{13}	6, 3; 3+3 3 3 3+3 3 3+3	flat-nosed square antiprism	PG_{27}	3, 4, 3 4 4 3 4 3	four-slope dome
PG_{14}*	3, 3; 3 3 3 5 3 5	tanbulan	PG_{28}	3 4 3 4 3 4 4 3 4 3	five-slope dome
PG_{29}	0, 6; 0 8 8 0 6 0	truncated cuboctahedron	PG_{33}	3+3, 3, 3+3 3 3 3 4 3	belted two-section
PG_{30}	0, 6; 0 6 6 0 6 0	truncated octahedron	PG_{34}	0 6 0 6 0 6 6 0 6 0	truncated icosahedron
PG_{31}	0, 6; 0 10 10 0 6 0	truncated icosadodecahedron	PG_{35}*	6, 3 4 3 3 3 6 3	……………………
PG_{32}	3+3, 3; 3+3 3 3 3 4 3	flattened great wedgecrown	PG_{36}	3 3 3 3+3 3 3+3	flat-nosed biclinoid

Of the polyhedrons listed in Table 5, 2 polyhedrons (PG_1 and PG_{24}) are regular, 9 known semiregular polyhedrons (PG_2, PG_{21} – PG_{23}, PG_{25}, PG_{29} – PG_{33}), 2 pyramids (PG_4, PG_6), 6 polyhedrons (PG_5, PG_{16} – PG_{19}, PG_{27}) are parts of the well-known semiregular polyhedrons, 6 polyhedrons (PG_7, PG_{12}, PG_{13}, PG_{19}, PG_{20}, PG_{32}) out of 9 declared for the first time by V.A. Zalgaller, and 9, indicated in the table with asterisks, new polyhedrons (PG_3, PG_8 – PG_{11}, PG_{14}, PG_{15}, PG_{26}, PG_{34}), two of which (PG_3, PG_{26}) are new semiregular polyhedrons. It follows from Table 5 that the same motives, even with the same insertions between the sides of the motive, can lead to different polyhedrons (for example, PG_{19}, PG_{20}). This is due in these cases to the possibility of various options for further continuation of the construction of closed figures. It should be noted that the same polyhedron, if it is not regular or semi-regular, can be obtained from different initial motives. However, the set of motives leading to convex simple regularly faceted polyhedrons is a closed finite set. The names of the new regularly faceted polyhedrons are absent in Table 5 due to the sufficient conditionality of these names, except for the names of two new semiregular polytopes associated with the known semiregular polytopes.

As an example, we give a detailed proof of the existence of two of the regular-faceted polytopes PG_3 and PG_{14}. The existence of other regularly faceted polyhedrons is proved in a similar way. Details of these proofs can be found in the author's monograph (Zhizhin, 2009), specially devoted to this issue. In addition, we give a proof of the impossibility of the existence of the M23 figure, declared by Zalgaller, as a regular polyhedron.

Figure PG_3

Each vertex of a semi-regular polyhedron of a flat-nosed cube (Figure 19) is a gonohedron 12_4 in a symmetric state. However, there are two options for the arrangement of squares and triangles on the surface of a polyhedron, each vertex of which is a gonohedron 12_4 in a symmetric state. In the known version, which is a flat-nosed cube, each motif of triangles a) in Figure 19 is surrounded by complexes 3 + 4 in Figure 18. In the new version of the arrangement of squares and triangles on the surface of a polyhedron, each motif of triangles a) is surrounded by complexes 4 + 3 in Figure 18. As a result, we have another polyhedron that cannot be reduced to the first by any motion. The existence of the second (new) polyhedron, as well as the first, is beyond doubt, since in this new polyhedron all the angles between flat faces are the same as in the well-known first polyhedron (they are defined when analyzing the gonohedron 12_4 in a symmetric state). The difference between them consists in the different order of alternation of triangles and squares. If we associate the triangle in the 3 + 4 complex with the direction of the arrow (since the triangle represents the pointed part of this complex), then in the known version the bypass of motif a) in Figure 19 occurs from right to left. Therefore, the well-known version of the flat-nosed cube can be called left. In the new polyhedron, motive a) in Figure 19 is traversed from left to right (Figure 19). Therefore, the new version of the flat-nosed cube can be called right. You will notice that both variants of the flat-nosed cube are related by a mirror image. However, since they are thus incompatible with each other no movement (like the right and left hand), then they should be considered different figures.

Figure PG_{14}

The figure PG14 involves (see Figure 19) two hard trihedral gonohedrons 6 with pentagons and triangles ($6_{3,5}$). The vertices of these gonohedrons have two vertices in each pentagon. The rest of the vertices of the pentagons are pairs of vertices of the gonohedrons 11_5, 12_5 and $16_{4,5}$. The remaining two vertices of the squares in the gonohedrons $16_{4,5}$ are the vertices of the gonohedrons 11_4. Two gonohedrons close the surface 3. To prove the statement that the figure PG_{14} is a regularly faceted polyhedron, we introduce a spatial coordinate system, giving the figure an orientation more convenient for the proof (Figure 20).

Figure 20. Polyhedron PG_{14}

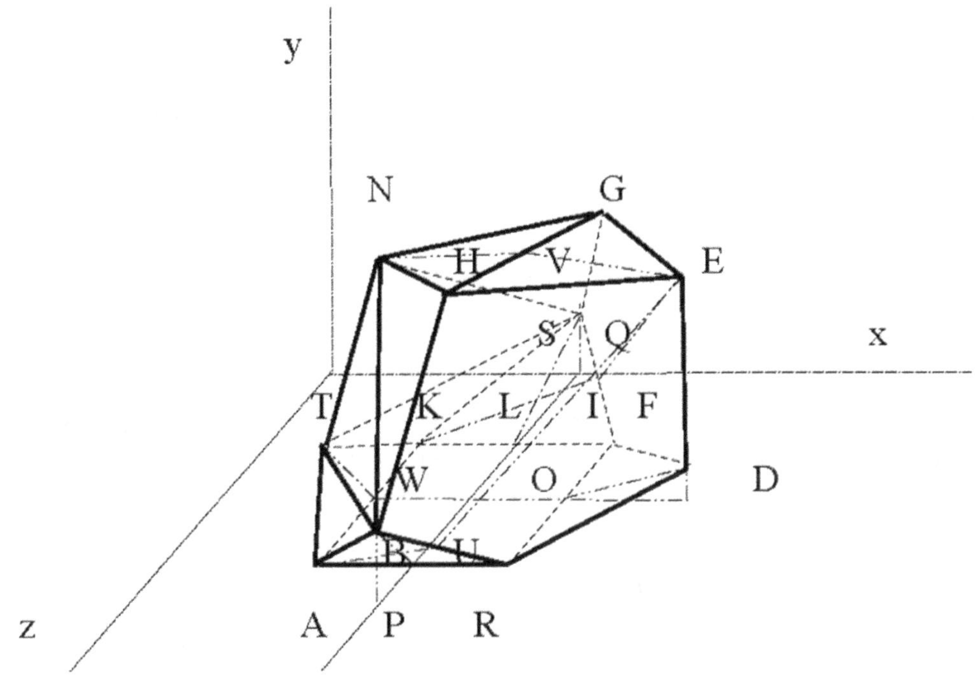

The polyhedron PG_{14} has a plane of symmetry passing through the common edge *AU* of the pentagons in gonohedrons $6_{3,5}$ and the common edge *TN* in gonohedrons 3. We denote the angles of inclination of the *ABC, LQF* triangles, symmetrically adjacent to the sides of the square α, and the angle of inclination of the triangle *CFD*, adjacent to one from the other sides of the square β (Figure 20). It follows from the triangles *BPR, ODM* that

$$PB = (\sqrt{3}/2)a\sin\alpha, RP = (\sqrt{3}/2)a\cos\alpha, OM = (\sqrt{3}/2)a\cos\beta, MD = (\sqrt{3}/2)a\sin\beta .$$

This allows you to express the coordinates of points *Q, B, D, C* in terms of corners α and β:

$$x_Q = x_S, y_Q = (\sqrt{3}/2)a\sin\alpha, z_Q = 0; x_B = x_S, y_B = (\sqrt{3}/2)a\sin\alpha,$$
$$z_B = (\sqrt{3}/2)a\cos\alpha, x_D = x_S + a/2 + (\sqrt{3}/2)a\cos\beta, y_D = (\sqrt{3}/2)a\sin\beta,$$
$$z_D = a/2 + (\sqrt{3}/2)a\cos\alpha; x_C = x_S + a/2, y_C = 0, z_C = a + (\sqrt{3}/2)a\cos\alpha.$$

Since the dihedral angle φ between the pentagons in the trihedral gonohedron $63_{,5}$ is constant, is found from (2.11) and is equal to 63,4347, then the distance between the vertices of the pentagons BQ=2a(cos54+sin72)sin(φ/2)=1.618028a.. Butfromthefigure(2.20)itfollowsthat $BQ = a(1+\sqrt{3}\cos\alpha)$. Comparing the last two equalities, we find the angle α=69.10.The distance BD can be calculated using the characteristics of the pentagon and the coordinates of the points B, *D.* As a result, we have equality.

$$BD^2 = 4a^2 \sin^2 54 =$$
$$[a/2+(\sqrt{3}/2)a\cos\beta]^2+[(\sqrt{3}/2)a\sin\alpha-(\sqrt{3}/2)a\sin\beta]^2+[(\sqrt{3}/2)a\cos\alpha+1/2]^2.$$

From this equality we find that the angle β=20.90· Having designated the coordinates of the point E, $x_E, y_E, z_e = a/2+(\sqrt{3}/2)a\cos\alpha$ we express the squares of the distances *QE, CE* in terms of the coordinates of the corresponding points and the characteristics of the pentagon.

$$QE^2 = 4a^2\sin^2 54 = (x_Q - x_E)^2 + (y_Q - y_E)^2 + (z_Q - z_E)^2,$$
$$CE^2 = 4a^2\sin^2 54 = (x_C - x_E)^2 + (y_C - y_E)^2 + (z_C - z_E)^2.$$

Resolving this system of equations, taking into account the known values of the coordinates, we find the unknown values of the coordinates $y_E/a = (x_E - x_S)/a = 1.30902^0$.
Taking into account that

$$ED = y_E - DM, DM = (\sqrt{3}/2)a\sin\beta = 0{,}309,$$

we get that $ED = a$. This is how it should be. But the fact that this result was obtained by calculation confirms the correctness of the calculation method. In addition, it turned out that $y_E = y_D$. This means that the edge *ED* is perpendicular to the horizontal coordinate plane.

Let's denote the coordinates of the point *T*: $x_T, y_T, z_T = a/2+(\sqrt{3}/2)a\cos\alpha$. Consider similar equations for squared distances *AT, TB:*

$$AT^2 = a^2 = (x_A - x_T)^2 + (y_A - y_T)^2 + (z_A - z_T)^2,$$
$$TB^2 = a^2 = (x_T - x_B)^2 + (y_T - y_B)^2 + (z_T - z_B)^2.$$

Solving this system of equations, we find the values of the coordinates of the point *T:* $x_T = x_S - 0.585414a$, $y_T = 0.861803a$, $z_T = 0.809014a$. Having written down the coordinates of the point H ($x_H = x_S + a/2$, $y_H = 1.618034a$, $z_H = 1.309014a$), we calculate the distance *TH*

$$TH = \sqrt{(x_T - x_H)^2 + (y_T - y_H)^2 + (z_T - z_H)^2} = 1{,}41421a.$$

Since *TH* is less than twice the height of the triangle, equal to $\sqrt{3}a$, then the edge *NB* is not conditional. Taking into account the symmetry of the surface of two gonohedrons 3 of the figure PG_{14}, it can be argued that no two triangles on this surface form a conditional edge (the calculation shows that TN = NA). The rest of the dihedral angles in the PG_{14} polyhedron differ significantly from 180 degrees and do not require numerical verification. Thus, it has been proved that the figure PG_{14} is a regularly faceted polyhedron.

To check the existence of Figure 20, its spatial model was built from paper. This made it possible to find that it has two sections in the form of a heptagonal polygon ODEVNTW (plane of symmetry) and an irregular pentagon RIEUA.

Figure M23

Let us prove that one of the eight polytopes declared by Zalgaller as new simple regularly faceted polyhedron (included in the set of suspicious regularly faceted polyhedrons) is not really one. This polyhedron is designated by Zalgaller as M 23. Its appearance is shown in Figure 21.

Figure 21. Figure M23

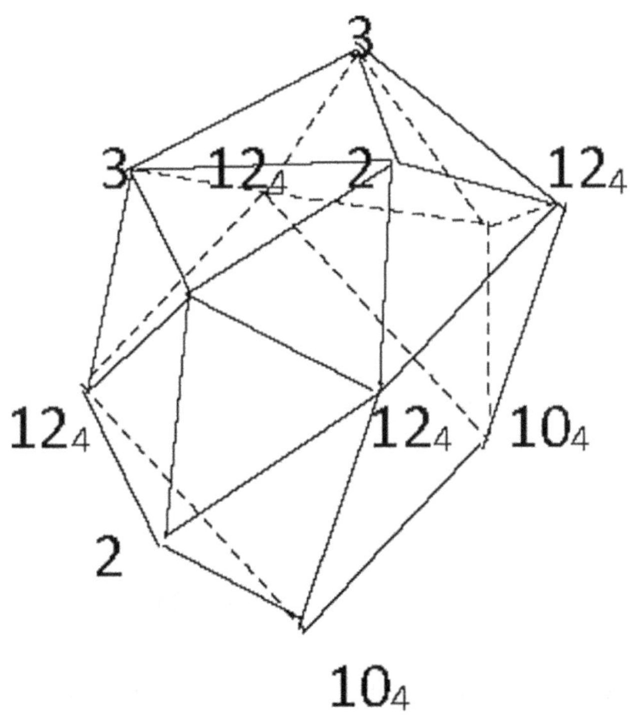

Figure M23 has two planes of symmetry intersecting at right angles. The sections of the figure with these planes are shown in Figure 22 a, b.

Figure 22. Sections of Figure M23

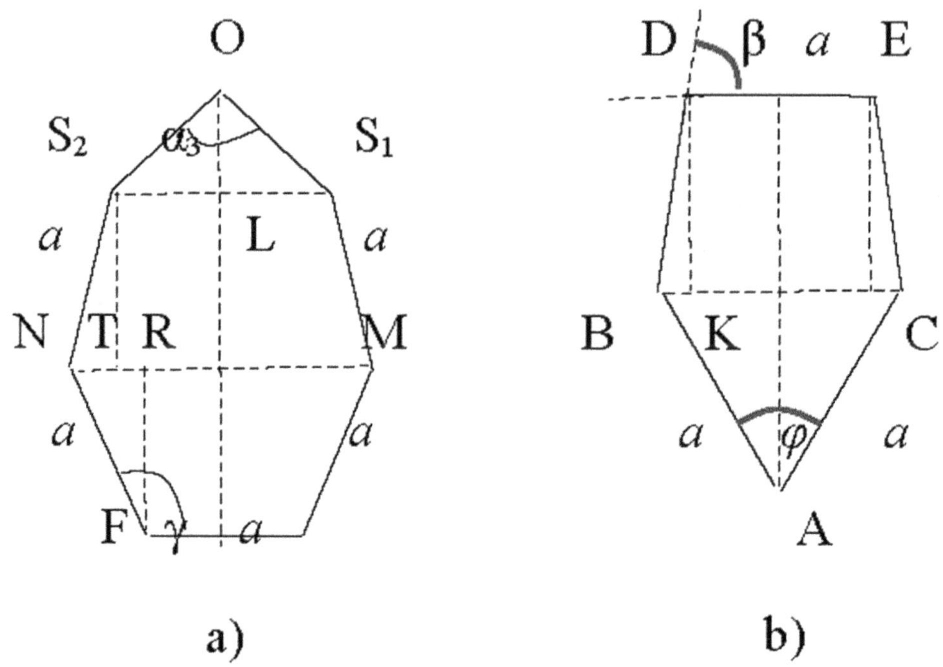

From Figure 22a it follows that

$$NM = a + 2\cos(\pi - \gamma), S_1S_2 = \sqrt{3}a\sin(\alpha_3 / 2), S_2T = \sqrt{a^2 - NT^2},$$
$$NT = (1/2)(NM - S_1S_2) = (a/2)[1 - 2\cos\gamma - \sqrt{3}\sin(\alpha_3 / 2)].$$

Thus, Figure 22a allows you to express the height of the figure M23:

$$H_1 = a\sin\gamma + (\sqrt{3}/2)a\cos(\alpha_3/2) + a\sqrt{1 - 0,25[1 - 2\cos\gamma - \sqrt{3}\sin(\alpha_3/2)]^2}.$$

From Figure 22b it is follows that BC=2$a$sin(φ/2), BK=0.5(BC–a).
In this Figure also can express the height of the figure M23

$$H_2 = a[\cos(\varphi/2) + \sqrt{1/2 - \sin^2(\varphi/2) + \sin(\varphi/2)}].$$

Since H_1=H_2, we have the equation for calculating the angle $\alpha 1$

$$\cos(\varphi/2) + \sqrt{1/2 - \sin^2(\varphi/2) + \sin(\varphi/2)} =$$
$$\sin\gamma + (\sqrt{3}/2)\cos(\alpha_3/2) + \sqrt{1 - 0,25[1 - 2\cos\gamma - \sqrt{3}\sin(\alpha_3/2)]}.$$

In this equation, the angle $\alpha 3$ is a function of the angle $\alpha 1$ according to equation (2.28). The angle $\alpha 1$ is determined from the value of the angle β (equation (2.21)). Since

$$\mu = \pi / 2 - \beta, \sin \mu = [2\sin(\varphi / 2) - 1] / \sqrt{3},$$

the value of the angle $\alpha 1$ determines the value of the angle φ. The angle γ depends on the angle $\alpha 1$ through the angles φ according to the formula:

$$\cos \gamma = -(\sqrt{3}/2)\sqrt{1 - \frac{1}{3\sin^2 \varphi}}.$$

However, the calculation showed that this equation has no solution. Therefore, it should be considered that figure M23 does not exist.

CONVEX COMPOSITE REGULARLY FACETED POLYHEDRON

Regularly faceted polyhedrons will be called state polyhedrons if they can be composed of simple regularly faceted polyhedrons, including prisms and antiprisms:

Theorem 2.3. There are 81 compound regularly faceted polyhedrons (excluding prisms and antiprisms), and 2 of them are regular polyhedrons, 4 semiregular polyhedrons, 75 irregular polyhedrons:

1. octahedron ($2PG_6$),
2. icosahedron ($PG_5 + 3PG_4$),
3. cuboctahedron ($2PG_{16}$),
4. icosadodecahedron ($2PG_{19}$),
5. rhombicuboctahedron ($2PG_{27} + P_8$),
6. rhomboicosadodecahedron ($PG_{27} + 3PG_{28}$),
7. elongated triangular pyramid ($PG_1 + P_3$),
8. elongated quadrangular pyramid ($PG_6 + P_4$),
9. elongated pentagonal pyramid ($PG_4 + P_5$),
10. twisted elongated quadrangular pyramid ($PG_6 + A_4$),
11. twisted elongated pentagonal pyramid ($PG_6 + A_5$),
12. triangular bipyramid ($2PG_1$),
13. pentagonal bipyramid ($2PG_4$),
14. elongated triangular bipyramid (2PG1 + P3),
15. elongated quadrangular bipyramid ($2P\Pi_6 + P_4$),
16. elongated pentagonal bipyramid ($2PG_4 + P_5$),
17. twisted elongated quadrangular bipyramid ($2PG_6 + A_4$),
18. elongated three-slope dome ($PG_{16} + P_6$),
19. elongated hipped roof ($PG_{27} + P_8$),
20. elongated five-slope dome ($PG_{28} + P_{10}$),
21. elongated five-slope rotunda ($PG_{19} + P_{10}$),
22. twisted elongated tricycle dome ($PG_{16} + A_6$),
23. twisted elongated four-slope dome ($PG_{28} + A_8$),

24. twisted elongated five-slope dome (PG_{28} + A_{10}),
25. twisted elongated five-slope rotunda (PG_{19} + A_{10}),
26. two-sided rotated bikupol ($2P_3$),
27. three-slope straight bikupol ($2PG_{16}$),
28. four-slope straight bikupol ($2PG_{17}$),
29. four-slope rotated bicupol ($2PG_{17}$),
30. five-slope straight bikupol ($2PG_{28}$),
31. five-slope rotated bikupol ($2PG_{28}$),
32. five-slope straight dome-rotunda (PG_{28} + PG_{19}),
33. five-slope rotated dome rotunda (PG_{28} + PG_{19}),
34. five-slope straight birotonda ($2PG_{19}$),
35. elongated three-slope straight bikupol ($2PG_{16}$ + P_6),
36. elongated three-slope rotated bikupol ($2PG_{16}$ + P_6),
37. elongated four-slope rotated bikupol ($2PG_{27}$ + P_8),
38. elongated five-slope straight bikupol ($2PG_{28}$ + P_{10}),
39. elongated five-slope rotated bikupol ($2PG_{28}$ + P_{10}),
40. elongated five-slope straight dome-rotunda (PG_{28} + PG_{19} + P_{10}),
41. elongated five-slope rotated dome-rotunda (PG_{28}+PG_{19}+P_{10}),
42. elongated five-slope straight birotunda ($2PG_{19}$+P_{10}),
43. elongated five-slope rotated birotunda ($2PG_{19}$+P_{10}),
44. twisted elongated tree-slope bikupol ($2PG_{16}$+A_6),
45. twisted elongated four-slope bikupol ($2PG_{27}$+A_8),
46. twisted elongated five-slope bikupol (PG_{28}+A_6+A_8),
47. twisted elongated five-slope dome-rotunda (PG_{28}+PG_{19}+A_{10}),
48. twisted elongated five-slope birotunda ($2PG_{19}$+A_{10}),
49. extended triangular prism (PG_6+P_3),
50. twice extended triangular prism ($2PG_6$+P_3),
51. three times extended triangular prism ($3PG_6$+P_3),
52. extended pentangular prism (PG_6+P_5),
53. twice extended pentangular prism ($2PG_6$+P_5),
54. extended sexangular prism (PG_6+P_6),
55. twice opposite extended sexangular prism ($2PG_6$+P_6),
56. twice obliquely extended sexangular prism ($2PG_6$+P_6),
57. three times extended sexangular prism ($3PG_6$+P_6),
58. extended dodecahedron (PG_{24}+PG_4),
59. twice opposite extended dodecahedron (PG_{24}+$2PG_4$),
60. twice obliquely extended dodecahedron (PG_{24}+$2PG_4$),
61. three times extended dodecahedron (PG_{24}+$3PG_4$),
62. twice obliquely truncated icosahedron (PG_4+PG_5),
63. three times extended truncated icosahedron (PG_1+PG_5),
64. extended truncated tetrahedron (PG_{21}+PG_{16}),
65. extended truncated cube (PG_{22}+PG_{27}),
66. twice extended truncated cube (PG_{22}+$2PG_{27}$),
67. extended truncated dodecahedron (PG_{28}+PG_{23}),

68. twice opposite extended truncated dodecahedron ($2PG_{28}+PG_{23}$),
69. twice obliquely extended truncated dodecahedron ($2PG_{28}+PG_{23}$),
70. three times extended dodecahedron ($3PG_{28}+PG_{23}$),
71. rotated romboicosododecahedron ($2PG_{28}+PG_{18}$),
72. twice opposite rotated romboicosododecahedron ($2PG_{28}+PG_{18}$),
73. twice obliquely rotated romboicosododecahedron ($3PG_{28}+PG_{18}$),
74. three times rotated romboicosododecahedron ($3PG_{28}+PG_{18}$),
75. truncated romboicosododecahedron ($2PG_{28}+PG_{18}$),
76. opposite rotated truncated romboicosododecahedron ($PG_{28}+PG_{18}$),
77. obliquely rotated truncated romboicosododecahedron ($2PG_{28}+PG_{17}$),
78. twice obliquely rotated truncated romboicosododecahedron ($2PG_{28}+PG_{17}$),
79. twice obliquely truncated romboicosododecahedron ($PG_{28}+PG_{17}$),
80. twice rotated truncated romboicosododecahedron ($PG_{28}+PG_{17}$),
81. ($PG_{11}+PG_6$).

In the list of polyhedrons, P_n and A_n denote a prism and an antiprism, respectively, with a base in the form of a regular n-gon.

Evidence

Regularly faceted polyhedrons listed in the list with numbers 1 and 2 are regular polyhedrons, with numbers from 3 to 6 there are known semiregular polyhedrons, the remaining 76 polyhedrons are irregular polyhedrons. Moreover, polyhedrons from the 1st to the 80th number consist of the known regular and semiregular polyhedrons and their parts. They were found by a simple search and their existence does not require special proof. The names of these polyhedrons are rather arbitrary, although they reflect some of their features (Zalgaller, 1966). In particular, with the same composition (indicated in parentheses), you can get different regularly faceted polyhedrons by arranging the component parts in different ways relative to each other. The new compound regularly faceted polyhedron is polytope 81. It includes the new simple regularly faceted polyhedron PG_{11}. At its base lies a pentagon (see Figure 18 (11) and Figure 21), forming two gonohedrons $17_{4,5}$ with adjoining squares and 11_5 gonohedron with adjacent triangles. A regularly faceted polyhedron PG_4 can be attached to this pentagon, and the dihedral angles between the triangles PG_4 and the polygons surrounding the pentagon of the figure PG_{11} turn out to be significantly less than 180 degrees (see descriptions of the polyhedrons PG_4 and PG_{11}). This means that the newly formed figure is convex and has no conditional edges. Thus, it is a compound regular polyhedron. Q.E.D.

Consequence

From Theorems 2.2 and 2.3 it follows that the total number of regularly faceted polyhedrons (with the exception of prisms and antiprisms) includes 117 polyhedrons, of which 36 are simple regularly faceted polyhedrons and 81 compound regularly faceted polyhedrons. If we exclude regular and well-known semiregular polytopes from the indicated total number of regularly faceted polyhedrons, then 100 polytopes remain, and not 92, as Johnson assumed (Johnson, 1960).

CONCLUSION

Convex polyhedrons with faces from regular polygons were investigated using 1) constructing their three-dimensional images by methods of descriptive geometry, 2) establishing a functional relationship between dihedral angles between faces (the same or different) when deforming polyhedral angles, 3) solving systems of equations describing a set of polyhedral angles of a polyhedron. A new way of searching for polyhedrons with faces from regular polygons is proposed, including the material construction of polyhedrons with a list of possible motives (the simplest combinations of regular polygons) and options for their continuation to the construction of closed convex surfaces. It was found that the well-known classification of convex polyhedrons with faces of regular polygons (Zalgaller, 1966), obtained using topological methods, claiming to enumerate all possible polyhedrons of this type, is far from complete. The existence of 9 new simple convex polyhedrons with faces of regular polygons is proved. In particular, the existence of modifications of the well-known semi-regular polyhedrons flat-nosed cube and flat-nosed dodecahedron was proved, differing from them in the alternation of polygons on the surface of polyhedrons and associated with them by mirroring, but not compatible with them by any movement (like the right and left hand). It is proved that the total number of convex polyhedrons with faces of regular polygons (simple and composite) is (excluding prisms and antiprisms) 117 polyhedrons (36 simple and 81 composite). Moreover, it was proved that some polyhedrons, previously declared as convex polyhedrons with faces of regular polygons (Zalgaller, 1966), do not actually exist. The corresponding systems of equations describing their set of polyhedral angles have no solution, i. e. the values of their characteristic angles cannot be determined. If we exclude regular and well-known semiregular polytopes from the indicated total number of regular polytopes, then exactly 100 polytopes remain, and not 92, as Johnson previously assumed (Johnson, 1960).

A compact classification of all polyhedral angles with faces of regular polygons (gonohedrons) is carried out. This not only simplified the search for new convex regular-faceted polyhedrons, but also made it possible to single out those gonohedrons that do and do not participate in the creation of these polyhedrons. Table 6 (columns 1) lists all simple regularly faceted polyhedrons of PG_i, indicating the value of the index. And for each polyhedron (columns 2) the numbers of the gonohedrons participating in the formation of this polyhedron and their number (in brackets) are given.

Table 7 lists the gonohedrons involved in the formation of simple regularly faceted polyhedrons (columns 1) and the number of different simple regularly faceted polyhedrons in which each of these gonohedrons is used (columns 2). In particular, from the data in Table 7, it follows that the pentahedral gonohedrons 3 and 12_4 (9 polyhedrons) are most used in the creation of various simple regularly faceted polyhedrons. Less widespread are the tetrahedral gonohedrons 10_4, 11_4 (6 polyhedrons), 9_4 (5 polyhedrons), 11_5 and $16_{4,5}$ (4 polyhedrons). The least common are trihedral gonohedrons. Trihedral gonohedrons with faces of squares (4), gonohedrons $6_{3,i}$ at i = 4,7, 8, 9, 11, gonohedrons $6_{4,I}$ at i=5, 7, gonohedron 7, gonohedrons $8_{3,I}$ at i=6,7, gonohedrons $8_{4,I}$ at i > 4, gonohedrons $8_{5,I}$ at i=6,7,9, gonohedrons $13_{3,4,j}$ at j=5,7,9,11, gonohedrons $13_{3,5,j}$ at j from 6 to 9, from 11 to 29, gonohedrons $13_{3,I,j}$ at i >5and any j, gonohedrons $13_{4,5,j}$ at j from 6 to 9, from 11 to 19, gonohedrons $13_{4,6,j}$ at j=7,9,11, gonohedrons $13_{4,I,j}$ at i =7 and any j, gonohedrons $13_{5,i,j}$ at any i and j, gonohedrons 11_i at i>6, gonohedrons 15, gonohedrons $16_{4,j}$ at j from 6 to 11, gonohedrons $16_{5,j}$ at any j, gonohedrons $17_{4,j}$ at j > 6, gonohedrons $17_{5,j}$ at any j do not take part in the creation of regularly faceted polyhedrons.

It follows from the data in Table 6 that the largest number of different gonohedrons (6) in one polyhedron is observed in the regularly faceted polyhedron PG_{14} discovered in this study. This polyhedron is distinguished by the greatest deviation from semiregular polyhedrons (Archimedes bodies), each of which is composed of gonohedrons of the same type). Due to the exclusivity of this polyhedron, we will assign it a special name polytypichedron-max. However, we will leave other regularly faceted polyhedrons discovered in this study without a name for now, confining ourselves to a letter designation. Five different gonohedrons are found in three regularly faceted polyhedrons discovered in this study. They also differ significantly from the bodies of Archimedes.

The conducted research is caused by the needs of the modern level of development of science, which focuses on the regularities of the formation of dissipative systems, the elementary cells of which are convex polyhedrons with faces of regular polygons. Examples include structures that arise in biological tissues, the shape of cells of which approaches the shape of convex regular-faceted polyhedrons, structures of clusters in nanotechnology, patterns of distribution of matter in the Universe, etc. In recent years, the author's studies on higher dimensions in geometric structures in chemical, physical and biological systems also indicate the importance of determining the geometry of three-dimensional polyhedrons, since they are always constituent parts of multidimensional structures. In this regard, the question of possible forms of convex polyhedrons with faces of regular polygons is of general scientific interest.

Table 6. The composition of gonohedrons in simple regularity faceted polyhedrons

1 *i*	2	1*i*	2	1*i*	2	1*i*	2	1*I*	2
1	1(4)	9*	3(4)	13	3(8)	19	9_5(10)	30	$6_{4,6}$(6)
2	12_4(24)		10_4(2)		12_4(8)		$13_{3,5,10}$(10)	31	$13_{4,6,10}$(30)
3*	12_4(24)		12_4(12)	14**	3(2)	20	9_5(3)	32	3(6)
4	3(1)	10*	11_4(2)		$6_{3,5}$(2)		11_5(3)		10_4(3)
	$8_{3,5}$(5)		11_5(1)		11_4(2)		$16_{4,5}$(6)		12_4(8)
5	$6_{3,5}$(6)		12_4(2)		11_5(2)		$17_{4,6}$(6)	33	3(4)
	10_5(3)		$16_{4,5}$(4)		12_5(2)	21	$6_{3,6}$(12)		10_4(4)
6	$8_{3,4}$(4)		$17_{4,5}$(4)		$16_{4,5}$(2)	22	$6_{3,6}$(12)		12_4(8)
	2(1)	11*	9_4(3)	15*	9_4(5)	23	$6_{3,10}$(60)		$6_{5,6}$(60)
7	3(4)		10_4(2)		10_4(4)	24	5(20)	34	9_4(2)
	10_4(2)		11_5(3)		11_4(2)	25	12_5(60)	35*	$6_{3,6}$(2)
8*	11_4(4)		12_4(4)	16	9_4(3)	26*	12_5(60)		11_6(2)
	3(6)		$17_{4,5}$(2)		$13_{3,4,6}$(6)	27	11_4(4)		$16_{4,6}$(4)
	9_4(1)	12	$6_{3,5}$(4)	17	$14_{4,5}$(12)		$13_{3,4,8}$(8)		$17_{4,6}$(8)
	11_4(2)		9_5(2)		$13_{4,5,10}$(30)	28	$14_{4,5}$(5)	36	2(4)
	12_4(4)		$16_{4,5}$(8)	18	$14_{4,5}$(30)		$13_{3,4,10}$(10)		3(4)
					$13_{4,5,10}$(20)	29	$13_{4,6,8}$(48)		

Table 7. Frequency of use of gonohedrons in simple regularly faceted polyhedrons

1	2	1	2	1	2	1	2	1	2
1	1	$6_{5,6}$	1	10_5	1	$13_{4,5,10}$	2	$16_{4,5}$	4
2	2	$6_{4,6}$	1	11_4	6	$13_{3,5,10}$	1	$17_{4,5}$	2
3	9	$8_{3,5}$	1	11_5	4	$13_{3,4,8}$	1	$17_{4,6}$	2
5	1	$8_{3,4}$	1	11_6	1	$13_{3,4,10}$	1		
$6_{3,5}$	3	9_4	5	12_4	9	$13_{4,6,8}$	1		
$6_{3,6}$	3	9_5	3	12_5	3	$13_{4,6,10}$	1		
$6_{3,10}$	1	10_4	6	$13_{3,4,6}$	1	$14_{4,5}$	3		

In Table 6, an asterisk denotes nine convex polyhedrons with faces of regular polygons discovered in this study. Two of them, 3* and 26*, can be attributed to the newly discovered bodies of Archimedes. The rest of the polyhedrons with this designation cannot be attributed to the bodies of Archimedes (and of course to the bodies of Plato). The polyhedron denoted by two asterisks 14** (plytypichedron-max), which has in its composition the maximum possible number of different types gonohedrons (6), is most different from the bodies of Archimedes (and the bodies of Plato). Interestingly, this figure has two sections in the form of irregular polygons (pentagon and heptagon), while the faces of this figure are constructed from regular polygons. Thus, this figure can be an example of the connection between symmetry and asymmetry (symmetry gives rise to asymmetry). The latter is typical for wildlife.

REFERENCES

Alexandrov, A. D. (1950). *The convex polyhedrons*. Gostechizdat.

Cauchy, O. (1944). Research on polyhedrons. *Advances in Mathematical Sciences*, *10*, 5–17.

D'Arsy, T. (1942). *On Growth and Form*. Cambridge University Press.

Euler, L. (1736). Solutio problematisad geometriam situs pertinentis. *Comment Acad Sci I Petropolitanae*, *8*, 128–140.

Fedorov, E.S. (1885). Beginning of the teachings on the figures. *Notes of the St. Petersburg Mineralogical Society*, *21*.

Fedorov, E.S. (1889). Symmetry of finite figures. *Notes of the St. Petersburg Mineralogical Society*, *25*, 1 - 52.

Fedorov, E.S. (1891). Symmetry of the correct systems of figures. *Notes of the St. Petersburg Mineralogical Society*, *28*.

Hadamard, J. (1938). *Elementary geometry. Part 2*. Uchpedgiz.

Johnson, W. W. (1960). Convex polyhedrons with regular faces. *Notices of the American Mathematical Society*, *7*, 952.

Petrov, O. V. (2007). *Dissipative of structure of the Earth as display of fundamental wave properties of a matter*. A.P. Karpinsky Russian Geological Research Institute.

Zalgaller, V. A. (1966). *Convex polyhedrons with correct sides*. Science.

Zalgaller, V.A. (1966). *List of all convex polyhedrons with correct sides*. The International Congress of Mathematicians.

Zhihin, G.V. (2005a). Correct both semiregular polyhedrons and filling of space with them. *Life and Safety*, *1-2*, 551-559.

Zhizhin, G. V. (2005 b). *Dissipative of structures in chemical, geological and ecological systems*. Science.

Zhizhin, G. V. (2006). Mathematical of a basis economically and ecologically rational use building and finishing materials. Architecture, Construction, Ecology, 215-262.

Zhizhin, G. V. (2008). To a question on possible mesh large-scale structure of the universe. *Proceedings of the International Conference "Problems of Practical Cosmology"*.

Zhizhin, G. V. (2009). *Convex right-faceted polyhedrons*. Polytechnika-Service.

Zhizhin, G. V. (2014). *4D world.* St. Petersburg: Polytechnica - Service.

Zhizhin, G. V. (2019). *The Geometry of Higher - Dimensional polytopes*. IGI Global.

Zhizhin, G. V. (2021). *Nanotechnologies and Clusters in the Spaces of Higher Dimension: Emerging Research and Opportunities*. IGI Global.

KEY TERMS AND DEFINITIONS

Convex Regularly Faceted Polyhedrons: Convex three-dimensional polytopes whose faces are regular polygons.

Gonohedron: A polyhedral corner whose faces are regular polygons.

Mechanical Operations on a Polyhedron: Conceivable transformations of a polyhedron, including cutting a polyhedron (cutting off vertices), deforming a section of a cut to obtain a regular polygon in a section, rotating parts of a polyhedron after cutting relative to each other.

Regular Polyhedron: Convex polyhedron with faces of regular polygons of one type.

Semiregular Polyhedrons: Convex polyhedrons with faces of regular polygons of two or more types, if all gonohedrons of the polyhedron have the same composition and arrangement of faces.

Chapter 3
Necessary and Sufficient Conditions for Proving the Existence of Polytopes of Higher Dimension

ABSTRACT

Based on the results of the study of the geometry of polytopes of higher dimension, a number of conditions necessary for the existence of polytopes have been formulated. Each of these conditions and their combination are necessary, but not sufficient for the existence of a polytope of the highest dimension. A sufficient condition for the existence of a polytope of the highest dimension is its complete construction (i.e., determination of its complete structure with a listing of all elements of the polytope and their contacts with each other). The importance of the concept of incidence in the polytope of the highest dimension is emphasized on the basis of which the law of conservation of incidents is formulated, which has a fundamental information value.

INTRODUCTION

When passing from three-dimensional polytopes (i.e., polyhedrons) to polytopes of higher dimensions, the question of the necessary and sufficient conditions for the existence of these polytopes should be raised. To answer this question, it is required to take into account the properties of multidimensional figures and refrain from attempts to mechanically transfer the properties of three-dimensional figures to the properties of multidimensional figures, arguing, as is often done, that when moving to a higher-dimensional space, nothing changes significantly, only the number of spatial coordinates increases. Such statements lead to important negative consequences. An example is the history of attempts to solve the 18th problem of Hilbert (Hilbert, 1901; Alexsandrov, 1969) on the construction of a space of higher dimension from congruent figures. The answer to this proposal was a series of works by various researchers in subsequent years, unfortunately, the properties of figures of higher dimensions were

DOI: 10.4018/978-1-7998-8374-6.ch003

not taken into account (Minkowski, 1911; Delaunay, 1929,1937, 1961; Delaunay, & Sandakova, 1961; Voronoi, 1952; Alexsandrov, 1934, 1954; Venkov, 1954, 1959; Stogrin, 1973).

In particular, when analyzing four-dimensional parallelotopes (Delone, 1929), a table of parallelohedrons of 51 figures was compiled. Later, another figure was added to this table (Shtogrin, 1973). However, it turned out (Zhizhin, 2019 b, 2021) that none of these figures was tested for the fulfillment of the Euler-Poincaré equation, which is necessary to prove the existence of a polytope of higher dimension (Poincaré, 1895). The Euler-Poincaré equation is the result of the continuation of the Euler equation, which is valid in three-dimensional space, to a space of higher dimension. A recent check (Zhizhin, 2019b 2021) showed that this equation does not hold for Delaunay parallelotopes. In this regard, the theory of four-dimensional stereohedrons developed by Delaunay (Delaunay, 1961; Delaunay, Sandakova, 1961), which also does not use the Euler-Poincaré equation, remains incorrect. It should be noted that in many works on polytopes of higher dimensions (Coxeter, 1961, 1973), when proving the existence of these polytopes, they are limited to calculating the possibility of locating a certain number of figures in a polytope around an element of a certain dimension, for example, an edge. Schläfli's symbols are built on this basis (Schläfli, 1901). But it cannot be argued that if a polytope satisfies the numbers in Schläfli's symbols, then it exists. This condition can be considered only as the minimum necessary condition for the existence of a polytope. The proof of the existence of a polytope of higher dimension can only be its construction with a list of all the elements included in it and with an indication of all contacts of these elements with each other. This is how the figures in three-dimensional space were studied in the works of Fedorov (Fedorov, 1885, 1889, 1891).

In recent studies by the author (Zhizhin, 2019 a, 2020), it was found that between elements of different dimensions of polytopes of the highest dimension, there are certain quantitative relationships between the incidence values of elements of one dimension in relation to elements of another dimension. The fulfillment of these quantitative relationships is also necessary for the existence of the polytope. Moreover, even if the Euler-Poincaré equation for a given figure is satisfied, and the quantitative relations between incidents are not satisfied, then the polytope cannot exist.

A sufficient condition for the existence of a higher-dimensional polytope is its complete construction, which includes recalculation of all elements of various dimensions included in the polytope with an indication (numerical) of the order of contact of a given element with all elements in its environment.

It should be noted the great importance of the concept of incidence in the structure of a polytope of the highest dimension (Zhizhin, 2019 a, 2020, 2021). On the basis of this concept, a new law was formulated that is fulfilled in polytopes of the highest dimension: the law of conservation of incidents. This law states that the integral incidence flow from low-dimensional elements to higher-dimensional elements is equal to the integral incidence flow from higher-dimensional elements to low-dimensional elements.

REGULAR POLYTOPES AND SCHLAFLI`S SYMBOLS

A regular polyhedron is a polyhedron with congruent regular polygons as faces, arranged in the same way around identical vertices. Also, it is a polyhedron whose symmetry group acts transitively on its flags; a regular polyhedron is a highly symmetric structure, being vertex-transitive, edge-transitive and face-transitive (Coxeter, 1963; Schulte, 2014).

The property of having a similar arrangement of faces around each vertex is equivalent with the following conditions: the vertices of the polyhedron all lie on a sphere; all the dihedral angles of the polyhedron are equal; all its vertex figures are regular polygons; all the solid angles are congruent (Cromwell, 1997).

There are five regular polyhedrons, known as Platonic polyhedral solids (see Chapter 1): tetrahedron, cube, octahedron, dodecahedron and icosahedron can be written as {3,3}; {4, 3}; {3, 4}; {5, 3}and {3, 5} by using the basic Schläfli (1901) symbols {p, q} where p is the number of vertices in a given face while q is the number of faces containing a given vertex. Their pair duals are (cube & octahedron) and (dodecahedron & icosahedron) while the tetrahedron is selfdual.

The ancient Greeks had good knowledge on the Platonic solids. Theaetetus, an Athenian contemporary of Plato, gave a mathematical description of all the five regular solids and may be the first known proof that no other convex regular polyhedrons exist. Plato wrote about the five regular solids in the dialogue Timaeus, about 360 B.C. Evclid gave a complect in his Elements (Heals, 1981).

For any three-dimensional polyhedron with the number of vertices v, edges e, and faces f of a convex polyhedron are related by the polyhedral formula (Euler, 1752)

$$v - e + f = \chi = 2(1 - g),$$

where χ is the Euler characteristic and g is the genus (i.e. the number of holes performed in a plastic sphere to make it homeomorphic to that surface). Euler characteristic is a topological invariant that describes the shape of a structure regardless of that way it is bent. A topological proof, showing that no other (out of the five Platonic) convex regular polyhedrons exit, can be derived from the Euler relations:

$$d_i pf - qv = 2e, \quad v - e + f = 2$$

Combining these equations, we obtain: $2e/q–e+2e/p=2$. Then, $1/q+1/p=1/e+1/2$ and, since e is strictly positive, the following condition comes out:

$$1/q + 1/p > 1/2. \tag{3.1}$$

Because p and q must be at least three, then (3.1) accepts only five solutions for the pair {p, q}:

{3, 3}, {4, 3}, {3, 4}, {5, 3}, {3, 5}.

Regular 4-polytope, written as {p, q, r}, have cells of the type {p, q}, face {p}, edges figure {r} and vertex figures {q, r}. It can also be red as: r-polyhedron (of the type {p, q}) meet at any edge of the polytope. In case of a 4-polytope, the vertex figure is a polyhedron formed by the neighboring vertices around a given vertex; the edges figure is a polygon.

The existence of a regular 4-polytope {p, q, r} is constrained by the existence of regular polyhedron {p, q} and {q, r}, which are its cells; if p, q, r, are integers, the following relation is obeyed (Coxeter, 1974)

$$p + 2q + r - 4/r - 4/p < 12. \tag{3.2}$$

The six regular 4-polytopes are: 5-Cell {3, 3, 3}; 8-Cell {4, 3, 3}; 16-Cell {3, 3, 4}; 24-Cell {3, 4, 3}; 120-Cell {5, 3, 3} and 600-Cell {3, 3, 5). Five of them can be associated to the Platonic solids but the sixth, the 24-cell has no close 3D equivalent. Among them, 5-Cell and 24-Cell are selfduals while the other are pairs: (8-Cell & 16-Cell); (120-Cell & 600-Cell). The symmetry groups of these 4-polytopes are all finite Coxeter groups (Coxeter, 1934, 1940, 1963; Davis, 2007).

In five dimensions, a regular 5-polytope is written as {p, q, r, s}, where {p, q, r} is a 4-face type, {p, q} is the cell type, {p} is the face type, {s} is the face figure, {r, s} is the edge figure and {q, r, s} is the vertex figure. The vertex figure is a 4-polytope (arrangement of neighboring vertices to each vertex), the edge figure is a polyhedron (seen as the arrangements of faces around each edge) while the face figure is a polygon (seen by the arrangements of sell around each face).

The condition reflected by the Schläfli symbols, as noted in the introduction, is a weak condition for the necessity of the existence of a polytope of higher dimension and does not prove its existence. Therefore, when formulating the conditions of necessity, one should turn to the Euler-Poincaré equation (Poincaré, 1895). Let us present the derivation of this equation in a modern presentation.

THE EULER - POINCARÉ EQUATION

So, let V is a closed manifold of n dimensions. We subdivide it into a certain number of manifolds v_n, each in n dimensions. These manifolds v_n will not be closed and their boundaries will be formed by a number of manifolds v_{n-1}, each of which has n - 1 dimensions. In turn, the boundaries of v_{n-1} will be formed by a certain number of manifolds v_{n-2} each dimension n - 2, and so on. This will lead to a certain number of manifolds of one dimension, the boundaries of which will be a certain number of isolated points or manifolds of zero dimension, which we denote v_0. Let the manifolds v_n, $v_{n-1},\dots,v_1$ be connected. The word "manifold", which Poincaré uses, we can now understand as "polytope", (Grunbaum, 1967), if its subdivision into a certain number of components is given.

Theorem 3.1. For any closed polytope with dimension n, if all elements of any dimension included in it are connected, the equality must be satisfied:

$$\sum_{i=0}^{n-1}(-1)^i f_i(n) = N = 1+(-1)^n, \tag{3.3}$$

in (3.3) $f_i(n)$ is the number of elements of a polytope P of dimension n having dimension i.

Proof. If two polytopes are obtained by dividing a manifold in two different ways, we will call them congruent. Let now be given a polytope P formed by the manifold V, regions v_n and their successive boundaries v_n, $v_{n-1},\dots,v_1$, v_0. Let's subdivide v_n it into smaller areas $v_n^{/}$. Full boundaries v_n will consist of a number of new areas $v^{//}_{n-1}$ and, in addition, from areas $v_{n-1}^{/}$ obtained by division v_{n-1}, and so on. Full boundaries $v_{n-1}^{/}$ and $v_{n-1}^{//}$ will consist of a number of new areas $v^{//}_{n-2}$ and, in addition, from areas $v_{n-2}^{/}$ obtained by division v_{n-2}, and so on. Finally, we come to $v_1^{/}$ and $v_1^{//}$, the full boundaries of which will consist of a certain number of new points $v_0^{//}$ and, in addition, from v_0. The resulting polytope $P^{/}$ will be called the derivative of the polytope P.

Now let there be given two congruent polytopes P and $P^{/}$ obtained using two different methods of decomposing some manifold V. There will always be such a polytope $P^{//}$ that will be simultaneously a derivative of P and $P^{/}$ and which will be obtained by combining both methods of decomposition.

If we denote by $v_n, v_n^/, v_n^{//}$ the subdivisions V in three decomposition methods that correspond to three polytopes $P, P^/, P^{//}$, then a necessary and sufficient condition for two points to simultaneously belong to the same region $v_n^{//}$ is that they must also simultaneously belong to the same region $v_n^/$, and the same area v_n.

It is required to establish that the number *N* is the same for two congruent polytopes. Since we have just found out that two congruent polytopes always have a common derivative polytope, it is enough for us to show the identity of the number *N* for the polytope and all its derivatives.

Let us investigate one of the regions v_{n-1} in polytope *P*. It will always belong to two and only two areas v_n that it separates from one another. On the contrary, the region v_{n-2} will belong, in general, to more than two regions v_n and more than two regions v_{n-1}. Areas v_{n-2} that will only belong to two areas v_{n-1} will be called singular. Let a singular region v_{n-2} be given that belongs to two regions v_{n-1}, which we denote $v_{n-1}^/$ and $v_{n-1}^{//}$. It is quite clear that $v_{n-1}^{//}$ will share the same two areas v_n that are separated by $v_{n-1}^{//}$ and, thus, v_{n-2} will also belong to only two areas v_n.

Quite also, we will say that a polytope v_q is singular if it belongs to only two polytopes v_{q+1}, and in this case both polytopes v_{q+1} to which it belongs v_q will belong to the same v_{q+2}, the same v_{q+3}, ..., the same v_n. Thus, the discarding of v_q and the mutual joining of the two v_{q+1} will not make any change in v_{q+2}, nor in v_{q+3}, ... nor in v_n.

Now let the polytope v_h be given. A certain number of polytopes v_{h-1} belongs to this polytope. If one of the latter is singular, then we call the polytope v_h irregular; otherwise, it will be regular.

Consider a polytope *P* with regions $v_n, v_{n-1}, \dots$ and derived from it polytope *P*/ with regions $v_n^/, v_{n-1}^/, \dots$. Let's try to rise from the polytope *P*/ to polytope *P*. Let two areas α and β be given in $v_n^/$. We assume that they are separated from one another by the $v_{n-1}^/$ area, which we will designate as γ. Hence the areas α and β are contiguous and are part of the same area v_n. We exclude the region γ, which serves as the boundary between α and β, and join region α to region β. Thus, we exclude one region $v_n^/$ and one region $v_{n-1}^/$. On the other hand, we do not exclude any region $v_{n-2}^/$ if the region γ is regular. Indeed, if none of the regions $v_{n-2}^/$ is singular, then each will belong to at least three regions $v_{n-1}^/$, and after exclusion γ it will belong to at least two regions $v_{n-1}^/$. Likewise, every region $v_q^/$ (where $q < n - 2$) that is a part of γ, will belong to at least three regions $v_{h-1}^/$ and, after exclusion γ, will still belong to at least two regions $v_{n-1}^/$. The exclusion γ, therefore, does not exclude any of the areas $v_q^/$, therefore, it does not change the value of the number *N*.

If, on the contrary, the region γ is irregular, we will no longer have the right to exclude it, since then there will be a region $v_{n-2}^/$ belonging only to γ and another region $v_{n-1}^/$. After exclusion γ, it will only belong to one area $v_q^/$, which is unacceptable.

In this case, the region γ does not form the only boundary between α and β. Indeed, since γ is irregular, there will be a singular region $v_{n-2}^/$, which can be denoted by δ and which will belong to γ and another region $v_{n-1}^/$, which we denote by γ/. This region γ/, according to what we saw above, separates the same regions as γ, i. e. α and β.

If the domain δ is regular, then we can exclude it and combine γ and γ′. Then the domain $\gamma' + \gamma$ will separate α from β. We would thus decrease γ_{n-1} and γ_{n-2} by one, while the other numbers γ_i would not change. The *N* number would remain unchanged.

If δ is irregular, then one region v_{n-3}' will be singular. Let's designate it ε. It will separate δ from another are δ′. We exclude ε and add δ to δ′, etc.

In the same way, we can exclude an area v_q' that separates two areas v_{q+1}' from one another, and join one of these areas v_{q+1}' to the second, but under two conditions:

1) if q is less than n - 1, the region v_q' must be singular;
2) and in any case, it should be regular.

After this assumption, we will reason as follows: In order to return from the polytope *P′* to the polytope *P*, we can assume that the polytope *P* does not include a singular region, but the polytope *P′* and intermediate polytopes can include such.

With the help of a whole series of exceptions and additions, we will return from polytope *P′* to polytope *P*, passing through a series of intermediate polytopes:

$$P_0 = P', P_1, P_2, \ldots, P_{m-1}, P_m.$$

How can one go from the polytope P_i to the polytope P_{i+1} ? If P_i includes singulars at v_0', one of them can be excluded. If they are absent, then all v_1' will be regular; if there are singular v_1', then one of them can be excluded.

If there are no singular v_1', all v_2' will be regular; if there are singular v_2', one of them can be excluded. Etc.

And finally, there will not be a single singular v_{n-2}', all v_{n-1}' will be regular and we will have the right to exclude any of them. If one of the v_n areas is divided into several louse areas v_n', you can select two of these areas v_n', which will be adjacent and separated from one another by the v_{n-1}' area, which will be their common border. Join one region to another, excluding this border. One should not stop until there is no singular region left and, on the other hand, no region is further v_n divided into several regions v_n'. Then we come to *P*.

None of these operations can change the number *N*. Consequently, this number has the same meaning for *P* and for *P′*, which was required to be proved. Therefore, to prove Theorem 3.1, it suffices to determine the number *N* for an arbitrary connected polytope. Let us do this for the simplest n - dimensional polytope of the simplex type.

The simplest n - dimensional polytope is the n - simplex, i. e. a figure with the number of vertices n + 1, in which each vertex is connected by an edge to any other vertex. In accordance with this definition in this polytope

$$f_0(n) = n + 1. \tag{3.4}$$

The number of edges in this polytope is equal to the number of combinations of the total number of vertices in 2, i.e.

$$f_1(n) = C_{n+1}^2 = \frac{(n+1)n}{2}. \tag{3.5}$$

All two - dimensional faces in an n - simplex are 2 - simplexes, i. e. triangles. The number of two - dimensional faces is:

$$f_2(n) = C_{n+1}^3 = \frac{(n+1)n(n-1)}{3!}. \tag{3.6}$$

Three - dimensional faces are composed of four two - dimensional faces (i.e., they have four vertices - tetrahedrons). Therefore, the number of three - dimensional faces is

$$f_3(n) = C_{n+1}^4 = \frac{(n+1)n(n-1)(n-2)}{4!}. \tag{3.7}$$

And go on. Thus, for the number of elements with dimension j in the n - simplex we have

$$f_j(n) = C_{n+1}^{j+1}. \tag{3.8}$$

We calculate the amount:

$$\sum_{j=0}^{n-1}(-1)^j f_j(n) = \sum_{j=0}^{n-1}(-1)^j C_{n+1}^{j+1} = (-1)^0 C_{n+1}^1 + (-1)^1 C_{n+1}^2 + (-1)^2 C_{n+1}^3 + \ldots + (-1)^{n-1} C_{n+1}^n. \tag{3.9}$$

Taking into account that $C_{n+1}^k = C_{n+1}^{n-k}$, we find that on the right - hand side of (3.9) with even n, all terms are pairwise reduced, i. e. the amount $\sum_{j=0}^{n-1}(-1)^j f_j(n)$ is zero. If n is odd, then all terms on the right - hand side of (3.9) are added up in pairs with the exception of the central term located in place $\frac{n+1}{2}$. In this case, the right - hand side after the transformation takes the form:

$2[1+\frac{n!}{n!}(1-1)]$, i. e. equals 2.

This proves the Euler – Poincaré formula for a simplex of dimension n, regardless of the general Poincaré proof for unknown functions $f_j(n)$.

It can also be noted that the right - hand side of equation (3.9) is the term in the expansion of Newton's binomial:

$$(1-1)^{n+1} = 1-(-1)^n - \sum_{j=0}^{n-1}(-1)^j C_{n+1}^{j+1}.$$

Since in this case Newton's binomial is obviously equal to zero, this proves that the right - hand side of (3.9) is equal to zero for even n and equal to 2 for odd n.

This completes the proof of Theorem 3.1.

Comment. It should be emphasized that the proof used the condition that polytopes are closed and connected. Therefore, Theorem 3.1 is applicable for a wide range of different types of polytopes: regular, irregular, convex, non-convex. This circle is much wider than the possibilities of using Schläfli symbols, limited by the obligatory condition of the correctness of polytopes and their relatively small dimension. Moreover, the fulfillment of Schläfli's symbols does not guarantee the fulfillment of Theorem 3.1, which is obligatory for the existence of polytopes.

Let us verify the fulfillment of Theorem 3.1 for some more complex polytopes.

The Number of Faces of Various Dimensions in the n-Prismahedrons

Multiply a segment (one - dimensional segment) by another segment (not necessarily equal to this). The rules for multiplying polytopes by each other can be found in the monograph (Zhizhin, 2019 a). As a result, we will have a rectangle. Multiply this rectangle by another segment. We get a parallelepiped, i. e. three - dimensional polyhedron. Multiply this box again by a certain segment. We get a four - dimensional prismahedron. If all segments have the same length, then this 4 - prismahedron is a 4 - cube. At each step of multiplication by a segment, the number of vertices in a figure multiply on 2. When multiplying m times, the number of vertices is equal 2^{m+1} and the dimension of the figure $n = 1 + m$, or $f_0(n)=2^n$. At each step of the multiplication, each vertex generates an edge, each edge generates a rectangle, each rectangle generates a parallelepiped, etc. The process of calculating the numbers of elements of different dimensions can be described by recurrence relations:

$$\begin{aligned}
&f_{0(n)} = 2f_{0(n-1)},\\
&f_{1(n)} = 2f_{1(n-1)} + f_{0(n-1)},\\
&f_{2(n)} = 2f_{2(n-1)} + f_{1(n-1)},\\
&\vdots\\
&f_{n(n)} = 1.
\end{aligned} \tag{3.10}$$

Here, the first lower index indicates the dimension of the face, and the second lower index indicates the dimension of the space. Substituting successively recurrence relations into each other, taking into account that $f_{0(2)}=4$, $f_{1(2)}=4$, $f_{2(2)}=1$, we obtain formulas more convenient for calculations for any finite figure dimension:

$$f_0 = 2^n C_n^0, f_1 = 2^{n-1} C_n^1, f_2 = 2^{n-2} C_n^2, \ldots, f_n = 2^{n-n} C_n^n = 1.$$

Thus, the number of elements of dimension i in the n - prismahedron is expressed by the formula:

$$f_i(n) = 2^{n-i} C_n^i, i = 0,1,...,n. \tag{3.11}$$

It is possible to precisely prove the fulfillment of the Euler - Poincaré condition for the n - prismahedron if we represent equation (3.3) in the form:

$$\sum_{i=0}^{n}(-1)^i f_{i(n)} = 1. \tag{3.12}$$

Using recurrence relations (3.10), we will successively lower the dimension of polytopes on the left side of (3.3). Then we get

$$\sum_{i=0}^{n}(-1)^i f_{i(n)} = (-1)^0 f_{0(2)} + (-1)^2 f_{1(2)} + (-1)^2 = 4 - 4 + 1 = 1.$$

Q.E.D.

The Number of Faces of Various Dimensions in the n - Polytopes with Anti – parallel Edges (n – cross - polytopes)

Let we have some segment AB and some point O on the plane outside the segment AB. Rotate the segment AB around the point O on 180 degrees. We get the segment $B'A'$ antiparallel to the segment AB. Connect the edges vertices indicated different letters: B with A', A with B' . We get a quadrangle with antiparallel sides. In this figure, four vertices formed from two vertices of the initial segment, i. e. vertices doubled (Figure 1).

Figure 1. A quadrangle with antiparallel edges

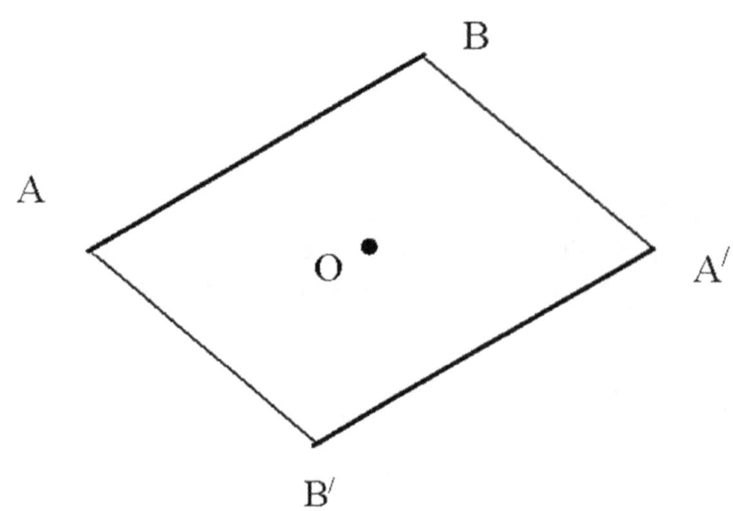

Four edges were formed from one edge of the initial segment: the original *AB*, new $B'A'$, and one edge were formed from each vertex of the original segment *AB*. In addition, a two -dimensional region was formed.

The movement of triangles along a helix leads to the formation of polytopes with antiparallel edges. Consider an arbitrary triangle *ABC* on the plane. Choose some point *O'* on the plane outside the triangle to his left. Let this point be the base of the axis of the helix passing through the triangle. Rotate the *ABC* triangle 180 degrees to the right, moving it up the helix, parallel to the initial plane. In the projection on the plane, both triangles *ABC* and the displaced triangle $A'B'C'$ will be located as shown in Figure 2.

Figure 2. Polytopes of dimension 3 with anti – parallel edges

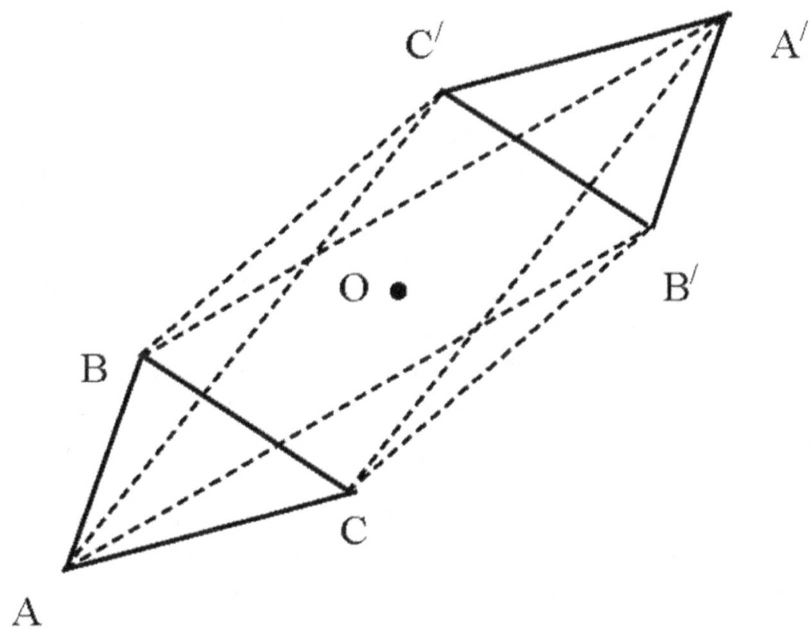

It is easy to see that the edges of the triangle *ABC* and $A'B'C'$ are antiparallel. It can now connect in space the vertices of the triangle *ABC* with the vertices of the triangle $A'B'C'$ so that there are no connections of the vertices with the same letters. In a projection on the plane the connection is represented by dotted segments. It can be seen that the connecting segments also break up into pairs of anti - parallel segments. Let us now verify that the image *ABCA'B'C'* in Figure 2, along with the dotted segments, is a projection of a three - dimensional convex polytope. One can use the Euler – Poincaré equation (3.3) for this aim. In Figure 2 *ABCA'B'C'* six vertices formed from three vertices of the initial triangle, i. e. vertices doubled (f_0=3+3=6). In the figure *ABCA'B'C'* three more edges of the new triangle are added to the three edges of the original triangle after the rotation of the original triangle. In addition, two edges began to additionally proceed from each vertex of the original triangle. Thus, the number of edges in the figure *ABCA'B'C'* is 12 (f_1=3+3+3•2=12). In Figure 2, another two - dimensional face of the triangle is added to one two - dimensional face of the original triangle after its rotation. In addition, at each vertex of the original triangle, two a two - dimensional faces were additionally formed

$ABC', ACB', BCA', A'B'C, A'C'B, B'C'A$.

Thus, the total number of two - dimensional faces is 8 (f_2=2+6=8). Substituting these values of elements of different dimensions into equation (3.3), can find:

$6 - 12 + 8 = 2$,

the Euler – Poincaré equation holds in this case for $n = 3$. This proves that the resulting figure is a convex polytope of dimension 3 (if the figure were not convex, the Euler – Poincaré equation would be violated).

The point O' in Figure 2 coincides with the center of the three - dimensional figure $ABCA'B'C'$ as diagonal figures pass through it and point O' located on the axis of the helix. Point O' can be considered as the origin of three - dimensional space. Coordinate axes x, y, z in this direction emanate from directions AA', BB', CC' respectively. Three pairs of these axes define the coordinate planes of the space of this shape.

Choose some point O'' on the plane outside the triangle below it. Let this point be the base of the axis of the helix passing through the triangle. Rotate the ABC triangle 180 degrees to the left, moving it up the helix, parallel to the initial plane. In the projection on the plane, both triangles ABC and the displaced triangle $A''B''C''$ will be located as shown in Figure 2.

It is easy to see that the edges of the triangle ABC and $A''B''C''$ are antiparallel. It can now connect in space the vertices of the triangle ABC with the vertices of the triangle $A''B''C''$ so that there are no connections of the vertices with the same letters. In a projection on the plane the connection is represented by dotted segments. It can be seen that the connecting segments also break up into pairs of anti - parallel segments. Let us now verify that the image $ABCA'B'C'$ in Figure 2, along with the dotted segments, is a projection of a three - dimensional convex polytope. Obviously, a shape $ABCA''B''C''$ has as many vertices, edges, and flat faces as a shape $ABCA'B'C'$. Therefore, it satisfies the Euler – Poincaré equation (3.3) with dimensionality $n = 3$, i. e. it is a convex three - dimensional polyhedron. The point O'' in Figure 2 coincides with the center of the three - dimensional figure $ABCA''B''C''$ as diagonal figures pass through it and point O'' located on the axis of the helix. Point O'' can be considered as the origin of three - dimensional space. Coordinate axes x, y, z in this direction emanated from directions AA'', BB'', CC'' respectively. The order of the coordinate axes x, y, z in the figures $ABCA'B'C'$ and $ABCA''B''C''$, as can be seen from Figure 2, coincide. This suggests that both figures $ABCA'B'C'$ and $ABCA''B''C''$are topologically one and same figure - the wrong octahedron.

Interestingly, to transform an arbitrary tetrahedron $ABCD$ into a tetrahedron $A'B'C'D'$ with anti - parallel edges, it is not enough to rotate it along a helix by 180 degrees. To do this, you must turn the helix together with the tetrahedron and move the tetrahedron along the helix in the opposite direction, also rotating it 180 degrees. In the initial state, the tetrahedron on the initial helix and the tetrahedron on the reversed helix, after its rotation by 180 degrees, will have anti -parallel edges. Both tetrahedrons can $ABCD$ and $A'B'C'D'$ be present by the Figure 3 for rotation to the right. On Figure 3 the point O' is the point of rotation.

Now connect the vertices of the tetrahedrons so that the connecting edges (dotted segments) do not have the same letters. In the resulting figure (along with dotted edges) the number of vertices doubled ($f_0 = 8$). The number of the edges completed six edges initial tetrahedron and six edges final tetrahedron. In addition, three more edges are additionally formed at each vertex of the initial tetrahedron. Thus, the

Figure 3. Polytope of dimension 4 with anti - parallel edges

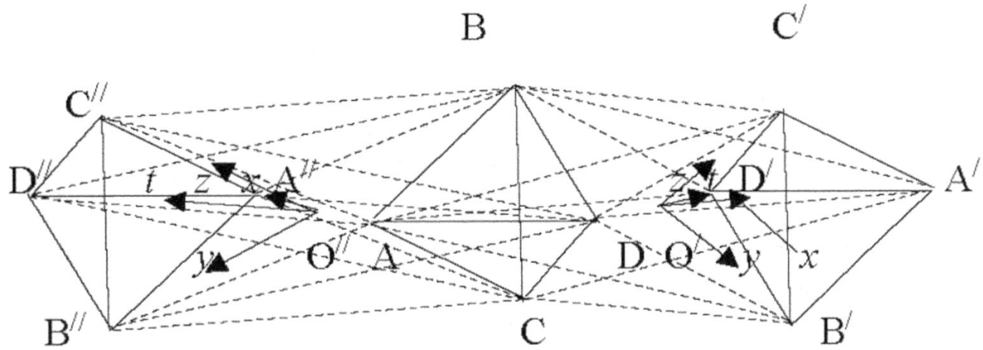

total number of edges is 24 (f_1=6+6+3•4=24). The flat faces of the figure include 4 faces of the original tetrahedron and 4 faces of the final tetrahedron. In addition, at each vertex of the initial tetrahedron, six flat faces are additionally formed. Thus, the total number of flat faces of the figure is 32 (f_2=4+4+4•6=32). The initial and final tetrahedra are included in the figure as three - dimensional faces. All tetrahedra do not have vertices of the same name by construction. Therefore, there are still 4 tetrahedra with three vertices of the original tetrahedron and one vertex of the final tetrahedron

$$ABDC', ACDB', BCDA', ABCD'.$$

In addition, there are 4 tetrahedra with one of the vertices of the initial tetrahedron and three vertices of the final tetrahedron

$$A'B'D'C, A'C'D'B, B'C'D'A, A'B'C'D.$$

There are also tetrahedrons of a tetrahedron with two vertices of the initial tetrahedron and with two vertices of the final tetrahedron

$$ABDC', ACDB', DCDA', ABCD'.$$

Thus, the total number of tetrahedrons in the figure is 16 (f_3=2+4+4+6=16). Substituting these values into equation (3.3), can find:

8 - 24 + 24 - 8 = 0.

Consequently, the Euler – Poincaré equation is satisfied in this case for n = 4. Thus, the polytope $ABCD\,A'B'C'D'$ in Figure 3 has dimension 4. Comparing figure $ABCD\,A'B'C'D'$ with Figure 1.3 (Zhizhin, 2021) of 4-cross-polytope from Stringham's work (Stringham, 1880), it is easy to see that these figures are topologically equivalent. Thus, figure $ABCD\,A'B'C'D'$ is a 4 - cross - polytope. Unlike Stringham's work, the resulting figure is not necessarily correct. Thus, it was found that a 4 – cross -

polytope is obtained as a result of rotation of the tetrahedron (3 - simplex) around the helicoid axis. Note that, as a result of the analysis, the octahedron can be considered as a 3 – cross - polytope.

The point O' in Figure 3 coincides with the center of the fourth - dimensional figure $ABCD\,A'B'C'D'$ as diagonal figures pass through it and point O' located on the axis of the helix. Point O' can be considered as the origin of forth - dimensional space. Coordinate axes *x, y, z, t* in this direction emanate from directions AA', BB', CC', DD' respectively. Six pairs of these axes define the coordinate planes of the space of this shape.

Choose some point O'' on the plane outside the tetrahedron *ABCD* below it. Let this point be the base of the axis of the helix passing through the tetrahedron. Rotate the *ABCD* tetrahedron 180 degrees to the left, moving it up the helix, parallel to the initial plane. In the projection on the plane, both tetrahedrons *ABCD* and the displaced tetrahedron $A''B''C''D''$ will be located as shown in Figure 3.

It is easy to see that the edges of the tetrahedrons *ABCD* and $A''B''C''D''$ are antiparallel. It can now connect in space the vertices of the tetrahedron *ABCD* with the vertices of the tetrahedron $A''B''C''D''$ so that there are no connections of the vertices with the same letters. In a projection on the plane the connection is represented by dotted segments. It can be seen that the connecting segments also break up into pairs of anti - parallel segments. Let us now verify that the image $ABCD\,A''B''C''D''$ in Figure 3, along with the dotted segments, is a projection of a fourth - dimensional convex polytope. Obviously, a shape $ABCD\,A''B''C''D''$ has as many vertices, edges, and flat faces as a shape $ABCD\,A'B'C'D'$. Therefore, it satisfies the Euler – Poincaré equation (3.3) with dimensionality $n = 4$. Thus, the polytope $ABCD\,A''B''C''D''$ in Figure 3 has dimension 4. It is easy to see (Zhizhin, 2018) that this too is 4 - cross - polytope.

The point O'' in Figure 3 coincides with the center of the 4 – cross - polytope as diagonal figures pass through it and point O'' located on the axis of the helix. Point O'' can be considered as the origin of forth - dimensional space. Coordinate axes *x, y, z, t* in this direction emanated from directions AA'', BB'', CC'', DD'' respectively. From Figure 3 it follows that the sequence of alternation of coordinate axes *x, y, z, t* in a 4 - cross – polytope $ABCD\,A'B'C'D'$ differs from the sequence of alternation of coordinate axes *x, y, z, t* in a 4 - cross – polytope $ABCD\,A''B''C''D''$. Thus, a surprising fact emerged: the figures $ABCD\,A'B'C'D'$ and $ABCD\,A''B''C''D''$, being 4 - cross - polytopes, are topologically different from each other. It is impossible to move from one of them to another by continuous transformation, since they have a different order of alternation of vertices.

Interestingly, to transform an arbitrary 4 – simplex *ABCDE* into a 4 - simplex $A'B'C'D'E'$ with anti - parallel edges, it is not enough to rotate it along a helix by 180 degrees. To do this, you must turn the helix together with the 4 - simplex and move the 4 - simplex along the helix in the opposite direction, also rotating it 180 degrees. In the initial state, the 4 - simplex on the initial helix and the 4 - simplex on the reversed helix, after its rotation by 180 degrees, will have anti -parallel edges. Both 4 - simplexes can *ABCDE* and $A'B'C'D'E'$ be present by the Figure 4 for rotation to the right. On Figure 4 the point *O* is the point of rotation.

Now connect the vertices of the 4 - simplexes so that the connecting edges (dotted segments) do not have the same letters. In the resulting figure (along with dotted edges) the number of vertices doubled (f_0 = 10). The number of the edges completed ten edges initial 4 - simplex and ten edges final 4 - simplex. In addition, four more edges are additionally formed at each vertex of the initial 4 - simplex. Thus, the total number of edges is 40 (f_1=10+10+5•4=40). The flat faces of the figure include 10 faces of the

Figure 4. Polytope of dimension 5 with anti - parallel edges

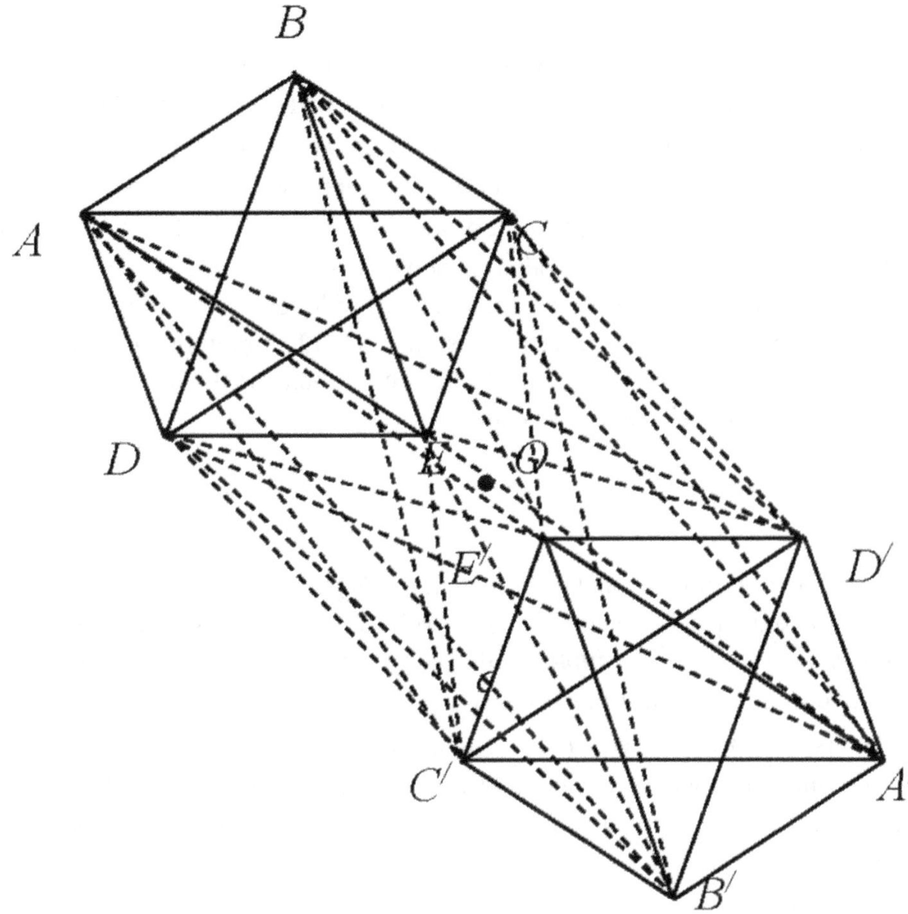

original 4 - simplex and 10 faces of the final 4 - simplex. In addition, from the connections of each of the vertices of a finite 4 - simplex with some two vertices of the original 4 - simplex, 30 flat faces in the form of triangles are formed additionally:

$$ABC', ABD', ABE', ACB', ACD', ACE', ADB', ADC', ADE', AEB', AEC', AED',$$
$$BCA', BCD', BCE', BDA', BDC', BDE', BEA', BEC', BED', CDA', CDB', CDE', \quad .$$
$$CEA', CEB', CED', DEA', DEB', DEC'.$$

Thus, the total number of flat faces of the Figure 4 is 80 (f_2=10+10+30+30=80). As three -dimensional faces, Figure 4 includes five tetrahedrons of the initial 4 - simplex and five tetrahedrons of the final 4 - simplex. In addition, tetrahedrons are formed from the connection of vertices from the initial 4 - simplex and the final 4 - simplex. Since there are 5 vertices in the 4 -simplex and 4 vertices in the tetrahedron, the following combinations are possible for the formation of tetrahedra with vertices from the initial 4 - simplex (letters without strokes without repetition) and from the final 4 - simplex (other letters with strokes). If there is only one vertex with a stroke, then the number of such combinations is equal to the number of combinations of five by three (C_5^3) multiplied by 2. This is because when choosing three letters from five for vertices without strokes, there are two more options for choosing letters with strokes for vertices of the final 4 - simplex ($2C_5^3$). This number must also be multiplied by 2, taking into account the possibility of completely replacing letters with strokes with letters without strokes and vice versa, replacing letters without strokes with letters with strokes ($2 \cdot 2C_5^3$). If there are two vertices with strokes in the tetrahedron, and, therefore, there are also two vertices without strokes, then the number of such letter combinations equals the product $C_5^2 C_{5-2}^2$. This is because after choosing two letters without strokes from five letters, it remains to choose two letter options from the remaining letters (5-2) for vertices with strokes. Since in this case the number of letters with strokes is equal to the number of letters without strokes, replacing letters with strokes with letters without strokes and vice versa does not give new options. Thus, the total number of three - dimensional faces (tetrahedrons) from the vertices of the initial and final 4 -simplexes is equal

$$f_3 = 5 + 5 + 4C_5^3 + C_5^2 C_3^2 = 10 + 40 + 30 = 80 .$$

When calculating the four -dimensional faces included in Figure 4, it is necessary to take into account the initial 4 - simplex and the final 4 - simplex (2), as well as 4 - simplexes formed from the vertices of the initial 4 -simplex and the final 4 - simplex at the same time. The number of variants of four letters without strokes with one letter with a stroke is equal C_5^4 . The number of variants of three letters without strokes and two letters with strokes is equal C_5^2 . You must add options to these options when replacing letters with strokes with letters without strokes and vice versa replacing letters without strokes with letters with strokes. Thus, the total number of four - dimensional figures included in the general Figure 4 is equal f_4=2+5+10+5+10=32.
Substituting values f_i into equation (3.1), can find:

10 - 40 + 80 – 80+32 = 2.

Consequently, the Euler – Poincaré equation is satisfied in this case for $n = 5$. Thus, the polytope in Figure 4 is obtained as a result of rotation of the 4 - simplex around the helicoid axis has dimension 5. Considering that a figure with antiparallel edges is formed in this case, we conclude that this figure is a 5 - cross - polytope. A comparison of the image of this figure with the image of a 5 – cross - polytope obtained earlier (Zhizhin, 2014, 2019) in a different way confirms their topological equivalence.

Now, from considering a sequential series of special cases, we can proceed to the general case. Let there be an (n -1) - simplex. We select the point O outside the simplex and rotate it 180 degrees along the helix with the axis passing through the point O. As a result, we get a (n -1) – simplex′ with antiparallel edges. Connect the non - homonymous vertices of these simplexes with edges. We obtain a polytope in which the number of vertices doubled. Since the number of vertices in the (n - 1) - simplex is equal C_n^1 (see formula (3.8), the number of vertices in the new polytope is equal $f_0 = 2C_n^1 = 2C_n^{n-1}$. The dimension of this polytope is n. The number of edges in the new polytope is the sum of the edges in two (n -1) – simplexes $2C_n^2$ (see formula (3.8)). In addition, (n – 1) edges additionally emanate from each vertex of the original (n -1) - simplex. Thus, the total number of edges in the new polytope is equal:

$$f_1 = 2C_n^2 + n(n-1) = 2\frac{n!}{2!(n-2)!} + n(n-1) = 2(n-1)n = 2^2 C_n^{n-2}.$$

The number of planar faces in the new polytope is formed from the sum of the planar faces in the initial and final polytopes (n -1) – simplexes $2C_n^3$. In addition, planar faces are formed due to the union of non - homonymous vertices of both (n - 1) - simplexes into triangles $2C_n^2(n-2)$. Thus, the total number of flat faces in the new polytope is equal:

$$f_2 = 2C_n^3 + 2C_n^2(n-2) = 2\frac{n!}{(n-3)!3!} + 2\frac{n!(n-2)}{(n-2)!2!} = \frac{n!(2+3!)}{(n-3)!3!} = 2^3 C_n^{n-3}.$$

The output of the values of the numbers of elements of dimension 1, 2, 3 in the new polytope already shows a general formula for calculating these numbers for any dimension j

$$f_j(n) = 2^{j+1} C_n^{n-1-j}. \qquad (3.13)$$

Calculations by the formula (3.13) of the concrete examples of the numbers of elements with dimensions j =1, 2, 3, 4 already considered confirm the values of these numbers obtained by direct determination. Comparing formulas (3.11) and (3.13), we conclude that the values $f_i(n)$ in formula (3.11) coincide with the values $f_j(n)$ in formula (3.13) for $j=n-i-1$. This means that the n – cross - polytope and the n - prismahedron are in relation to each other with respect to inclusion inversion. In this case, the number of faces of dimension $i = n - 1$ in the n - prismahedron is equal to the number of faces of dimension 0 in the n – cross - polytope, the number of faces of dimension $i = n$ - 2 in the n - prismahedron is equal to the number of faces of dimension 1 in the n – cross - polytope, etc. The number of faces of dimension 0 in the n - prismahedron is equal to the number of faces of dimension n - 1 in the n - prismahedron.

Thus, it is proved that, as well as for the n - prismahedron, the Euler - Poincaré equation is also valid for the n – cross - polytope.

It should be noted that the obtained formulas for the numbers of faces of different dimensions for the n - simplex, n - prismahedron and n – cross - polytope, respectively, coincide with the previously obtained formulas for the numbers of faces of different dimensions in the correct simplex, cube, and correct cross - polytope. It is also interesting to note the result obtained, that any cross - polytope is the result of a simplex rival along the helicoid axis.

The performed consideration of various polytopes of arbitrary dimension confirmed the fulfillment of the Euler-Poincaré for these polytopes, which corresponds to Theorem 3.1.

INCIDENCE IN THE HIGHER DIMENSIONAL POLYTOPES

Let P is an n – polytope, i. e. a convex polytope of dimension n. We denote by B (P) the boundary complex P, that is, a complex consisting of all faces of P with dimension n - 1 or less (Grunbaum, 1967). The points not belonging to B (P) will be called interior points of P. The set of interior points of the polytope P will be denoted by V. Thus, we have equality $P = B(P) \bigcup V$. Since all the faces of a polytope from vertices (which in chemical compounds are images of atoms, molecules or functional groups) to facets enter the boundary complex, it is of interest to establish relationships between faces of different dimensions in the boundary complex *B* (*P*).

For each element b_j of the boundary complex *B* (*P*) (here *j* is the serial number of the element in the boundary complex) can introduce of the notation its dimension $i_{b_j}, 0 \le i_{b_j} < n$.

There is denoted $k_{b(i_1)b(i_2)}$ the incidence value of the element with dimension i_1 of the boundary complex of polytope *P* with dimension *n* to the elements of the boundary complex having dimension i_2(0£i_1<i_2£*n*). This value is equal to the number of elements of the boundary complex of this polytope having dimension i_2 which the given element with dimension i_1 belongs.

Consider simplex polytopes of various dimensions.

Let a polytope of simplex type have dimension 2 (triangle). For this polytope, the incidence coefficients have the following values:

$$k_{b(0)b(1)} = C_2^1 = 2; k_{b(0)b(2)} = C_2^2 = 1;$$
$$k_{b(1)b(2)} = C_1^1 = 1.$$

Let a polytope of simplex type have dimension 3 (tetrahedron). For this polytope, the incidence coefficients have the following values:

$$k_{b(0)b(1)} = C_3^1 = 3; k_{b(0)b(2)} = C_3^2 = 3; k_{b(0)b(3)} = C_3^3 = 1;$$
$$k_{b(1)b(2)} = C_2^1 = 2; k_{b(1)b(3)} = C_2^2 = 1;$$
$$k_{b(2)b(3)} = C_1^1 = 1.$$

Let a polytope of simplex type have dimension 4 (4 - simplex, Figure 5).

Figure 5. The polytope 4 - simplex

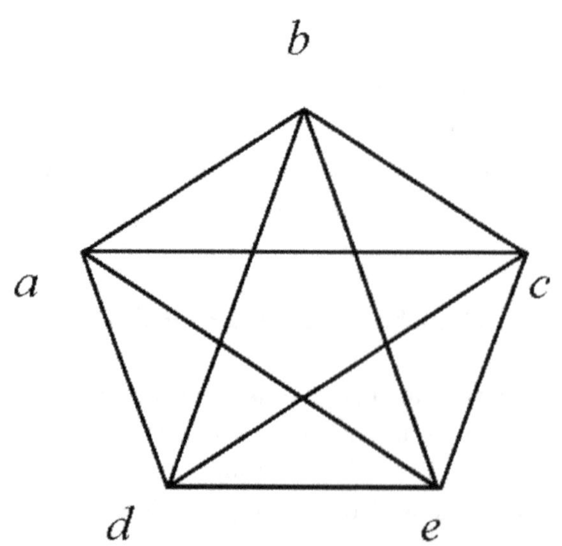

For this polytope, the incidence coefficients have the following values:

$$k_{b(0)b(1)} = C_4^1 = 4; k_{b(0)b(2)} = C_4^2 = 6; k_{b(0)b(3)} = C_4^3 = 4; k_{b(0)b(4)} = C_4^4 = 1;$$
$$k_{b(1)b(2)} = C_3^1 = 3; k_{b(1)b(3)} = C_3^2 = 3; k_{b(1)b(4)} = C_3^3 = 1;$$
$$k_{b(2)b(3)} = C_2^1 = 2; k_{b(2)b(4)} = C_2^2 = 1;$$
$$k_{b(3)b(4)} = C_1^1 = 1.$$

Let a polytope of simplex type have dimension 5 (5 - simplex, Figure 6).

Figure 6. The polytope 5 - simplex

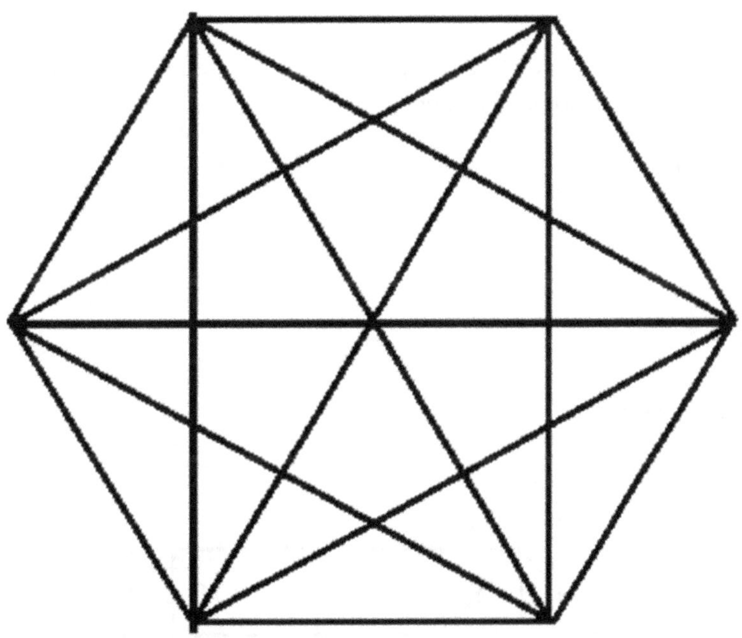

For this polytope, the incidence coefficients have the following values:

$$k_{b(0)b(1)} = C_5^1 = 5; k_{b(0)b(2)} = C_5^2 = 10; k_{b(0)b(3)} = C_5^3 = 10; k_{b(0)b(4)} = C_5^4 = 5; k_{b(0)b(5)} = C_5^5 = 1;$$
$$k_{b(1)b(2)} = C_4^1 = 4; k_{b(1)b(3)} = C_4^2 = 6; k_{b(1)b(4)} = C_4^3 = 4; k_{b(1)b(5)} = C_4^4 = 1;$$
$$k_{b(2)b(3)} = C_3^1 = 3; k_{b(2)b(4)} = C_3^2 = 3; k_{b(2)b(5)} = C_3^3 = 1;$$
$$k_{b(3)b(4)} = C_2^1 = 2; k_{b(3)b(5)} = C_2^2 = 1;$$
$$k_{b(4)b(5)} = C_1^1 = 1.$$

Thus, it follows from the analysis that the general formula for calculating the incidence coefficient of an element of dimension i_1 with respect to elements of dimension i_2 in a polytope of type simplex of dimension n has the form:

$$k_{b(i_1)b(i_2)}(n) = C_{n-i_1}^{i_2-i_1}, n \geq i_2 > i_1. \tag{3.14}$$

Consider n – prismahedron (or n - cube) polytopes of various dimensions.

Let a polytope of prismahedron type have dimension 2 (square). For this polytope, the incidence coefficients have the following values:

$$k_{b(0)b(1)} = 2; k_{b(0)b(2)} = 1;$$
$$k_{b(1)b(2)} = 1.$$

Let a polytope of prismahedron type have dimension 3 (parallelepiped, cube). For this polytope, the incidence coefficients have the following values:

$$k_{b(0)b(1)} = 3; k_{b(0)b(2)} = 3; k_{b(0)b(3)} = 1;$$
$$k_{b(1)b(2)} = 2; k_{b(1)b(3)} = 1;$$
$$k_{b(2)b(3)} = 1.$$

Figure 7. The polytope 4 - cube

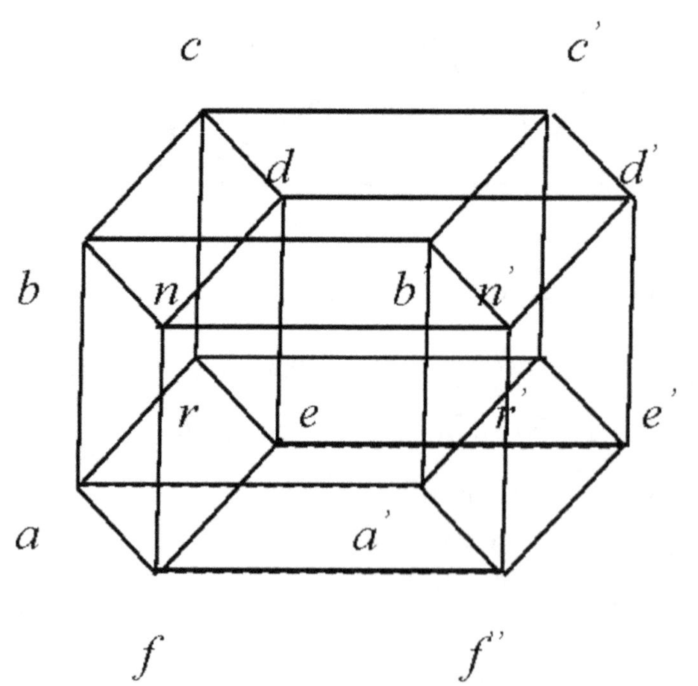

Let a polytope of prismahedron type have dimension 4 (4 - prismahedron, 4 - cube, Figure 7). For this polytope, the incidence coefficients have the following values:

$$k_{b(0)b(1)} = 4; k_{b(0)b(2)} = 6; k_{b(0)b(3)} = 4; k_{b(0)b(4)} = 1;$$
$$k_{b(1)b(2)} = 3; k_{b(1)b(3)} = 3; k_{b(1)b(4)} = 1;$$
$$k_{b(2)b(3)} = 2; k_{b(2)b(4)} = 1;$$
$$k_{b(3)b(4)} = 1.$$

From this analysis it follows that the values of the incidence coefficients of polytopes of the n -prismahedron type (n - cube) coincide with the values of the incidence coefficient of the n -simplex for the same dimension of polytopes n. Therefore, formula (3.14) obtained for incidence coefficients in the case of an n - simplex simultaneously refers to the incidence coefficients of elements of different dimensions in n - prismahedron. It can be shown that the right-hand side of formula (3.14) is expressed in a uniform manner through the number of elements of different dimensions of the polytopes of the n - simplex and n - prismahedron. It is argued that the n -simplexes and n - prismahedrons are valid the following formula:

$$k_{b(i)b(j)}(n) = \frac{f_j(n)}{f_i(n)} f_i(j), n \geq j > i. \tag{3.15}$$

To prove this assertion in the case of an n - simplex, we express the functions on the right - hand side using equality (3.8)

$$k_{b(i)b(j)}(n) = \frac{!_{n+1}^{j+1}}{C_{n+1}^{i+1}} C_{j+1}^{i+1} = \frac{(n-i)!}{(n-j)!(j-i)!} = C_{n-i}^{j-i}.$$

In the case of a n – prismahedron we express the functions on the right – hand side of (3.15) using equality (3.11)

$$k_{b(i)b(j)}(n) = \frac{2^{n-j} !_n^j}{2^{n-i} C_n^i} 2^{j-i} C_j^i = \frac{(n-i)!}{(n-j)!(j-i)!} = C_{n-i}^{j-i}.$$

Q.E.D.

It follows directly from the universal formula (3.15) (the law of incidence) that the values of the incidence coefficients between elements of different dimensions in n - simplexes and in n - prismahedrons increase significantly with increasing dimension of polytopes. if we recall that the tops of polytopes contain material bodies (atoms, molecules, functional groups), then the incidence coefficients (belonging) characterize the exchange of information between material bodies. Consequently, the result indicates that with an increase in the dimension of polytopes, the intensity of the exchange of information between elements of polytopes sharply increases. This result does not follow from the Euler – Poincaré equation. It must be taken into account when constructing polytopes of higher dimension (in particular, parallelohedra). If there is no necessary increase in the incidence coefficients in the considered polytopes, then they cannot exist even if the Euler – Poincaré equation is satisfied. This newly discovered fact was not taken into account, for example, in the construction of parallelohedra of higher dimension earlier (Delaunay, 1937, 1961; Delaunay & Sandakova, 1961; Venkov, 1954, 1959; Alexandrov, 1934, 1954; Voronoi, 1952).

In particular, it follows from formula (3.15) that in n - simplices and in n – prismahedrons, the number of three - dimensional bodies based on one two - dimensional face increases linearly with increasing dimension n:

$$k_{b(2)b(3)}(n) = C_{n-2}^{1} = n-2.$$

Consider n – cross - polytopes of various dimensions.
Let a cross - polytope have dimension 3 (octahedron, Figure 2).

$$k_{b(0)b(1)} = 4; k_{b(0)b(2)} = 4; k_{b(0)b(3)} = 1;$$
$$k_{b(1)b(2)} = 2; k_{b(1)b(3)} = 1;$$
$$k_{b(2)b(3)} = 1.$$

For this polytope, the incidence coefficients have the following values:

Let a cross - polytope have dimension 4 (4 – cross - polytope, Figure 3). For this polytope, the incidence coefficients have the following values:

$$k_{b(0)b(1)} = 6; k_{b(0)b(2)} = 12; k_{b(0)b(3)} = 6; k_{b(0)b(4)} = 1;$$
$$k_{b(1)b(2)} = 4; k_{b(1)b(3)} = 4; k_{b(1)b(4)} = 1;$$
$$k_{b(2)b(3)} = 2; k_{b(2)b(4)} = 1;$$
$$k_{b(3)b(4)} = 1.$$

Let a cross - polytope have dimension 5 (5 – cross - polytope, Figure 4). For this polytope, the incidence coefficients have the following values:

$$k_{b(0)b(1)} = 8; k_{b(0)b(2)} = 20; k_{b(0)b(3)} = 20; k_{b(0)b(4)} = 8; k_{b(0)b(5)} = 1;$$
$$k_{b(1)b(2)} = 6; k_{b(1)b(3)} = 12; k_{b(1)b(4)} = 6; k_{b(1)b(5)} = 1;$$
$$k_{b(2)b(3)} = 4; k_{b(2)b(4)} = 4; k_{b(2)b(5)} = 1;$$
$$k_{b(3)b(4)} = 2; k_{b(3)b(5)} = 1;$$
$$k_{b(4)b(5)} = 1.$$

From the analysis of n – cross - polytopes, it follows that the values of incidence coefficients in cross - polytopes are significantly higher than the values of incidence coefficients in n -simplexes and

n - prismahedrons with the same dimension of polytopes. In addition, the incidence coefficients in n – cross - polytopes increase more sharply with increasing polytope dimensions compared to the incidence coefficients in n - simplices and n - prismahedrons.

The Local Law of Conservation of Incidents for Convex Polytopes

Theorem 3.2. The sum of the incidence coefficients from the elements of lower dimension to the elements of high dimension in any convex polytope is equal to the sum of the incidence coefficients from the elements of high dimension to the elements of low dimension. It depends on the type of polytope and its dimension.

Proof. From the studies of incidents carried out in this chapter between elements of different dimensions in the n - simplex, n - cross polytope, n - cube, n – hyper - rhombohedron, clusters of chemical compounds, it follows that the sum of the incidence coefficients between elements of arbitrary dimension j and elements of another arbitrary the dimensions of h in the forward and reverse directions are always equal to each other. This means that the sum of the incidence coefficients of all elements of dimension j in a given polytope with respect to all elements of dimension h is equal to the sum of the incidence coefficients of all elements of dimension h with respect to all elements of dimension j in this polytope. In this case, the incidence coefficient of a given element of dimension j with respect to an element of dimension h is generally not equal to the incidence coefficient of an element of dimension h with respect to an element of dimension j. Here are some examples.

In 4 - cross - polytopes, the sum of incidence coefficients from elements of zero dimension to elements of dimension 1 is equal:

$$\sum_{i=1}^{8} k_{d_0^i d_1} = k_{d_0^i d_1} \cdot i = 6 \cdot 8 = 48.$$

At the same time, the sum of incidence coefficients from elements of dimension 1 to elements of dimension 0 is:

$$\sum_{i=1}^{24} k_{d_0 d_1^i} = k_{d_0 d_1^i} \cdot i = 2 \cdot 24 = 48.$$

The sum of incidence coefficients from elements of dimension 3 to elements of dimension 4 is equal

$$\sum_{i=1}^{16} k_{d_3^i d_4} = k_{d_3^i d_4} \cdot i = 1 \cdot 16 = 16.$$

At the same time, of incidence coefficient from element of dimension 4 to elements of dimension 3 is $k_{d_3 d_4} = 16$.

In 4 – cube (or 4 – hyper – rhombohedron), the sum of incidence coefficients from elements of zero dimension to elements of dimension 1 is equal:

$$\sum_{i=1}^{16} k_{d_0^i d_1} = k_{d_0^i d_1} \cdot i = 4 \cdot 16 = 64.$$

At the same time, the sum of incidence coefficients from elements of dimension 1 to elements of dimension 0 is equal:

$$\sum_{i=1}^{32} k_{d_0 d_1^i} = k_{d_0 d_1^i} \cdot i = 2 \cdot 32 = 64.$$

The sum of incidence coefficients from elements of dimension 3 to elements of dimension 4 is equal:

$$\sum_{i=1}^{8} k_{d_3^i d_4} = k_{d_3^i d_4} \cdot i = 1 \cdot 8 = 8.$$

At the same time, of incidence coefficient from element of dimension 4 to elements of dimension 3 is $k_{d_3 d_4} = 8$.

In the cluster of Mackay, the sum of incidence coefficients from elements of zero dimension to elements of dimension 1 is equal (Zhizhin, 2020):

$$\sum_{i=1}^{24} k_{d_0^i d_1} = k_{d_0^i d_1} \cdot i = 6 \cdot 24 = 144.$$

At the same time, the sum of incidence coefficients from elements of dimension 1 to elements of dimension 0 is equal:

$$\sum_{i=1}^{72} k_{d_0 d_1^i} = k_{d_0 d_1^i} \cdot i = 2 \cdot 72 = 144.$$

The sum of the coefficients from elements of dimension 1 to elements of dimension 3 consists of two parts: for edges of icosahedrons and edges connecting icosahedrons. For the edges of icosahedrons, this part is equal to

$$\sum_{i=1}^{60} k_{d_1^i d_3} = k_{d_1^i d_3} \cdot i = 3 \cdot 60 = 180.$$

For the edges connecting icosahedrons, this part is equal:

$$\sum_{i=1}^{12} k_{d_1^i d_3} = k_{d_1^i d_3} \cdot i = 5 \cdot 12 = 60.$$

Thus, the total sum of incidence coefficients from elements of dimension 1 to elements of dimension 3 in Mackay clusters is:

$$\sum_{i=1}^{60} k_{d_1^i d_3} + \sum_{i=1}^{12} k_{d_1^i d_3} = 180 + 60 = 240.$$

At the same time the sum of incidence coefficients from elements of dimension 3 to elements of dimension 1 in the cluster of Mackay consist of two parts too: for edges of icosahedrons and edges connecting icosahedrons. For the edges of icosahedrons, this part is equal to

$$\sum_{i=1}^{60} k_{d_1 d_3^i} = k_{d_1 d_3^i} \cdot i = 3 \cdot 60 = 180.$$

For the edges connecting icosahedrons, this part is equal:

$$\sum_{i=1}^{12} k_{d_1 d_3^i} = k_{d_1 d_3^i} \cdot i = 5 \cdot 12 = 60.$$

Thus, the total sum of incidence coefficients from elements of dimension 3 to elements of dimension 1 in Mackay clusters is:

$$\sum_{i=1}^{60} k_{d_1 d_3^i} + \sum_{i=1}^{12} k_{d_1 d_3^i} = 180 + 60 = 240.$$

As we can see, the sums of incidence coefficients from low - dimensional elements to high -dimensional elements and from high - dimensional elements to low - dimensional elements consist of terms linking elements of different dimensions, and these terms are equal for the same dimension pair. Therefore, the total amounts in the direct (from the smaller to the larger) and the inverse (from the larger to the smaller) are equal to each other.

Obviously, such equalities hold for any convex polytope. This reflects the general property of belonging as a contact: if an element with a certain dimension contacts in some extent with an element of another dimension, then the latter contacts in this extent with the first.

It is also obvious from the studies that the total sum of the incidence coefficients calculated over the entire set of elements of different dimensions of a polytope depends on the type of polytope and its dimension.

The proof of Theorem 3.2 is completed.

In this case, the law of conservation of incidents is called local, since it establishes a balance of incidents between a set of elements of one dimension and a set of elements of another dimension, that is, only between elements of some two dimensions.

CONCLUSION

Necessary conditions for the existence of higher-dimensional polytopes are investigated. In the case of regular polytopes, they should include the need for equality of the number of elements of a certain dimension in the vicinity of elements of a different dimension to certain numbers, i. e. satisfy Schläfli symbols. However, this is the weakest condition of necessity and it has a narrow field of application. A broader, one might say, universal necessary condition for the existence of polytopes of higher dimension is the condition to satisfy the Euler-Poincaré equation. This condition is applicable to any closed connected polytopes of any dimension and any kind. For the first time, the derivation of the Euler-Poincaré equation was repeated in general form using modern concepts and notations. The fulfillment of the Euler-Poincaré equation for polytopes of the highest dimension of different classes is checked. Another necessary condition for the existence of higher-dimensional polytopes is associated with the need to correspond to certain incidence ratios between elements of different dimensions in a polytope. Analytical formulas for these relations are given. Thus, several conditions are necessary for the existence of higher-dimensional polytopes. Each of these conditions is necessary, but not sufficient. We can say that a polytope of the highest dimension exists only when it is constructed, i.e. all elements of various dimensions included in the polytope are listed, and a specific order of arrangement of elements relative to each other is indicated with a list of their contacts. In the next chapter it will be shown that the absence of this sufficient condition leads to the impossibility of the existence of a number of higher-dimensional polytopes, which were previously considered to exist. Let us emphasize the importance of the concept of incidence in the process of studying the geometry of higher-dimensional polytopes. On the basis of this concept, the law of the conservation of incidents was formulated, which has important informational value, for example, in the theory of heredity (Zhizhin, 2019 c, e).

REFERENCES

Alexandrov, A. D. (1934). The derivation of four - dimensional abnormal parallelohedrons. *Proceedings of the U.S.S.R. Academy of Sciences. Atmospheric and Oceanic Physics*, *6*, 803–817.

Alexandrov, A. D. (1954). On filling space with polyhedrons. *Bulletin of the Leningrad University. Mathematics Series.*, *1*, 33–43.

Alexandrov, P. S. (Ed.). (1969). *Hilbert's problems*. Science.

Coxeter, Y. S. M. (1934). Discrete groups generated by reflections. *Annals of Mathematics*, *35*(3), 588–621. doi:10.2307/1968753

Coxeter, Y. S. M. (1940). Regular and semi-regular polytopes. *J Math Zeit*, *46*(1), 380–407. doi:10.1007/BF01181449

Coxeter, Y. S. M. (1961). *Introduction to Geometry*. John Wiley & Sons, INC.

Coxeter, Y. S. M. (1963). Regular Polytopes. Academic Press.

Coxeter, Y. S. M. (1974). *Regular Complex Polytopes*. Cambridge University Press.

Cromwell, P. R. (1977). *Polyhedrons*. Cambridge University Press.

Davis, M. W. (2007). The geometry and topology of coxeter groups. London Mathematical Society, 32.

Delaunay, B.N. (1929). Sur la partition ré gulière de l'espace A4 dimensions. *Bulletin de L'Académie des Sciences de L'Urss., 1*, 79 - 110, 147 - 164.

Delaunay, B. N. (1937). Geometry of positive quadratic forms. *Progresses of Mathematical Sciences, 3*, 16–62.

Delaunay, B. N. (1961). Proof of the main theorem of the theory of stereohedrons. *Reports. Academy of Sciences of the USSR, 138*(6), 1270–1272.

Delaunay, B. N., & Sandakova, N. N. (1961). Theory of stereohedrons. *Proceeding Mathematical In. name V.A. Steklov of Academy Sciences of the USSR, 64*, 28–51.

Euler, L. (1752-1753). Elementa doctrinae solidorum. *Novi Comment Acad Sci I Petropolitanae, 4*, 376–386.

Fedorov, E.S. (1885). Beginning of the teachings on the figures. *Notes of the St. Petersburg Mineralogical Society, 21.*

Fedorov, E.S. (1889). Symmetry of finite figures. *Notes of the St. Petersburg Mineralogical Society, 25*, 1 - 52.

Fedorov, E.S. (1891). Symmetry of the correct systems of figures. *Notes of the St. Petersburg Mineralogical Society, 28.*

Grünboum, B. (1967). *Convex Polytopes*. Springer.

Heath, T. L. (1981). A History of Greek Mathematics. Academic Press.

Hilbert, D. (1901). Gesammelte Abhandlangen. *Archiv f. Mathematics and Physics, 3*(1), 44–63, 213–237.

Minkowski, H. (1911). Diskontinuitätshereich für arithemetische Äquivalenz. *Gesammelte Abhandlangen, II*, 70–72.

Poincaré, A. (1895). Analysis situs. *J. de é Ecole Polytechnique, 1.*

Schläfli, L. (1901). *Theorie der vielfachen Kontinuität Zürcher und Furrer*. doi:10.1007/978-3-0348-5118-3

Schulte, E. (2014). Polyhedra, complexes, nets and symmetry. *Acta Crystallographica. Section A, Foundations of Crystallography, 70*(3), 203–216. doi:10.1107/S2053273314000217 PMID:24815970

Stringham, W. I. (1880). Regular figures in *n* - dimensional space. *American Journal of Mathematics, 3*(1), 1–14. doi:10.2307/2369441

Venkov, B. A. (1954). On a class of Euclidean polyhedrons. *Bulletin of the Leningrad University Mathematics, 2*, 11–31.

Venkov, B.A. (1959). On the design of parallelohedra. *Mathematical Collection, 49*(2), 207 -224.

Voronoi, G. F. (1952). *Studies on primitive parallelohedra. In Collected Works, 2.* Publishing House of the Academy of Sciences of the Ukrainian SSR.

Zhizhin, G. V. (2018). *Chemical Compound Structures and the higher Dimension of Molecules: Emerging research and Opportunities.* IGI Global. doi:10.4018/978-1-5225-4108-0

Zhizhin, G. V. (2019a). *The Geometry of Higher - Dimensional polytopes.* IGI Global. doi:10.4018/978-1-5225-6968-8

Zhizhin, G. V. (2019b). On the Structure of Parallelohedrons of Higher Dimension: Hilbert's 18th Problem. *International Journal of Chemoinformatics and Chemical Engineering, 8*(2), 69–96. doi:10.4018/IJCCE.20190701.oa2

Zhizhin, G. V. (2019c). *Attractors and Higher Dimensions in Population and Molecular Biology: Emerging Research and Opportunities.* IGI Global. doi:10.4018/978-1-5225-9651-6

Zhizhin, G. V. (2019d). The Law of Conservation of Incidents in the space of Nanoworld. *International Journal of Chemoinformatics and Chemical Engineering, 8*(1), 25–46. doi:10.4018/IJCCE.2019010103

Zhizhin, G. V. (2019e). The Polytope of Hereditary Information the Structure, Location, Signification. *Biochemistry and Modern Applications, 2*(1), 56–62. doi:10.33805/2638-7735.124

Zhizhin, G. V. (2020). *Nanotechnologies and Clusters in the Spaces of Higher Dimension: Emerging Research and Opportunities.* IGI Global.

Zhizhin, G. V. (2021). *Normal Partitions and Hierarchical Fillings of N-Dimensional Spaces.* IGI Global. doi:10.4018/978-1-7998-6768-5

KEY TERMS AND DEFINITIONS

Dimension of the Space: The member of independent parameters needed to describe the change in position of an object in space.

Incidence Coefficients of Elements of Lower Dimension With Respect to Elements of Higher Dimension: The number of elements of a certain higher dimension to which the given element of a lower dimension belongs.

Incidence Coefficients of Elements of Higher Dimension With Respect to Elements of Lower Dimension: The number of elements of a given lower dimension that are included in a particular element of a higher dimension.

N–Cross-Polytope: The convex polytope of dimension n in which opposite related of centrum edges not have connection of edge.

N–Cube: The convex polytope of dimension *n* in which each vertex incident to *n* edges.

N-Simplex: The convex polytope of dimension n in which each vertex is joined by edges with all remain vertices of polytope.

Polytope: Polyhedron in the space of higher dimension.

Chapter 4
Regular and Semi-Regular Polytopes of Higher Dimension

ABSTRACT

A direct construction of regular polytopes of dimension four was carried out by connecting three-dimensional figures along whole flat faces included in the polytope. It was found that the images of a 24-cell cell known from Coxeter's works do not correspond to reality. It is proved that the 600-cell and 120-cell polytopes cannot exist, since the process of their construction leads to a contradiction with the necessary conditions for the existence of polytopes. The existence of a new class of polytopes has been discovered: polyincidental quasi-regular polytopes having edges with different incidence values within the same polytope. Semi-regular four-dimensional polytopes are constructed using the operations of cutting off the neighborhood of the vertices of regular polytopes. The images of semi-regular polytopes of the highest dimension are presented.

INTRODUCTION

In the first chapter, regular and semi-regular three-dimensional polyhedrons (Plato's bodies and Archimedes' bodies) were considered. As the dimension increases, the class of semi-regular polytopes changes sharply and the composition of regular polytopes changes. This chapter is devoted to the study of these polytopes. Moreover, in accordance with the conditions necessary for the existence of polytopes, considered in the previous chapter, an analysis was made of the possibility of the existence of four-dimensional polytopes, for which a complete proof of the existence has not been previously carried out. Direct construction of the 24-cell, with a list of all its components, proved the possibility of its existence. Direct construction of a 600-cell led to a contradiction with the necessary conditions for the incidence of elements of different dimensions in this polytope. This proves that the 600 - cell cannot exist. Consequently, the 120-cell, which is dual to it, cannot exist either. Semi-regular polytopes of higher dimension are obtained from regular polytopes of higher dimension as a result of single and double cutting of the vertices of the polytopes. Images of a series of semi-regular polytopes of dimension 4 are shown.

DOI: 10.4018/978-1-7998-8374-6.ch004

REGULAR POLYTOPES OF HIGHER DIMENSION

Regular Polytopes of Four Dimension

Regular four-dimensional polytopes consist of regular three-dimensional shapes adjacent to each other along two-dimensional whole faces. Following Schläfli (Schläfli, 1901), each regular polytope should be assigned three digits {p, q, r}, where *p* is the number of two-dimensional faces of the polytope, *q* is the number of flat faces incident to one vertex, *r* is the number of three-dimensional faces incident to one edge of the polytope. By virtue of the assumption of correctness, all vertices of the polytope are equivalent, i. e. each of them is incident to the same number of edges. A regular two-dimensional face in a regular polytope can be a regular triangle, a square and a regular pentagon, since in order to form a three-dimensional angle in a three-dimensional cell of a four-dimensional polytope, flat faces can be either three triangles, or four triangles, or five triangles, or three squares, or three pentagons.

Construction of a Polytope Simplex of Dimension 4

Let's say we have a regular tetrahedron *abcd* (Figure 1).

Figure 1. The regular tetrahedron

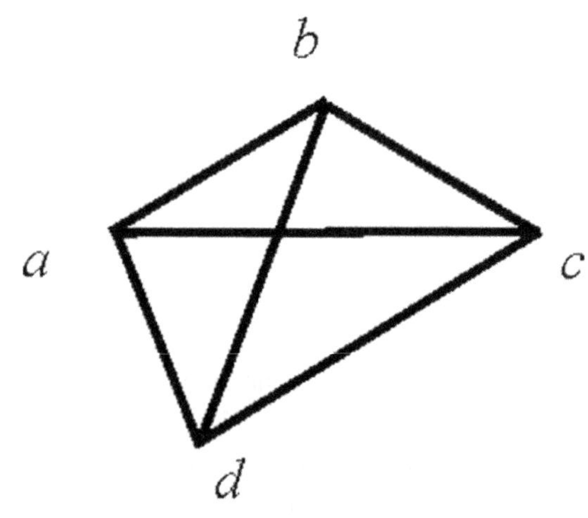

Figure 1 is a projection of a three-dimensional figure *abcd* of a tetrahedron onto a two-dimensional plane. Let's attach to the face *acd* of this tetrahedron one more same tetrahedron *acde* (Figure 2).

If now the vertex *b* of the initial tetrahedron is connected by an edge with the vertex *d* of the second tetrahedron, then in the projection onto the two-dimensional plane we will have a figure *abcde* (Figure 3).

Consider this figure. In it you can see projections of five tetrahedrons. If the length of an edge in the original tetrahedron is equal to *A*, then the values of the lengths of all edges are also equal to *A*. This leads to a figure of five regular tetrahedrons with an edge length *A*. Moreover, any pair of five tetrahedrons is attached to each other along an integral triangular face, i. e. each triangular face has two tetrahedrons attached. It is easy to determine that the number of edges in Figure 3 is 10: *ab, ac, ae, ad, bc, be, bd,cd,*

Figure 2. Two regular tetrahedrons attached to the face acd

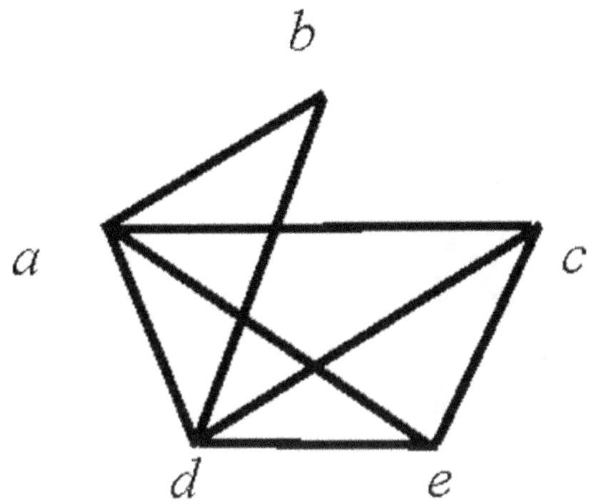

Figure 3. The 4 - simplex

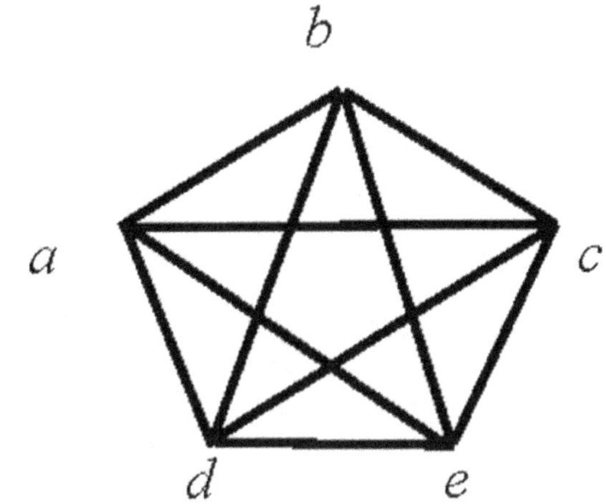

ce, de. The number of flat faces in Figure 3 is also 10: *abc, abd, bce, dbc, bde, ace, abe, adc, ade, dce*. The dimension of any polytope can be determined by the Euler-Poincaré equation (Poincaré, 1895):

$$\sum_{i=0}^{n-1}(-1)^i f_i(n)=1+(-1)^n, \qquad (4.1)$$

in (4.1) $f_i(n)$ is the number of elements of a polytope P of dimension n having dimension i.

In this case f_0=5, f_1=10, f_2=10, f_3=5. Substituting these values into equation (4.1), we have

$$5 - 10 + 10 - 5 = 0.$$

Thus, we see that the Euler-Poincaré equation (4.1) holds in this case for $n = 4$. This proves that Figure 3 has dimension 4. This figure is called a simplex of dimension 4 or 4-simplex. In Figure 3, the number of vertices incident to one 2D face is 3; the number of flat faces incident to one vertex is 3; the number of three-dimensional faces incident to one edge is 3. Therefore, the Schläfli symbol for a regular 4-simplex has the form {3, 3, 3}. Any vertex of a 4-simplex can be considered as a vertex of a pyramid with a tetrahedron base built on the other vertices of a 4-simplex. All vertices of a 4-simplex are the same. They are incident to the same number of edges, the same number of two-dimensional faces, the same number of three-dimensional faces.

Construction of a Polytope Cube of Dimension 4

Let there be a cube *abcdefrn* (Figure 4).

Let's attach two more cubes to the cube faces *cdre, ndfe* so that the applied cubes also have one common face (Figure 5).

Figure 4. The tree-dimensional cube

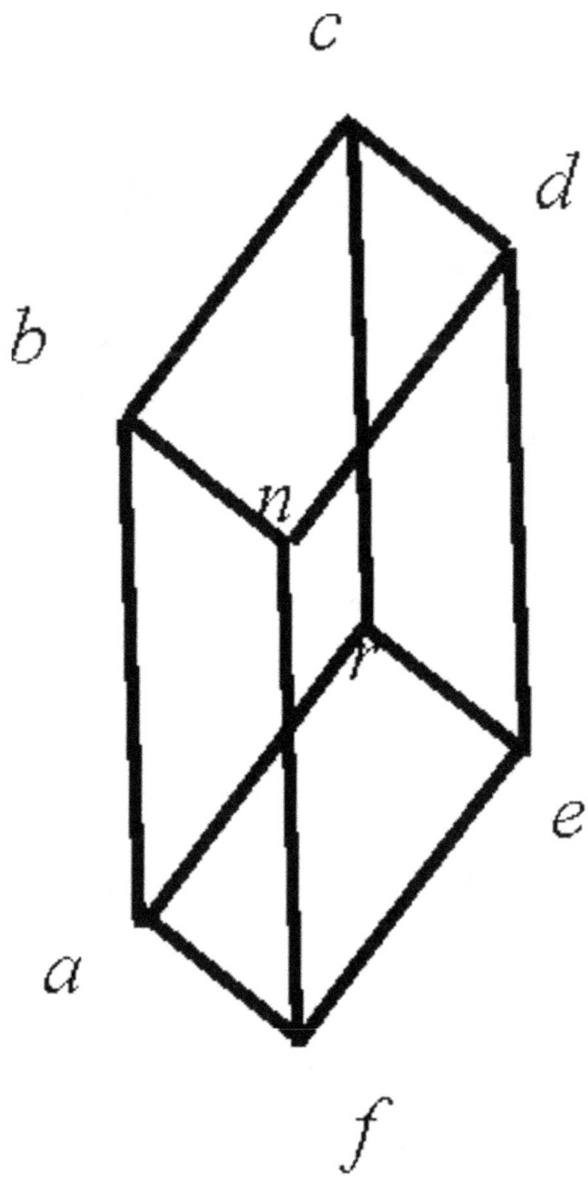

Let's insert into the figure in Figure 5 another cube that has one common face with the cubes just connected (Figure 6) and connect the vertices a and a' in the louse, as well as the vertices b and b' with the edges.

Consider Figure 6 In it you can see the projection of eight cubes:

abcdrefn, a′b′c′d′r′e′f′n′, cdbnc′a′b′n′, arefa′r′e′f′, cdrec′d′r′e′, bnafb′n′a′f′, ndfen′d′f′e, bcarb′c′a′r.

Each flat face of any of the cubes is common to two of the eight cubes.

Figure 5. The tree cube with common faces

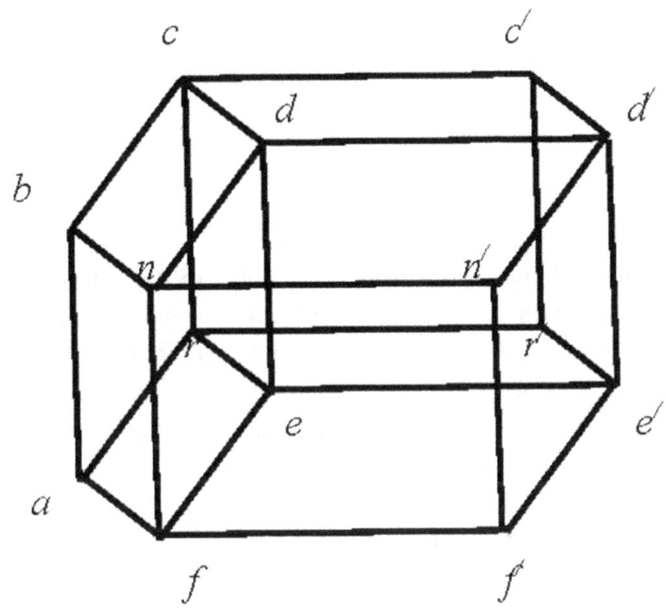

Figure 6. The polytope 4- cube

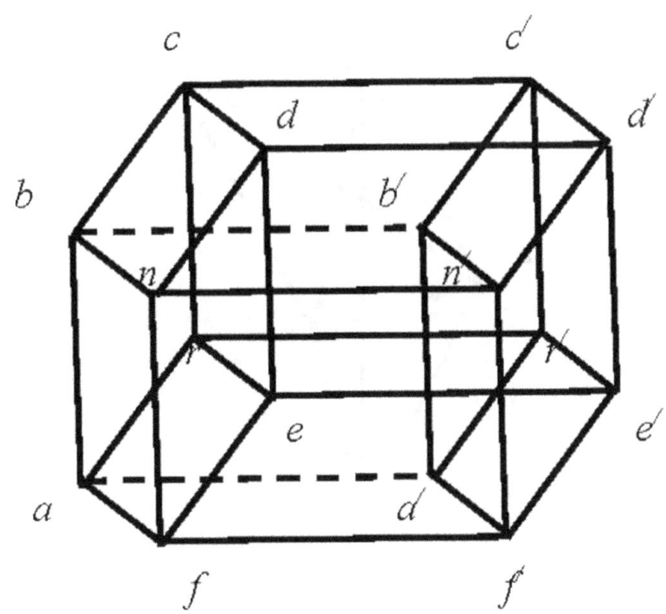

The Figure 6 has 16 vertices; 32 edges *ab, af, ar, aa′, bn, bc, bb′,·cd, cc′, cr, dn, de, dd′, ef, ee′, er, rr′, nn′,nf, ff′, a′b′, a′f′, a′r′, b′c′, b′n′, c′d′, c′r′,d′e′, d′n′, e′r′, e′f′,f′n′;* 24 a flat faces *abbf, cdre, bncd, afre, arbc, bfde, a′b′n′f′, c′d′r′e′, b′n′c′d′,a′f′r′e′, a′r′b′c′, n′f′d′e′, cdc′d′, bb′nn′, bcb′c′, ndn′d′, rer′e′, afa′f′, rar′a′, fef′e′, nfn′f′, ded′e′, crc′r′, aba′b′.*
In this case f_0=16, f_1=32, f_2=24, f_3=8. Substituting these values into equation (4.1), we have

$$16 - 32 + 24 - 8 = 0.$$

Thus, we see that the Euler-Poincaré equation (4.1) holds in this case for n = 4. This proves that Figure 6 has dimension 4. This figure is called a cube of dimension 4 or 4-cube. In Figure 6, the number of vertices incident to one 2D face is 4; the number of flat faces incident to one vertex is 5; the number of three-dimensional faces incident to one edge is 3; the number of two-dimension faces incident to one edge is 3. The flat faces all are squares, all vertices and edges are equal and, therefore, the polytope is regular. The Schläfli symbol for a regular 4-cube has the form {4, 3, 3}.

Construction of a Cross - Polytope of Dimension 4

Let's say we have a regular tetrahedron *abcd* (Figure 1). We added to it also a same tetrahedron, but not face *dac* (as it was on Figure 2), and to face *abd* (Figure 7).

Figure 7. The two tetrahedrons with common face abd

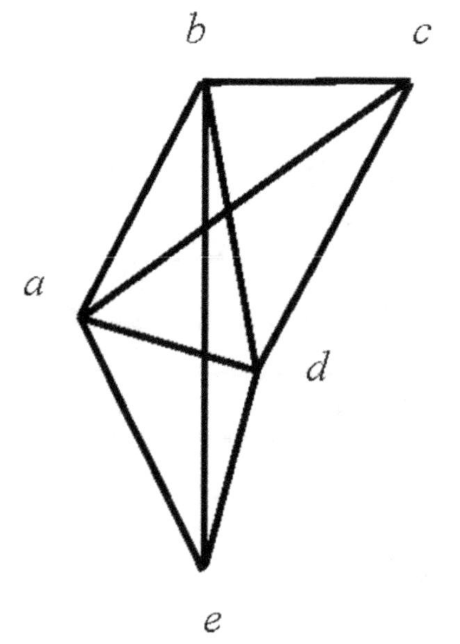

After this we added to face *ade* also same tetrahedron *adef* (Figure 8).

Figure 8. The tree tetrahedrons with common faces dac and abd

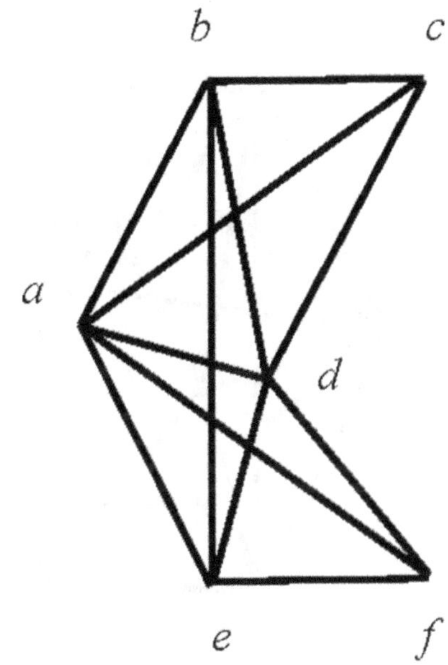

Now we added to faces *adc* and *adf* again one same tetrahedron *adcf.* So will add an edge *cf* (Figure 9).

Figure 9. The four tetrahedrons with common faces abd, ade, adc, adf

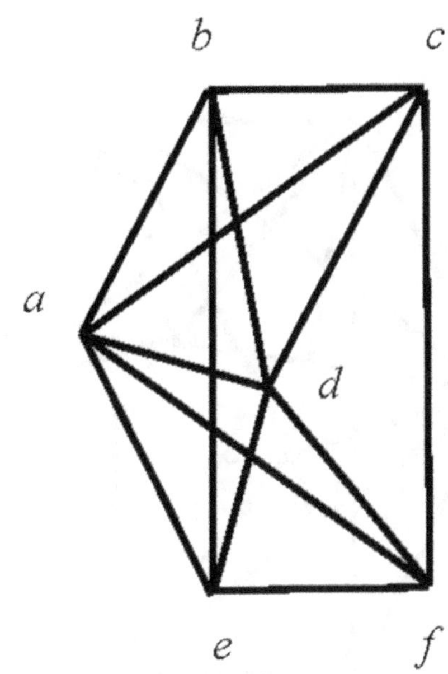

To a face *abc* add another same tetrahedron *abcg* (Figure 10).

Figure 10. The five tetrahedrons with common faces abd, ade, adc, adf, abc

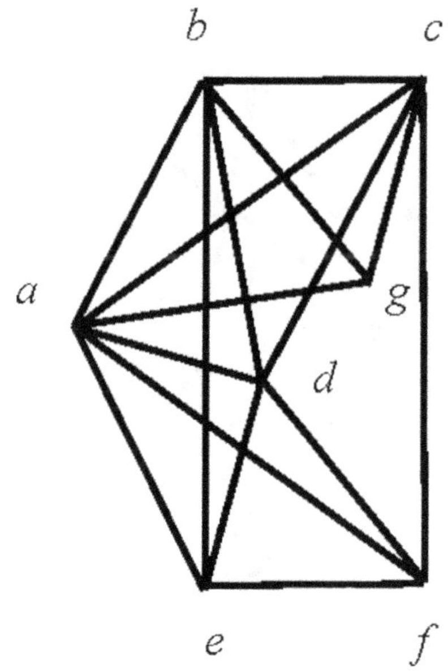

This will create the edges *ag, bg, cg* and vertex *g*. To face *bcg* add tetrahedron *bcgs* (Figure 11).

Figure 11. The six tetrahedrons with common faces abd, ade, adc, adf, abc, bcg

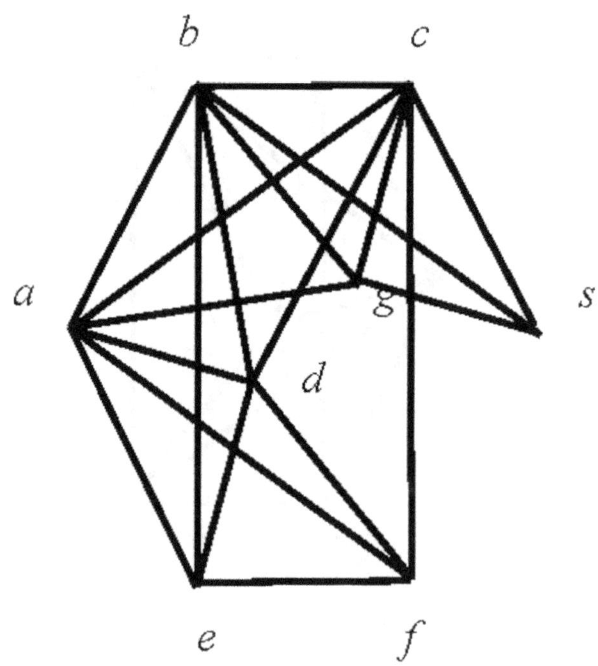

To face *cgs* add tetrahedron *cgsf* (Figure 12). In this case tetrahedron *agcf* are formed complement.

Figure 12. The eight tetrahedrons with common faces

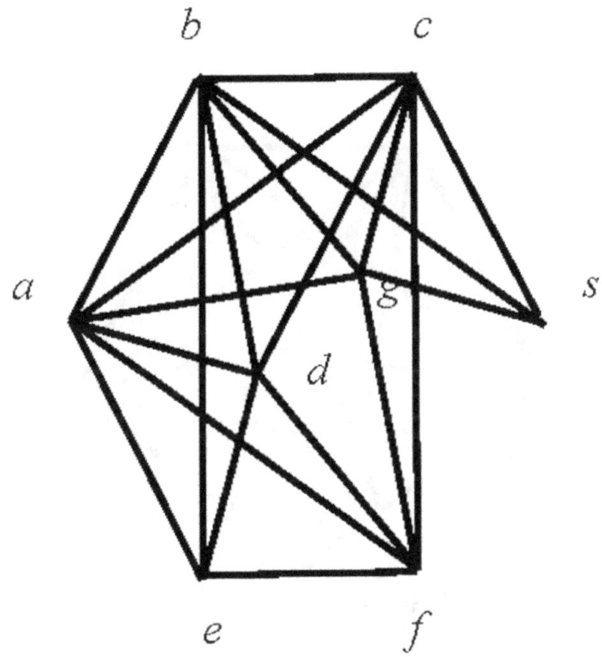

To face *edf* add tetrahedron *edsf (*Figure 13). In this case tetrahedrons *afdc, bdef, agcf* are formed complement.

Figure 13. The twelve tetrahedrons with common faces

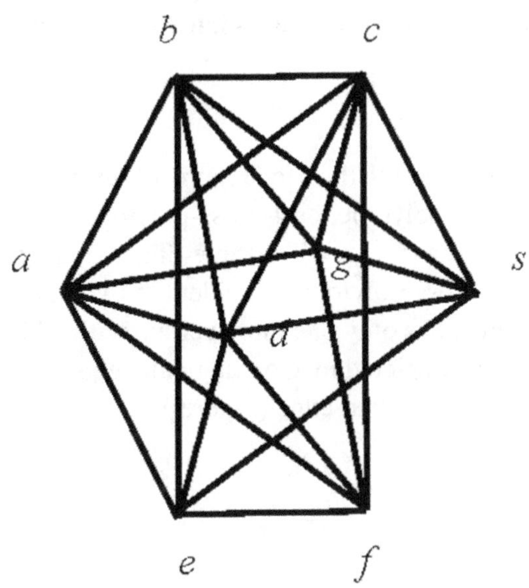

In this case, tetrahedrons *afdc, agcf, bdef* are additionally formed.

Let's connect the vertices *e* and *g* with an edge. In this case, tetrahedrons *agef, abge, acfg, dfsg* are formed (Figure 14).

Figure 14. The sixteen tetrahedrons with common faces

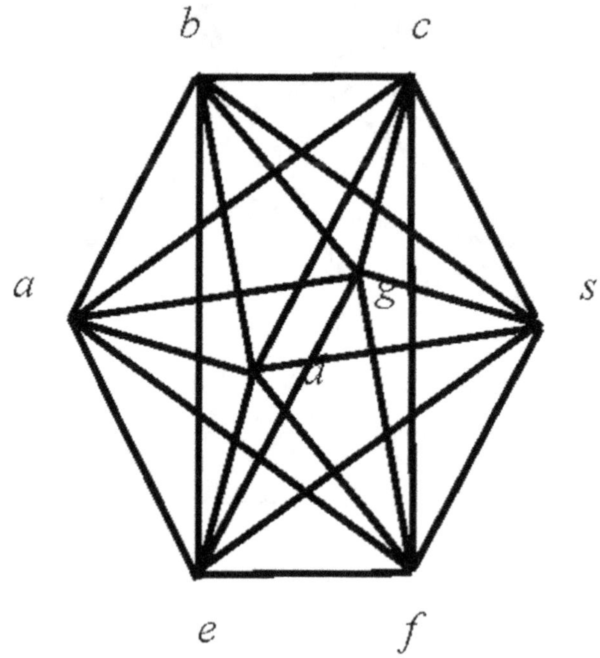

The polytope in Figure 14 has 8 vertices, f_0=8 ; 24 edges *ad, bc, ab, bd, dc, eb, ac, ae, ed, df, af, ef, cf, bg, gc, ag, bf, cf, gf, gs, sf, ds, es, eg,* f_1=24; 32 triangular faces *bcd, acd, abd, abc, edb, abe, aed, adf, edf, aef, acf, dcf, abg, bcg, acg, dsg, bcs, gcs, agf,gcf, csf, gsf, eds, bsd, dcs,sdf, csf, bes, age, egf,* f_2=32; 16 tetrahedrons, f_3=16. Substituting the obtained values into the Euler-Poincaré equation (4.1), we obtain

$$8 - 24 + 32 - 16 = 0.$$

Thus, the Euler-Poincaré equation is fulfilled for Figure 14 with $n = 4$. This proves that this figure has dimension 4. It is called a cross-polytope, or 4-cross-polytope. This figure is correct, since all its vertices are equivalent. Each of them has 6 edges. The three-dimensional figures included in it are regular tetrahedrons. Each flat face of a 4-cross-polytope is incident to two 3D figures. A distinctive feature of the cross-polytope is that each vertex has only one more vertex, with which there is no edge connection. Therefore, a 4-cross-polytope can be depicted in a topologically equivalent form, where each vertex has an opposite vertex, between which there is no edge (Figure 15).

Figure 15. The 4 – cross – polytope

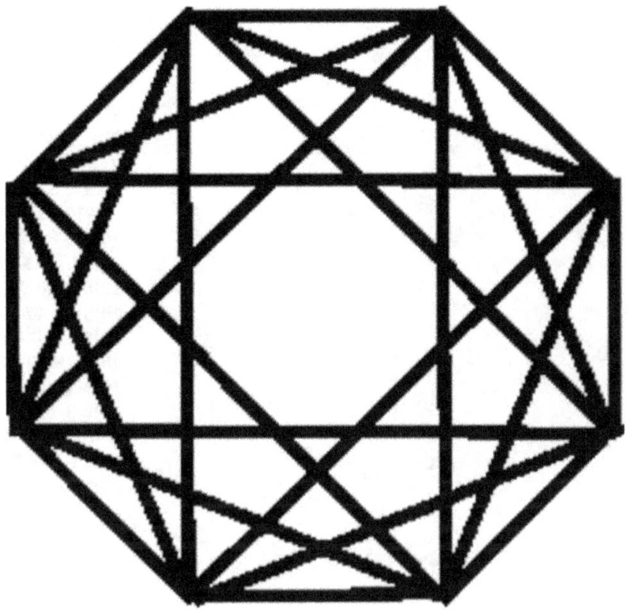

Construction of a Polytope 24 – Cell

The 24-Cell polytope is supposed to consist of 24 regular octahedrons (Coxeter, 1963, 1974). An octahedron can be obtained from a tetrahedron using the clipping operation. In this operation, the edges of the polytope are shortened. According to the operation t, all the edges of the polyhedron are shortened from both sides by one third. The second cutting operation t^2 consists in shortening all the edges of the polyhedron on both sides by half. Then all the edges of the original polytope turn into vertices, and if these vertices are connected by edges a new polyhedron is formed. Let us choose 4-cross-polytope as the initial polytope. It has 24 edges. As a result of applying the t^2 cut operation to each tetrahedron in the 4-cross-polytope, all edges of the 4-cross-polytope will turn into vertices, all tetrahedrons will turn into regular octahedrons. Since there are 16 tetrahedrons in the 4-cross-polytope, the t^2 operation will lead to the formation of 16 octahedrons. However, between the formed octahedrons, empty volumes are formed, which, as can be seen by the construction, also have the shape of an octahedron. Thus, an additional 8 octahedrons are formed additionally. The total number of octahedrons in the new polytope will be 24. Make sure that the formed figure is closed, and all octahedrons are adjacent to each other along whole flat faces. The result of the described construction (Zhizhin, 2013, 2014) is shown in Figure 16 in projection onto a two-dimensional plane.

All newly formed peaks are numbered. They are located on the edges of the 4-cross-polytope, the edges of which are indicated by the dotted line. All octahedrons are listed according to the vertex numbering (next to the figure). The resulting polytope is a 24-cell or in Schläflis ‘ notation {3,4,3}. An image of polytope {3, 4, 3} suggested by Coxeter (Coxeter, 1963, 1974) is shown in Figure 17.

However, these works do not list the 24 octahedrons included in this figure. You can see that it is impossible to find 24 octahedrons in this figure. Therefore, Figure 17 cannot be recognized as depicting a 24-cell polytope.

Figure 16. The {3, 4, 3} polytope

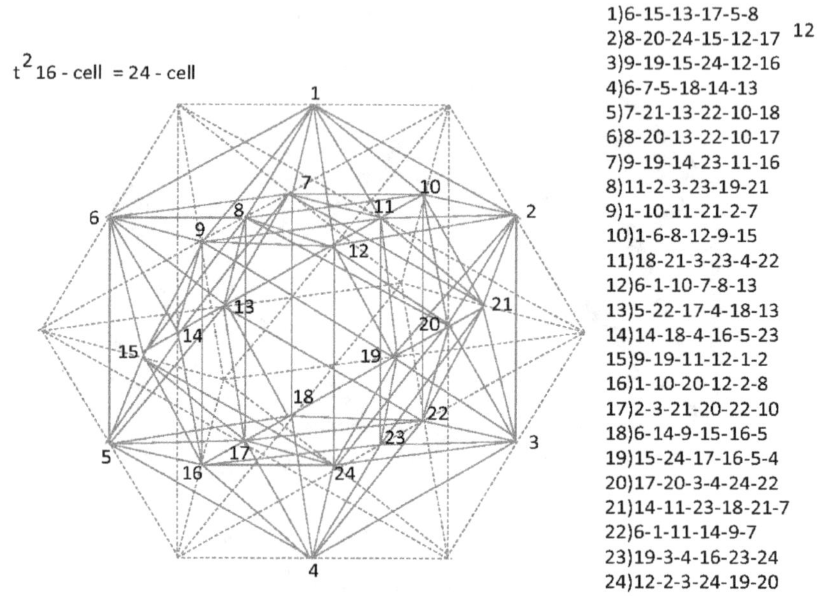

Figure 17. The {3, 4, 3} polytope at Coxeter

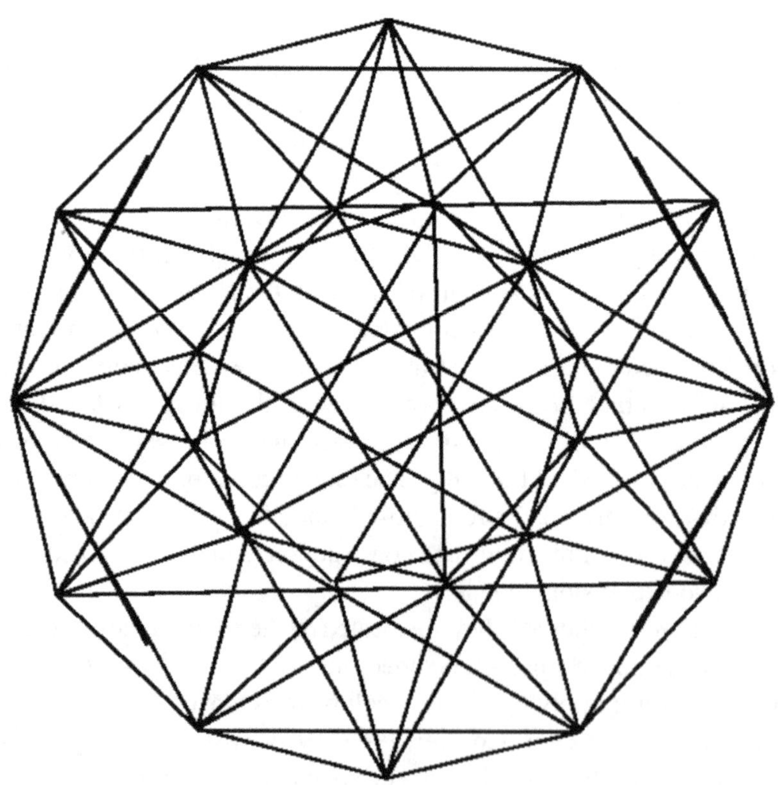

Let us make sure that Figure 16 and the set of 24 octahedrons indicated next to the figure represent a closed polytope. To do this, you need to make sure that each flat face of an octahedron from the list on Fig 4.16 is closed by some other unique face of another octahedron from this list. The results of this analysis are shown in Table1. It follows from it that all the flat faces of the octahedrons included in the polytope are covered by other flat faces of the octahedrons in the singular. The entire set of 24 octahedra is closed.

It should be noted that edge 13 - 10 of octahedron 12 in projection onto a two-dimensional plane lies on the same straight line as edge 14 - 11 of octahedron 22 and edge 15 - 12 of octahedron 10. This leads to this. that at vertices 11 - 14 the apparent incidence is 9. The same situation exists with edges 21 - 18, 19 - 16, 17 - 20, respectively, of octahedra 11, 23, 20. Therefore, the apparent incidence at vertices 17 - 20 is also 9. The edge 15 - 12, connecting vertices 15 and 12, does not connect them in construction with vertices 14 and 13. Edge 13 - 10, connecting vertices 13 and 10, does not connect these vertices in construction with vertices 12 and 11. Edge 14 - 11, connecting vertices 14 and 11, does not connect them with vertices 13 and 12. Therefore, the incidence of vertices 11 - 14 is 8, as well as for vertices 17 - 20 and all other vertices of the polytope. The location of vertices 11 - 14 on one straight line should be considered the result of projection from four-dimensional space (the same applies to vertices 17 - 20). Thus, each vertex of polytope {3, 4, 3} is incident to eight edges. All 96 polytope edges were divided into 12 groups, each of which contains the same number of parallel edges - 8. All this proves that the polytope in Figure 16 is a correct polytope.

Table 1. The harmony of the faces of octahedrons in the polytope {3,4,3}

Number of octahedron	Faces of octahedron number of the contacting octahedron along the corresponding face							
1	6-5-15	15-5-17	17-5-13	6-5-13	6-8-15	15-8-17	17-8-13	6-8-13
	18	19	13	4	10	2	6	12
2	8-12-15	8-12-20	15-12-24	24-12-20	8-17-15	8-17-20	15-17-24	24-17-20
	10	16	3	24	1	6	19	20
3	9-15-16	9-15-12	16-15-24	24-15-12	9-19-16	9-19-12	16-19-24	24-19-12
	18	10	19	2	7	15	23	24
4	6-14-5	6-14-7	5-14-18	7-14-18	6-13-5	6-13-7	5-13-18	7-13-18
	18	22	14	21	1	12	13	5
5	7-13-18	7-13-10	18-13-22	22-13-10	7-21-18	7-21-10	18-21-22	22-21-10
	4	12	13	6	21	9	11	17
6	8-20-17	8-20-10	17-20-22	22-20-10	8-13-17	8-13-10	22-13-17	22-13-10
	2	16	20	17	1	12	13	5
7	9-14-16	9-14-11	16-14-23	23-14-11	9-19-16	9-19-11	16-19-23	23-19-11
	18	22	14	21	3	15	23	8
8	11-19-23	11-19-2	23-19-3	3-19-2	11-21-23	11-21-2	23-21-3	3-21-2
	7	15	23	24	21	9	11	17
9	1-7-11	1-7-10	11-7-21	21-7-10	1-2-11	1-2-10	11-2-21	21-2-10
	22	12	21	5	15	16	8	17

Continued on following page

Table 1. Continued

Number of octahedron	Faces of octahedron number of the contacting octahedron along the corresponding face							
10	6-1-9	6-1-8	9-1-12	12-1-8	6-15-9	6-15-8	9-15-12	12-15-8
	22	12	15	16	18	1	3	2
11	18-22-4	18-22-21	4-22-3	3-22-21	18-23-4	18-23-21	4-23-3	3-23-21
	13	5	20	17	14	21	23	8
12	6-1-8	6-1-7	8-1-10	10-1-7	6-13-8	6-13-7	8-13-10	10-13-7
	10	22	16	9	1	4	6	5
13	13-5-17	17-5-4	4-5-18	13-5-18	13-22-17	17-22-4	4-22-18	13-22-18
	1	19	14	4	6	20	11	5
14	14-5-16	14-5-18	16-5-4	4-5-18	14-23-16	14-23-18	16-23-4	4-23-18
	18	4	19	13	7	21	23	11
15	9-1-12	9-1-11	12-1-2	2-1-11	9-19-12	9-19-11	12-19-2	2-19-11
	10	22	16	9	3	7	15	8
16	1-2-12	1-2-10	12-2-20	20-2-10	1-8-12	1-8-10	12-8-20	20-8-10
	15	9	24	17	10	12	2	6
17	10-2-20	10-2-21	20-2-3	2-3-21	10-22-20	10-22-21	20-22-23	3-22-21
	16	9	24	8	6	5	20	11
18	6-5-15	6-5-14	15-5-16	16-5-14	6-9-15	6-9-14	15-9-16	16-9-14
	1	4	19	14	10	22	3	7
19	15-5-16	15-5-17	16-5-4	4-5-17	15-24-16	15-24-17	16-24-4	4-24-17
	18	1	14	13	3	2	23	20
20	17-24-4	17-24-20	4-24-3	3-24-20	17-22-4	17-22-20	4-22-3	3-22-20
	19	2	23	24	13	6	11	17
21	14-7-18	14-7-11	18-7-21	21-7-11	14-23-18	14-23-11	18-23-21	21-23-11
	4	22	5	9	14	7	11	8
22	6-9-1	6-9-14	14-9-11	11-9-1	6-7-1	6-7-14	14-7-11	11-7-1
	10	18	7	15	12	4	21	9
23	19-3-24	19-3-23	24-3-4	4-3-23	19-16-24	19-16-23	24-16-4	4-16-23
	24	8	20	11	3	7	19	14
24	19-3-24	19-3-2	24-3-20	20-3-2	19-12-24	19-12-2	24-12-20	20-12-2
	23	8	20	17	3	15	2	16

Since the incidence of any vertex with respect to the edges is 8, it is easy to write down the entire list of edges using Figure 15. Obviously, each edge in this list (by vertices) will occur twice. Therefore, the total number of edges is f_1=24•8/2=96.

From Table1 it is easy to compile a list of all two-dimensional faces of the polytope {3, 4, 3}

6 – 5 – 15, 15 – 5 – 17, 17 – 5 – 13, 6 – 5 – 13, 6 – 8 – 15, 15 – 8 – 17, 17 – 8 – 13, 6 – 8 – 13, 8 – 12 – 15, 8 – 12 – 20, 12 – 15 – 24, 12 – 20 – 24, 8 – 17 – 20, 15 – 17 – 24, 17 – 24 – 20, 9 – 15 – 96, 9 – 12 – 15, 15 – 16 – 24, 9 – 16 – 19, 9 – 12 – 19, 16 – 19 – 24, 12 – 19 – 24, 5 – 6 – 14, 6 – 7 – 14,

5 – 14 – 18, 7 – 14 – 18, 6 – 7 – 13, 5 – 13 – 18, 7 – 13 – 18, 7 – 10 – 13, 13 – 18 – 22, 10 – 13 – 22, 7 – 21 – 18, 7 – 10 – 21, 18 – 21 – 22, 10 – 21 – 22, 8 – 10 – 20, 17 – 20 – 22, 10 – 2 – 22, 8 – 10 – 13, 13 – 17 – 22, 9 – 11 – 14, 14 – 16 – 23, 11 – 14 – 23, 9 – 11 – 19, 16 – 19 – 23, 11 – 19 – 23, 2 – 11 – 19, 3 – 19 – 23, 2 – 3 – 19, 11 – 21 – 23, 2 – 11 – 21, 3 – 21 – 23, 2 – 3 – 21, 1 – 7 – 11, 1 – 7 – 10, 7 – 11 – 21, 1 – 2 – 11, 1 – 2 – 10, 2 – 10 – 21, 1 – 6 – 9, 1 – 6 – 8, 1 – 9 – 12, 1 – 8 – 12, 6 – 9 – 15, 4 – 18 – 22, 3 – 4 – 22, 3 – 21 – 22, 4 – 18 – 23, 18 – 21 – 23, 3 – 4 – 23, 1 – 6 – 7, 1 – 8 – 10, 4 – 5 – 17, 4 – 5 – 18, 4 – 17 – 22, 5 – 14 – 16, 4 – 5 – 16, 14 – 18 – 23, 4 – 16 – 23, 1 – 9 – 11, 1 – 2 – 12, 2 – 12 – 19, 2 – 12 – 20, 2 – 10 – 20, 2 – 3 – 20, 20 – 22 – 23, 5 – 15 – 16, 6 – 9 – 14, 9 – 14 – 16, 4 – 16 – 24, 4 – 17 – 24, 3 – 4 – 24, 3 – 20 – 24, 3 – 20 – 22, 7 – 11 – 14, 3 – 19 – 24.

Thus, f_2=96, f_0=f_3=24.
Substituting the obtained values into the Euler-Poincaré equation (4.1), we obtain

$$24 - 96 + 96 - 24 = 0.$$

Thus, the Euler-Poincaré equation is fulfilled for Figure 15 with $n = 4$. This proves that polytope 24-cell has dimension 4.

About Polytopes {5, 3, 3} and {3, 3, 5}

The polytopes {5, 3, 3} and {3, 3, 5} are dual to each other. It is stated that the polytope {5, 3, 3} (600-cell) consists of 600 tetrahedrons, and each edge of any tetrahedron from this set has 5 tetrahedrons indented. The polytope {3, 3, 5} consists of 120 dodecahedrons, and each edge of any of the dodecahedrons is indented by 3 dodecahedrons from this set. The existence of these polytopes was first announced by W. Stringham (Stringham,1880). However, there is no integral depiction of each of these two polytopes in this work. Moreover, the proof of the existence of the regular polytope {5, 3, 3} Stringham is based on the initial assumption that the three-dimensional space around some point is filled with 20 regular tetrahedrons. However, it is not. It is possible to fill the three-dimensional space around a certain point only with irregular tetrahedrons, applying them to each other along whole faces around the selected point in space. The solid angle at the apex of a regular tetrahedron is 0.55129 sr. Therefore, 20 such angles are 11.0258 sr $< 2\pi^2$ »19.72 sr - the full solid angle in three-dimensional space. Therefore, it should be recognized that the existence of the polytope {5, 3, 3} and its dual polytope {3, 3, 5} has not yet been proven. In the works of G. Coxeter (Coxeter, 1963, 1974) there are images {3, 3, 5} and {5, 3, 3} obtained, according to the author, on the basis of the image of a 24-cell in Figure 16. However, as has just been shown, this image of the polytope {3, 4, 3} does not allow us to list all 24 octahedrons so this octahedrons on this figure are not. In the very images of polytopes {5, 3, 3} and {3, 3, 5} given of Coxeter, it is also impossible to list the declared 600 tetrahedrons and 120 dodecahedrons. Therefore, the question of the existence of polytopes {5, 3, 3} and {3, 3, 5} should be considered open.

Just as when considering the previous regular four-dimensional polytopes, let us try to check the possibility of the existence of a 600 -cell by construction, i. e. consistently applying three-dimensional figures to each other. It is claimed that: 600-cell consists of 600 tetrahedrons and each edge of any of the tetrahedra of this set is incidentally 5 tetrahedrons. Let us check this statement by constructing a projection of this figure on to a two-dimensional plane, applying tetrahedrons to each other along whole faces. We will number the vertices and write down the list of plotted tetrahedrons. Moreover, in the

construction we will strive to achieve this incident. The edge for which this incidence is reached will be outlined in red. Consistently reaching this incidence value, we will move from the center of construction to the periphery. Let's check if it is possible to achieve a figure of 600 tetrahedrons, in which all edges have an incidence of 5 in relation to three-dimensional figures, and the flat edges have an incidence of 2 in relation to three-dimensional figures.
Let there be a regular tetrahedron 1 - 2 - 3 - 4 (Figure 18).

Figure 18. The building of polytope 600-cell

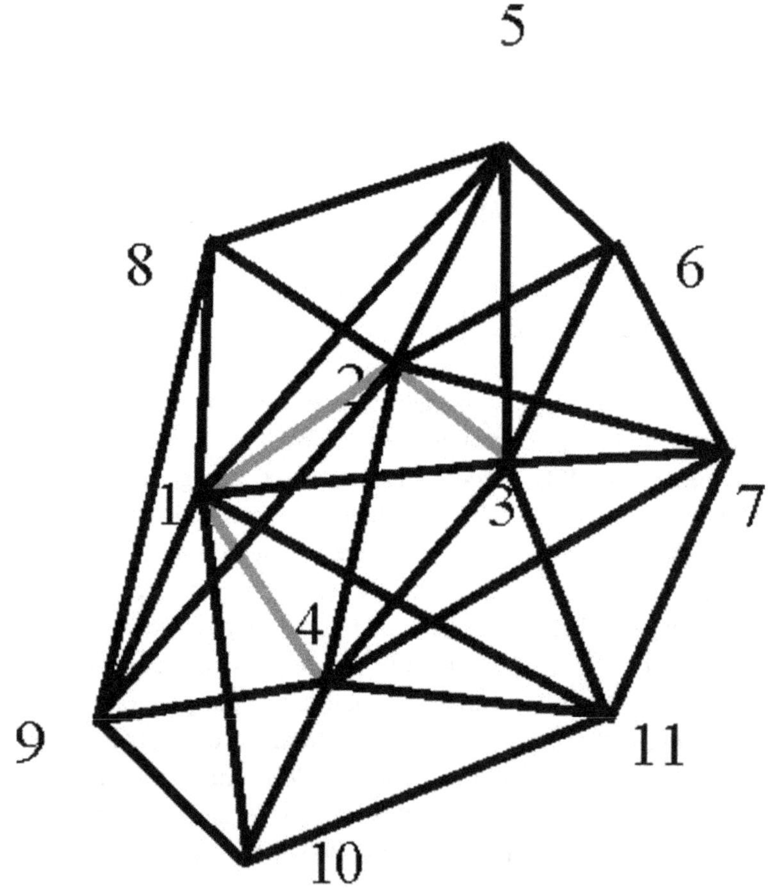

We will apply to its two-dimensional faces other tetrahedrons of the same size along whole faces. According to the accepted incidence condition, only one tetrahedron can be attached to each face. Let us apply tetrahedron 1 - 2 - 3 - 5 to face 1 - 2 - 3. The incidence of edge 2 - 3 with respect to tetrahedra in this case becomes equal to 2 (as well as edges 1 - 2, 1 - 3). To increase the incidence of an edge (for example, edges 2 - 3), apply to face 2 - 3 - 5 tetrahedron 1 - 2 - 3 - 5 another tetrahedron 2 - 3 - 5 - 6. The incidence of edge 2 - 3 became equal to 3. Apply to tetrahedron 2 - 3 - 5 - 6 along face 2 - 6 - 3 another tetrahedron 2 - 6 - 3 - 7. The incidence of edge 2 - 3 has become equal to 4. Let's connect vertices 4 and 7 with an edge. This is equivalent to joining tetrahedra 1 - 2 - 3 - 4, 2 - 6 -3 - 7 along the edges, respectively, 2 - 3 - 4 and 2 - 3 - 7 tetrahedron 2 - 3 - 4 - 7. The incidence of edge 2 - 3 became equal 5, and the incidence of edge 3 - 4 became equal to 2. Let's continue constructions to increase the incidence of edge 1 - 2. Add tetrahedron 1 - 2 - 3 - 5 to face 1 - 2 - 5 tetrahedron 1 - 2 - 5 - 8. The incidence of edge 1 - 2 becomes 3. Let's connect to tetrahedron 1 - 2 - 8 - 5 along the face 1 - 2 - 8 tetrahedron 1 - 2 - 8 - 9. The incidence of the edge becomes 4. Let's connect to tetrahedron 1 - 2 - 8 - 9 and tetrahedron 1 - 2 - 3 - 4, respectively, along faces 1 - 2 - 9 and 1 - 2 - 4 tetrahedron 1 - 2 - 4 - 9. This is equivalent to joining vertices 4 and 9 with an edge. Then the incidence of edge 1 - 2 becomes equal to 5, and the incidence of edge 1 - 4 becomes equal to 2. To increase the incidence of edges 1 - 4, connect the tetrahedron 1 - 2 - 4 - 9 to the tetrahedron 1 - 2 - 4 - 9 along the face 1 - 4 - 9 tetrahedron 1 - 4 - 9 - 10. The incidence of edge 1 - 4 becomes 3. Connect to tetrahedron 1 - 10 - 4 - 9 along the face 1 - 4 - 10 tetrahedron 1 - 4 - 10 - 11. The incidence of edge 1 - 4 becomes equal to 4. Connect the vertices with an edge 3 and 11. This is equivalent to joining tetrahedra 1 - 2 - 3 - 4 and 1 - 4 - 10 - 11, respectively, along edges 1 - 3 - 4 and 1 - 4 - 11 of tetrahedron 1 - 3 - 4 - 11. Incidence of edge 1 - 4 becomes 5, and the incidence of edge 3 - 4 becomes 3. Connect an edge to vertices 7 and 11. This is equivalent to creating a tetrahedron 3 - 4 - 7 - 11. The incidence of edge 3 - 4 becomes 4. To increase the incidence of edges 3 - 4, it would be possible to connect vertices 3 and 10 with an edge. This would lead to the creation of the tetrahedron 3 - 4 - 10 - 11. But in this case, the tetrahedron 1 - 4 - 3 - 10 was certainly formed, and this, in turn, would lead to an increase in the incidence of faces 1 - 3 - 4, 4 - 3 - 11, 1 - 4 - 10 to 3, which is impossible under the accepted conditions. To increase the incidence of edges 3 - 4, one could connect vertices 4 and 6 with an edge. This would lead to the creation of a tetrahedron 2 - 3 - 4 - 6 and to an increase in the incidence of edge 2 - 3 above 5 and to an increase in the incidence of edge 2 - 3 - 4 more than 2. And this is impossible by condition. The incidence of edges 3 - 4 could also be increased by attaching other tetrahedra to the flat faces that have this edge. But this would lead to an increase in incidence of flat faces more than 2. And this is impossible by condition. Thus, edge 3 - 4 cannot reach the required incidence 5 without violating the required conditions for incidence of other edges 5 and flat faces 2. By virtue of the arbitrariness and generality of the constructions, this proves that a polytope of tetrahedra with incidence conditions for edges and flat faces, respectively 5 and 2 does not exist. Consequently, there is no 600-cell polytope and its dual 120-cell.

The Integral Law of Conservation of Incidents for Regular Polytopes with Dimension 4

In Chapter 3, the local law of conservation of incidents in polytope was established. Now let us consider the law of conservation of incidents for elements of a polytope of all dimensions possible for a given polytope (Zhizhin, 2019 a, b, c). Since correct four-dimensional polytopes have just been investigated, it is natural to consider correct four-dimensional polytopes on this issue. Due to the fact that elements of all possible dimensions for a given polytope are considered, the law of conservation of incidents will be called integral.

Let us introduce the notation. Let $k_{d_j^i d_u}$ be the number of elements of dimension u, which include an element of dimension $j(u>j)$ with number i. Thus, $k_{d_j^i d_u}$ is the incidence coefficient of the element i with dimension j relative to elements with dimension u.

For the polytope 4-simplex (see Figure 3) we have:

The factors of incidents (from smaller dimension to larger, Figure 3) are:

$$k_{d_0^i d_1} = 4, i = 1 \div 5; k_{d_0^i d_2} = 5, i = 1 \div 5; k_{d_0^i d_3} = 4, i = 1 \div 5; k_{d_0^i d_4} = 1, i = 1 \div 5; k_{d_1^i d_2} = 3, i = 1 \div 10;$$
$$k_{d_1^i d_3} = 3, i = 1 \div 10; k_{d_1^i d_4} = 1, i = 1 \div 10; k_{d_2^i d_3} = 2, i = 1 \div 10; k_{d_2^i d_4} = 1, i = 1 \div 10; k_{d_3^i d_4} = 1, i = 1 \div 5.$$

Sum up the incidence coefficients for all vertices, edges, two - dimension faces and three - dimension faces of the 4 – simplex:

$$\sum_{i=1}^{5} k_{d_0^i d_1} = 4 \cdot 5 = 20, \sum_{i=1}^{5} k_{d_0^i d_2} = 5 \cdot 5 = 25, \sum_{i=1}^{5} k_{d_0^i d_3} = 4 \cdot 5 = 20, \sum_{i=1}^{5} k_{d_0^i d_4} = 5, \sum_{i=1}^{10} k_{d_1^i d_2} = 3 \cdot 10 = 30,$$
$$\sum_{i=1}^{10} k_{d_1^i d_3} = 3 \cdot 10 = 30, \sum_{i=1}^{10} k_{d_1^i d_4} = 10, \sum_{i=1}^{10} k_{d_2^i d_3} = 2 \cdot 10 = 20, \sum_{i=1}^{10} k_{d_2^i d_4} = 10, \sum_{i=1}^{5} k_{d_3^i d_4} = 1 \cdot 5 = 5.$$
$$\sum_{i=1}^{5} k_{d_0^i d_1} + \sum_{i=1}^{5} k_{d_0^i d_2} + \sum_{i=1}^{5} k_{d_0^i d_3} + \sum_{i=1}^{5} k_{d_0^i d_4} + \sum_{i=1}^{10} k_{d_1^i d_2} + \sum_{i=1}^{10} k_{d_1^i d_3} + \sum_{i=1}^{10} k_{d_1^i d_4} + \sum_{i=1}^{10} k_{d_2^i d_3} + \sum_{i=1}^{10} k_{d_2^i d_4} + \sum_{i=1}^{5} k_{d_3^i d_4} = 180.$$

The factors of incidents (from larger dimension to smaller, Figure 3) are

$$k_{d_0 d_1^i} = 2, i = 1 \div 10; k_{d_0 d_2^i} = 3, i = 1 \div 10; k_{d_0 d_3^i} = 4, i = 1 \div 5; k_{d_0 d_4} = 5; k_{d_1 d_2^i} = 3, i = 1 \div 10;$$
$$k_{d_1 d_3^i} = 6, i = 1 \div 5; k_{d_1 d_4} = 10; k_{d_2 d_3^i} = 4, \mathrm{i} = 1 \div 5; k_{d_2 d_4} = 10; k_{d_3 d_4} = 5.$$

Sum up the incidence coefficients for all elements of the 4 - simplex with dimension larger of zero:

$$\sum_{i=1}^{10} k_{d_0 d_1^i}{}_0 + \sum_{i=1}^{10} k_{d_0 d_2^i} + \sum_{i=1}^{10} k_{d_1 d_2^i} + \sum_{i=1}^{5} k_{d_0 d_3^i} + \sum_{i=1}^{5} k_{d_1 d_3^i} + \sum_{i=1}^{5} k_{d_2 d_3^i} + k_{d_0 d_4} + k_{d_1 d_4} + k_{d_2 d_4} + k_{d_3 d_4} = 180.$$

You can see that the sum of incidents in a 4 - simplex from elements with a lower dimension to elements with a higher dimension is equal to the sum of incidents from elements with a higher dimension to elements with a lower dimension. Thus, the sum of incidents retains its value when the direction of the relationship between the elements it is changes.
For polytope 4-cube
The factors of incidents (from smaller dimension to larger, Figure 6) are:

$$k_{d_0^i d_1} = 4, i = 1 \div 16; k_{d_0^i d_2} = 6, i = 1 \div 16; k_{d_0^i d_3} = 4, i = 1 \div 16; k_{d_0^i d_4} = 1, i = 1 \div 16; k_{d_1^i d_2} = 3, i = 1 \div 32;$$
$$k_{d_1^i d_3} = 3, i = 1 \div 32; k_{d_1^i d_4} = 1, i = 1 \div 32; k_{d_2^i d_3} = 2, i = 1 \div 24; k_{d_2^i d_4} = 1, i = 1 \div 24; k_{d_3^i d_4} = 1, i = 1 \div 8.$$

Sum up the incidence coefficients for all vertices, edges, two - dimension faces and three - dimension faces of the 4 – cube:

$$\sum_{i=1}^{16} k_{d_0^i d_1} = 4 \cdot 16 = 64, \sum_{i=1}^{16} k_{d_0^i d_2} = 6 \cdot 16 = 96, \sum_{i=1}^{16} k_{d_0^i d_3} = 4 \cdot 16 = 64, \sum_{i=1}^{16} k_{d_0^i d_4} = 16, \sum_{i=1}^{32} k_{d_1^i d_2} = 3 \cdot 32 = 96,$$

$$\sum_{i=1}^{32} k_{d_1^i d_3} = 3 \cdot 32 = 96, \sum_{i=1}^{32} k_{d_1^i d_4} = 32, \sum_{i=1}^{24} k_{d_2^i d_3} = 2 \cdot 24 = 48, \sum_{i=1}^{24} k_{d_2^i d_4} = 24, \sum_{i=1}^{8} k_{d_3^i d_4} = 1 \cdot 8 = 8.$$

$$\sum_{i=1}^{16} k_{d_0^i d_1} + \sum_{i=1}^{16} k_{d_0^i d_2} + \sum_{i=1}^{16} k_{d_0^i d_3} + \sum_{i=1}^{16} k_{d_0^i d_4} + \sum_{i=1}^{32} k_{d_1^i d_2} + \sum_{i=1}^{32} k_{d_1^i d_3} + \sum_{i=1}^{32} k_{d_1^i d_4} + \sum_{i=1}^{24} k_{d_2^i d_3} + \sum_{i=1}^{24} k_{d_2^i d_4} + \sum_{i=1}^{8} k_{d_3^i d_4} = 544.$$

The factors of incidents (from larger dimension to smaller, Figure 6) are

$$k_{d_0 d_1^i} = 2, i = 1 \div 32; k_{d_0 d_2^i} = 4, i = 1 \div 24; k_{d_0 d_3^i} = 8, i = 1 \div 8; k_{d_0 d_4} = 16; k_{d_1 d_2^i} = 4, i = 1 \div 24;$$
$$k_{d_1 d_3^i} = 12, i = 1 \div 8; k_{d_1 d_4} = 32; k_{d_2 d_3^i} = 6, \mathrm{i} = 1 \div 8; k_{d_2 d_4} = 24; k_{d_3 d_4} = 8.$$

Sum up the incidence coefficients for all elements of the 4 - cube with dimension larger of zero:

$$\sum_{i=1}^{32} k_{d_0 d_1^i} \sum_{0\ i=1}^{24} k_{d_0 d_2^i} + \sum_{i=1}^{24} k_{d_1 d_2^i} + \sum_{i=1}^{8} k_{d_0 d_3^i} + \sum_{i=1}^{8} k_{d_1 d_3^i} + \sum_{i=1}^{8} k_{d_2 d_3^i} + k_{d_0 d_4} + k_{d_1 d_4} + k_{d_2 d_4} + k_{d_3 d_4} = 544.$$

You can see that the sum of incidents in a 4 - cube from elements with a lower dimension to elements with a higher dimension is equal to the sum of incidents from elements with a higher dimension to elements with a lower dimension. Thus, the sum of incidents retains its value when the direction of the relationship between the elements it is changes.

For polytope 4 – cross - polytope

The factors of incidents (from smaller dimension to larger, Figure 15) are:

$$k_{d_0^i d_1} = 6, i = 1 \div 8; k_{d_0^i d_2} = 12, i = 1 \div 8; k_{d_0^i d_3} = 8, i = 1 \div 8; k_{d_0^i d_4} = 1, i = 1 \div 8; k_{d_1^i d_2} = 4, i = 1 \div 24;$$
$$k_{d_1^i d_3} = 4, i = 1 \div 24; k_{d_1^i d_4} = 1, i = 1 \div 24; k_{d_2^i d_3} = 2, i = 1 \div 32; k_{d_2^i d_4} = 1, i = 1 \div 32; k_{d_3^i d_4} = 1, i = 1 \div 16.$$

Sum up the incidence coefficients for all vertices, edges, two - dimension faces and three - dimension faces of the 4 – cross – polytope:

$$\sum_{i=1}^{8} k_{d_0^i d_1} + \sum_{i=1}^{8} k_{d_0^i d_2} + \sum_{i=1}^{8} k_{d_0^i d_3} + \sum_{i=1}^{8} k_{d_0^i d_4} + \sum_{i=1}^{24} k_{d_1^i d_2} + \sum_{i=1}^{24} k_{d_1^i d_3} + \sum_{i=1}^{24} k_{d_1^i d_4} + \sum_{i=1}^{32} k_{d_2^i d_3} + \sum_{i=1}^{32} k_{d_2^i d_4} + \sum_{i=1}^{16} k_{d_3^i d_4} = 544.$$

The factors of incidents (from larger dimension to smaller, Figure 15) are

$$k_{d_0 d_1^i} = 2, i = 1 \div 24; k_{d_0 d_2^i} = 3, i = 1 \div 32; k_{d_0 d_3^i} = 4, i = 1 \div 16; k_{d_0 d_4} = 8; k_{d_1 d_2^i} = 3, i = 1 \div 32;$$
$$k_{d_1 d_3^i} = 6, i = 1 \div 16; k_{d_1 d_4} = 24; k_{d_2 d_3^i} = 4, \mathrm{i} = 1 \div 8; k_{d_2 d_4} = 32; k_{d_3 d_4} = 16.$$

Sum up the incidence coefficients for all elements of the 4 - cube with dimension larger of zero

$$\sum_{i=1}^{24} k_{d_0 d_1^i}{}_0 + \sum_{i=1}^{32} k_{d_0 d_2^i} + \sum_{i=1}^{32} k_{d_1 d_2^i} + \sum_{i=1}^{16} k_{d_0 d_3^i} + \sum_{i=1}^{16} k_{d_1 d_3^i} + \sum_{i=1}^{16} k_{d_2 d_3^i} + k_{d_0 d_4} + k_{d_1 d_4} + k_{d_2 d_4} + k_{d_3 d_4} = 544.$$

You can see that the sum of incidents in a 4 – cross - polytopes from elements with a lower dimension to elements with a higher dimension is equal to the sum of incidents from elements with a higher dimension to elements with a lower dimension. Thus, the sum of incidents retains its value when the direction of the relationship between the elements it is changes.
For polytope 24-cell
The factors of incidents (from smaller dimension to larger, Figure 16) are:

$$k_{d_0^i d_1} = 8, i = 1 \div 24; k_{d_0^i d_2} = 12, i = 1 \div 24; k_{d_0^i d_3} = 6, i = 1 \div 24; k_{d_0^i d_4} = 1, i = 1 \div 24; k_{d_1^i d_2} = 3, i = 1 \div 96;$$
$$k_{d_1^i d_3} = 3, i = 1 \div 96; k_{d_1^i d_4} = 1, i = 1 \div 96; k_{d_2^i d_3} = 2, i = 1 \div 96; k_{d_2^i d_4} = 1, i = 1 \div 96; k_{d_3^i d_4} = 1, i = 1 \div 24.$$

Sum up the incidence coefficients for all vertices, edges, two - dimension faces and three - dimension faces of the 24 - cell

$$\sum_{i=1}^{24} k_{d_0^i d_1} + \sum_{i=1}^{24} k_{d_0^i d_2} + \sum_{i=1}^{24} k_{d_0^i d_3} + \sum_{i=1}^{24} k_{d_0^i d_4} + \sum_{i=1}^{96} k_{d_1^i d_2} + \sum_{i=1}^{96} k_{d_1^i d_3} + \sum_{i=1}^{96} k_{d_1^i d_4} + \sum_{i=1}^{96} k_{d_2^i d_3} + \sum_{i=1}^{96} k_{d_2^i d_4} + \sum_{i=1}^{24} k_{d_3^i d_4} = 1632.$$

The factors of incidents (from larger dimension to smaller, Figure 16) are

$$k_{d_0 d_1^i} = 2, i = 1 \div 96; k_{d_0 d_2^i} = 3, i = 1 \div 96; k_{d_0 d_3^i} = 6, i = 1 \div 24; k_{d_0 d_4} = 24; k_{d_1 d_2^i} = 3, i = 1 \div 96;$$
$$k_{d_1 d_3^i} = 12, i = 1 \div 24; k_{d_1 d_4} = 24; k_{d_2 d_3^i} = 8, \mathrm{i} = 1 \div 24; k_{d_2 d_4} = 96; k_{d_3 d_4} = 24.$$

Sum up the incidence coefficients for all elements of the 24 - cell with dimension larger of zero:

$$\sum_{i=1}^{96} k_{d_0 d_1^i}{}_0 + \sum_{i=1}^{96} k_{d_0 d_2^i} + \sum_{i=1}^{96} k_{d_1 d_2^i} + \sum_{i=1}^{24} k_{d_0 d_3^i} + \sum_{i=1}^{24} k_{d_1 d_3^i} + \sum_{i=1}^{24} k_{d_2 d_3^i} + k_{d_0 d_4} + k_{d_1 d_4} + k_{d_2 d_4} + k_{d_3 d_4} = 1632.$$

You can see that the sum of incidents in a 24 - cell from elements with a lower dimension to elements with a higher dimension is equal to the sum of incidents from elements with a higher dimension to elements with a lower dimension. Thus, the sum of incidents retains its value when the direction of the relationship between the elements it is changes.

Comment. When the dimension of polytopes is more than four, there are three types of regular polytopes: regular n-simplex, regular n-cube, regular n-cross-polytope. In Chapter 3, analytical formulas were obtained for calculating the number of faces of different dimensions in an n-simplex, n-prismahedron, n-cross-polytope in the general case without using the requirement that the polytopes are correct. Obviously, these formulas are fully applicable for regular polytopes of this type, since the derivation of the formulas was purely topological (metric parameters and dependencies were not used). The necessary ratios between the incidences of elements of different dimensions obtained in Chapter 3 are also fully applicable to regular polytopes with a dimension greater than four.

THE CONVEX SEMI – REGULAR POLYTOPES OF HIGHER DIMENSION

Chapter 2 covered in detail semi-regular polyhedrons, i.e. polytopes of dimension 3, in which different polygons participate in the formation of the surface, but all vertices are compatible with motion (Zhizhin, 2009; Zalgaller, 1966). With the transition to a space of higher dimension, the concept of a semi-regular polytope becomes more complex. If we envisage the possibility of participation of different polygons as two-dimensional faces, then we must also take into account that in this case, the polytope must also include different faces of dimension greater than two. If we restrict ourselves to four-dimensional polytopes for now, then this means that the polytopes include various three-dimensional polytopes with different flat faces. However, in the concept of a semiregular polytope, we must leave the requirement of equivalence of the vertices of the four-dimensional polytope, i. e. compatibility with their movement.

For the first time, there was announced the existence of convex semi-regular polytopes in four-dimensional space in Gosset's work (Gosset, 1900). It was a short note, containing only a statement about the existence of three four-dimensional semi-regular finite polytopes with an indication of their composition. Elte (Elte, 1912) came to the same results. Only in 1988 the validity of Gosset's statements was publicly mathematically confirmed (Makarov, 1988). In all the works mentioned, polytopes were considered in which all two-dimensional faces are the same - regular triangles, but the three-dimensional faces in the same polytope can be different. In 1900 and 1910 Stott's work (Stott, 1900; Stott, 1910) was published, in which, regardless of Gosset's work, semi-regular polytopes were considered, and the presence of a set of two-dimensional faces in the same polytope of different regular polygons was allowed. It is significant that in all these works no images of some semi-regular polytopes were shown. Images of semi-regular polytopes appeared in print only after 2013 in the works of Zhizhin (Zhizhin, 2013, 2014, 2018).

Chapter 2 introduced two operations on three-dimensional polytopes. The operation t, consisting in cutting off the polyhedral angles of a regular polyhedron with their replacement by regular polygons with the same lengths of the sides as the edges of the remaining parts of the polyhedron. The operation t^2 consisting in repeating the operation t of cutting out the polyhedral angles of the obtained semi-regular polyhedron after performing the operation t. In this case, the necessary deformation of the dihedral angles and the lengths of the sides of the polygons is used in order to obtain closed bodies with the same lengths of sides in the whole body.

The operations *t* and can use for obtaining the semi-regular polytopes of higher dimension. The semi-regular polytopes in this case will have hyper–faces of the regular and semi-regular polytopes of the different form. The edges of the semi-regular polytopes will be uniform, i. e. they have equal meaning incidence of the tree-dimension polytopes throughout polytope of the higher dimension.

For notation of the semi-regular polytopes will be used the products of notation operation by notation of the regular polytopes of the symbol Schläfli (Schläfli, 1901). For the four-dimension polytopes the symbol Schläfli has look $\{i_1,i_2,i_3\}$, where i_1 is the number of vertices incident one flat face of the polytope, i_2 is the number of the flat faces of the polytope incident one vertex of the polytope, $i_3$is the number tree-dimension faces of the polytope incident one the edge.

Obtaining semi-regular polytopes is reasonable to begin with the analysis of the simplest regular polytope of a four-dimensional regular simplex {3,3,3}. Applying the clipping operation *t*, i. e. replacing the edges by one third of the edge in its middle, we obtain a semi-regular polytope *t*{3,3,3} whose set of hyper-faces consists of 5 tetrahedrons and 5 truncated tetrahedrons (Figure 19).

Figure 19. The four-dimensional semi-regular polytope t{3,3,3}

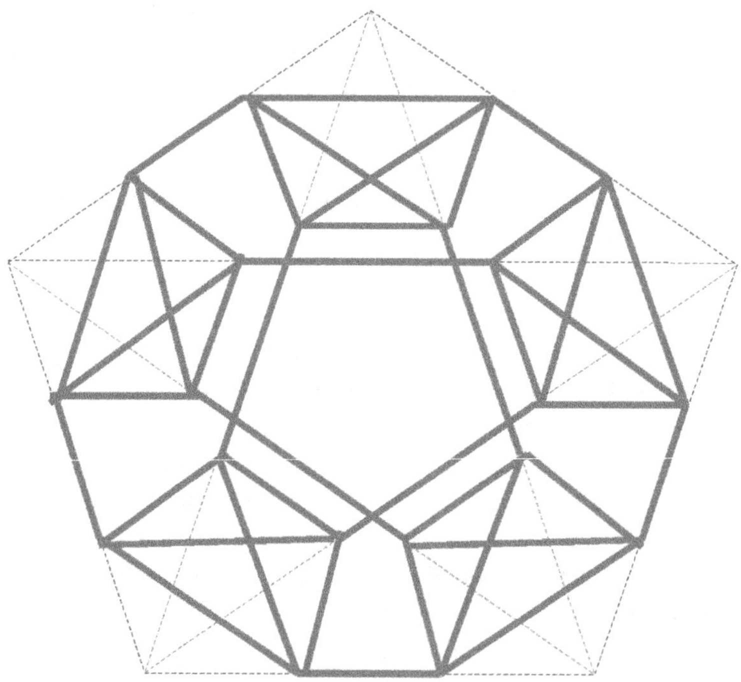

Applying the clipping operation *t* to hyper-cube {4,3,3}, we obtain a semi-regular polytope *t*{4,3,3} whose set of hyper-faces consists of 16 tetrahedrons and 8 truncated cube (Figure 20).

Applying the clipping operation t to 4-cross-polytope {3,3,4}, we obtain a semi-regular polytope t {3,3,4} whose set of hyper-faces consists of 8 octahedrons and 16 truncated tetrahedrons (Figure 21).

Figure 20. The four-dimensional semi-regular polytope t{4,3,3}

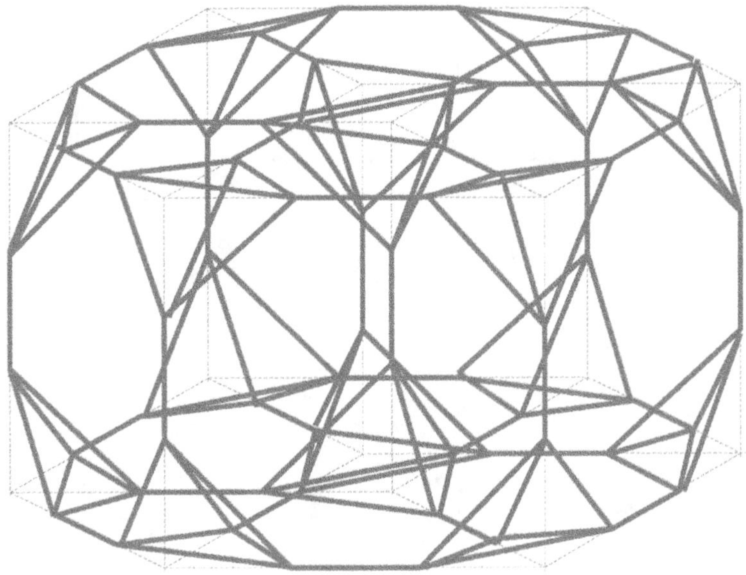

Applying the clipping operation t^2 to regular simplex {3,3,3}, we obtain a semi-regular polytope t^2{3,3,3} whose set of hyper-faces consists of 5 octahedrons and 5 tetrahedrons (Figure 22).

Figure 21. The four-dimensional semi-regular polytope t{3,3,4}

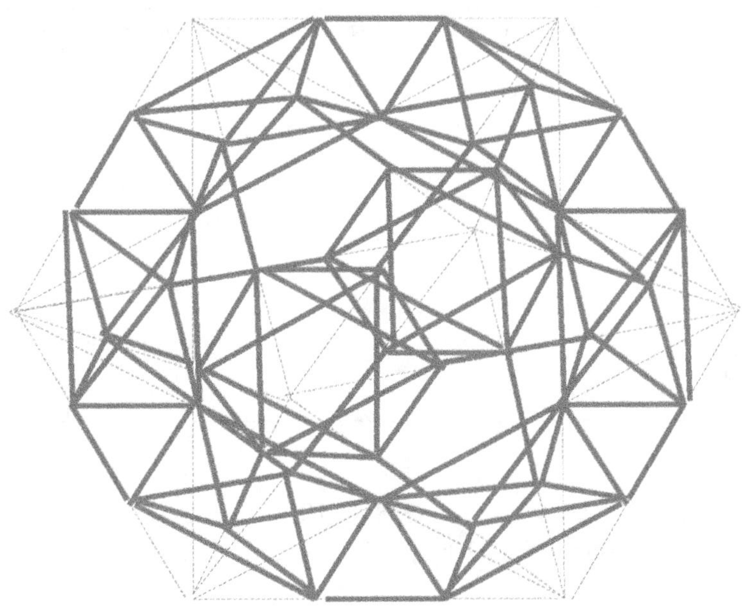

In view of the fact that in the projection in Figure 22 a part of the edges overlap, it is difficult to trace the existence of 5 octahedrons and 5 tetrahedrons in the figure. In order to see them enough to slightly to dismiss the coincidence vertices. This is done in Figure 23.

Figure 22. The four-dimensional semi-regular polytope $t^2\{3,3,3\}$

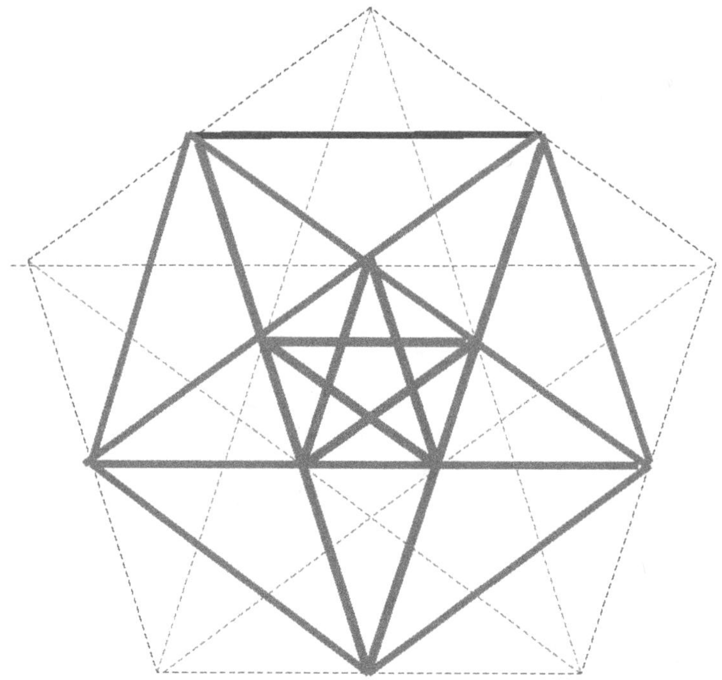

Table 2. The convex semi-regular four-dimension polytopes

N	Types and number of the hyper-faces semi-regular polytopes
1	*O*(5), *T*(5)
2	*I*(24), *T*(120)
3	*I*(120), *O*(600)
4	*tT*(5), *T*(5)
5	*tC*(8), *T*(16)
6	*tT*(16), *O*(8)
7	*tO*(24), *C*(24)
8	*tD*(120), *T*(600)
9	*tT*(600), *I*(120)
10	*CO*(8), *T*(16)
11	*tI*(120), *tT*(600)
12	*CO*(24), *C*(24)
13	*ID*(120), *T*(600)
14	*I*(10), *T*(85)

Then the octahedrons are figures 4, 6, 1, 10, 5, 8; 4, 10, 7, 9, 5, 3;10, 7, 2, 8, 1, 3; 4, 9, 2, 8, 3, 6; 5, 6, 2, 7, 1, 9. There are tetrahedrons 1, 5, 7, 10; 1, 6, 2, 8; 9, 2, 7, 3; 10, 8, 3, 4; 5, 4, 9, 6.

Figure 23. The four-dimensional semi-regular polytope $t^2\{3,3,3\}$ with dismissal vertices

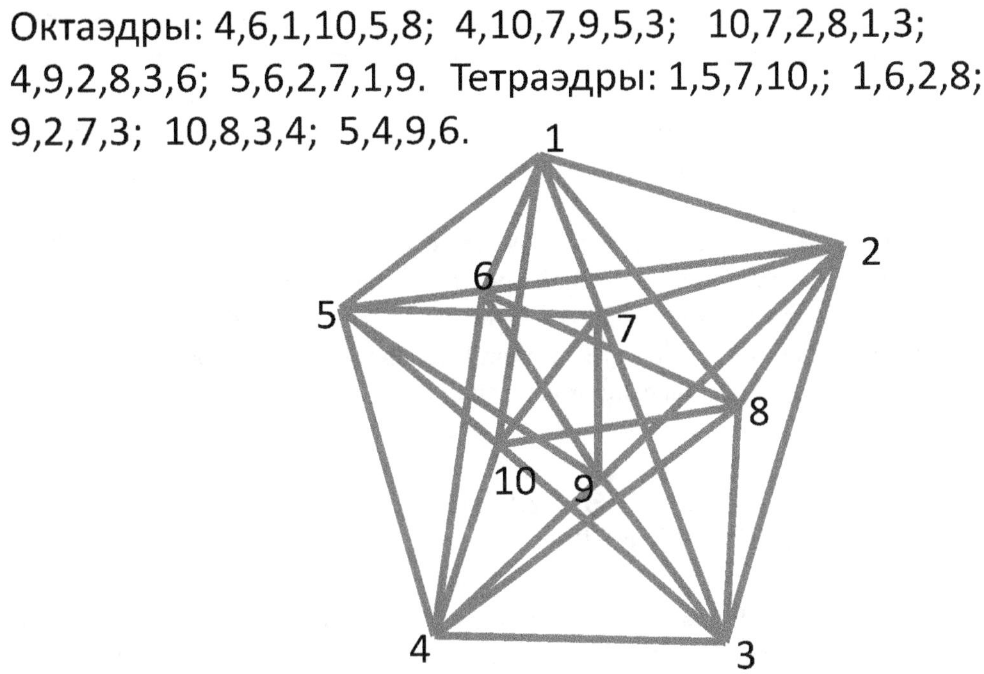

Applying the clipping operation t^2 to hyper-cube {4,3,3}, we obtain a semi-regular polytope $t^2\{4,3,3\}$ whose set of hyper-faces consists of 16 tetrahedrons and 8 cube-octahedrons (Figure 24).

Figure 24. The four-dimensional semi-regular polytope $t^2\{4,3,3\}$

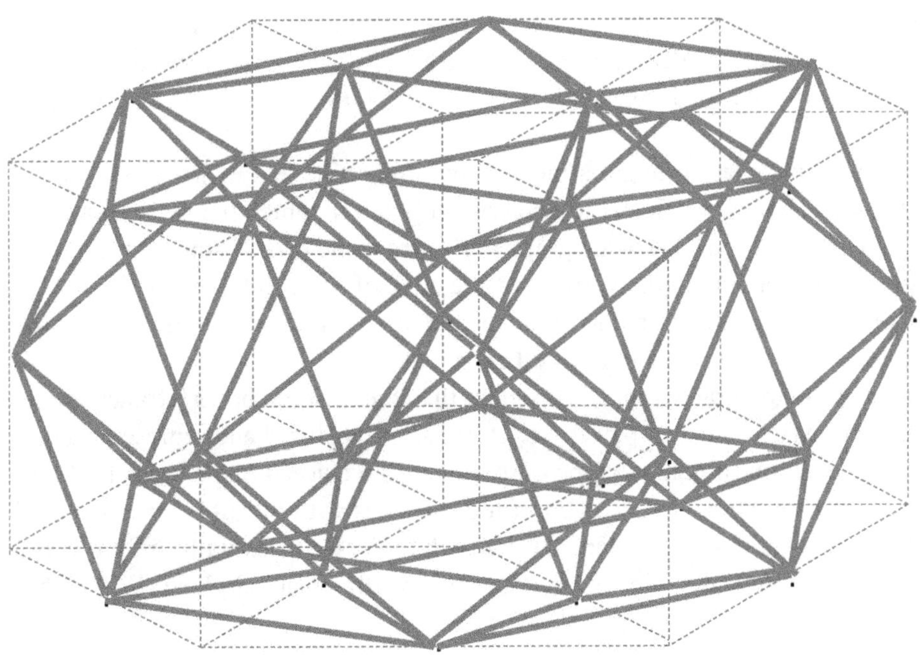

If we denote *O* - octahedron, *T* - tetrahedron, *I* - icosahedron, *D* - dodecahedron, *C* - cube, *ID* – icosadodecahedron, *CO* - cuboctahedron, *tT* – truncated tetrahedron, *tC* – truncated cube, *tO* – truncated octahedron, *tD* – truncated dodecahedron, *tI* – truncated icosahedron, then it is possible to enumerate all the obtained convex semi - regular four-dimensional polytopes. Some of these polytopes are not represented in the drawings because of their cumbersomeness. A list of all semi-regular polytopes obtained is given in Table2, indicating the composition of the set of hyper-faces (the number of definite types of the hyper - faces is indicated in parentheses).

In Table2 the first three semi-regular polytopes were established by Gosset (Gosset,1900). In addition to the operations t and t^2 for obtaining semi-regular polytopes, the known (Coxeter, 1963) transformation of an octahedron into an icosahedron was used. Obviously, by known projections on a two-dimensional plane of regular polytopes of dimension greater than four, applying the introduced operations one can obtain semi-regular polytopes of dimension greater than four.

CONVEX POLYINCIDENTAL QUASI-REGULAR POLYTOPES OF HIGHER DIMENSION

In 2014, the author discovered the existence of another class of higher-dimensional polytopes: polyincidental quasi-regular polytopes (Zhizhin, 2014). In these polytopes, as well as in semiregular polytopes, several types of regular flat and three-dimensional faces are simultaneously involved in the construction of the polytope. However, the edges of a polytope can have different incidence values within the same polytope in relation to, for example, three-dimensional faces. Moreover, at different vertices the edges with different incidence values are located in different ways. Therefore, despite the fact that the number of edges outgoing from each vertex is the same for all vertices, it is impossible to assume that these vertices are equivalent and compatible with motion. In connection with the listed properties, these polytopes are called polyincidental quasi-regular polytopes.

As a proof of the existence of regular-faceted convex polytopes of higher dimension, we consider, for example, the following 9 metrically equal regular tetrahedrons with the vertices indicated in brackets: 1(1, 2, 3, 4), 2(1, 2, 3, 5), 3(1, 2, 5, 6), 4(1, 2, 4, 6), 5(2, 3, 4, 6), 6(2, 3, 4, 5), 7(1, 4, 5, 6), 8(1, 3, 4, 5), 9(3, 4, 5, 6).

It is easy to verify that every triangular face of any of the nine tetrahedrons is common to two tetrahedrons from the list. There is sufficient to construct a graph in which the vertices are the tetrahedrons from the indicated list, and each edge of the graph will correspond to the common triangular face between two tetrahedrons corresponding to the vertices of the edge of the graph (Figure 25).

Table 3 lists the edges of the graph and the corresponding triangular faces of the nine tetrahedrons along which these tetrahedrons contact each other.

Since each face of nine tetrahedrons is closed by the face of some other tetrahedron of the indicated set of tetrahedrons, the figure made up of these tetrahedrons is closed and represents a certain polytope. Constructing the projection of this polytope onto a two-dimensional plane leads to the proof that each of the six vertices of nine tetrahedrons is connected by edges with the all other vertices. Therefore, the projection onto the two-dimensional plane of this polytope coincides with the projection of the five-dimensional regular simplex (Zhizhin, 2014) onto the two-dimensional plane (Figure 26).

Figure 25. The graph of set nine listed tetrahedrons

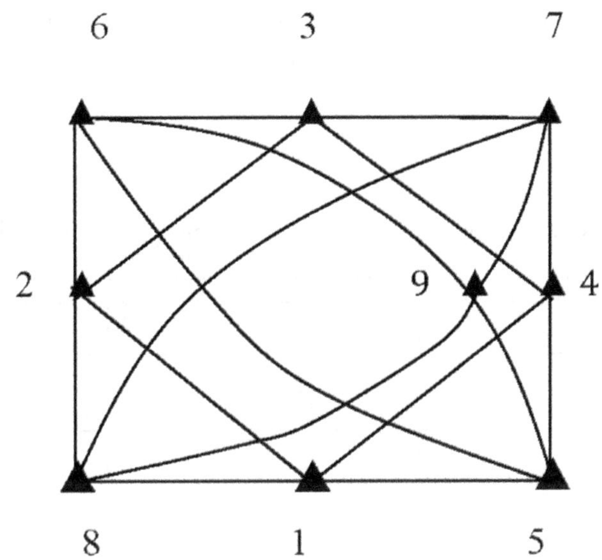

Figure 26. Polytope with hyper- faces of tetrahedrons 1, 2, 3, 4, 5, 6, 7, 8, 9

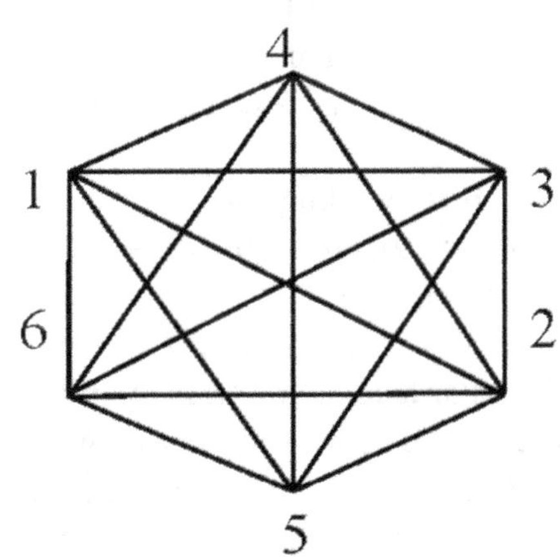

Table 3. Correspondence of the edges of the graph in Figure 25 to triangular faces of tetrahedrons

Edge of graph	Face of tetrahedron	Edge of graph	Face of tetrahedron	Edge of graph	Face of tetrahedron
1-2	123	1-5	243	1-4	124
1-8	143	2-8	135	2-3	125
2-6	235	3-6	625	3-4	126
3-7	156	4-5	246	4-7	146
6-9	356	6-5	263	5-9	364
7-8	154	7-9	564	8-9	345

It follows from Figure 26 and Table3 that the constructed polytope has 6 vertices (f_0=6), 15 edges (f_1=15), 18 flat faces (f_2=18), 9 three-dimensional faces (f_3=9). Substituting these values f_i, i=1,2,3, into the Euler-Poincaré equation (4.1), we find that the Euler-Poincaré equation holds for dimension $n = 4$

$$6 - 15 + 18 - 9 = 0.$$

Table 4. Incidence of the edges of polytopes on Figure 26 to tetrahedrons of the polytopes

Edge of polytope	Incident tetrahedrons	Edge of polytope	Incident tetrahedrons	Edge of polytope	Incident tetrahedrons
25	2, 3, 6	16	3, 4, 7	24	1, 4, 5
13	1, 2, 8	45	7, 8, 9	36	5, 6, 9
12	1, 2, 3, 4	46	4, 5, 7, 9	35	2, 6, 8, 9
23	1, 2, 5, 6	14	1, 4, 7, 8	15	2, 3, 7, 8
56	3, 6, 7, 9	26	3, 4, 5, 6	34	1, 5, 8, 9

This proves that the polytope constructed has dimension 4. Thus, removing from a 5-dimensional regular simplex of six tetrahedrons is not changed its projection onto a two-dimensional plane, but led to a decrease in the dimension of the polytope, while preserving its closure and convexity.

It follows from Figure 26 and Table3 that each of the six edges of the polytope (out of 15) is incidentally to the three tetrahedrons of the polytope, and 9 edges of the polytope are incidentally to the four tetrahedrons of the polytope (Table4).

We introduce 25 tetrahedrons with corresponding vertices: 1(1, 2, 3, 4), 2(1, 2, 3, 5), 3(1, 2, 5, 6), 4(1, 2, 6, 7), 5(1, 2, 4, 7), 6(2, 4, 7, 8), 7(4, 7, 8, 9), 8(4, 7, 9, 10), 9(4, 7, 1, 9), 10(2, 3, 4, 8), 11(3, 4, 8, 10), 12(3, 4, 9, 10), 13(3, 4, 1, 9), 14(2, 3, 5, 8), 15(3, 5, 8, 10), 16(3, 5, 9, 10), 17(3, 5, 1, 9), 18(2, 5, 6, 8), 19(5, 6, 8, 10), 20(5, 6, 9, 10), 21(1, 5, 6, 9), 22(2, 6, 7, 8), 23(6, 7, 8, 10), 24(6, 7, 9, 10), 25(1, 6, 7, 9).

It is easy to verify that every triangular face of any from the twenty-five tetrahedrons is common to two tetrahedrons from the list. There is sufficient to construct a graph in which the vertices are the tetrahedrons from the indicated list, and each edge of the graph will correspond to the common triangular face between two tetrahedrons corresponding to the vertices of the edge of the graph (Figure 27).

Table5 lists the edges of the graph and the corresponding triangular faces of the twenty-five tetrahedrons along which these tetrahedrons contact each other.

Since each face of twenty-five tetrahedrons is closed by the face of some other tetrahedron of the indicated set of tetrahedrons, the figure made up of these tetrahedrons is closed and represents a certain polytope.

It follows from Figure 27 and Table5 that the constructed polytope has 10 vertices (f_0=10), 35 edges (f_1=35), 50 flat faces (f_2=50), 25 three-dimensional faces (f_3=25). Substituting these values f_i, i=1,2,3, into the Euler-Poincaré equation (4.1), we find that the Euler-Poincaré equation holds for dimension $n = 4$

$$10 - 35 + 50 - 25 = 0.$$

This proves that the polytope constructed has dimension 4.

Figure 27. The graph of set twenty-five listed tetrahedrons

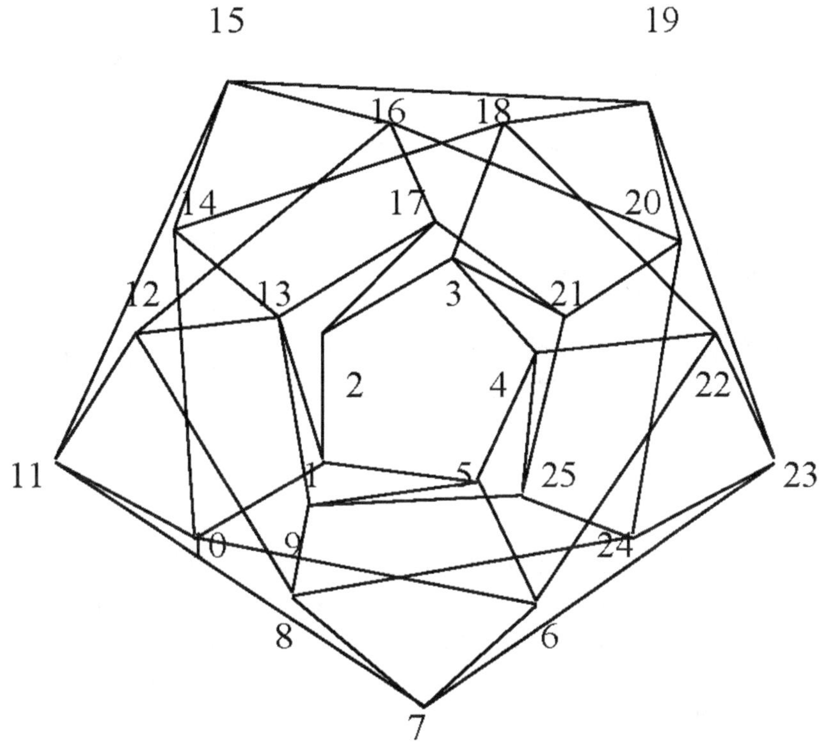

Table 5. Correspondence of the edges of the graph in Figure 14 to triangular faces of tetrahedrons

Edge of graph	Face of tetrahedron	Edge of graph	Face of tetrahedron	Edge of graph	Face of tetrahedron
1-2	1, 2, 3	1-5	1, 2, 4	1-10	3, 2, 4
1-13	1, 4, 3	2-3	1, 2, 5	2-14	3, 2, 5
2-17	1, 3, 5	3-21	6, 1, 5	3-4	1, 2, 6
3-18	2, 5, 6	4-5	2, 1, 7	4-22	2, 7, 6
4-25	1, 7, 6	6-5	2, 4, 7	5-9	1, 7, 4
6-7	7, 8, 4	6-10	2, 8, 4	6-22	2, 7, 8
7-8	4, 7, 10	7-11	4, 8, 10	7-23	7, 8, 10
8-9	4, 9, 7	8-12	4, 9, 10	8-24	7, 9, 10
9-13	1, 4, 9	9-25	1, 7, 9	10-14	2, 3, 8
10-11	3, 4, 8	11-12	3, 4, 10	11-15	3, 8, 10
12-13	3, 4, 9	12-16	3, 9, 10	13-17	1, 3, 9
14-15	3, 5, 8	14-18	2, 5, 8	15-16	3, 5, 10
15-19	5, 8, 10	16-17	3, 5, 9	16-20	5, 9, 10
17-21	1, 5, 9	18-19	5, 6, 8	18-22	2, 6, 8
19-20	5, 6, 10	19-23	6, 8, 10	20-21	5, 6, 9
20-24	6, 9, 10	21-25	1, 6, 9	22-23	6, 7, 8
23-24	6, 7, 10	24-25	6, 7, 9		

It follows from Figure 27 and Table5 that each of the ten edges of the polytope (out of 35) is incidentally to the five tetrahedrons of the polytope, and 25 edges of the polytope are incidentally to the four tetrahedrons of the polytope (Table 6).

Table 6. Incidence of the edges of polytopes with the graph on Figure 14 to tetrahedrons the polytopes

Edge of polytope	Incident tetrahedrons	Edge of polytope	Incident tetrahedrons	Edge of polytope	Incident tetrahedrons
2-3	1, 2, 6, 5, 7	1-4	1, 11, 16, 4, 21	1-5	2, 3, 12, 17, 22
2-6	3, 4, 8, 9, 10	5-7	5, 8, 13, 18, 22	7-8	14, 6,19, 9, 24
4-8	7, 10, 15,20, 25	6-9	11, 12, 13, 14,1 5	9-10	16, 17, 18, 19,20
10-3	21, 22, 23, 24,25	1-2	2, 3, 4, 1	1-3	1, 2, 21, 22
4-2	1, 7, 4, 10	4-3	1, 7, 21, 25	2-5	5, 2, 8, 3
3-5	2, 5, 22, 23	1-6	3, 4, 11, 12	5-6	3, 8, 12, 13
4-6	4, 10, 11, 15	2-7	5, 6, 8, 9	3-7	5, 6, 23, 24
2-8	6, 7, 9, 10	3-8	6, 7, 24, 25	6-7	8, 9, 13, 14
6-8	9, 10, 14, 15	1-9	11, 12, 16, 17	4-9	11, 15, 16, 20
5-9	12, 13, 17, 18	7-9	13, 14, 18, 19	8-9	14, 15, 19, 20
1-10	16, 17, 21, 22	4-10	16, 20, 21, 25	5-10	17, 18, 22, 23
7-10	18, 19, 23, 24	8-10	19, 20, 24, 25		

CONCLUSION

For the first time, the proof of the existence of a number of polytopes of higher dimension is carried out by their direct construction. First of all, we are talking about regular four-dimensional polytopes, the existence of which was assumed based on the satisfaction of their symbols by Schläfli. The construction of polytopes was carried out by attaching to each other along the whole flat faces of three-dimensional faces included in the polytope. This is how polytopes 4 - simplex, 4 - cube, 4 - cross-polytopes are constructed. When constructing, the necessary conditions for the existence of polytopes were checked: the Euler-Poincaré equation, the incidence relationship between elements of different dimensions. In the same way, a 24-cell was built, consisting of 24 octahedrons. It is shown that the image of this polytope given in the works of Coxeter does not correspond to the actual image. This actual image lists all the elements of various dimensions included in the polytope and indicates the specific order of their contacts with each other. The possibility of the existence of 600-cell and 120-cell polytopes widespread in the literature, i. e. four-dimensional polytopes, consisting, respectively, of 600 tetrahedrons and 120 dodecahedrons. Their existence was justified by the satisfaction of their symbols of the Shläfli, while there were no specific images of these polytopes with a list of the figures included in them in the literature. The depictions of these polytopes in Coxeter's works were purely artistic. An attempt to directly construct a 600-cell unit with successive laying to each other along the whole faces of the tetrahedra showed that the process of applying tetrahedra leads to a contradiction with the necessary conditions for incidence of edges and flat faces. Therefore, the 600-cell polytope cannot exist, as well as the 120-cell polytope, which in theory is dual to the 600-cell.

A mechanism is considered for obtaining semiregular polytopes of higher dimension using the operations of cutting off the neighborhood of vertices of regular polytopes. the images of four-dimensional semiregular polytopes obtained by this method are presented. The existence of a new class of polytopes, which are called polyincidental quasi-regular polytopes, is revealed. In four-dimensional polytopes of this class, regular two- and three-dimensional figures appear with flat and three-dimensional faces. However, their edges have different meanings of incidence in relation to 3D shapes. This violates the condition of equivalence of vertices in the volume of a polytope, which is required for semiregular polytopes.

The considered examples of deviations from the correctness conditions for polytopes of higher dimension show the possibility of a wide variety of deviations from the correctness conditions for large polytope dimensions.

REFERENCES

Coxeter H.S.M. (1963). *Regular Polytopes.* John Wiley & Sons, Inc.

Coxeter, Y. S. M. (1974). *Regular Complex Polytopes.* Cambridge University Press.

Elte, E. L. (1912). *The semi-regular polytopes of the hyperspace.* University of Groningen.

Gosset, T. (1900). On the regular and semi-regular figures in space of *n* dimensions. *Messenger of Mathematics, 29,* 43–48.

Makarov, M. V. (1988). On the derivation of four-dimensional semiregular polyhedrons. *Problems of Discrete Geometry, 103,* 139–150.

Poincaré A. (1895). Analysis situs. *J. de é Ecole Polytechnique, 1.*

Schläfli, L. (1901). *Theorie der vielfachen Kontinuität Zürcher und Furrer.* doi:10.1007/978-3-0348-5118-3

Stott A. B. (1900). On sertain series of sections of the regular four-dimensional hyper-solids. *Verhandelingen der Koninklijke Akademie van Wetenschappen te Amsterdam,* 3-7.

Stott A. B. (1910). Geometrical deduction of semiregular from regular polytopes and space fillings. *Verhandelingen der Koninklijke Akademie van Wetenschappen te Amsterdam,* 1 – 11.

Stringham, W. I. (1880). Regular figures in *n*-dimensional space. *American Journal of Mathematics, 3*(1), 1–14. doi:10.2307/2369441

Zalgaller, V. A. (1966). *Convex polyhedrons with correct sides.* Science.

Zhizhin, G. V. (2009). *Convex right-faceted polyhedrons.* Polytechnika-service.

Zhizhin, G. V. (2013). Relations for the numbers of faces of various dimensions in the tower of n - dimensional convex regular polytopes. In *Proceedings of the 9th All - Russian School "Mathematical Studies in the Natural Sciences".* Apatity: Geological Institute, Russian Academy of Sciences.

Zhizhin, G. V. (2014). *World - 4D.* Polytechnic Service.

Zhizhin, G. V. (2018). *Chemical Compound Structures and the Higher Dimensional of Molecules: Emerging Research and Opportunities.* IGI Global. doi:10.4018/978-1-5225-4108-0

Zhizhin, G. V. (2019 a). *The Geometry of Higher - Dimensional Polytopes.* IGI Global. doi:10.4018/978-1-5225-6968-8

Zhizhin, G. V. (2019 b). *Attractors and Higher Dimensions in Population and Molecular Biology: Emerging Research and Opportunities.* IGI Global. doi:10.4018/978-1-5225-9651-6

Zhizhin, G. V. (2019 c). The Law of Conservation of Incidents in the Space of Nanoworld. *International Journal of Chemoinformatics and Chemical Engineering*, *8*(1), 25–46. doi:10.4018/IJCCE.2019010103

KEY TERMS AND DEFINITIONS

Dimension of the Space: The member of independent parameters needed to describe the change in position of an object in space.

Incidence Coefficients of Elements of Lower Dimension With Respect to Elements of Higher Dimension: The number of elements of a certain higher dimension to which the given element of a lower dimension belongs.

Incidence Coefficients of Elements of Higher Dimension With Respect to Elements of Lower Dimension: The number of elements of a given lower dimension that are included in a particular element of a higher dimension.

N–Cross-Polytope: The convex polytope of dimension n in which opposite related of centrum edges not have connection of edge.

N–Cube: The convex polytope of dimension n in which each vertex incident to n edges.

N–Simplex: The convex polytope of dimension n in which each vertex is joined by edges with all remain vertices of polytope.

Nanocluster: A nanometric set of connected atoms, stable either in isolation state or in building unit of condensed matter.

Polytope: Polyhedron in the space of higher dimension.

Chapter 5
Irregular Polytopes of Higher Dimension

ABSTRACT

Methods for the formation of irregular polytopes of the highest dimension and their properties are investigated. It was found that a distinctive feature of irregular polytopes of higher dimension is the existence of a continuous outer boundary of the polytope and a jump in dimension on this boundary. This feature makes it possible to consider irregular polytopes of higher dimension as carriers of higher dimension in low-dimensional space, for example, in three-dimensional space. The possibility of creating objects with a steadily increasing dimension has been opened when deepening into the object. Various options for the implementation of the creation of incorrect polytopes of the highest dimension with the calculation of the dimension in specific cases are considered in detail.

INTRODUCTION

Many people know about the existence of five regular three-dimensional polyhedrons (Plato's bodies) (Coxeter, 1963; Elte, 1912; Gosset, 1900). Some of these people know (Grünboum, 1967; Stringham, 1880; Schläfli, 1901) that in a high-dimensional space (more than four) there are only three regular polytopes (n-simplex, n-cube, n-cross-polytope). What other figures can be in high-dimensional space for many people remains unknown. The author's studies in chemistry, physics and biology have shown (Zhizhin, 2013; 2018; 2019 a, b, c; 2020; 2021) that the structural units of most material bodies are objects of high dimension. As a rule, they differ in shape from regular polytopes n-simplex, n - cube, n - cross-polytope. As follows from the previous chapter, the differences can be insignificant, retaining the basic properties of regular bodies, and then they are called semi-regular (Zalgaller, 1966; Zhizhin, 2009). However, the differences can be significant, violating the basic requirements for correctness. First of all, such differences can be attributed to metric violations of correctness: different lengths of edges or angles in one object. In this case, the topological structure can be close to the structure of regular bodies. Such differences will not change the dimensions of the object, since the Euler-Poincaré equation, which determines the dimension of the object, does not include metric characteristics. Therefore, irregular

DOI: 10.4018/978-1-7998-8374-6.ch005

polytopes of the highest dimension, in which differences from regular bodies are associated only with metric violations of correctness, will not interest us in this chapter special. Of interest are significant differences from the correct figures associated with structural changes in objects. In this engraving, an attempt is made to formulate variants of such structural changes, leading to the formation of irregular polytopes of the highest dimension. Explore these structures and determine their dimensions. Moreover, the options for such structural changes will be selected on the basis of previous studies of the structures of real chemical compounds in physical and biological objects. In subsequent chapters, these options will be illustrated with specific examples from the fields of chemistry, physics, and biology. Abstract options for changing structures (Diudea, 2018; McMullen & Schulte, 2002) will not be considered.

ADDING A CENTER TO SIMPLEX POLYTOPES

In chemical compounds, tetrahedral coordination is widespread. It's easy to see how the higher dimensional space appears when you add a vertex to the center of the tetrahedron.

Suppose we have an arbitrary tetrahedron (correct or incorrect, it does not matter). Place a vertex in the center of the tetrahedron and connect it with edges to other vertices of the tetrahedron (Figure 1).

Figure 1. Polytope a tetrahedron with center

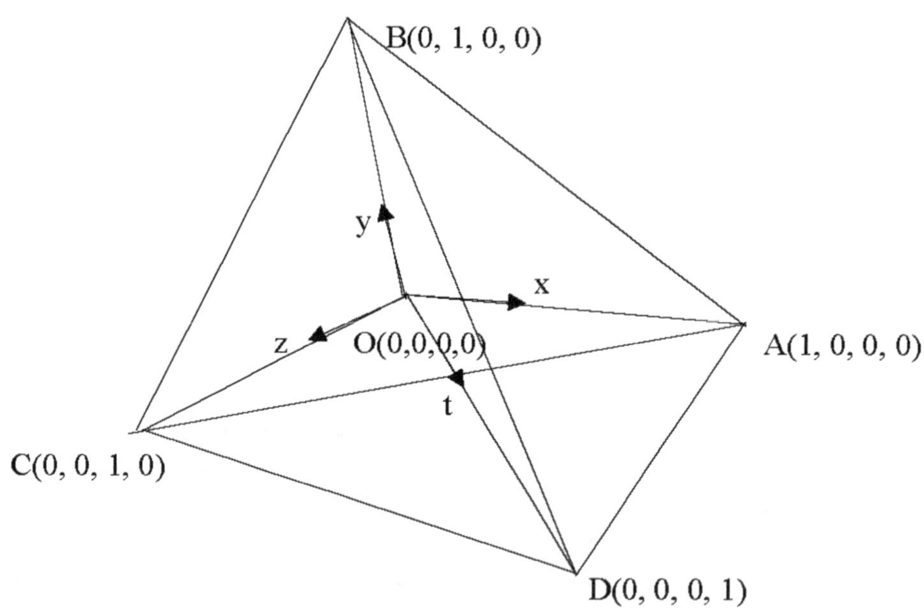

To the center of the tetrahedron, i. e. we place the origin of coordinates (x, y, z, t) at the vertex and direct the coordinate axes to the other four vertices of the tetrahedron. We take the distance from the center of the tetrahedron to the vertices as unity. Then the coordinates of the vertices have integer values O (0, 0, 0, 0), A (1, 0, 0, 0), B (0, 1, 0, 0), C (0, 0, 1, 0), D (0, 0, 0, 1). Let's define the dimension of this figure according to the Euler-Poincaré equation (Poincaré, 1895)

$$\sum_{i=0}^{n-1}(-1)^i f_i(n) = 1 + (-1)^n, \tag{5.1}$$

in (5.1) $f_i(n)$ is the number of elements of a polytope P of dimension n having dimension i.

In this case a polytope P has five vertices *O, A, B, C, D*. Then we have f_0=5. The polytope P has 10 edges *OA, OB, OC, OD, AB, AD, AC, BD, BC, CD*. Then we have f_1=5. The polytope P has 10 two-dimensional faces *OAB, OBC, OCD, OBD, OCA, OAD, ABC, CBD, CAD, CBA*. Then we have f_2=10. The polytope P has 5 three-dimensional faces (three tetrahedrons) *OBAD, OCBD, OCBA, OCAD, ABCD*. Then we have f_3=5. Substituting values f_i, i=0÷3 into equation (5.1), we have

$$5 - 10 + 10 - 5 = 0.$$

Thus, we see that the Euler-Poincaré equation (5.1) holds in this case for $n = 4$. This proves that Figure 1 has dimension 4. This figure is a simplex of dimension 4 or 4-simplex. The polytope on Figure 1 topological equal to polytope on Figure 4.3. The number of vertices incident to one 2D face is 3; the number of flat faces incident to one vertex is 3; the number of three-dimensional faces incident to one edge is 3; the number of edges incident to one vertex is 4.

It should be noted that the polytope in Figure 1 is an incorrect polytope, since by construction it is clear that the edges outgoing from the center of the tetrahedron and the edges of the original ABCD tetrahedron have different lengths, even if the original tetrahedron was correct. This leads to the fact that all other elements are metrically different. In this case, an essential point arises for what follows. When depicting polytope P (4-simplex) in the form of a tetrahedron with a center, it turns out that it has an external three-dimensional boundary in the form of a surface composed of two-dimensional faces. This does not contradict the geometry of Riemann (Riemann, 1854) according to which the space of the highest dimension is not infinite, it has a boundary. In this case, the boundary is a three-dimensional surface. Therefore, objects of the highest dimension can be located inside three-dimensional space, in this case a four-dimensional simplex.

Let's turn again to the image of the 4-simplex polytope shown in Figure 4.3. Let's add the center now to this polytope and connect the center with all other vertices with edges (Figure 2).

Figure 2. The 4 –simplex with centrum.

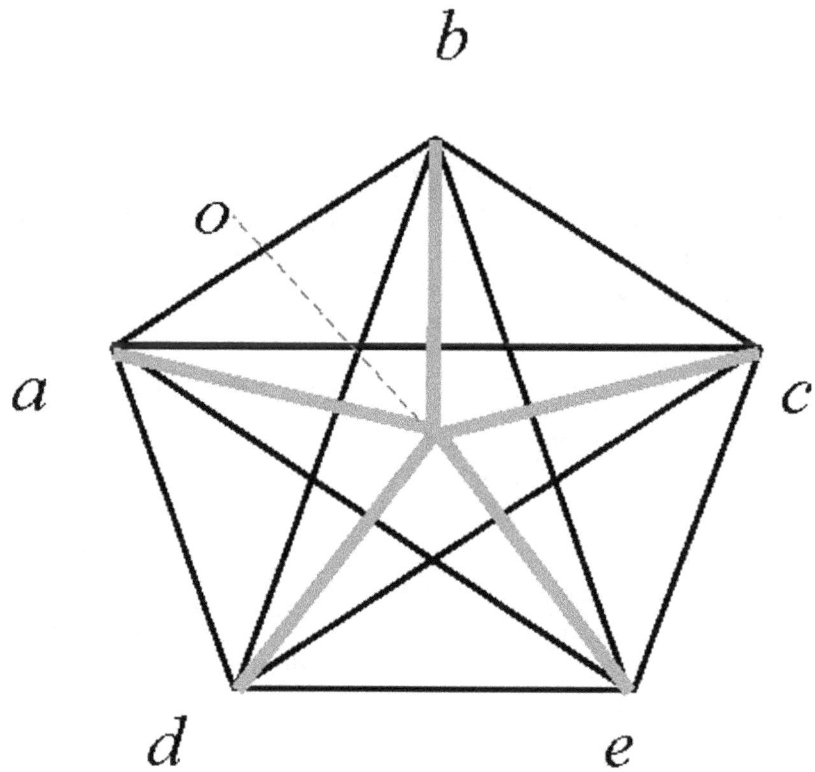

In Figure 2 can indicate 6 vertices *o, a, b, c, d, e* ($f_0 = 6$); 15 edges

ab, bc, ce, de, ad, ac, ae, be, bd, cd, ob, oc, oa, od, oe

($f_1 = 15$); 20 trigonal faces

(*abc, aeb, abo, abd, bcd, bco, bce, aeo, aed, aec, edo, edc, edb, dco, dca*),

$f_2 = 20$; 15 tetrahedrons

(*abed, abec, abcd, dbce, aecd, obcd, oecd, aoed, aoeb, aobc, boed, coae, doeb, eobc, aocd*),

$f_3 = 15$; 6 4-simplexes

(*abcde, abedo, abeco, abcdo,dbceo, aecdo*), $f_4 = 6$.

In this polytope, each vertex is connected by an edge to all other vertices.

Substituting values f_i, i=0÷4 into equation (5.1), we find that the Euler-Poincaré equation is satisfied for $n = 5$

6 – 15 + 20 – 15 + 6 = 2.

This proves that a 4-simplex with center has dimension 5 and is polytope irregular 5 - simplex.

It can be noted that in this case, too, the original 4-simplex *abcde* includes a 5-simplex for which the 4-simplex polytope is the outer four-dimensional boundary.

Obviously, this process can be continued. Then, in the general case, adding a vertex to the center of an n-simplex polytope will lead to the formation of an irregular polytope (n + 1) - simplex for which the original n-simplex polytope is an external n-dimensional boundary.

ADDING A CENTER TO A CROSS-POLYTOPES

A spatial figure of minimum dimension, representing a cross-polytope, is a regular octahedron - one of the five Plato bodies. Add a vertex to the center of the octahedron and connect it with edges to the vertices of the octahedron (Figure 3).

Figure 3. Octahedron with a center

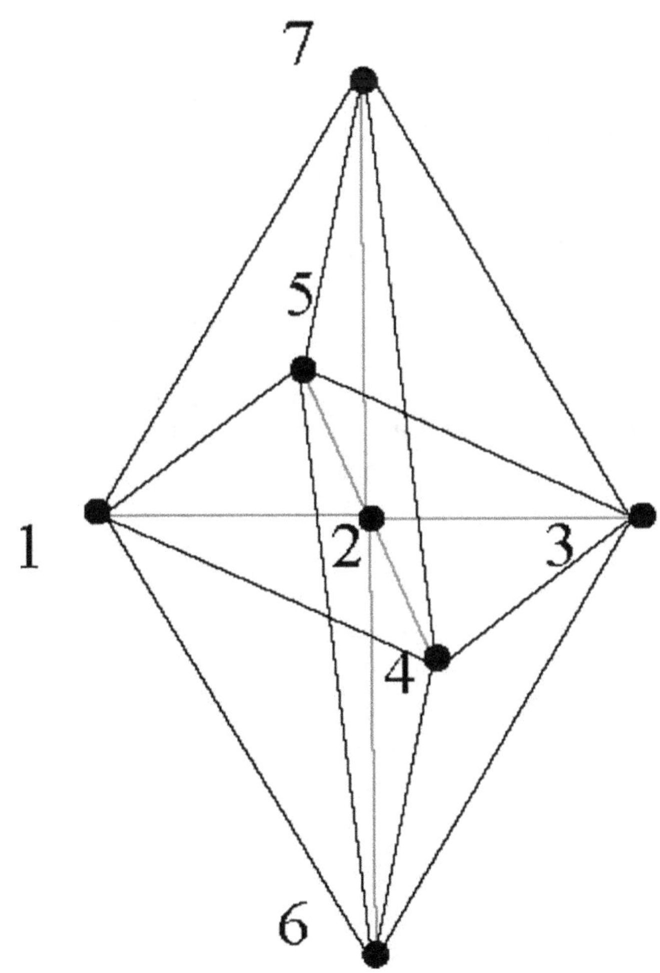

Polytope on Figure 3 has 7 vertices (f_0=7), 18 edges (f_1=18)

1 - 2, 2 - 3, 2 - 5, 2 - 4, 1 - 5, 5 - 3, 3 - 4, 1 - 4, 2 - 6, 2 - 7, 1 - 7, 5 - 7, 3 - 7, 7 - 4, 1 - 6, 5 - 6, 4 - 6, 3 - 6.

Polytope composed of eight tetrahedrons

1 - 2 - 4 - 6, 2 - 4 - 3 - 6, 2 - 5 - 3 - 6, 1 - 2 - 5 - 6, 1 - 2 - 5 - 7, 2 - 5 - 3 - 7, 2 - 4 - 3 - 7, 1 - 2 - 4 - 7.

To these three - dimensional figures you need to add an octahedron without a center

1 - 5 - 3 - 7 - 9 - 11.

Thus, the total number of three - dimensional figures is equal to 9 (f_3=9). Each flat face of any of these nine polyhedrons is simultaneously one of the flat faces of some other of these nine polyhedrons. Therefore, the total number of two - dimensional faces in the polytope in Figure 3 is 16 (f_2=16):

1 - 4 - 6, 4 - 3 - 6, 5 - 3 - 6, 1 - 5 - 6, 5 - 3 - 7, 1 - 5 - 7, 1 - 7 - 4, 7 - 3 - 4, 2 - 4 - 6, 2 - 3 - 6, 1 - 2 - 6, 5 - 2 - 6, 1 - 2 - 7, 2 - 5 - 7, 2 - 3 - 7, 2 - 6 - 4, 1 - 2 - 5, 2 - 5 - 3, 2 - 4 - 3, 1 - 2 - 4.

Define dimension polytope in Figure 3 on the Euler - Poincaré equation (Poincaré, 1895) (5.1).

Substituting the obtained values f_i(i=0,1,2,3) in the equation (5.1) you can see that the Euler – Poincaré equation is satisfied in this case with a value of n equal to 4. This proves that the dimension of an octahedron with center is 4.

By construction, it is clear that an octahedron with a center is an irregular polytope. In particular, its center is incident to six edges, and the remaining vertices are incident to five edges. Moreover, the outer boundary of this four-dimensional polytope is the three-dimensional surface of a regular octahedron.

Add a center in 5-cross-polytope.

We will use the image of the 5-cross-polytope obtained in the monograph (Zhizhin, 2014), and add to this image the center and edges going from the center (green). Then we get the following image of a 5-cross-polytope with a center (Zhizhin, 2019) (Figure 4)

Figure 4. The 5 –cross-polytope with centrum.

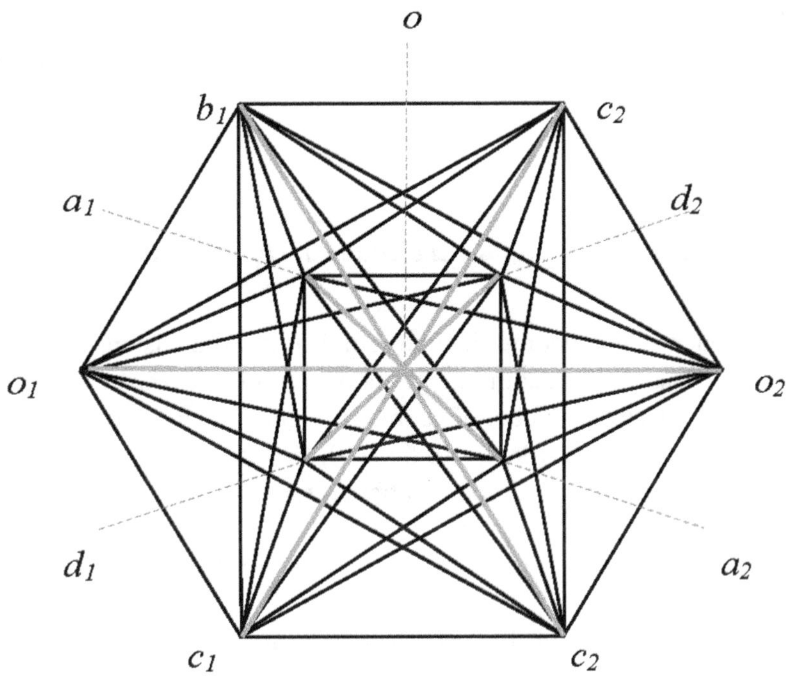

As follows from Zhizhin (2014) 5 cross-polytope has 10 vertices (f_0=10), 40 edges (f_1= 40), 80 triangular faces (f_2= 80), 80 tetrahedrons (f_3= 80), 32 4-cross-politopes (f_4= 32). The introduction of the center into the 5-cross-polytop adds, according to Figure 4 10 edges

$$(oa_1, ob_1, oc_1, od_1, oo_1, ob_2, oa_2, oc_2, od_2, oo_2),$$

24 triangular faces

$$\left\{ \begin{matrix} o_1b_1o, b_1a_1o, b_1oa_2, b_1od_2, b_1oc_2, b_1oc_1, c_2d_2o, c_2od_1, c_2oa_2, c_2oo_2, o_2d_2o, o_2oa_2, o_2ob_2, b_2a_2o, b_2od_2, \\ b_2oc_2, b_2oc_1, b_2d_1o, c_1od_1, c_1od_2, c_1oa_2, c_1oo_1, a_1o_1o, o_1d_1o \end{matrix} \right\},$$

28 tetrahedrons

$$\left(\begin{matrix} b_1od_2a_2, b_1c_1oo_1, b_1d_1oa_1, b_1a_1d_2o, b_1od_1a_2, b_1oa_2c_2, b_1od_1c_1, c_2oa_1d_1, c_2b_2oo_2, \\ c_2a_2od_2, c_2a_1d_2o, c_2oa_2d_1, c_2od_1b_1, c_2oa_2b_2, c_1oa_2d_2, c_1a_1od_1, c_1d_1a_2o, c_1oa_1d_2, \\ c_1od_2b_2, c_1od_1b_1, o_1a_1d_1o, b_2od_1a_1, b_2d_2oa_2, b_2a_2d_1o, b_2od_2a_1, b_2oa_1c_1, b_2od_2c_2, o_2d_2a_2o \end{matrix} \right),$$

18 four-dimensional simplexes

$$\begin{pmatrix} b_1a_1od_2c_2, b_1c_1d_1a_1o, b_1o_1a_1oc_2, b_1od_2o_2c_2, b_1od_2a_2o_2, c_1d_1oa_2b_2, c_1o_1d_1b_2o, c_1oa_2o_2b_2, c_1oo_2d_2o_2, \\ c_2b_2a_2d_2o, c_2o_2d_2ob_1, c_2oa_1o_1b_1, c_2oa_1d_1o_1, b_2a_2od_1c_1, b_2od_1o_1c_1, b_2od_1a_1o_1, o_2c_2od_2a_2b_2 \end{pmatrix},$$

6 five- dimensional simplexes

$$(o_1b_1a_1od_1c_1, c_2od_2a_2o_2b_2, c_1od_2a_2o_2b_2, b_2oa_1o_1d_1c_1, c_2od_1a_1o_1b_1, b_1oa_2o_2d_2c_2).$$

Adding the obtained quantities of geometric figures of different dimensions connected with the center of the 5-cross polytope to the corresponding numbers of figures not connected with the center of the 5-cross polytope, we obtain the total number of geometric figures of different dimensions in the 5-cross polytope with center: f_0=11, f_1=50, f_2=104, f_3=108, f_4=50, f_5=7. Substituting these values into equation (5.1), we find that the Euler-Poincare equation is satisfied for $n = 6$

11 – 50 + 104 – 108 + 50 -7 =0.

This proves that a 5-cross-polytope with center has dimension 6.

By construction, it is clear, that a 5 – cross – polytope with a center is an irregular polytope. In particular, a center of polytope is incident to ten edges, and the remaining vertices are incident to nine edges. Moreover, the outer boundary of this six-dimensional polytope is the five-dimensional surface of a 5 – cross - polytope.

It is obvious that an increase in the dimension of the cross-polytope will not change the noted pattern: adding a center to the n-cross-polytope increases the dimension of the polytope by one, the polytope becomes irregular and has an external border in the form of an n-cross-polytope.

Comment. Consideration of the process of adding a center to simplex and cross-polytopes of arbitrary dimension leads to interesting conclusions. The resulting polytope in all cases has a dimension greater than the initial polytope by one. The resulting polytope in all cases becomes incorrect, and, unlike regular polytopes, it has an external boundary with a dimension less than the dimension of the resulting polytope by one. It should be assumed that the dimension of the boundary of the polytope coincides with the dimension of the surrounding space in which the polytope is located, which has a large dimension. This means that it is possible to locate, inside a polytope of a given dimension, another polytope that is not associated with the surrounding polytope by chemical bond and has a dimension greater by one. For example, in three-dimensional space, there is a four-dimensional object with a three-dimensional border. Inside a four-dimensional object, a five-dimensional object with a four-dimensional border can be located, etc. The dimension of objects moving inside the structure can constantly increase.

Insert Shapes into Each to Other

Insert a Tetrahedron into a Chemically Bonded Tetrahedron

The result of inserting a tetrahedron into a tetrahedron is shown in Figure 5.

Figure 5. The polytope tetrahedron in tetrahedron.

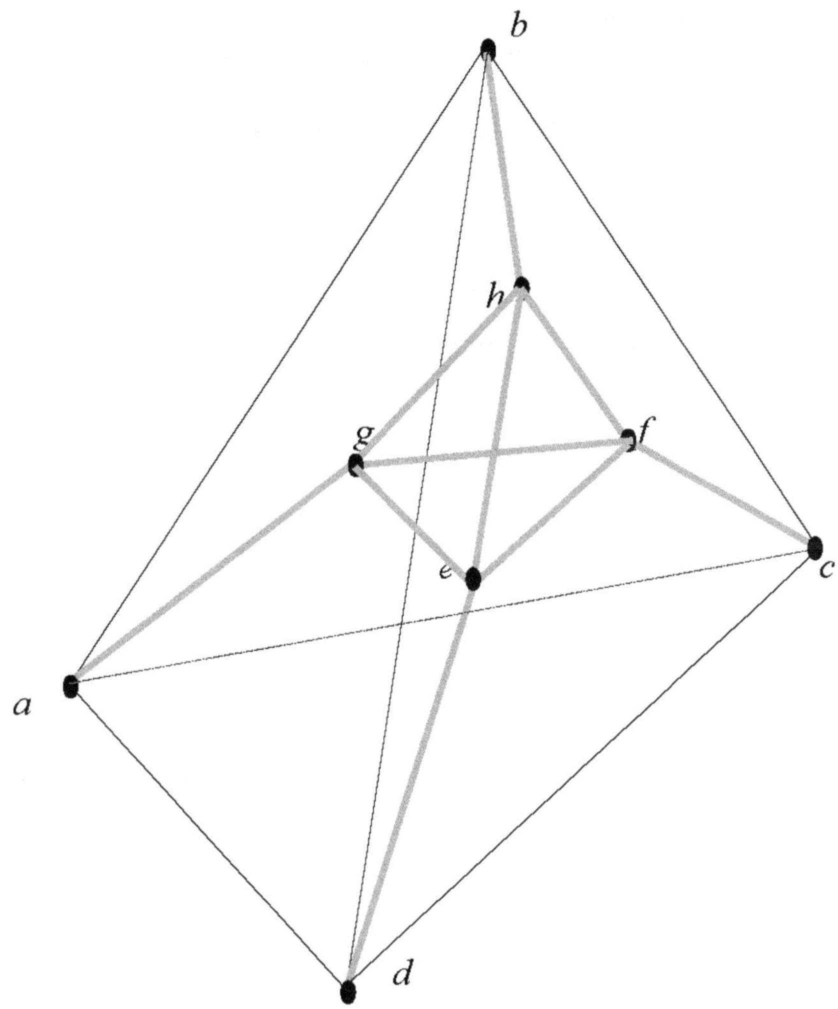

In Figure 5 there is eight vertices, $f_0 = 8$. The number of edges is 16 (*ab, bc, cd, ad, bd, ac, gh, hf,ef, he, gf, eg, bh, fc, ed, ag*), $f_1 = 16$. The number of elements of dimension 2 is 14 (triangles *abd, bcd, abc, acd, ghe, hef, ghf, gfe* and quadrangles *aghb,aged, hbed, hbfc, efcd, hfbc*), $f_2 = 14$. The number of elements of dimension 3 is 6 (tetrahedrons *abcd, ghef* and prismatoides *abhged, hbedbfc, aghbfc, gefadc*), $f_3 = 6$. On Figure 5 for example the edges correspondent of chemical bounds is indicated brown, remain edges (black) it is need for creating convex body. Substituting

the values of the number of elements of different dimension in the equation Euler –Poincaré (5.1) we obtain:

$$8 - 16 + 14 - 6 = 0.$$

We find that it holds for $n = 4$. This proves that the figure who projection is shown in Figure 5 there is polytope of dimension 4.

By construction, it is clear, that a polytope tetrahedron in tetrahedron is an irregular polytope. In particular, a polytope included irregular polyhedrons prismatoides. Moreover, the outer boundary of this four-dimensional polytope is the three-dimensional surface of a tetrahedron.

Insert an Octahedron into a Chemically Bonded Octahedron

The result of inserting an octahedron into an octahedron is shown in Figure 6.

Figure 6. The polytope octahedron in octahedron.

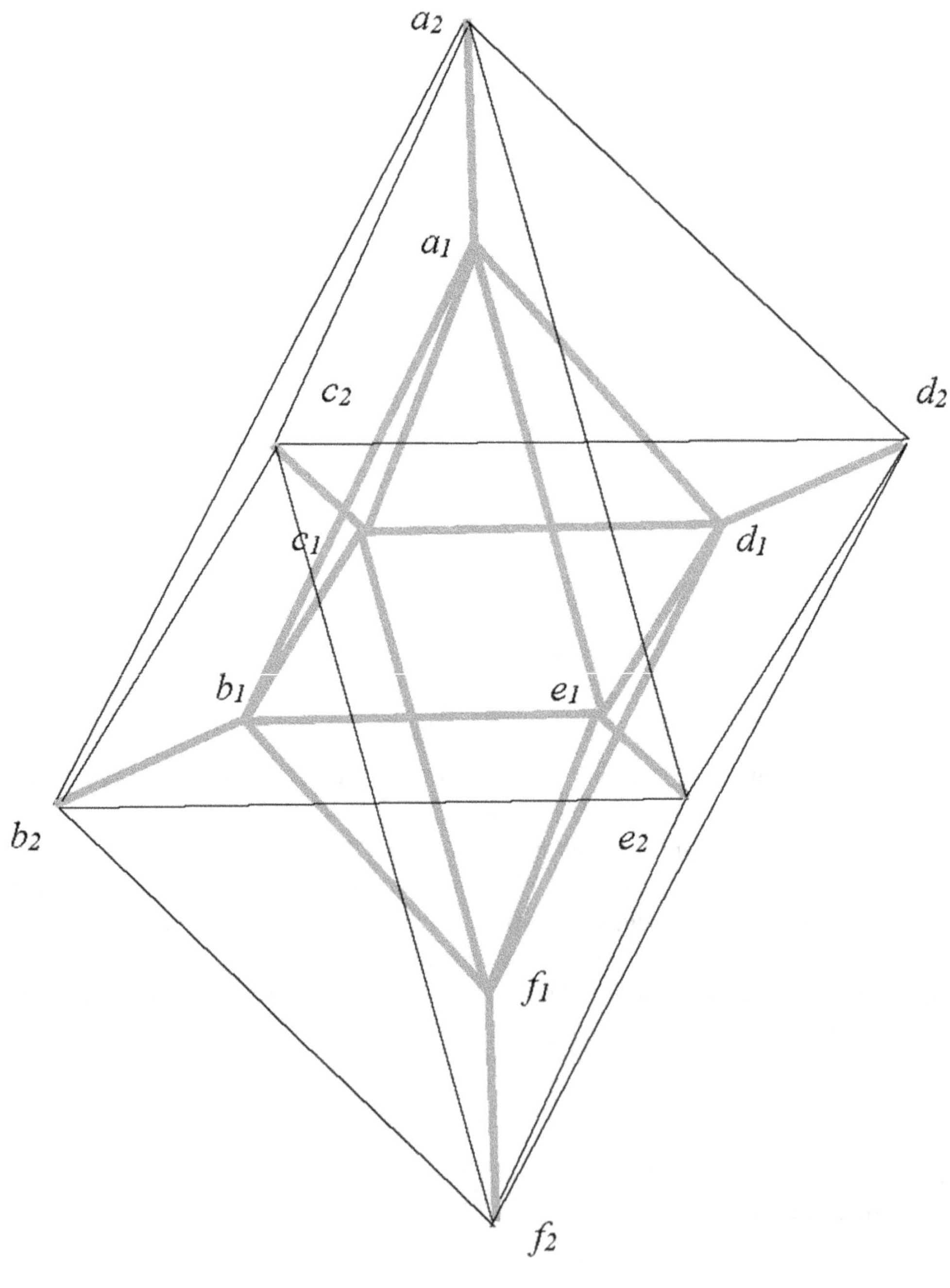

In this case the number of elements of zero dimension is $f_0 = 12$. The number of elements of dimension one is $f_1 = 12 + 12 + 6 = 30$. The number of elements of dimension 2 is sum of the number small triangles 8 and big triangles 8, add 12 quadrangles, i. e. $f_2 = 28$. The number of elements of dimension 3 is sum two octahedrons and 8 prisms, i. e. $f_3 = 10$. Substituting the values of numbers of elements of different dimensions in the equation (5.1), we obtain

$$12 - 30 + 28 - 10 = 0.$$

We find that it holds for $n = 4$. This proves that the Figure 6 is polytope of dimension 4.

By construction, it is clear, that a polytope octahedron in octahedron is an irregular polytope. In particular, the polytope included irregular polyhedrons prismatoides. Moreover, the outer boundary of this four-dimensional polytope is the three-dimensional surface of an octahedron.

Insert an Icosahedron into a Chemically Bonded Icosahedron

The result of inserting an icosahedron into an icosahedron is shown in Figure 7.

Figure 7. The polytope icosahedron in icosahedron.

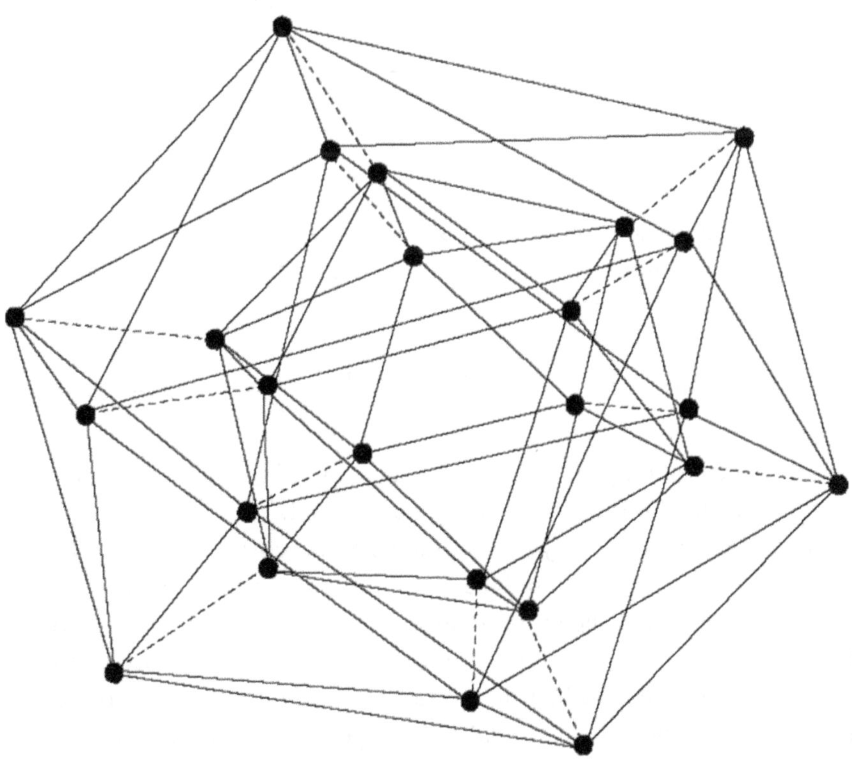

On Figure 7 connect the edges of the vertices of both icosahedrons lying on the same line passing through the common center of the icosahedrons (dotted lines in Figure 7). This shows that the space between the icosahedrons is completely filled with triangular prisms, the bases of which are the triangular faces of the smaller and larger icosahedrons (Figure 8).

Figure 8. The triangular prism

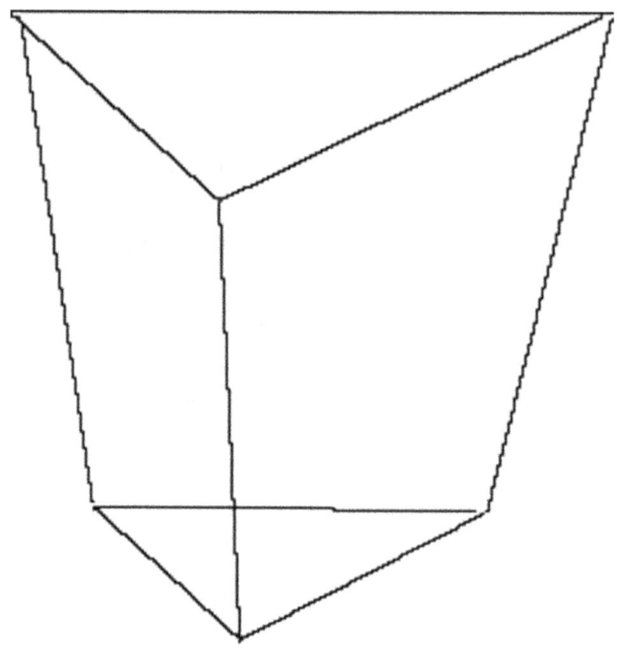

These prisms are adjacent to each other along flat quadrilateral side faces. The number of these prisms is equal to the number of triangular faces of the icosahedron - 20. Since the icosahedron has 12 vertices and 30 edges for a given design, we have:

$$f_0 = 2 \cdot 12 = 24, f_1 = 2 \cdot 30 + 12 = 72, f_2 = 2 \cdot 20 + 30 = 70, f_3 = 20 + 2 = 22.$$

Substituting these values into equation (5.1), we have 24 – 72 + 70 – 22 = 0. This proves that the dimension of polytope icosahedron in icosahedron is equal to 4.

By construction, it is clear, that a polytope icosahedron in icosahedron is an irregular polytope. In particular, the polytope included irregular polyhedrons triangular prisms. Moreover, the outer boundary of this four-dimensional polytope is the three-dimensional surface of an icosahedron.

Comment. It can be shown that connecting one more third icosahedron to two connected icosahedrons will increase the dimension of the icosahedron cluster by one. In this case, the dimension of the boundary surface remains the same, in this case equal to 3. In general, the dimension of a system of several connected shells grows linearly with the number of shells (Zhizhin, 2019). This way one can achieve a large difference between the dimension of the system of connected shells and the dimension of the outer boundary set. Obviously, this is typical not only for the system of icosahedrons with a common center, but also for any system of any polyhedrons or polytopes with a common center. Adding a vertex to the center of a system of connected shells also increases the dimension of the system by one. We can say that adding a center and adding shells increases the deviation of the system from correctness (the number of signs of deviations from correctness increases). Thus, we can conclude that the more deviations from correctness in the shell system, the greater the dimension of this system. If the shell itself has the highest dimension, then adding shells to the system will lead to a significant increase in the dimension of the system. In this case, the boundary set will also have the highest dimension.

Polytope of Polyhedrons with Common Vertices

There is another way to form a higher-dimensional irregular polytope found in chemical structures - to connect the polyhedrons at one or more vertices.

The Center of one Tetrahedron is the Vertex of another Tetrahedron

The projection onto a two-dimensional plane of a polytope consisting of two tetrahedra, the vertex of one of which is the center of the other, is shown in Figure 9.

Figure 9. The polytope from two tetrahedrons, the vertex of one of which is the center of the other.

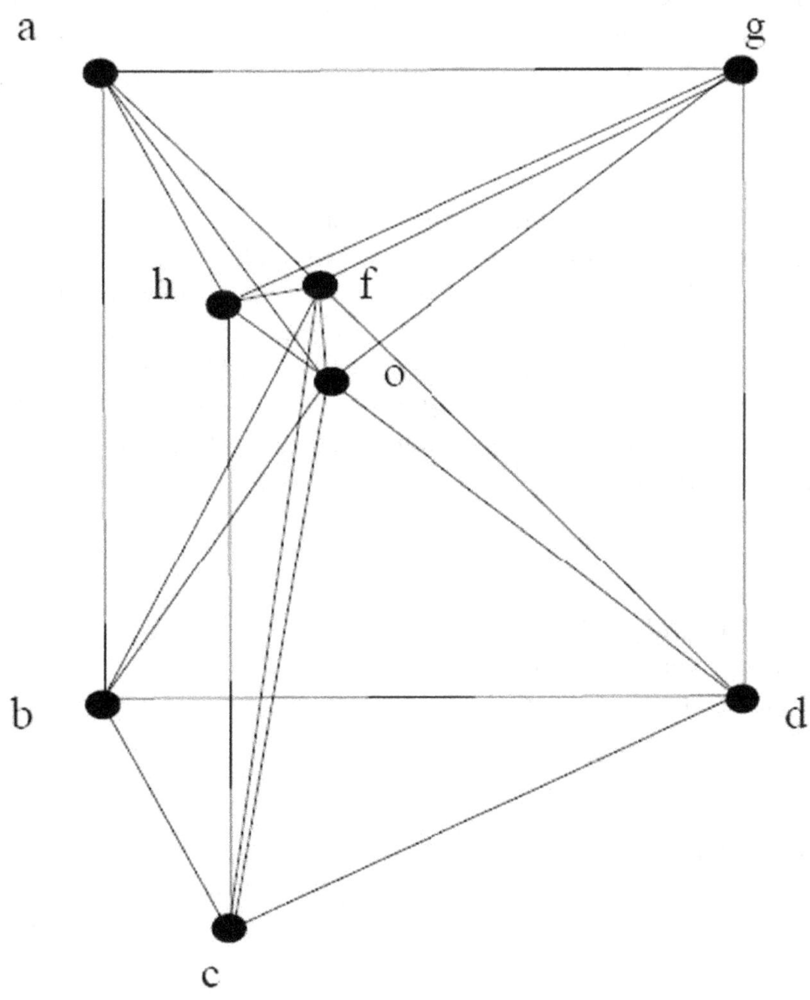

In this figure there is a tetrahedron *bcdf* with center *o* and a tetrahedron *oahg* with center *f*. Each a tetrahedron with a center is a polytope of dimension 4.

The vertex *f* of the first tetrahedron is the center of the second tetrahedron, and the vertex *o* of the second tetrahedron is the center of the first tetrahedron. It is necessary to determine the dimension of the polytope *bcdfahgo*. The polytope in Figure 9 has 8 vertices ($f_0 = 8$), 22 edges (*ab, ag, af, ao, ah, gh, gf, go, gd, bf, bo, bd, bc, df, do, dc, hc, hf, ho, cf, co, fo*). Therefore, $f_1 = 22$. The polytope in Figure 9 has 29 planar faces, of which 26 are the triangles (*aho, afo, ahf, afg, ahg, aog, aob, afb, bfo, bco, bod, bfd, bfc, bcd, ghf, gho, gfo, god, gfd, dfo, dco, dfc, cof, chf, cho, hfo*) and 3 quadrangles (*abdg, hcgd, abhc*). Therefore, $f_2 = 29$. The polytope in Figure 9 has 20 three - dimensional figures, of which 13 are tetrahedrons (*bcfd, ahog, dcfo, bcdo, bfdo, cfdo, ahof, ahgf, hogf, aogf, fgod, fohc, foab*), 6 pyramids (*agbdo, ahbcf, ahbco, agbdf, Chgdf, chgdo*) and one (*ahgbcd*) prism. Therefore, $f_3 = 20$. It follows from the construction of the polytope in Figure 9 that it includes two tetrahedrons with the center *bcdfo* and *oahgf*. In addition to these two polytopes with dimension 4, five 4 - polytopes also appear in the polytope in Figure 9. Three of these polytopes have as their base three rectangular faces of the prism *ahgbcd*, whose vertices are connected with the vertices *f, o* located inside the prism. To prove their 4 - dimensionality, consider one of these polytopes *abhcfo*, since the proofs for the other two polytopes are similar. This polytope has 6 vertices ($f_0 = 6$); 13 edges (*ab, ah, hc, bc, af, hf, bf, cf, ho, ao, bo, co, fo*), $f_1 = 13$; 13 two - dimensional faces (*ahf, aho, abo, abf, afo, bfo, boc, ahbc*), $f_2 = 13$; 6 three -dimensional faces (*hfoc, abof, bfoc, afho, ahcbf, ahcbo*), $f_3 = 13$. Substituting the obtained values of the numbers of faces of different dimensions into equation (5.1) can find that equation (5.1) is satisfied for $n = 4$

$$6 - 13 + 13 - 6 = 0.$$

It is proved by the 4 - dimensionality of the polytope *abhcfo*.

The two polytopes of dimension 4 there are formed by the *ahgbcd* prism with the vertex *f* or *o* inside its. Consider the prism *ahgbcd* with the vertex *f* (the proof for the prism with vertex *o* is similar). The polytope *ahgbcdf* has 7 vertices, $f_0 = 7$; 15 edges (*ah, hg, ag, bd, bc, cd, ab, hc, gd, af, fh, fg, bf, fc, fd*), $f_1 = 15$; 14 two - dimensional faces (*ahg, bdc, ahf, hfg, afg, bfc, fcd, bfd, fhc, afb, fgd, ahbc, hcgd, agbd*), $f_2 = 14$; 6 three - dimensional faces (*ahgbcd, ahgf, bcdf, abdgf, hgcdf, ahbcf*), $f_3 = 6$. Substituting the values of the numbers of faces of various dimensions obtained for the polytope *ahgbcdf* into equation (5.1) can find that it is satisfied for $n = 4$

$$7 - 15 + 14 - 6 = 0.$$

This proves that the polytope *ahgbcdf* has a dimension of 4.

Thus, for the polytope in Figure 9 are $f_0 = 8, f_1 = 22, f_2 = 29, f_3 = 20, f_4 = 7$. Substituting these values into equation (5.1) can find that it is satisfied for $n = 5$

$$8 - 22 + 29 - 20 + 7 = 2.$$

This proves that the polytope in Figure 9 has dimension 5.

It is clear, that the polytope on Figure 9 is irregular polytope, so into it are the irregular polyhedrons. Surprisingly, the outer boundary of this five-dimensional polytope is a three-dimensional figure - a triangular prism.

The Center of One Octahedron is the Vertex of another Octahedron

The projection onto a two-dimensional plane of a polytope consisting of two octahedrons, one of vertex of each octahedron is the center of the other octahedron, is shown in Figure 10.

Figure 10. The polytope consisting of two octahedrons, one of vertex of each octahedron is the center of the other octahedron.

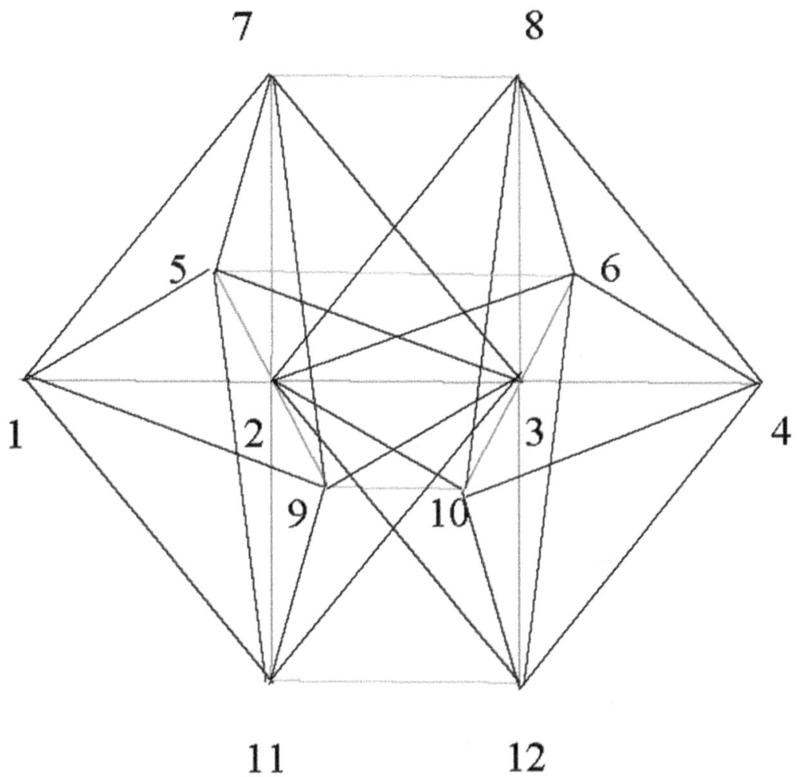

In Figure 10, the region in which the octahedrons enter each other 5 - 7 - 8 - 6 - 10 - 12 - 9 - 11 - 2 - 3 should be distinguished. As well as two regions in which only of the octahedron is present 1 - 5 - 7 - 9 - 11 - 2, 4 - 8 - 6 - 10 - 12 - 3 should be distinguished. An area in which only one of the octahedrons is present, for example, 1 - 5 - 7 - 9 - 11 - 2 (Figure 11) has 6 vertices (f_0=6), 13 edges (f_1=13)

1 - 5, 1 - 7, 1 - 2, 1 - 9, 1 - 11, 5 - 7, 7 - 9, 9 - 11, 5 - 11, 2 - 5, 2 - 7, 2 - 9, 2 - 11,

12 flat two - dimensional faces (f_2=12)

1 - 7 - 5, 1 - 5 - 2, 1 - 7 - 2, 1 - 9 - 7, 1 - 2 - 9, 1 - 9 - 11, 1 - 2 - 11, 1 - 5 - 11, 2 - 5 - 7, 2 - 7 - 9, 2 - 11 - 9, 5 - 2 - 11,

5 three - dimensional areas (f_3=5)

1 - 5 - 2 - 7, 1 - 2 - 7 - 9, 1 - 2 - 9 - 11, 1 - 5 - 2 - 11, 1 - 7 - 9 - 5 - 11 - 2.

Substituting the obtained values f_i(i=0,1,2,3) in the equation (5.1), you can see that the Euler - Poincaré equation is satisfied in this case with a value of n equal to 4:

$6 - 13 + 12 - 5 = 0.$

Figure 11.

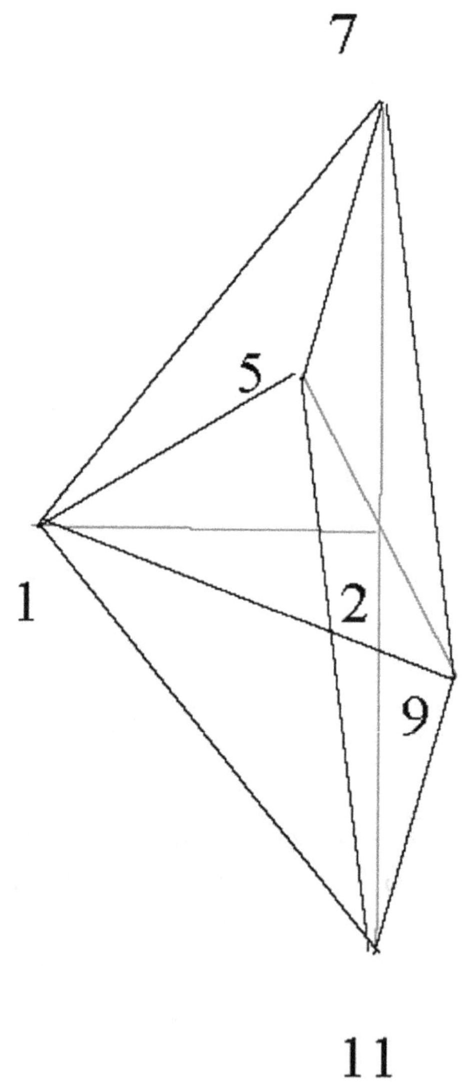

This proves that each of these regions has dimension 4. It should be noted that the region 1 - 7 - 9 - 5 - 11 - 2 does not contain edge 1 - 2, therefore it has dimension 3. As soon as edge 1 - 2 is added, region has dimension 4. The base of this figure is constructed from four triangles that are faces of other three - dimensional figures in this polytope.

The region 5 - 7 - 8 - 6 - 10 - 12 - 9 - 11 - 2 - 3 in which the octahedrons penetrate into each other is quite complicated. However, here we must bear in mind that the edges of the octahedrons in this region, which are not valence bonds, were necessary for the creation of convex figures (octahedrons) when considering these figures separately. When the general region, which includes octahedrons, is considered, their purpose is lost, due to the introduction of other edges that create a unifying closed convex figure. Therefore, in this region one should leave only the edges corresponding to the valence bonds and the edges forming the boundary of the region. Other edges 7 - 3, 8 - 2, 5 - 3, 2 - 6, 2 - 10, 3 - 9 must be removed. In this case, the region takes the form shown in Figure 12.

Figure 12. Simplified region penetrated by adjacent octahedrons in chains

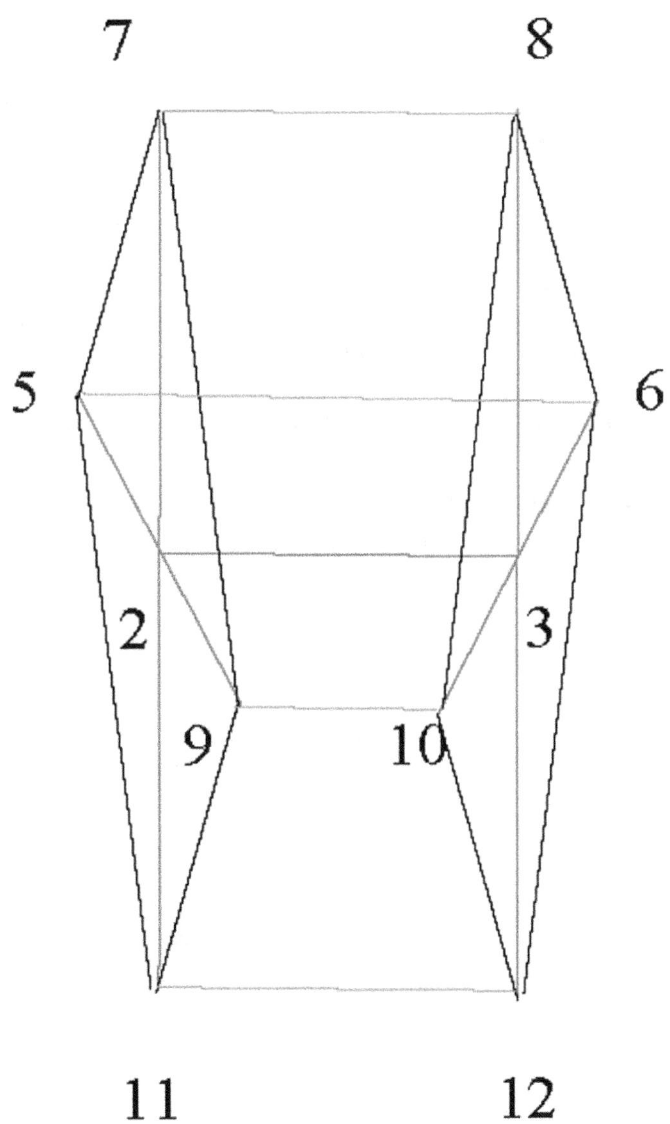

The polytope on Figure 12 has 10 vertices (f_0=10), 21 edges (f_1=21)

7 - 8, 6 - 5, 2 - 3, 9 - 10, 11 - 12, 5 - 7, 7 - 9, 9 - 11, 11 - 5, 2 - 5, 2 - 7, 2 - 9, 2 - 11, 8 - 6, 6 - 12, 12 - 10, 10 - 8, 3 - 8, 3 - 6, 3 - 12, 3 - 10,

16 two - dimensional faces (f_2=16)

5 - 7 - 8 - 6, 5 - 2 - 6 - 3, 7 - 8 - 2 - 3, 7 - 8 - 9 - 10, 9 - 10 - 11 - 12, 2 - 3 - 11 - 12, 2 - 3 - 9 - 10, 2 - 3 - 5 - 6, 2 - 5 - 7, 2 - 7 - 9, 2 - 9 - 11, 2 - 5 - 11, 3 - 6 - 8, 3 - 6 - 12, 3 - 10 - 12, 3 - 8 - 10,

5 tree - dimensional faces (f_3=5)

2 - 5 - 7 - 3 - 8 - 6, 2 - 7 - 9 - 3 - 8 - 10, 2 - 9 - 11- 3 - 10 - 12, 2 - 5 - 11 - 3 - 6 - 12, 5 - 7 - 8 - 6 - 9 - 10 - 11 - 12 - 2 - 3 (not contain edge 2 - 3).

Substituting the obtained values f_i(i=0,1,2,3) in the equation (5.1) you can see that the Euler - Poincaré equation is satisfied in this case with a value of n equal to 4

$$10 - 21 + 16 - 5 = 0.$$

This proves that polytope on Figure 12 has dimension 4.

Thus, in this case Figure 10 is chain of the polytopes with dimension 4 connected each other on flat faces. Moreover, the outer boundary of the chain of these polytopes is a three-dimensional surface.

Three Tetrahedrons with the Center, Having one Common Vertex

The projection onto a two-dimensional plane of a polytope consisting of three tetrahedrons with the center, having one common vertex, is shown in Figure 13.

Figure 13. The polytope from three tetrahedrons with the center and one common vertex.

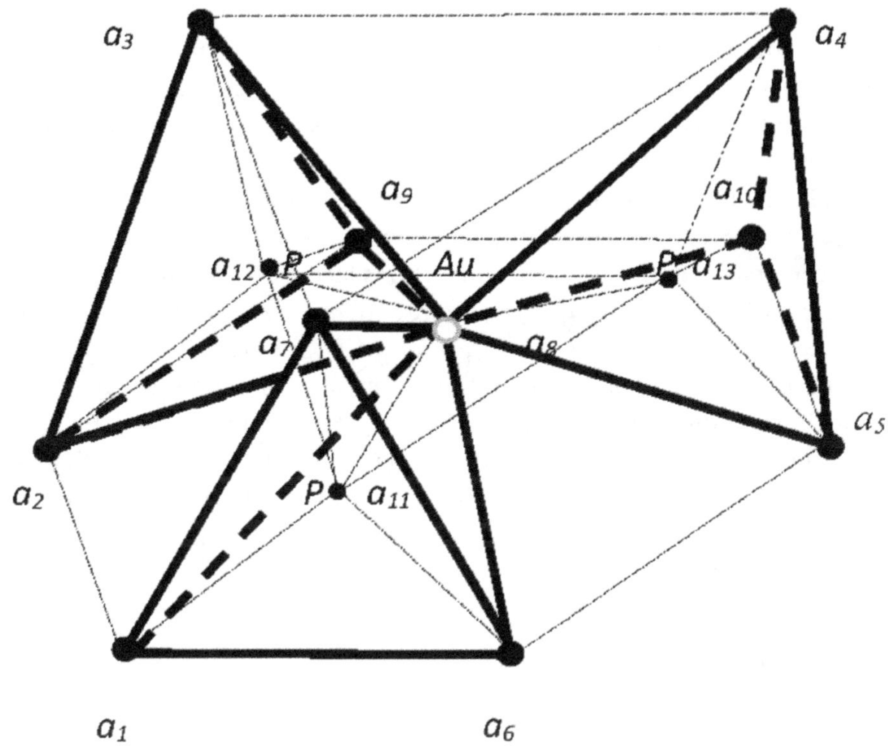

The dimension of each tetrahedron with center is still equal to 4.

To prove the necessity of the three tetrahedrons with a common vertex to form a convex shape. Connect the vertices of a_3, a_4, a_7 by line segments, forming a triangle $a_3\,a_4\,a_7$. Connect also the three centers of the tetrahedrons with each other, forming a triangle $a_7\,a_{11}a_{13}$. Connect the center of the tetrahedrons with vertices corresponding of the tetrahedrons and vertices in the grounds of the tetrahedrons, forming a hexagon $a_1\,a_2\,a_9\,a_{10}\,a_5\,a_6$ (thin lines on Figure 13). Define dimension polytope in Figure 13 on the Euler- Poincare equation (5.1) (Poincare, 1895)

To calculate the number of elements of large dimensions we turn first to a simple polytope, a part of a polytope in Figure Temporarily excluded from Figure 13 the centers of the tetrahedrons – a_{11}, $a_{12,}$ a_{13}, and all edges emanating from these vertices. Then, the polytope has 13 vertices, i. e.f_0=10. The number of edges of the polytope is sum the number edges of three tetrahedrons (6•3=18), the number of edges connecting tetrahedrons at the base figure (3), the number of the edges connecting vertices of the tetrahedrons at top of Figures 13. Thus, the number of edges polytope without centers of the tetrahedrons equal 24, i.e.f_1=24. The number of flat elements is sum of the flat faces of tetrahedrons (4•3=12), 1 hexagon, 3 triangles between tetrahedrons at base of figure, 3 lateral tetragons, 4 triangles of tetrahedron at top of figure. Thus, the number of flat elements is 23, i. e.f_2=23. The number of three-dimension elements is sum of 4 tetrahedrons, 3 figures between tetrahedrons, 1 hexagon at base and figure composed from boundary flat faces. Thus, the number of three-dimension elements is 9, i. e.f_3=9. Substituting the values f_i in equation (5.1), we see that it holds for n = 4

$10 - 24 + 23 - 9 = 0.$

Therefore, three tetrahedrons with common vertices is polytope with dimension 4.

For add centers in tetrahedrons the number of the vertices becomes equal 13, i. e. $(f_0)_c=13$. For this there add the number of the edges: 4•3=12 edges in tetrahedrons with centers, and 3 edges connecting centers. Thus, common number of edges on Figure 13 equal 39, i. e. $(f_1)_c=39$. The number flat faces there increase on 18 triangles in the tetrahedrons, 4 triangles in tetrahedron $a_{11}a_{12}a_{13}a_8$, 6 tetragons with vertices part which are centers of the tetrahedrons. Thus, common number of flat faces on Figure 13 equal 51, i. e. $(f_2)_c=51$. For add centers the number of three-dimensions faces increases on 4•3=12 tetrahedrons into tetrahedrons with centers, tetrahedron $a_{11}a_{12}a_{13}a_8$, figure $a_1a_2a_9a_{10}a_5a_6a_{11}a_{12}a_{13}a_8$, prism $a_{11}a_{12}a_{13}a_3a_4a_7$, 3 pyramids with vertex a_8 $(a_8a_1a_2a_{12}a_{11}$, $a_8a_5a_6a_{11}a_{13}$, $a_8a_9a_{10}a_{12}a_{13})$, 3 prism $(a_1a_2a_3a_7a_{11}a_{12}$, $a_5a_6a_4a_7a_{11}a_{13}$, $a_9a_{10}a_3a_4a_{12}a_{13})$. Thus, common number of three-dimension faces on Figure 13 equal 30, i. e. $(f_3)_c=30$.

It is known from the preceding that the Figure 13 has polytopes of dimension 4. Each tetrahedron with center there is polytope of dimension 4 and 3 tetrahedrons without center, but with a common vertex, there is a polytope of dimension 4. In addition, in Figure 13 between any two tetrahedrons with the center is polytope dimension 4. Obviously, such polytopes are 3. That to proof this statement we consider any polytope from them. For example, the polytope $a_4a_5a_6a_7a_{11}a_{13}$ (Figure 14).

Figure 14. The polytope $a_4a_5a_6a_7a_{11}a_{13}$

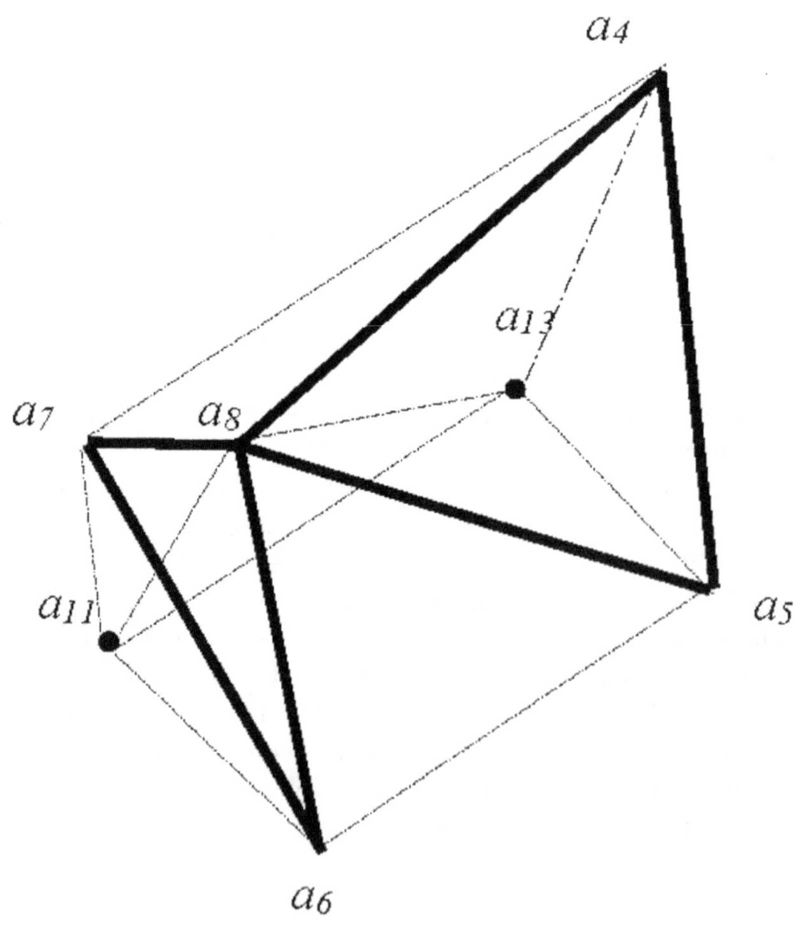

Figure 14 has 7 vertices (f_0=7); 15 edges

$a_{11}a_7, a_{11}a_8, a_{11}a_6, a_{11}a_{13}, a_7a_6, a_7a_8, a_7a_4, a_8a_4, a_8a_{13}, a_8a_5, a_6a_5, a_{13}a_5, a_4a_{13}, a_4a_5, a_8a_6$;

14 flat faces

$a_{11}a_8a_6, a_{11}a_7a_8, a_6a_7a_8, a_{11}a_7a_6, a_8a_5a_{13}, a_8a_{13}a_4, a_5a_{13}a_4, a_8a_5a_4, a_6a_8a_5, a_{11}a_8a_{13}, a_7a_8a_4, a_6a_{11}a_{13}a_5, a_6a_5a_7a_4, a_{11}a_7a_{13}a_4$;

6 three-dimension faces

$a_6a_{11}a_8a_7, a_8a_5a_{13}a_4, a_7a_8a_{11}a_{13}a_4, a_6a_{11}a_8a_{13}a_5, a_{11}a_6a_7a_{13}a_5a_4, a_6a_7a_8a_4a_5$.

Therefore, for Figure 14 are f_0=7, f_1=15, f_2=14, f_3=6. Substituting the values f_i in equation (5.1), we see that it holds for $n = 4$

$$7 - 15 + 14 - 6 = 0.$$

This proof that Figure 14 has dimension 4.

As each figure in polytope on Figure 13 is 3, so common number polytopes with dimension 4 in Figure 13 equal 7, i. e. $(f_4)_c$=7. Substituting the values $(f_i)_c$ in equation (5.1), we see that it holds for $n = 5$

$$13 - 39 + 51 - 30 + 7 = 2.$$

Thus, the polytope on Figure 13 is irregular polytope, so into it are the irregular polyhedrons. This polytope has dimension 5. The outer boundary of this five-dimensional polytope is a three-dimensional irregular figure - a prism with bases triangle and hexagon.

CONCLUSION

For the first time, irregular polytopes of higher dimension have been investigated in sufficient detail. Some methods of the formation of polytopes of the highest dimension in chemical structures in nature are considered. A characteristic feature of irregular polytopes of higher dimension has been found: the existence of a continuous outer boundary with a lower dimension of a given polytope. This feature is not possessed by regular polytopes of higher dimension, in which the boundary complex includes sets of faces of different dimensions, smaller than the dimension of the polytope (Grünbaum, 1967). This feature of irregular polytopes of higher dimension, which was not known before, makes it possible to consider irregular polytopes of higher dimension to be carriers of higher dimension in spaces of lower dimension, for example, in three-dimensional spaces. Moreover, irregular polytopes of the highest dimension, composed of chemically unbound shells, provide the possibility of the existence in nature of objects of a higher dimension with a steadily increasing dimension as the distance from the object boundary deeper into the object. A dimension jump on the outer boundary of irregular polytopes of higher dimension allows for contact between spaces of different dimensions. A number of methods for creating irregular polytopes of higher dimension are investigated, connected with the introduction of elements of lower dimension inside polytopes, contacts of polytopes of higher dimension with each other, etc. In subsequent chapters, these options for the formation of irregular polytopes of the highest dimension will be illustrated on specific material objects. More complex variants of the formation of irregular polytopes of the highest dimension are also possible, in comparison with those considered in this chapter. Some of these options will be explored with specific examples in subsequent chapters.

REFERENCES

Coxeter H.S.M. (1963). *Regular Polytopes.* John Wiley & Sons, Inc.

Elte, E. L. (1912). *The semi-regular polytopes of the hyperspace.* University of Groningen.

Gosset, T. (1900). On the regular and semi-regular figures in space of *n* dimensions. *Messenger of Mathematics, 29*, 43–48.

Grünboum, B. (1967). *Convex Polytopes.* Springer.

Poincaré A. (1895). Analysis situs. *J. de é Ecole Polytechnique, 1.*

Schläfli, L. (1901). *Theorie der vielfachen Kontinuität Zürcher und Furrer.* doi:10.1007/978-3-0348-5118-3

Stringham, W. I. (1880). Regular figures in *n*-dimensional space. *American Journal of Mathematics, 3*(1), 1–14. doi:10.2307/2369441

Zalgaller, V. A. (1966). *Convex polyhedrons with correct sides.* Science.

Zhizhin, G. V. (2009). *Convex right-faceted polyhedrons.* Polytechnika-service.

Zhizhin, G. V. (2013). Relations for the numbers of faces of various dimensions in the tower of n - dimensional convex regular polytopes. In *Proceedings of the 9th All - Russian School "Mathematical Studies in the Natural Sciences"*. Apatity: Geological Institute, Russian Academy of Sciences.

Zhizhin, G. V. (2018). *Chemical Compound Structures and the higher Dimension of Molecules: Emerging research and Opportunities.* IGI Global. doi:10.4018/978-1-5225-4108-0

Zhizhin, G. V. (2019 a). *The Geometry of Higher - Dimensional polytopes.* IGI Global. doi:10.4018/978-1-5225-6968-8

Zhizhin, G. V. (2019 b). On the Structure of Parallelohedrons of Higher Dimension: Hilbert's 18th Problem. *International Journal of Chemoinformatics and Chemical Engineering, 8*(2), 69–96. doi:10.4018/IJCCE.20190701.oa2

Zhizhin, G. V. (2019 c). *Attractors and Higher Dimensions in Population and Molecular Biology: Emerging Research and Opportunities.* IGI Global. doi:10.4018/978-1-5225-9651-6

Zhizhin, G. V. (2020). *Nanotechnologies and Clusters in the Spaces of Higher Dimension: Emerging Research and Opportunities.* IGI Global.

Zhizhin, G. V. (2021). *Normal Partitions and Hierarchical Fillings of N-Dimensional Spaces*. IGI Global. doi:10.4018/978-1-7998-6768-5

KEY TERMS AND DEFINITIONS

Dimension of the Space: The member of independent parameters needed to describe the change in position of an object in space.

Incidence Coefficients of Elements of Lower Dimension With Respect to Elements of Higher Dimension: The number of elements of a certain higher dimension to which the given element of a lower dimension belongs.

N–Cross-Polytope: The convex polytope of dimension n in which opposite related of centrum edges not have connection of edge.

N–Cube: The convex polytope of dimension n in which each vertex incident to n edges.

N–Simplex: The convex polytope of dimension n in which each vertex is joined by edges with all remain vertices of polytope.

Polytope: Polyhedron in the space of higher dimension.

Chapter 6
Higher Dimensional Polytopes That Are Products of Lower Dimensional Polytopes

ABSTRACT

The geometry of polytopes, which are products of other polytopes, is investigated. It is proved that for the existence of a product of polytopes, as a polytope, it is necessary that the Euler-Poincaré equation be fulfilled for its factors. Then the Euler-Poincaré equation must be satisfied for the product of these polytopes. It is proved that the products of polytopes for any factors are incorrect polytopes. Thus, as previously established by the author, the possibility of constructing an n-dimensional space using the product of polytopes is carried out by incorrect polytopes. It is shown that the incorrectness of the product of polytopes, as a polytope, leads to the formation of a continuous closed boundary surface in a polytope with dimension one less than the dimension of the product of polytopes. This, in turn, leads to the fundamental possibility of creating multi-shell systems from unrelated products of polytopes with a steady increase in dimension as we go deeper into the system.

INTRODUCTION

The products of polytopes form an important class of polytopes with important properties. The products of polytopes are not simplicial polytopes, even if the factors of this product are simplices; therefore, their studies are of considerable interest (Ziegler, 1995). The peculiarities and properties of the product of polytopes made it possible to solve the 18th problem of Hilbert (Hilbert, 1901), posed 120 years ago, of constructing n - dimensional spaces from congruent figures (Zhizhin, 2019 b, 2021). Given the importance of this class of polytopes, a more precise and detailed study of these properties is required. In particular, the author's previous works (Zhizhin, 2019 a, 2021) did not formulate in a general analytical form the necessary condition for the existence of a product of polytopes as polytopes of higher dimension (an analogue of the Euler - Poincaré equation, Poincaré, 1895). It is required to define in more detail the conditions of incidence of faces in the products of polytopes. The products of polytopes are generally

DOI: 10.4018/978-1-7998-8374-6.ch006

referred to as irregular polytopes with one exception: an n - cube can be considered as a product of cubes of lower dimension. In this regard, it is necessary to take into account the results of the study of the properties of irregular polytopes obtained in the previous chapters of this work and previously unknown. In addition, to prove the theorem on the necessary conditions for the existence of a product of polytopes, it is necessary to improve the notation system. This chapter is devoted to solving all of the above issues.

THE STRUCTURE OF POLYTOPES WHICH ARE PRODUCTS OF POLYTOPES

There is the structure of product of polytopes having different structures of their factors is determined.

Theorem 6.1. If there are convex polytopes of dimensions n and m, respectively denoted P_n and Q_m (or simply P and Q), then their product $P_n \times Q_m$, (or simply ×, when it is clear what polytopes are multiplied) has a face $F_k^{\times}$ with numbers

$$f_k^{\times} = \sum_{i=0}^{j} f_{k-i}^{P_n} f_i^{Q_m}, \tag{6.1}$$

j=*k*, if 0£*k*<*m*; *j*=*m*, if *m*£*k*£*n*+*m*; *n*³*m*.

Here the symbol f indicates the number of faces, the lower index of f and F indicates the dimension of faces, the superscript indicates belonging of a face to the respective polytope.

Proof. To find the product of polytopes P_n and Q_m one fined the product of geometric elements of different dimensions of one of the polytopes by geometric elements of different dimensions of another polytope. According to the definition of polytopes product (Ziegler, 1995), the product of vertices (elements of zero - dimension) of one of the polytopes by the vertices of another polytope is a set of vertices, the number of which is equal to the product of the number of vertices of one of the polytopes by the number of vertices of another polytope:

$$f_0^{\times} = f_0^{P_n} f_0^{Q_m}.$$

The number of elements dimension 1 of polytopes two product is sum from two parts: a product of the number of elements dimension 1 of one polytope by the number of elements dimension 0 of another polytope and a product of the number of elements dimension 0 of one polytope by the number of elements dimension 1 of another polytope:

$$f_1^{\times} = f_1^{P_n} f_0^{Q_m} + f_0^{P_n} f_1^{Q_m}.$$

The number of elements dimension 2 of polytopes two product is sum from three parts:

a product of the number of elements dimension 2 of one polytope by the number of elements with dimension 0 of another polytope, a product of the number of elements of dimension 0 of one polytope by the number of elements dimension 1 of another polytope, and a product of the number of elements dimension 0 of one polytope by the number of elements of dimension 2 of another polytope

$$f_2^{\times} = f_2^{P_n} f_0^{Q_m} + f_1^{P_n} f_1^{Q_m} + f_0^{P_n} f_2^{Q_m} .$$

Continuing this process, can obtain the expression for the number of elements of dimension k in the product of polytopes P_n and Q_m

$$f_k^{\times} = f_k^{P_n} f_0^{Q_m} + f_{k-1}^{P_n} f_1^{Q_m} + \ldots + f_0^{P_n} f_k^{Q_m} . \tag{6.2}$$

The specific form of equation (6.2) depends on the ratio of k with the dimensions of polytopes P_n and Q_m. If k is greater than m, i. e. the dimension of the polytope Q_m, then in the series (6.2) several last items are lost because of the absence of faces of the polytope Q_m with dimension greater than m. If k is greater than n, i. e. the dimension of the polytope P_n, in the series (6.2) several first items are lost because of the absence of faces of the polytope P_n with the dimension greater than n.
If $k < m$ the series (6.2) can be written in the form

$$f_k^{\times} = \sum_{i=0}^{k} f_{k-i}^{P_n} f_i^{Q_m} . \tag{6.3}$$

When $k^3 m$ summation by i in the series (6.3) is carried out only up to the value $i = m$

$$f_k^{\times} = \sum_{i=0}^{m} f_{k-i}^{P_n} f_i^{Q_m} . \tag{6.4}$$

At that, in the row (6.4) the items corresponding to the values of $k - i > n$ are equal to zero. Equations (6.3) and (6.4) can be written as a single equation:

$$f_k^{\times} = \sum_{i=0}^{j} f_{k-i}^{P_n} f_i^{Q_m} , \tag{6.5}$$

where $j=k$, if 0£k<m; $j=m$, if m£k£$n+m$; $n^3 m$.
Theorem 6.1 is proved.

Theorem 6.1 determines the number of figures of different dimensions in the product of polytopes, but it cannot identify in general terms what is the form of these figures. It follows from theorem 6.1 that the form of the figures of polytopes product depends on the form of the polytopes, which are the factors of this product.

We consider the sum of the alternating series of the number of faces $\sum_{i=0}^{n+m} (-1)^i f_i^{\times}$ of the product of polytopes P_n and Q_m .

Theorem 6.2. The sum of the alternating series of the number of faces of the product of polytopes P_n and Q_m is equal to the product of the sums of the alternating series of the numbers of the faces of the factors, i. e. polytopes P_n and Q_m:

$$\sum_{i=0}^{n+m}(-1)^i f_i^{\times} = \sum_{k=0}^{m}(-1)^k f_k^{Q_m} \cdot \sum_{j=0}^{n}(-1)^j f_j^{P_n} . \tag{6.6}$$

Proof. Let us expand the expressions for the terms of the sum on the left-hand side of (6.6), taking into account (6.5), as well as the conditions that polytopes P_n cannot have elements of dimension greater than n, and polytopes of Q_m cannot have elements of dimension greater than m. Then the left-hand side of (6.6) has the following term-by-term general form

$$\begin{aligned}\sum_{i=0}^{n+m}(-1)^i f_i^{\times} &= f_0^{P_n} f_0^{Q_m} - (f_1^{P_n} f_0^{Q_m} + f_0^{P_n} f_1^{Q_m}) + (f_2^{P_n} f_0^{Q_m} + f_1^{P_n} f_1^{Q_m} + f_0^{P_n} f_2^{Q_m}) \\ &+ \ldots + (-1)^m (f_m^{P_n} f_0^{Q_m} + f_{m-1}^{P_n} f_1^{Q_m} + \ldots + f_0^{P_n} f_m^{Q_m}) + (-1)^{m+1} (f_{m+1}^{P_n} f_0^{Q_m} + f_m^{P_n} f_1^{Q_m} + \ldots + f_1^{P_n} f_m^{Q_m}) \\ &+ (-1)^{m+2} (f_{m+2}^{P_n} f_0^{Q_m} + f_{m+1}^{P_n} f_1^{Q_m} + \ldots + f_2^{P_n} f_m^{Q_m}) + \ldots + (-1)^n (f_n^{P_n} f_0^{Q_m} + f_{n-1}^{P_n} f_1^{Q_m} + \ldots + f_{n-m}^{P_n} f_m^{Q_m}) \\ &+ (-1)^{n+1} (f_n^{P_n} f_1^{Q_m} + f_{n-1}^{P_n} f_2^{Q_m} + \ldots + f_{n-m+1}^{P_n} f_m^{Q_m}) + (-1)^{n+2} (f_n^{P_n} f_2^{Q_m} + f_{n-1}^{P_n} f_3^{Q_m} + \ldots + f_{n-m+2}^{P_n} f_m^{Q_m}) \\ &+ \ldots + (-1)^{n+m-1} (f_n^{P_n} f_{m-1}^{Q_m} + f_{n-1}^{P_n} f_m^{Q_m}) + (-1)^{n+m} f_n^{P_n} f_m^{Q_m} .\end{aligned} \tag{6.7}$$

Let us group on the right-hand side of (6.7) the terms with factors $f_0^{Q_m}, f_1^{Q_m}, \ldots, f_m^{Q_m}$. Then equation (6.7) takes the form

$$\begin{aligned}\sum_{i=0}^{n+m}(-1)^i f_i^{\times} &= (-1)^0 f_0^{Q_m}[(-1)^0 f_0^{P_n} + (-1)^1 f_1^{P_n} + \ldots + (-1)^n f_n^{P_n}] + (-1)^1 f_1^{Q_m}[(-1)^0 f_0^{P_n} \\ &+ (-1)^1 f_1^{P_n} + \ldots + (-1)^n f_n^{P_n}] + (-1)^2 f_2^{Q_m}[(-1)^0 f_0^{P_n} + (-1)^1 f_1^{P_n} + \ldots + (-1)^n f_n^{P_n}] \\ &+ \ldots + (-1)^m f_m^{Q_m}[(-1)^0 f_0^{P_n} + (-1)^1 f_1^{P_n} + \ldots + (-1)^n f_n^{P_n}] = [(-1)^0 f_0^{Q_m} + (-1)^1 f_1^{Q_m} \\ &+ \ldots + (-1)^m f_m^{Q_m}][(-1)^0 f_0^{P_n} + (-1)^1 f_1^{P_n} + \ldots + (-1)^n f_n^{P_n}].\end{aligned}$$

Q.E.D.

Theorem 6.3. For the product of polytopes to satisfy the Euler-Poincaré equation, it is necessary that the factors of the product satisfy the Euler-Poincaré equation.

Proof. Theorem 6.3 immediately follows from Theorem 6.2. Indeed, according to the Euler-Poincaré equation, we have

$$\sum_{k=0}^{m}(-1)^k f_k^{Q_m} = \sum_{k=0}^{m-1}(-1)^k f_k^{Q_m} + (-1)^m f_m^{Q_m}, \sum_{j=0}^{n}(-1)^j f_j^{P_n} = \sum_{j=0}^{n-1}(-1)^k f_j^{P_n} + (-1)^n f_n^{P_n} .$$

Substituting these expressions into the right-hand side of equation (6.6), taking into account that $f_m^{Q_m} = f_n^{P_n} = 1$, we obtain

$$\sum_{i=0}^{n+m-1} (-1)^i f_i^{\times} = 1-(-1)^{n+m}.\text{Q.E.D.} \quad (6.8)$$

Now the results obtained by the author earlier on the works of various specific polytopes are presented in an updated form.

THE POPYTOPES WHICH ARE THE PRODUCTS OF POLYTOPES BY ONE – DIMENSIONAL SEGMENT

It is obvious that the product of a point *a* by a one - dimensional segment is the same segment with point on end coinciding with point *a*. The product of segment by segment is a quadrangle. The product triangle on the segment is a triangular prism (Zhizhin, 2017 a) with the bases equal to the triangle and generating lines equal to the segment.

The Polytope which is the Product of a Tetrahedron by a Segment

For the designation of a polytope let's agree to indicate not only its dimension but the number of its vertices by lower index of polytope symbol. Besides, one will also indicate the number of its facets in brackets after the polytope sign, dimension of facets by lower index of facet symbol and the number of vertices by upper index of facets symbol. Since, in the general case, any face of a polytope is also a polytope, the polytope and its faces are denoted by the same letters *P*, *Q* etc. they differ in the corresponding dimensions. For example, a tetrahedron in this record has the form $P_3^4(4P_2^3)$. Let's call such a record a structural polytope formula. One need to get from structural formulas of polytopes as factors the structural formula of the product polytopes. Thus, in this case, we consider the product $P_3^4(4F_2^3)\times Q_1^2$. According to theorem 6.1 and the ratios (6.5) there are:

$$f_0^{\times} = f_0^{P_3} f_0^{Q_1} = 4 \cdot 2 = 8,$$

$$f_1^{\times} = f_1^{P_3} f_0^{Q_1} + f_0^{P_3} f_1^{Q_1} = 6\cdot 2+4\cdot 1=16.$$

When calculating the number of faces of dimension greater than 1 one will indicate in brackets after the sign of the faces number the designation of the corresponding face. This will give us an opportunity to get the composition of faces in the product polytopes. Thus,

$$f_2^{\times} = f_2^{P_3} f_0^{Q_1} + f_1^{P_3} f_1^{Q_1} = 4\cdot 2+6\cdot 1=14. \quad (6.9)$$

For receiving of the composition of the faces in the dimension 2 polytopes product one multiply the faces in brackets in the product of the numbers of faces, observing the same rules as at the definition of the product polytopes, i. e. dimensions are added, and the numbers of vertices are multiplied. At that, indexes P and Q are replaced by index $\times$. Then 14 faces obtained in (6.9), are deciphered so

$$14 = 8P_2^{\times,3} + 6P_2^{\times,4}. \tag{6.10}$$

This means that 14 faces of dimension 2 in the product include eight faces $P_2^{\times,3}$ (i.e., triangles) and six faces $P_2^{\times,4}$ (i.e., quadrangles). In future, equations similar to (6.10), (6.9) will be recorded in a single line.

For the faces of dimension 3, there is:

$$f_3^{\times} = f_3^{P_3} f_0^{Q_1} + f_2^{P_3} f_1^{Q_1} = 1\cdot 2 + 4\cdot 1 = 6 = 2P_3^{\times,4} + 4P_3^{\times,6}.$$

Thus, 6 faces of dimension 3 in the product consist of two tetrahedrons and four triangular prisms $P_3^6(3P_2^4, 2P_2^3)$.

For the faces of dimension 4, there is:

$$f_4^{\times} = f_3^{P_3} f_1^{Q_1} = 1\cdot 1 = 1 = P_4^{\times,8}.$$

According to the preceding equation, can obtain the structural formula of the product:

$$P_3^4(4F_2^3) \times Q_1^2 = P_4^8[4P_3^6(3P_2^4, 2P_2^3), 2P_3^4(4P_2^3)]. \tag{6.11}$$

Therefore, the product of the tetrahedron by the segment is polytope $4D$, which includes four triangular prisms and 2 tetrahedrons. You can call a polytope (6.11) tetrahedral prism. The picture of the polytope (6.11) is shown in Figure 1.

Figure 1. The tetrahedral prism

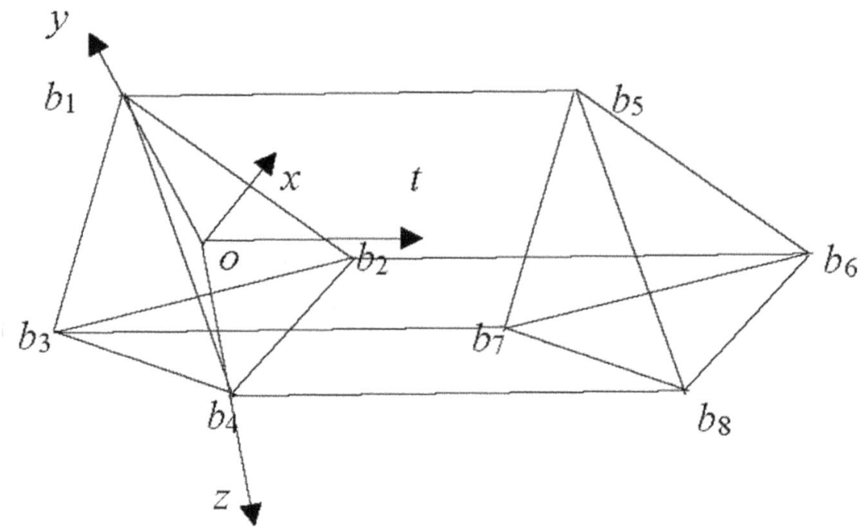

It is easy to verify that the equation (6.8) for the product polytopes (6.11) is satisfied:

$8 - 16 + 14 - 6 = 0.$

In future, the verification of the Euler – Poincaré (Poincaré, 1895) equation satisfaction is omitted, as a reader can easily do it on his own.

The product of polytopes (6.11) is an irregular polytope, since it contains irregular polyhedrons, trigonal prisms and tetrahedrons. You can see that this four-dimensional polytope is limited three-dimensional surface.

The Polytope which is the Product of a Cube by a Segment

The structural formula of a cube is $P_3^8(6P_2^4)$. It is necessary to find a structural formula of the product $P_3^8(6P_2^4)\times Q_1^2$. According to theorem 6.1 and formulas (6.5), can find:

$$f_0^{\times} = f_0^{P_3} f_0^{Q_1} = 8 \cdot 2 = 16,$$

$$f_1^{\times} = f_1^{P_3} f_0^{Q_1} + f_0^{P_3} f_1^{Q_1} = 12 \cdot 2 + 8 \cdot 1 = 32,$$

$$f_2^{\times} = f_2^{P_3} f_0^{Q_1} + f_1^{P_3} f_1^{Q_1} = 6 \cdot 2 + 12 \cdot 1 = 24 = 24P_2^{\times,4}.$$

$$f_3^{\times} = f_3^{P_3} f_0^{Q_1} + f_2^{P_3} f_1^{Q_1} = 1 \cdot 2 + 6 \cdot 1 = 8 = 8P_3^{\times,8},$$

$$f_4^{\times} = f_3^{P_3} f_1^{Q_1} = 1 \cdot 1 = 1 = P_4^{\times,16}(8P_3^{\times,8}). \tag{6.12}$$

Thus, easy to see that the product of a cube by a segment if the length of a segment equal of the length of edge a cube is 4 – cube (hyper - cube) containing 8 three-dimension cubes. If the length of a segment not equal of length of a cube, so polytope (6.12) is irregular polytope.

The Polytope which is the Product of an Octahedron by a Segment

The structural formula of an octahedron is $P_3^6(8P_2^4)$. It is necessary to find a structural formula of the product $P_3^6(8P_2^4)\times Q_1^2$. According to theorem 6.1 and formulas (6.5), can find:

$$f_0^{\times} = f_0^{P_3} f_0^{Q_1} = 6 \cdot 2 = 12,$$

$$f_1^{\times} = f_1^{P_3} f_0^{Q_1} + f_0^{P_3} f_1^{Q_1} = 12\cdot 2 + 8\cdot 1 = 32,$$

$$f_2^{\times} = f_2^{P_3} f_0^{Q_1} + f_1^{P_3} f_1^{Q_1} = 6\cdot 2 + 12\cdot 1 = 24 = 24P_2^{\times,4}.$$

$$f_3^{\times} = f_3^{P_3} f_0^{Q_1} + f_2^{P_3} f_1^{Q_1} = 1\cdot 2 + 8\cdot 1 = 10 = 8P_3^{\times,6}(3P_2^{\times,4}, 2P_2^{\times,3}) + 2P_3^{\times,6}(8P_2^{\times,3}),$$

$$f_4^{\times} = f_3^{P_3} f_1^{Q_1} = 1\cdot 1 = 1 = P_4^{\times,12}.$$

According to the last equality can obtain the structural formula of the product:

$$P_3^6(8P_2^4)\times Q_1^2 = P_4^{\times,12}[8P_3^{\times,6}(3P_2^{\times,4}, 2P_2^{\times,3}), 2P_3^{\times,6}(8P_2^{\times,3})]. \tag{6.13}$$

Thus, the product of an octahedron by a segment is a polytope *4D,* which includes 8 triangular prisms and two octahedrons. One can call the polytope (6.13) an octahedral prism.

The polytope (6.13) is an irregular four-dimensional polytope, since it contains various three-dimensional and two-dimensional figures and, therefore (see Chapter 5), it is limited to a three-dimensional surface.

The Polytope which is the Product of an Icosahedron by a Segment

The structural formula of an icosahedron is $P_3^{12}(20P_2^3)$. It is necessary to find the structural formula of the product $P_3^{12}(20P_2^3)\times Q_1^2$. According to theorem 6 and formulas (6.5), one can find

$$f_0^{\times} = f_0^{P_3} f_0^{Q_1} = 12\cdot 2 = 24,$$

$$f_1^{\times} = f_1^{P_3} f_0^{Q_1} + f_0^{P_3} f_1^{Q_1} = 30\cdot 2 + 12\cdot 1 = 72,.$$

$$f_2^{\times} = f_2^{P_3} f_0^{Q_1} + f_1^{P_3} f_1^{Q_1} = 20\cdot 2 + 30\cdot 1 = 70 = 40P_2^{\times,3} + 30P_2^{\times,4},$$

$$f_3^{\times} = f_3^{P_3} f_0^{Q_1} + f_2^{P_3} f_1^{Q_1} = 1 \cdot 2 + 20 \cdot 1 = 22 = 20P_3^{\times,6}(3P_2^{\times,4}, 2P_2^{\times,3}) + 2P_3^{\times,12},$$

$$f_4^{\times} = f_3^{P_3} f_1^{Q_1} = 1 \cdot 1 = 1 = P_4^{\times,12}.$$

According to the last equality, one can to obtain the structural formula of the product

$$P_3^{12}(20P_2^3) \times Q_1^2 = P_4^{\times,24}[20P_3^{\times,6}(3P_2^{\times,4}, 2P_2^{\times,3}), 2P_3^{\times,12}(20P_2^{\times,3})]. \quad (6.14)$$

Thus, the product of an icosahedron by a segment is a polytope 4*D*, which includes 20 triangular prisms and 2 icosahedrons. One can call a polytope (6.14) icosahedral prism.

The polytope (6.14) is an irregular four-dimensional polytope, since it contains various three-dimensional and two-dimensional figures and, therefore (see Chapter 5), it is limited to a three-dimensional surface.

The Polytope which is the Product of a Dodecahedron on a Segment

The structural formula of a dodecahedron is $P_3^{20}(12P_2^5)$. It is necessary to find the structural formula of product $P_3^{20}(12P_2^5) \times Q_1^2$. According to theorem 6.1 and formulas (6.5), one can find:

$$f_0^{\times} = f_0^{P_3} f_0^{Q_1} = 20 \cdot 2 = 40,$$

$$f_1^{\times} = f_1^{P_3} f_0^{Q_1} + f_0^{P_3} f_1^{Q_1} = 30 \cdot 2 + 20 \cdot 1 = 80,.$$

$$f_2^{\times} = f_2^{P_3} f_0^{Q_1} + f_1^{P_3} f_1^{Q_1} = 12 \cdot 2 + 30 \cdot 1 = 54 = 24P_2^{\times,5} + 30P_2^{\times,4},$$

$$f_3^{\times} = f_3^{P_3} f_0^{Q_1} + f_2^{P_3} f_1^{Q_1} = 1 \cdot 2 + 12 \cdot 1 = 14 = 12P_3^{\times,10}(5P_2^{\times,4}, 2P_2^{\times,5}) + 2P_3^{\times,20}(12P_2^{\times,5}),$$

$$f_4^{\times} = f_3^{P_3} f_1^{Q_1} = 1 \cdot 1 = 1 = P_4^{\times,40}.$$

According to the last equality, one can to obtain the structural formula of the product

$$P_3^{20}(12P_2^5) \times Q_1^2 = P_4^{\times,40}[12P_3^{\times,10}(5P_2^{\times,4}, 2P_2^{\times,5}), 2P_3^{\times,20}(12P_2^{\times,5})]. \quad (6.15)$$

Thus, the product of a dodecahedron by a segment is a polytope 4*D* including 12 pentagonal prisms and 2 dodecahedrons. One can call the polytope (6.15) a dodecahedral prism.

The polytope (6.15) is an irregular four-dimensional polytope, since it contains various three-dimensional and two-dimensional figures and, therefore (see Chapter 5), it is limited to a three-dimensional surface.

The Polytope which is the Product of an Egyptian Pyramid by a Segment

The structural formula of an Egyptian pyramid is $P_3^5(4P_2^3, P_2^4)$. It is necessary to find the structural formula of product $P_3^5(4P_2^3, P_2^4)\times Q_1^2$. According to theorem 6.1 and formulas (6.5), one can find:

$$f_0^\times = f_0^{P_3} f_0^{Q_1} = 5 \cdot 2 = 10,$$

$$f_1^\times = f_1^{P_3} f_0^{Q_1} + f_0^{P_3} f_1^{Q_1} = 5\cdot 1 + 8\cdot 2 = 21,.$$

$$f_2^\times = f_2^{P_3} f_0^{Q_1} + f_1^{P_3} f_1^{Q_1} = 5\cdot 2 + 8\cdot 1 = 18 = 8P_2^{\times,3} + 10P_2^{\times,4},$$

$$f_3^\times = f_3^{P_3} f_0^{Q_1} + f_2^{P_3} f_1^{Q_1} = 1\cdot 2 + 5\cdot 1 = 7 = 2P_3^{\times,5} + 4P_2^{\times,6} + P_3^{\times,8},$$

$$f_4^\times = f_3^{P_3} f_1^{Q_1} = 1\cdot 1 = 1 = P_4^{\times,10}.$$

According to the last equality, one can obtain the structural formula of the product

$$P_3^5(4P_2^3, P_2^4)\times Q_1^2 = P_4^{\times,10}[4P_3^{\times,6}(3P_2^{\times,4}, 2P_2^{\times,3}), P_3^{\times,8}(6P_2^{\times,4}), 2P_3^{\times,5}(4P_2^{\times,3}, P_2^{\times,4})]. \quad (6.16)$$

Thus, the product of an Egyptian pyramid by segment is a polytope $4D$ including four triangular prisms, 1 quadrangular prism and 2 Egyptian pyramids (Figure 2). One can call the polytope a pyramidal prism.

Figure 2. The pyramidal prism

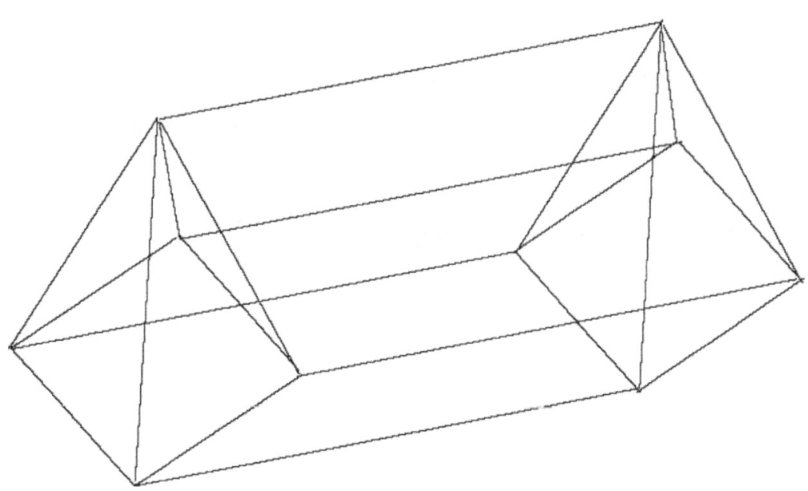

The polytope (6.16) is an irregular four-dimensional polytope, since it contains various three-dimensional and two-dimensional figures and, therefore (see Chapter 5), it is limited to a three-dimensional surface.

It is interesting that the pyramids of ancient Egypt had a strict geometric form (the pyramids of Cheops, Khafre, Mikerin) and were built in groups not far from each other (Snegirev & Franzev, 1938). In aggregate, this led, according to the analysis carried out, to an increase in the dimensionality of the space in the vicinity of the pyramids.

The Polytope which is the Product of a 4 - Simplex by a Segment

The structural formula of a 4 - simplex is $P_4^5[5P_3^4(4P_2^3)]$. It is necessary to find the structural formula of the product $P_4^5[5P_3^4(4P_2^3)]\times Q_1^2$. According to theorem 6.1 and formulas (6.5), one can find:

$$f_0^{\times} = f_0^{P_4} f_0^{Q_1} = 5 \cdot 2 = 10,$$

$$f_1^{\times} = f_1^{P_4} f_0^{Q_1} + f_0^{P_4} f_1^{Q_1} = 10 \cdot 2 + 5 \cdot 1 = 25,.$$

$$f_2^{\times} = f_2^{P_4} f_0^{Q_1} + f_1^{P_4} f_1^{Q_1} = 10 \cdot 2 + 10 \cdot 1 = 30 = 20P_2^{\times,3} + 10P_2^{\times,4},$$

$$f_3^{\times} = f_3^{P_4} f_0^{Q_1} + f_2^{P_4} f_1^{Q_1} = 5 \cdot 2 + 10 \cdot 1 = 20 = 10P_3^{\times,4}(4P_2^{\times,3}) + 10P_3^{\times,6}(3P_2^{\times,4}, 2P_2^{\times,3}),$$

$$f_4^{\times} = f_4^{P_4} f_0^{Q_1} + f_3^{P_4} f_1^{Q_1} = 1 \cdot 2 + 5 \cdot 1 = 7$$
$$= 5P_4^{\times,8}[4P_3^{\times,6}(3P_2^{\times,4}, 2P_2^{\times,3}), 2P_3^{\times,4}(4P_2^{\times,3})] + 2P_4^{\times,5}[5P_3^{\times,4}(4P_2^{\times,3})],$$

$$f_5^{\times} = f_4^{P_4} f_1^{Q_1} = 1 \cdot 1 = 1 = P_5^{\times,10}.$$

According to the last equality, one can obtain the structural formula of the product

$$P_4^5[5P_3^4(4P_2^3)]\times Q_1^2 = P_5^{\times,10}\{5P_4^{\times,8}[4P_3^{\times,6}(3P_2^{\times,4}, 2P_2^{\times,3}), 2P_3^{\times,4}(4P_2^{\times,3})], 2P_4^{\times,5}[5P_3^{\times,4}(4P_2^{\times,3})]\}. \quad (6.17)$$

Thus, the product of a 4 - simplex by a segment is a polytope *5D* including 5 the tetrahedral prisms (6.11) and 2 the 4 - simplexes. One can call the polytope (6.17) a 5 – simplex - prism. The image of the 5 – simplex - prism is shown in Figure 3.

Figure 3. The 5 – simplex – prism

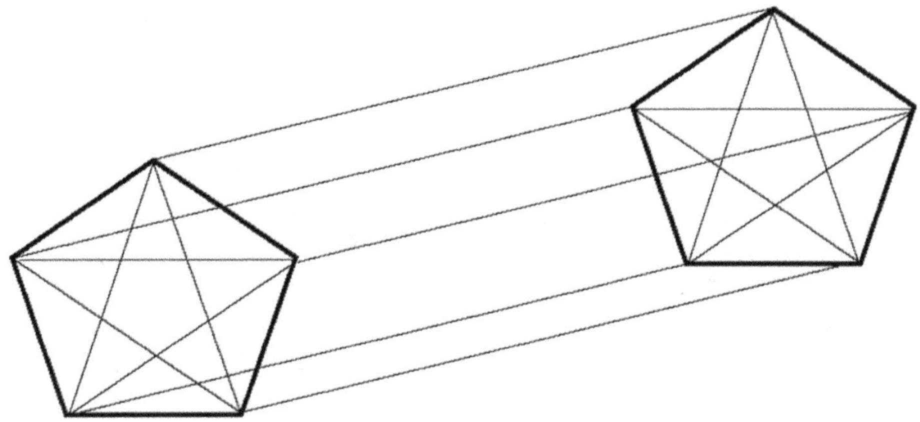

The polytope (6.17) is an irregular five-dimensional polytope, since it contains various four-dimensional, three-dimensional, and two-dimensional figures and, therefore (see Chapter 5), it is limited to a four-dimensional surface.

The Polytope which is the Product of a 4 – Cross - Polytope by a Segment

The structural formula of a 4 – cross - polytope is $P_4^8[16P_3^4(4P_2^3)]$. It is necessary to find the structural formula of the product $P_4^8[16P_3^4(4P_2^3)]\times Q_1^2$. According to theorem 6.1 and formulas (6.5), there are:

$$f_0^{\times} = f_0^{P_4} f_0^{Q_1} = 8 \cdot 2 = 16,$$

$$f_1^{\times} = f_1^{P_4} f_0^{Q_1} + f_0^{P_4} f_1^{Q_1} = 24 \cdot 2 + 8 \cdot 1 = 56,.$$

$$f_2^{\times} = f_2^{P_4} f_0^{Q_1} + f_1^{P_4} f_1^{Q_1} = 32 \cdot 2 + 24 \cdot 1 = 88 = 64P_2^{\times,6} + 24P_2^{\times,4},$$

$$f_3^{\times} = f_3^{P_4} f_0^{Q_1} + f_2^{P_4} f_1^{Q_1} = 16 \cdot 2 + 32 \cdot 1 = 64 = 32P_3^{\times,4}(4P_2^{\times,3}) + 32P_3^{\times,6}(3P_2^{\times,4}, 2P_2^{\times,3}),$$

$$f_4^{\times} = f_4^{P_4} f_0^{Q_1} + f_3^{P_4} f_1^{Q_1} = 1 \cdot 2 + 16 \cdot 1 = 18$$
$$= 16P_4^{\times,8}[4P_3^{\times,6}(3P_2^{\times,4}, 2P_2^{\times,3}), 2P_3^{\times,4}(4P_2^{\times,3})] + 2P_4^{\times,8}[16P_3^{\times,4}(4P_2^{\times,3})],$$

$$f_5^{\times} = f_4^{P_4} f_1^{Q_1} = 1 \cdot 1 = 1 = P_5^{\times,16}.$$

According to the last equality, we obtain the structural formula of the product

$$P_4^8[16P_3^4(4P_2^3)]\times Q_1^2 = P_5^{\times,16}\{16P_4^{\times,8}[4P_3^{\times,6}(3P_2^{\times,4}, 2P_2^{\times,3}), 2P_3^{\times,4}(4P_2^{\times,3})], 2P_4^{\times,8}[16P_3^{\times,4}(4P_2^{\times,3})]\}. \quad (6.18)$$

Thus, the product of a 4 – cross - polytope by a segment is a polytope $5D$ including 16 the tetrahedral prisms (6.11) and 2 the 4 – cross - polytopes. We can call the polytope (6.18) a 5 – cross - prism (Figure 4).

Figure 4. The 5 - cross – prism

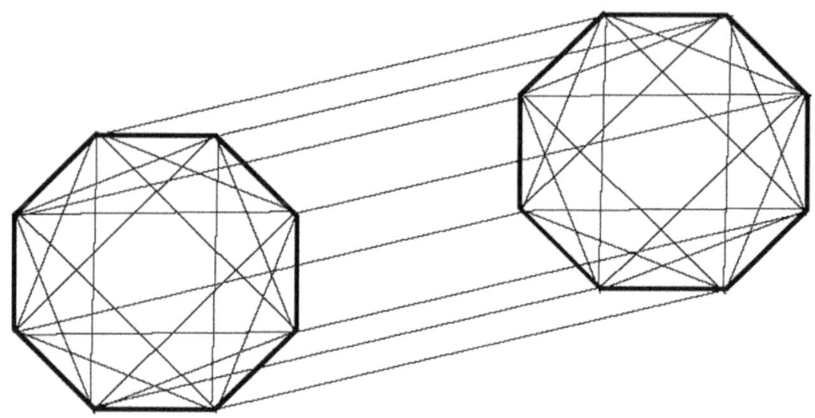

The polytope (6.18) is an irregular five-dimensional polytope, since it contains various four-dimensional, three-dimensional, and two-dimensional figures and, therefore (see Chapter 5), it is limited to a four-dimensional surface.

The General Structural Formula of Polytopes which are the Products Polytopes by a One-Dimensional Segment

One will call prisms with bases in the form of polytopes the polytopic prisms.

Theorem 6.4. The product of a polytope by a segment, if all its faces of dimension 2 are equal, is a polytopic prism with the structure defined by equation:

$$P_n^{K_n}\{f_{n-1}P_{n-1}^{K_{n-1}}...[f_3P_3^{K_3}(f_2P_2^{K_2})]...\}\times Q_1^2 = P_{n+1}^{2K_n}\{f_{n-1}P_n^{2K_{n-1}}...[f_3P_4^{2K_3}(f_2P_3^{2K_2},2P_2^{K_2})2P_3^{K_3}]...2P_n^{K_n}\}. \quad (6.19)$$

There $P_n^{K_n}$ is the polytope of dimension n with the number of vertices K_n;

$P_{n-1}^{K_{n-1}}$ are hyper - faces with the number f_{n-1}, each having K_{n-1} vertices; accordingly $P_2^{K_2}$ are hyper - faces of dimension 2 at faces $P_3^{K_3}$; f_2 is the number of faces $P_2^{K_2}$ with the number of vertices K_2 and so on; $P_{n+1}^{2K_n}$ is a polytopic prism of dimension $n+1$ with the number of vertices $2K_n$, the bases of which are the polytopes $P_n^{K_n}$. The polytopic prism $P_{n+1}^{2K_n}$ includes f_{n-1} polytopic prisms of dimension n and the number of the vertices $2K_{n-1}$, and also two polytopes $P_n^{K_n}$ which are the bases of polytopic prisms and so on, up to prisms $P_3^{2K_2}$ with a number f_2 and with bases $P_2^{K_2}$.

Proof. Theorem 6.1 follows from the carried out examinations of the product of polytopes of different dimensions by a segment. General laws which allowed formulate the theorem can be traced in all these products. Whatever the dimension n of the polytope was, at multiplying it by a segment from each vertex of the polytope this segment as if grows. This forms polytopic prism, the bases of which are two copies of the initial polytope. The dimension of this polytopic prism is by 1 more than the dimension of the initial polytope, i. e. $n + 1$. This polytopic prism includes polytopic prisms built on the hyper - faces of the initial polytope and their number is equal to the number of hyper-faces of the initial polytope f_{n-1}. The dimension of these politopic prisms is n, since the dimension of the initial polytope hyper - faces is equal to n - 1, but when multiplied by a segment the dimension is increased by 1. In addition, a polytopic prism of dimension $n + 1$ consists of two copies of the initial polytope of dimension n, as the bases of the prism. In its turn, the polytopic prisms of dimension n include the polytopic prisms of dimension n - 1, built on the hyper - faces of the polytope of dimension n - 1, and two copies of the polytope of dimension n - 1. This process of the polytopic prisms inclusion in each other happens until the formation of three - dimensional prisms built on two - dimensional facets of the initial polytope.

The equality (6.19) is an analytical expression of these constructions. This process of formation occurs when all two - dimensional faces of the initial polytope are the same, i. e. they are equal polygons. If the initial polytope has two - dimensional faces with different numbers of vertices so the analytical record of the product this polytope by a segment will be added by prisms built on different bases. The ratios (6.19) do not include such options of the polytopic prisms. Structural formulas for the products of irregular polytopes on a one-dimensional segment can be obtained using the proposed mechanism for calculating the product of polytopes, for example, as it is done in this chapter for the Egyptian pyramid.

Theorem 6.4 it is proved.

Supplement. Values of f_i in (6.19) depend on the form and dimension of the polytope. According to Zhizhin (2019a) the formulas for their calculation in the case of a simplex, a cube and a cross - polytope will be the following: for simplex $f_i(n) = C_{n+1}^{i+1}$; i=0,1,…,n; for cube $f_i(n) = 2^{n-i} C_n^i$; i=0,1,…,n; for cross – polytope $f_i(n) = 2^{1+i} C_n^{n-1-i}$, i=0,1,…,n.

Naturally, the results of research about products of polytopes by a segment are to be taken into account when analyzing the products of polytopes by more complex figures, since they all have edges are factors in these products.

The products of both regular and irregular polytopes on a one-dimensional segment are irregular polytopes. All of them have a boundary surface with a dimension equal to the dimension of the original polytope as a factor.

THE POLYTOPES WHICH ARE THE PRODUCTS OF POLYTOPES BY A TRIANGLE

The Polytope which is the Product of a Triangle by a Triangle

There we consider the product $P_2^3 \times Q_2^3$. According to theorem 6.1 and ratios (6.5) there are:

$$f_0^{\times} = f_0^{P_2} f_0^{Q_2} = 3 \cdot 3 = 9,$$

$$f_{\times}^1 = f_1^{P_2} f_0^{Q_2} + f_0^{P_2} f_1^{Q_2} = 3 \cdot 3 + 3 \cdot 3 = 18,$$

$$f_2^{\times} = f_2^{P_2} f_0^{Q_2} + f_1^{P_2} f_1^{Q_2} + f_0^{P_2} f_2^{Q_2} = 1 \cdot 3 + 3 \cdot 3 + 3 \cdot 1 = 15 = 6P_2^{\times,3} + 9P_2^{\times,4},$$

$$f_3^{\times} = f_2^{P_2} f_1^{Q_2} + f_1^{P_2} f_2^{Q_2} = 1 \cdot 3 + 3 \cdot 1 = 6 = 6P_3^{\times,6},$$

$$f_4^{\times} = f_2^{P_2} f_2^{Q_2} = 1 \cdot 1 = 1 = P_4^{\times,9}.$$

Thus, the product of a triangle by a triangle is 4 - polytope having 9 vertices and 6 triangle prisms. The existence of such polytope was proved of Ziegler (1995) independently.

In particular, the fundamental difference between this polytope and simplicial polytopes and between this polytope and three - dimensional polyhedrons because of its multidimensionality is indicated of Ziegler (1995). The structural formula of this polytope (the triangular prismahedron) has the form:

$$P_2^3 \times Q_2^3 = P_4^9[6P_3^6(3P_2^4, 2P_2^3)]. \tag{6.20}$$

The image of a triangular prismahedron (Zhizhin, 2017) is shown in Figure 5.

Figure 5. The triangular prismahedron

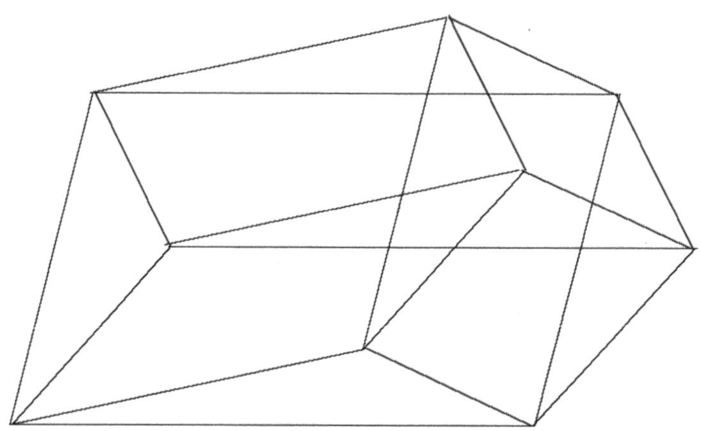

The four-dimensional polytope in Figure 5 is incorrect, since it includes various geometric shapes and, therefore, an external three-dimensional surface can be distinguished in it, for example, consisting of the outer border of two trigonal prisms attached to each other on a common base *abfcde* and *abcdhg*.

The Polytope which is the Product of a Polygon by a Triangle

According to theorem 6.1 and ratios (6.5) for the product $P_2^i \times Q_2^3$ (i is the number of the polygon sides) there are:

$$f_0^{\times} = f_0^{P_2} f_0^{Q_2} = i \cdot 3 = 3i,$$

$$f_1^{\times} = f_1^{P_2} f_0^{Q_2} + f_0^{P_2} f_1^{Q_2} = i \cdot 3 + 3 \cdot i = 6i,.$$

$$f_2^{\times} = f_2^{P_2} f_0^{Q_2} + f_1^{P_2} f_1^{Q_2} + f_0^{P_2} f_2^{Q_2} = 1 \cdot 3 + i \cdot 3 + i \cdot 1 = 3 + 4i = 3P_2^{\times,i} + 3iP_2^{\times,4} + iP_2^{\times,3},$$

$$f_3^{\times} = f_2^{P_2} f_1^{Q_2} + f_1^{P_2} f_2^{Q_2} = 1 \cdot 3 + i \cdot 1 = 3 + i = iP_3^{\times,6} + 3P_3^{\times,2i},$$

$$f_4^{\times} = f_2^{P_2} f_2^{Q_2} = 1 \cdot 1 = 1 = P_4^{\times,3i}.$$

Thus, the product of the polygon P_2^i by the triangle Q_2^3 is a 4 - polytope having $3i$ vertices, 3 prisms with bases P_2^i and *i* prisms with triangle bases. The structural formula of this product has the form:

$$P_2^i \times Q_2^3 = P_4^{\times 3i}[3P_3^{\times 2i}(\mathrm{i}\, P_2^{\times,4}, 2P_2^{\times,3}), \mathrm{i}\, P_3^{\times,6}(3P_2^{\times,4}, 2P_2^{\times,3})]. \quad (6.21)$$

Examples of products of a square, a pentagon and a hexagon by a triangle are shown in Figures 6 –8.

*Figure 6. The 4*3 – angular prismahedron*

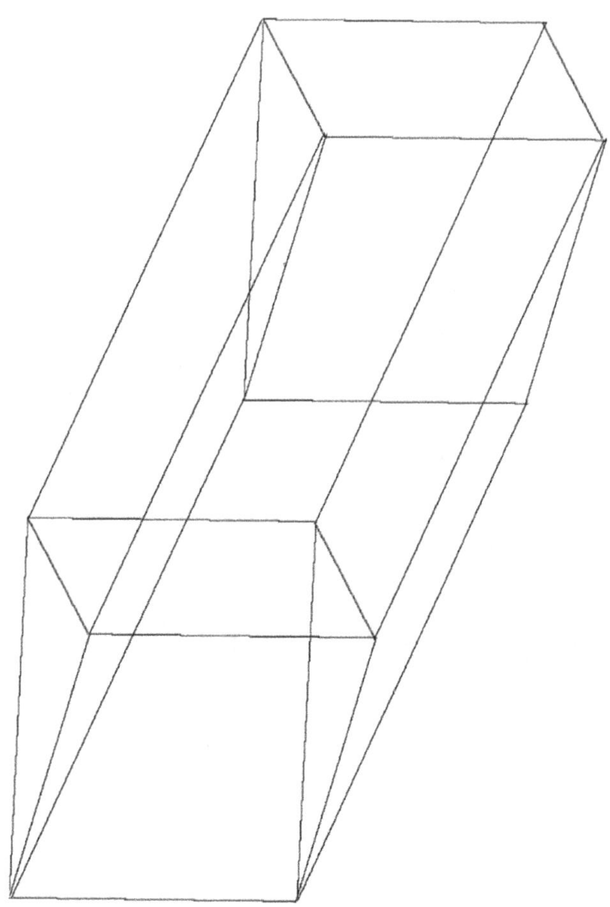

*Figure 7. The 5*3 – angular prismahedron*

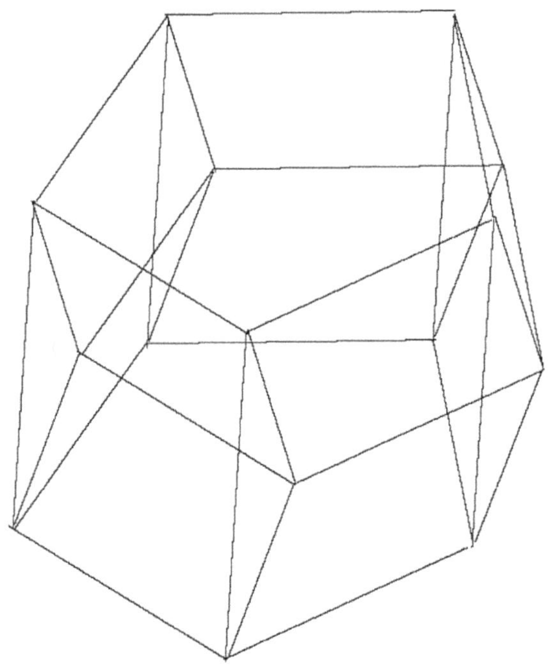

*Figure 8. The 6*3 – angular prismahedron*

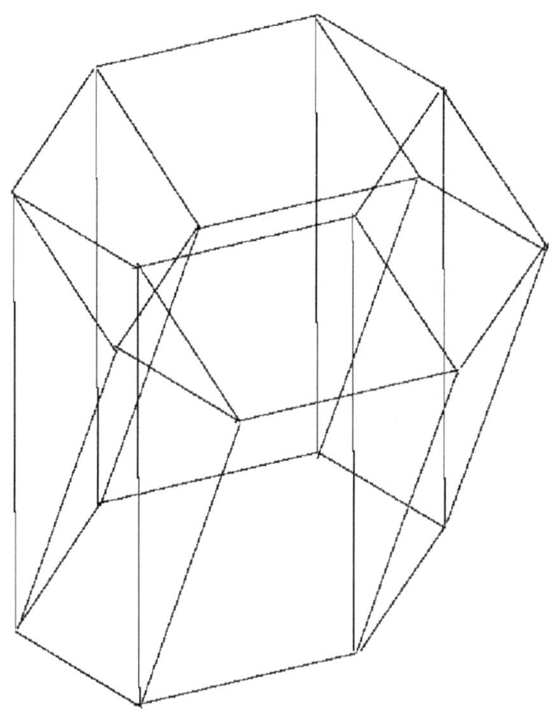

The polytope (6.21) is a generalization of the polytope (6.20) in the case when one of the multiplied polygons has the number of sides greater than 3. This polytope in the general case can be called 3 *i - angular prismahedron. Therefore, the polytope (6.21) can be described more accurately as 3*3 - angular prismahedron.

The four-dimensional polytopes in Figures 6 – 8 are incorrect, since it includes various geometric shapes and, therefore, an external three-dimensional surface can be distinguished in it.

The Polytope which is the Product of a Tetrahedron by a Triangle

According to theorem 6.1 and ratios (6.5) for the numbers of elements of different dimension in the product $P_3^4(4P_2^3)\times Q_2^3$ there are:

$$f_0^{\times} = f_0^{P_3} f_0^{Q_2} = 4 \cdot 3 = 12,$$

$$f_1^{\times} = f_1^{P_3} f_0^{Q_2} + f_0^{P_3} f_1^{Q_2} = 6\cdot 3 + 3\cdot 4 = 30,.$$

$$f_2^{\times} = f_2^{P_3} f_0^{Q_2} + f_1^{P_3} f_1^{Q_2} + f_0^{P_3} f_2^{Q_2} = 4\cdot 1 + 6\cdot 3 + 4\cdot 1 = 26 = 8P_2^{\times,3} + 18P_2^{\times,4},$$

$$f_3^{\times} = f_3^{P_3} f_0^{Q_2} + f_2^{P_3} f_1^{Q_2} + f_1^{P_3} f_2^{Q_2} = 1\cdot 3 + 4\cdot 1 + 6\cdot 1 = 21 = 18P_3^{\times,6} + 3P_3^{\times,4},$$

$$f_4^{\times} = f_3^{P_3} f_1^{Q_2} + f_2^{P_3} f_2^{Q_2} = 1\cdot 3 + 4\cdot 1 = 7 = 3P_4^{\times,8} + 4P_4^{\times,9},$$

$$f_5^{\times} = f_3^{P_3} f_2^{Q_2} = 1\cdot 1 = 1 = P_5^{\times,12}.$$

Thus, the product of a tetrahedron by a triangle is a 5-polytope consisting from 3 tetrahedron prisms (6.11) and 4 triangular prismahedrons (6.20). The structural formula of this polytope has the form:

$$P_3^4(4P_2^3)\times Q_2^3 = P_5^{12}\{4P_4^9[6P_3^6(3P_2^4,2P_2^3)],2P_4^8[4P_3^6(3P_2^4,2P_2^3),2P_3^4(4P_2^3)]\}. \tag{6.22}$$

One can call the five-dimensional polytope (6.22) a tetrahedral prismahedron. The image of this polytope on a two-dimension plane is shown in Figure 9.

Figure 9. The tetrahedral prismahedron

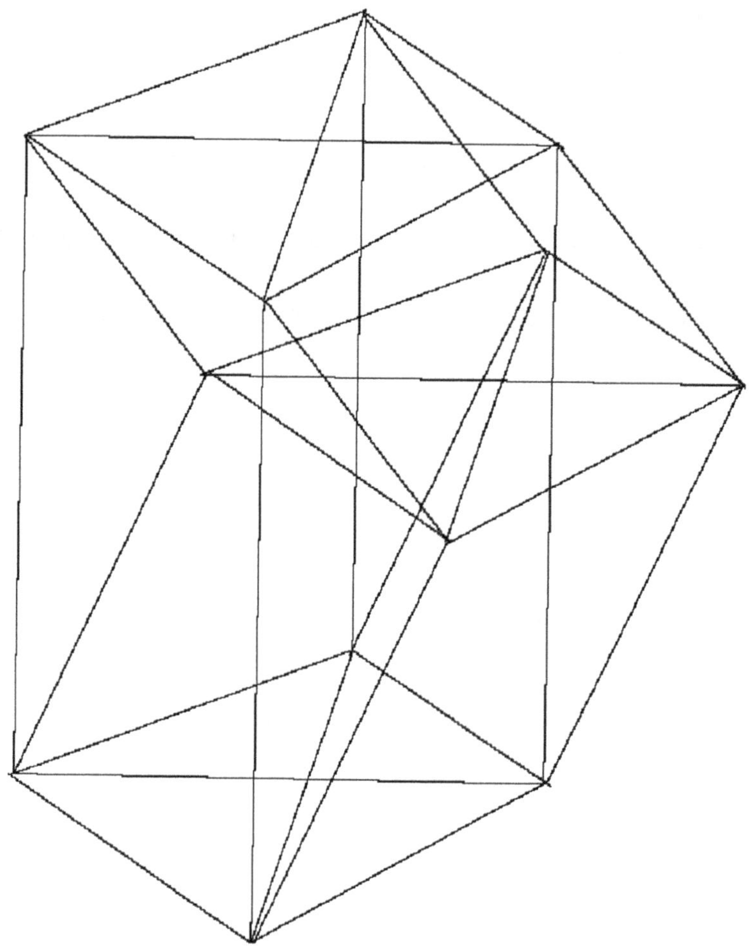

The five-dimensional polytope in Figure 9 is incorrect polytope since it includes various geometric shapes and, therefore, an external four-dimensional surface can be distinguished in it (see Chapter 5).

The Polytope which is the Product of an Octahedron by a Triangle

According to theorem 6.1 and ratios (6.5) for the numbers of elements of different dimension in the product $P_2^6(8P_2^3)\times Q_2^3$ there are:

$$f_0^{\times} = f_0^{P_3} f_0^{Q_2} = 6 \cdot 3 = 18,$$

$$f_1^{\times} = f_1^{P_3} f_0^{Q_2} + f_0^{P_3} f_1^{Q_2} = 12 \cdot 3 + 6 \cdot 3 = 54,.$$

$$f_2^{\times} = f_2^{P_3} f_0^{Q_2} + f_1^{P_3} f_1^{Q_2} + f_0^{P_3} f_2^{Q_2} = 8 \cdot 1 + 12 \cdot 3 + 6 \cdot 1 = 66 = 30P_2^{\times,3} + 36P_2^{\times,4},$$

$$f_3^{\times} = f_3^{P_3} f_0^{Q_2} + f_2^{P_3} f_1^{Q_2} + f_1^{P_3} f_2^{Q_2} = 1\cdot 3 + 8\cdot 3 + 12\cdot 1 = 39$$
$$= 36P_3^{\times,6}(3P_2^4, 2P_2^3) + 3P_3^{\times,6}(8P_2^{\times,3}),$$

$$f_4^{\times} = f_3^{P_3} f_1^{Q_2} + f_2^{P_3} f_2^{Q_2} = 1\cdot 3 + 8\cdot 1 = 11$$
$$= 3P_4^{\times,12}[8P_3^{\times,6}(3P_2^{\times,4}, 2P_2^{\times,3}), 2P_3^{\times,6}(8P_2^{\times,3})] + 8P_4^{\times,9}[6P_3^{\times,6}(3P_2^{\times,4}, 2P_2^{\times,3})],$$

$$f_5^{\times} = f_3^{P_3} f_2^{Q_2} = 1\cdot 1 = 1 = P_5^{\times,18}.$$

Thus, the product of an octahedron by a triangle is a 5 - polytope including 8 the 3*3 - angular prismahedrons (6.20) and 3 the octahedral prisms (6.13).
The structural formula of the polytope P_5^{18} has the form:

$$P_2^6(8P_2^3)\times Q_2^3 = P_5^{\times,18}\{8P_4^{\times,9}[6P_3^{\times,6}(3P_2^{\times,4}, 2P_2^{\times,3}), 3P_3^{\times,12}[8P_3^{\times,6}(3P_2^{\times,4}, 2P_2^{\times,3}), 2P_3^{\times,6}(8P_2^{\times,3})]\}. \tag{6.23}$$

One can call the polytope (6.23) an octahedral prismahedron. The image of the polytope P_5^{18} is shown in Figure 10.

Figure 10. The octahedral prismahedron

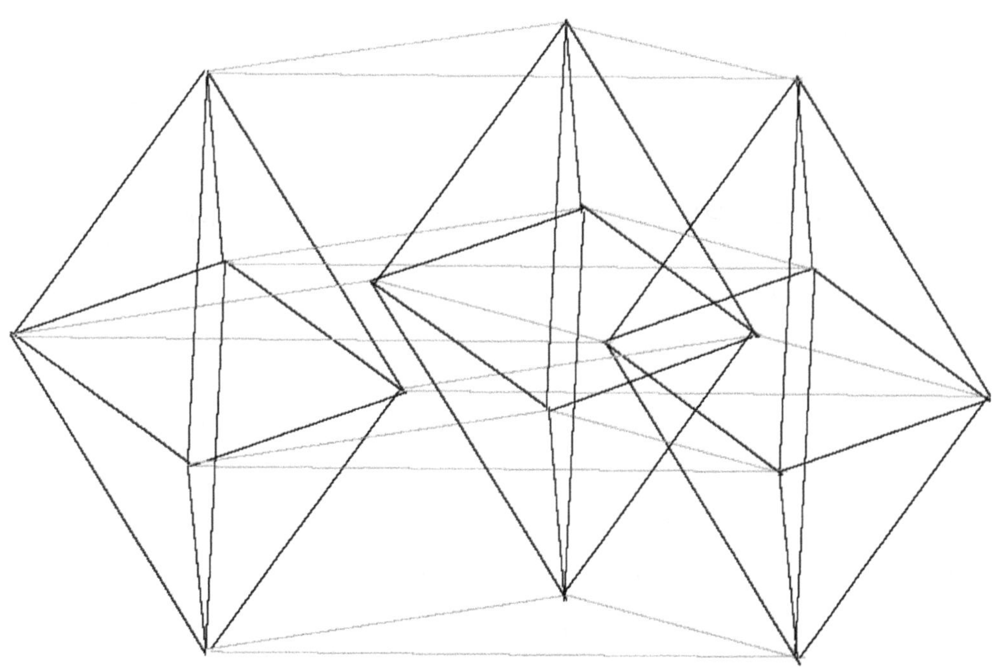

The five-dimensional polytope in Figure 10 is incorrect polytope since it includes various geometric shapes and, therefore, an external four-dimensional surface can be distinguished in it (see Chapter 5).

The Polytope which is the Product of a Cube by a Triangle

According to theorem 6.1 and ratios (6.5) for the numbers of elements of different dimension in the product $P_3^8(6P_2^4)\times Q_2^3$ there are:

$$f_0^\times = f_0^{P_3} f_0^{Q_2} = 8 \cdot 3 = 24,$$

$$f_1^\times = f_1^{P_3} f_0^{Q_2} + f_0^{P_3} f_1^{Q_2} = 12\cdot 3 + 8\cdot 3 = 60,.$$

$$f_2^\times = f_2^{P_3} f_0^{Q_2} + f_1^{P_3} f_1^{Q_2} + f_0^{P_3} f_2^{Q_2} = 6\cdot 3 + 12\cdot 3 + 8\cdot 1 = 62 = 8P_2^{\times,3} + 54P_2^{\times,4},$$

$$f_3^\times = f_3^{P_3} f_0^{Q_2} + f_2^{P_3} f_1^{Q_2} + f_1^{P_3} f_2^{Q_2} = 1\cdot 3 + 6\cdot 3 + 12\cdot 1 = 33 = 21P_3^{\times,8} + 12P_3^{\times,6},$$

$$f_4^\times = f_3^{P_3} f_1^{Q_2} + f_2^{P_3} f_2^{Q_2} = 1\cdot 3 + 6\cdot 1 = 9 = 3P_4^{\times,16} + 6P_4^{\times,12},$$

$$f_5^\times = f_3^{P_3} f_2^{Q_2} = 1\cdot 1 = 1 = P_5^{\times,24}.$$

Thus, the product of a cube by a triangle is a 5 - polytope including 3 the hyper - cubes (6.12) and 6 the 3*4 - angular prismahedrons (6.21).
The structural formula of the polytope P_5^{24} has the form:

$$P_3^8(6P_2^4)\times Q_2^3 = P_5^{\times,24}\{3P_4^{\times,16}(80P_2^{\times,4}), 6P_4^{\times,12}[3P_3^{\times,8}(4P_2^{\times,4}, 2P_2^{\times,3}), 4P_3^{\times,6}(3P_2^{\times,4}, 2P_2^{\times,3})]\}. \quad (6.24)$$

One can call the polytope (6.24) a cube - polytopic prismahedron. The image of the polytope P_5^{24} is shown in Figure 11.

Figure 11. The cube - polytopic prismahedron

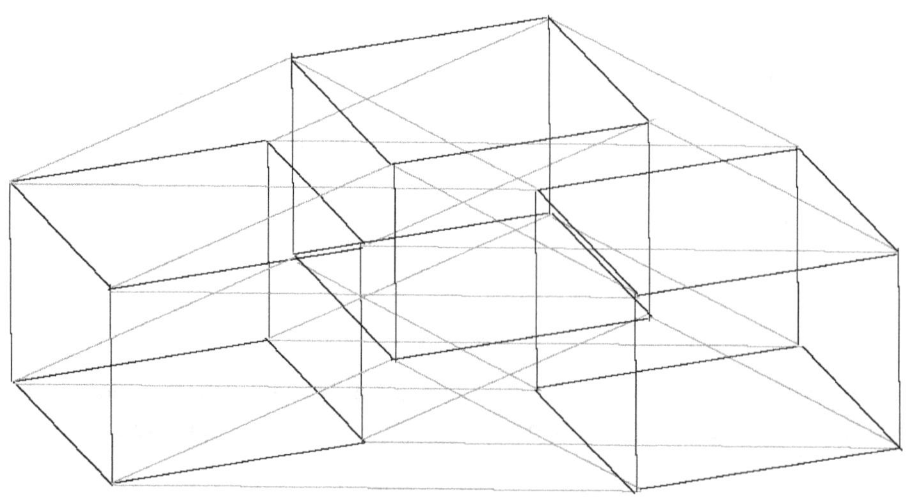

The five-dimensional polytope in Figure 11 is incorrect polytope since it includes various geometric shapes and, therefore, an external four-dimensional surface can be distinguished in it (see Chapter 5).

The Polytope which is the Product of a Dodecahedral by a Triangle

According to theorem 6.1 and ratios (6.5) for the numbers of elements of different dimension in the product $P_3^{20}(12P_2^5)\times Q_2^3$ there are:

$$f_0^{\times} = f_0^{P_3} f_0^{Q_2} = 20 \cdot 3 = 60,$$

$$f_1^{\times} = f_1^{P_3} f_0^{Q_2} + f_0^{P_3} f_1^{Q_2} = 30\cdot 3 + 20\cdot 3 = 150,$$

$$f_2^{\times} = f_2^{P_3} f_0^{Q_2} + f_1^{P_3} f_1^{Q_2} + f_0^{P_3} f_2^{Q_2} = 12\cdot 3 + 30\cdot 3 + 20\cdot 1 = 146 = 36P_2^{\times,5} + 90P_2^{\times,4} + 20P_2^{\times,3},$$

$$f_3^{\times} = f_3^{P_3} f_0^{Q_2} + f_2^{P_3} f_1^{Q_2} + f_1^{P_3} f_2^{Q_2} = 1\cdot 3 + 12\cdot 3 + 30\cdot 1 = 69 = 3P_3^{\times,20} + 36P_3^{\times,10} + 30P_3^{\times,6},$$

$$f_4^{\times} = f_3^{P_3} f_1^{Q_2} + f_2^{P_3} f_2^{Q_2} = 1\cdot 3 + 12\cdot 1 = 15 = 3P_4^{\times,40} + 12P_4^{\times,15},$$

$$f_5^{\times} = f_3^{P_3} f_2^{Q_2} = 1\cdot 1 = 1 = P_5^{\times,60}.$$

Thus, the product of a dodecahedron by a triangle is a 5 - polytope including 3 the dodecahedral prisms (6.15) and 12 the 3*5 - angular prismahedrons (6.21).
The structural formula of the polytope P_5^{60} has the form:

$$P_5^{\times,60}\{3P_4^{\times,40}[12P_3^{\times,10}(5P_2^{\times,4},2P_2^{\times,3}),2P_3^{\times,20}(12P_2^{\times,5})],12P_4^{\times,15}[3P_3^{\times,10}(5P_2^{\times,4},2P_2^{\times,5}),5P_3^{\times,6}(3P_2^{\times,4},2P_2^{\times,3})]\}. \quad (6.25)$$

One can call the polytope (6.25) a dodeca - polytopic prismahedron.
The five-dimensional polytope (6.25) is incorrect polytope since it includes various geometric shapes and, therefore, an external four-dimensional surface can be distinguished in it (see Chapter 5).

The Polytope which is the Product of an Icosahedron by a Triangle

According to theorem 6.1 and ratios (6.5) for the numbers of elements of different dimension in the product $P_3^{12}(20P_2^5)\times Q_2^3$ there are:

$$f_0^{\times} = f_0^{P_3} f_0^{Q_2} = 12 \cdot 3 = 36,$$

$$f_1^{\times} = f_1^{P_3} f_0^{Q_2} + f_0^{P_3} f_1^{Q_2} = 30\cdot 3 + 12\cdot 3 = 126,.$$

$$f_2^{\times} = f_2^{P_3} f_0^{Q_2} + f_1^{P_3} f_1^{Q_2} + f_0^{P_3} f_2^{Q_2} = 20\cdot 3 + 30\cdot 3 + 12\cdot 1 = 162 = 90P_2^{\times,4} + 72P_2^{\times,3},$$

$$f_3^{\times} = f_3^{P_3} f_0^{Q_2} + f_2^{P_3} f_1^{Q_2} + f_1^{P_3} f_2^{Q_2} = 1\cdot 3 + 20\cdot 3 + 30\cdot 1 = 93 = 3P_3^{\times,12} + 90P_3^{\times,6},$$

$$f_4^{\times} = f_3^{P_3} f_1^{Q_2} + f_2^{P_3} f_2^{Q_2} = 1\cdot 3 + 20\cdot 1 = 23 = 3P_4^{\times,24} + 20P_4^{\times,9},$$

$$f_5^{\times} = f_3^{P_3} f_2^{Q_2} = 1\cdot 1 = 1 = P_5^{\times,36}.$$

Thus, the product of an icosahedron by a triangle is a 5 - polytope including 3 the icosahedral prisms (6.14) and 20 the 3*3 - angular prismahedrons (6.21).
The structural formula of the polytope P_5^{36} has the form:

$$P_5^{\times,36}\{20P_4^{\times,9}[6P_3^{\times,6}(3P_2^{\times,4},2P_2^{\times,3})],3P_4^{\times,24}[20P_3^{\times,6}(3P_2^{\times,4},2P_2^{\times,3}),2P_3^{\times,12}(20P_2^{\times,3})]\}. \quad (6.26)$$

One can call the polytope (6.26) an icosa - polytopic prismahedron.

The five-dimensional polytope (6.26) is incorrect polytope since it includes various geometric shapes and, therefore, an external four-dimensional surface can be distinguished in it (see Chapter 5). This last statement is confirmed by the polytope structural formula (6.26).

The Polytope which is the Product of a 4 - Simplex by a Triangle

According to theorem 6.1 and ratios (6.5) for the numbers of elements of different dimension in the product $P_3^5[5P_3^4(4P_2^3)]\times Q_2^3$ there are:

$$f_0^{\times} = f_0^{P_4} f_0^{Q_2} = 5 \cdot 3 = 15,$$

$$f_1^{\times} = f_1^{P_4} f_0^{Q_2} + f_0^{P_4} f_1^{Q_2} = 10\cdot 3 + 5\cdot 3 = 45,.$$

$$f_2^{\times} = f_2^{P_4} f_0^{Q_2} + f_1^{P_4} f_1^{Q_2} + f_0^{P_4} f_2^{Q_2} = 10\cdot 3 + 10\cdot 3 + 5\cdot 1 = 65 = 30P_2^{\times,4} + 35P_2^{\times,3},$$

$$f_3^{\times} = f_3^{P_4} f_0^{Q_2} + f_2^{P_4} f_1^{Q_2} + f_1^{P_4} f_2^{Q_2} = 5\cdot 3 + 10\cdot 3 + 10\cdot 1 = 55 = 15P_3^{\times,4} + 40P_3^{\times,6},$$

$$f_4^{\times} = f_4^{P_4} f_0^{Q_2} + f_3^{P_4} f_1^{Q_2} + f_2^{P_4} f_2^{Q_2} = 1\cdot 3 + 5\cdot 3 + 10\cdot 1 = 28 = 3P_4^{\times,5} + 15P_4^{\times,8} + 10P_4^{\times,9},$$

$$f_5^{\times} = f_4^{P_4} f_1^{Q_2} + f_3^{P_4} f_2^{Q_2} = 1\cdot 3 + 5\cdot 1 = 8 = 3P_5^{\times,10} + 5P_5^{\times,12},$$

$$f_6^{\times} = f_4^{P_4} f_2^{Q_2} = 1\cdot 1 = 1 = P_6^{\times,15}.$$

Thus, the product of a 4 - simplex by a triangle is a 6-polytope including 3 the 5 – simplex - polytopic prisms (6.16) and 5 the tetrahedral prismahedrons (6.22).

The structural formula of the polytope P_6^{15} ,if we restrict ourselves to listing elements of the sixth, fifth, and fourth dimensions, has the form:

$$P_6^{\times,15}\{3P_5^{\times,10}(5P_4^{\times,8},2P_4^{\times,5}),5P_5^{\times,12}(4P_4^{\times,9},3P_4^{\times,8})\}. \quad (6.27)$$

One can call the polytope (6.27) a 6 – simplex - polytopic prismahedron. The image of this polytope is shown in Figure 12.

Figure 12. The 6 – simplex - polytopic prismahedron

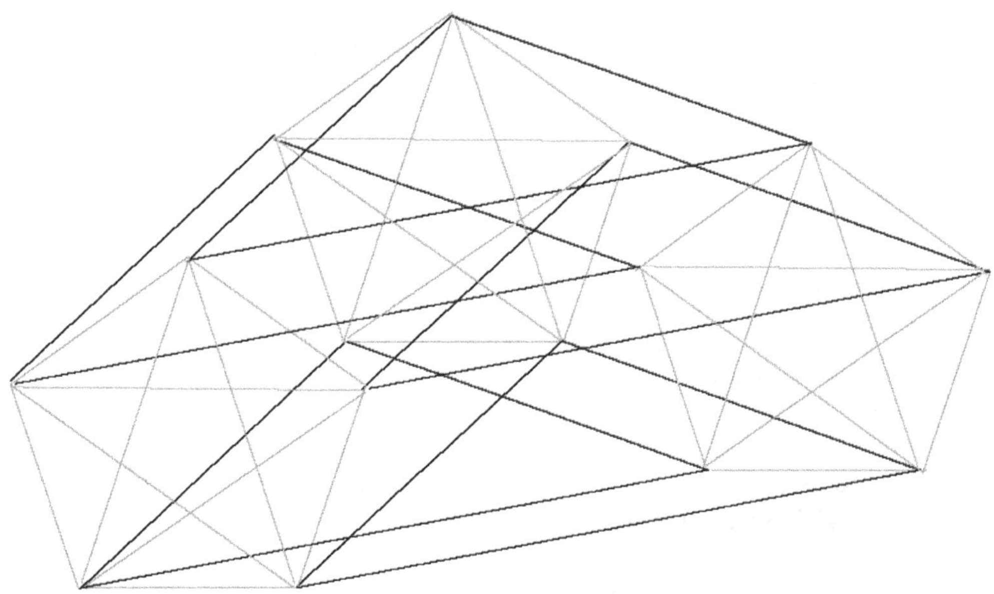

The six-dimensional polytope (6.27) is incorrect polytope since it includes various geometric shapes and, therefore, an external five-dimensional surface can be distinguished in it (see Chapter 5). This last statement is confirmed by the polytope structural formula (6.27).

The Polytope which is the Product of a 4 – Cross - Polytope by a Triangle

According to theorem 6.1 and ratios (6.5) for the numbers of elements of different dimension in the product $P_4^8[16P_3^4(4P_2^3)]\times Q_2^3$ there are:

$$f_0^{\times} = f_0^{P_4} f_0^{Q_2} = 8 \cdot 3 = 24,$$

$$f_1^{\times} = f_1^{P_4} f_0^{Q_2} + f_0^{P_4} f_1^{Q_2} = 24 \cdot 3 + 8 \cdot 3 = 96,$$

$$f_2^{\times} = f_2^{P_4} f_0^{Q_2} + f_1^{P_4} f_1^{Q_2} + f_0^{P_4} f_2^{Q_2} = 32 \cdot 3 + 24 \cdot 3 + 8 \cdot 1 = 176 = 72 P_2^{\times,4} + 104 P_2^{\times 3},$$

$$f_3^{\times} = f_3^{P_4} f_0^{Q_2} + f_2^{P_4} f_1^{Q_2} + f_1^{P_4} f_2^{Q_2} = 16 \cdot 3 + 32 \cdot 3 + 24 \cdot 1 = 168 = 48 P_3^{\times,4} + 120 P_3^{\times,6},$$

$$f_4^{\times} = f_4^{P_4} f_0^{Q_2} + f_3^{P_4} f_1^{Q_2} + f_2^{P_4} f_2^{Q_2} = 1 \cdot 3 + 16 \cdot 3 + 32 \cdot 1 = 83 = 51 P_4^{\times,8} + 32 P_4^{\times,9},$$

$$f_5^{\times} = f_4^{P_4} f_1^{Q_2} + f_3^{P_4} f_2^{Q_2} = 1 \cdot 3 + 16 \cdot 1 = 19 = 3 P_5^{\times,16} + 16 P_5^{\times 12},$$

$$f_6^{\times} = f_4^{P_4} f_2^{Q_2} = 1 \cdot 1 = 1 = P_6^{\times 24}.$$

Thus, the product of a 4 – cross - polytope by a triangle is a 6-polytope including 3 the 5 – cross -polytopic prisms (6.13) and 16 the tetrahedral prismahedrons (6.6).

The structural formula of the polytope P_{24}^6, we restrict ourselves to listing elements of the sixth, fifth, and fourth dimensions, has the form:

$$P_6^{\times,24}\{3P_5^{\times,16}(16P_4^{\times,8}, 2P_4^{\times,8}), 16P_5^{\times,12}(4P_4^{\times,9}, 3P_4^{\times,8})\}. \tag{6.28}$$

One can call the polytope a 6 - cross - polytopic prismahedron. The six-dimensional polytope (6.28) is incorrect polytope since it includes various geometric shapes and, therefore, an external five-dimensional surface can be distinguished in it (see Chapter 5). This last statement is confirmed by the polytope structural formula (6.28).

The General Structural Formula for Polytopes which are the Products of Polytopes by a Triangle

If we pay attention to a polytope taking into account its dimensions and the number of vertices and the number of its facets and the number of vertices of every facet, we can prove the general rule of changing these parameters of a polytope for its multiplication by a triangle.

Theorem 6.5. Let's assume that a convex polytope of dimension *n,* with the number of vertices *m*, has a number of facets ν with vertex number *k* for every facet. Then its product by a triangle is defined by the equation

$$P_n^m(\nu P_{n-1}^k) \times Q_2^3 = P_{n+2}^{3m}(\nu P_{n+1}^{3k}, 3P_{n+1}^{2m}). \tag{6.28}$$

Proof. When the polytope P_n^m is multiplied by the triangle Q_2^3, the number of vertices of the polytope is multiplied by 3, since the triangle has 3 edges and each edge at this multiplication translates polytope vertices in its direction. At that the dimension of the polytope naturally, according to the definition (Ziegler, 1995), is increased by two. Thus, one will get a polytope P_{n+2}^{3m}. If one multiply the polytope by the triangle, its facets P_{n-1}^k are multiplied by the triangle also. The number of these facets of the

product is equal to the number of facets of the polytope P_n^m , i. e. it is equal to ν. In addition, at the multiplication by the triangle the polytope P_n^m is multiplied by each edge (i. e. multiplication by Q_1^2 3 times). Therefore, 3 facets P_{n+1}^{2m} more appear as facets of the polytope P_{n+2}^{3m}. It is easy to make sure that all the considered above products of the polytopes by the triangle satisfy the equality (6.28).
Theorem 6.5 is proved.

Corollary. If the polygon Q_2^t with the number of vertices t is the polytope factor in (6.28), the equation (6.28) can be rewritten as

$$P_n^m(\nu P_{n-1}^k) \times Q_2^t = P_{n+2}^{tm}(\nu P_{n+1}^{tk}, tP_{n+1}^{2m}). \tag{6.29}$$

As an example, the result of the product of a tetrahedron by a square is given in Figure 13.

Figure 13. The product of a tetrahedron by a square

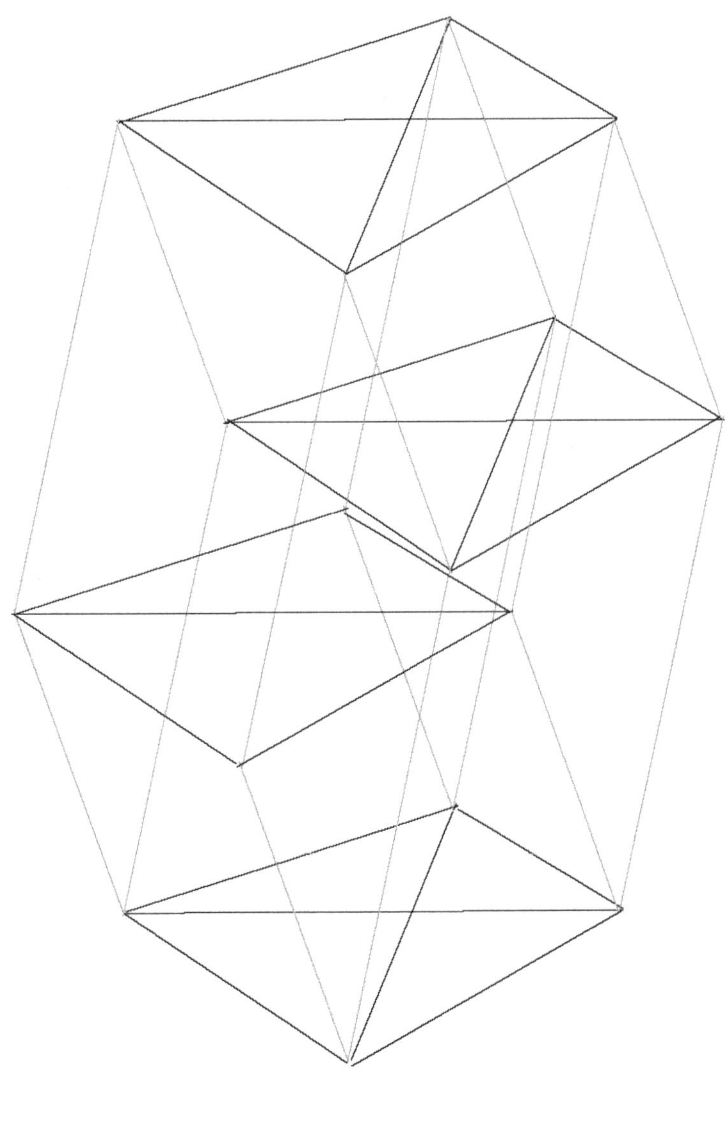

In all cases, the polytope (6.29) of dimension n +2 is incorrect and has a boundary surface of dimension n + 1.

THE POLYTOPES WHICH ARE THE PRODUCTS OF POLYTOPES BY A TETRAHEDRON

The Polytope which is the Product of a Tetrahedron by a Tetrahedron

According to theorem 6.1 and ratios (6.5) for the numbers of elements of different dimension in the product $P_3^4(4P_2^3)\times Q_3^4(4P_2^3)$ there are:

$$f_0^{\times} = f_0^{P_3} f_0^{Q_3} = 4 \cdot 4 = 16,$$

$$f_1^{\times} = f_1^{P_3} f_0^{Q_3} + f_0^{P_3} f_1^{Q_3} = 6\cdot 4 + 4\cdot 6 = 48,$$

$$f_2^{\times} = f_2^{P_3} f_0^{Q_3} + f_1^{P_3} f_1^{Q_3} + f_0^{P_3} f_2^{Q_3} = 4\cdot 4 + 6\cdot 4 + 4\cdot 4 = 68 = 36P_2^{\times,4} + 32P_2^{\times,3},$$

$$f_3^{\times} = f_3^{P_3} f_0^{Q_3} + f_2^{P_3} f_1^{Q_3} + f_1^{P_3} f_2^{Q_3} = 1\cdot 4 + 4\cdot 6 + 6\cdot 4 + 4\cdot 1 = 56 = 8P_3^{\times,4} + 48P_3^{\times,6},$$

$$f_4^{\times} = f_3^{P_3} f_1^{Q_3} + f_2^{P_3} f_2^{Q_3} + f_1^{P_3} f_3^{Q_3} = 1\cdot 6 + 4\cdot 4 + 6\cdot 1 = 28 = 12P_4^{\times,8} + 16P_4^{\times,9},$$

$$f_5^{\times} = f_3^{P_3} f_2^{Q_3} + f_2^{P_3} f_3^{Q_3} = 1\cdot 4 + 4\cdot 1 = 8 = 8P_5^{\times,12},$$

$$f_6^{\times} = f_3^{P_3} f_3^{Q_3} = 1\cdot 1 = 1 = P_6^{\times,16}.$$

Thus, the product of a tetrahedron by a tetrahedron is a 6-polytope including 8 the tetrahedral prismahedrons (6.22).
The short structural formula of the polytope P_6^{16} has the form:

$$P_6^{16}(8P_5^{12}). \qquad (6.30)$$

The image of this polytope is shown in Figure 14.

Figure 14. The product of a tetrahedron by a tetrahedron

One can call the polytope (6.30) a 6 - complex of tetrahedral prismahedrons. The six-dimensional polytope (6.30) is incorrect polytope since it includes various geometric shapes and, therefore, an external five-dimensional surface can be distinguished in it (see Chapter 5). This last statement is confirmed by the polytope structural formula (6.30).

The Polytope which is the Product of an Octahedron by a Tetrahedron

According to theorem 6.1 and ratio (6.5) for the numbers of elements of different dimension in the product $P_3^6(8P_2^3)\times Q_3^4(4P_2^3)$ there are:

$$f_0^{\times} = f_0^{P_3} f_0^{Q_3} = 6 \cdot 4 = 24,$$

$$f_1^{\times} = f_1^{P_3} f_0^{Q_3} + f_0^{P_3} f_1^{Q_3} = 12 \cdot 4 + 6 \cdot 6 = 84,$$

$$f_2^{\times} = f_2^{P_3} f_0^{Q_3} + f_1^{P_3} f_1^{Q_3} + f_0^{P_3} f_2^{Q_3} = 8 \cdot 4 + 12 \cdot 6 + 6 \cdot 4 = 128 = 72P_2^{\times,4} + 56P_2^{\times,3},$$

$$f_3^{\times} = f_3^{P_3} f_0^{Q_3} + f_2^{P_3} f_1^{Q_3} + f_1^{P_3} f_2^{Q_3} = 1 \cdot 4 + 8 \cdot 6 + 12 \cdot 4 + 6 \cdot 1 = 106 = 6P_3^{\times,4} + 100P_3^{\times,6},$$

$$f_4^{\times} = f_3^{P_3} f_1^{Q_3} + f_2^{P_3} f_2^{Q_3} + f_1^{P_3} f_3^{Q_3} = 1 \cdot 6 + 8 \cdot 4 + 12 \cdot 1 = 50 = 12P_4^{\times,8} + 32P_4^{\times,9} + 6P_4^{\times,12},$$

$$f_5^{\times} = f_3^{P_3} f_2^{Q_3} + f_2^{P_3} f_3^{Q_3} = 1 \cdot 4 + 8 \cdot 1 = 12 = 8P_5^{\times,12} + 4P_5^{\times,18},$$

$$f_6^{\times} = f_3^{P_3} f_3^{Q_3} = 1 \cdot 1 = 1 = P_6^{\times,24}.$$

Thus, the product of an octahedron by a tetrahedron is a 6 - polytope including 8 the tetrahedral prismahedrons (6.22) and 4 the octahedral polytopic prisms (6.23).
The short structural formula of the polytope P_6^{24} has the form:

$$P_6^{24}(4P_5^{18}, 8P_5^{12}). \qquad (6.31)$$

The image of this polytope is shown in Figure 15.

Figure 15. The product of an octahedron by a tetrahedron

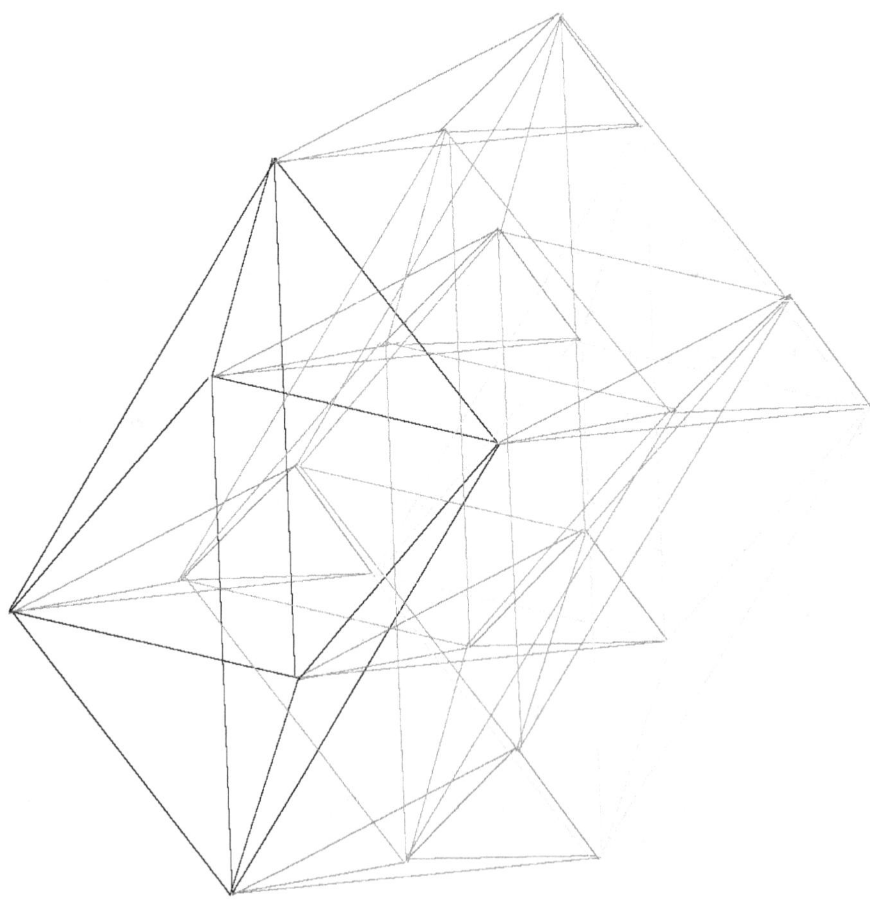

One can call the polytope P_6^{24} a 6 - complex of the tetrahedral prismahedrons and the octahedral polytopic prisms. The six-dimensional polytope (6.31) is incorrect polytope since it includes various geometric shapes and, therefore, an external five-dimensional surface can be distinguished in it (see Chapter 5). This last statement is confirmed by the polytope structural formula (6.31).

THE POLYTOPES WHICH ARE THE PRODUCTS OF POLYTOPES WITH UNIFORM FACETS

Comparing (6.30) and (6.31) with equalities defined by theorems 3, 4, we note their general structure. This allows us to formulate and prove a general theorem on the polytopes which are the products of the polytopes with uniform facets.

Theorem 6.6. Let's assume that there is a convex polytope P of dimension n, with the number of vertices m, which has ν facets with the number of vertices k for each facet. Let's also there is set a convex polytope Q of dimension π, with the number of vertices l, having μ facets with the number of vertices λ for every facet. Then the product of these polytopes is defined by equation

$$P_n^m(\nu P_{n-1}^k)\times Q_\pi^l(\mu P_{\pi-1}^\lambda)=P_{n+\pi}^{lm}(\nu P_{n+\pi-1}^{lk},\mu P_{n+\pi-1}^{\lambda m}). \tag{6.32}$$

Proof. For multiplication of the polytope P_n^m by the polytope Q_π^l, the number of the vertices of polytope P is multiplied by the number of the vertices of polytope Q since each vertex of the polytope P at multiplication will correspond to all vertices of the polytope Q. The dimension of the product polytope according to the definition (Ziegler, 1995) is the sum of the dimensions of polytopes-factors, i. e. we receive the polytope $P_{n+\pi}^{lm}$ as the result of the product. Multiplication of the polytope P by the polytope Q is accompanied also by multiplication of the polytope P facets by the polytope Q. That is why the result of this multiplication will be the facets of the polytopes product defined according to the same rules, i.e. $P_{n+\pi-1}^{lk}$. The number of these facets is equal to the number of facets of the polytope P, i. e. ν. Besides, in the product of the polytopes P and Q there are facets formed by the product of the facets of polytope Q by polytope P. These are facets $P_{n+\pi-1}^{lk}$, their number is equal to the number of facets of the polytope Q, i. e. μ. The theorem 6.6 is proved.

Let's use theorem 6.6 for the product of an octahedron by a tetrahedron:

$$P_3^6(8P_2^3)\times Q_3^4(4P_2^3)=P_6^{24}(4P_5^{18},8P_5^{12}).$$

The received formula coincides with (6.31) got after detailed derivation.

It's easy to verify that all the previous formulas about polytopes product are particular cases of the formula (6.32). Theorem 6.6 allows to write down at once the result of the product of any convex polytopes with uniform faces without detailed derivation of this result.

In principle, all these works are incorrect polytopes, regardless of the fact that their cofactors may be correct polytopes. And all of them, regardless of their dimension, have an external boundary in the form of a continuous surface with a dimension less by one than the dimension of the polytope, which is the product of polytopes.

THE VALUES OF THE INCIDENCE COEFICIENTS OF ELEMENTS WITH A LOWER DIMENSION IN RELATION TO ELEMENTS WITH A HIGHER DIMENSION IN POLITOPIC PRISMAHEDRONS.

In the third chapter of this work, for a number of polytopes, the systems of incidence coefficients of elements with a lower dimension in relation to elements with a higher dimension were given for each polytope from this series. It is of interest to write down the same systems of incidence coefficients for polytopes, which are products of polytopes of lower dimension.

Consider the product of a triangle and a one-dimensional segment. For this product, the incidence coefficients are:

$$k_{b(0)b(1)} = 3; k_{b(0)b(2)} = 3; k_{b(0)b(3)} = 1;$$
$$k_{b(1)b(2)} = 2; k_{b(1)b(3)} = 1;$$
$$k_{b(2)b(3)} = 1.$$

Consider the product of a tetrahedron and a one-dimensional segment (Figure 1). For this product, the incidence coefficients are:

$$k_{b(0)b(1)} = 4; k_{b(0)b(2)} = 6; k_{b(0)b(3)} = 4; k_{b(0)b(4)} = 1;$$
$$k_{b(1)b(2)} = 3; k_{b(1)b(3)} = 3; k_{b(1)b(4)} = 1;$$
$$k_{b(2)b(3)} = 2; k_{b(2)b(4)} = 1;$$
$$k_{b(3)b(4)} = 1.$$

It should be noted that the incidence values are independent of the belonging of an element to a tetrahedron or a prism as part of a polytopic prismahedron.

Consider the product of a triangle and a triangle (Figure 5). For this product, the incidence coefficients are:

$$k_{b(0)b(1)} = 4; k_{b(0)b(2)} = 5; k_{b(0)b(3)} = 3; k_{b(0)b(4)} = 1;$$
$$k_{b(1)b(2)} = 3; k_{b(1)b(3)} = 3; k_{b(1)b(4)} = 1;$$
$$k_{b(2)b(3)} = 2; k_{b(2)b(4)} = 1;$$
$$k_{b(3)b(4)} = 1.$$

Consider the product of a tetrahedron and a triangle (Figure 9). For this product, the incidence coefficients are:

$$k_{b(0)b(1)} = 5; k_{b(0)b(2)} = 7; k_{b(0)b(3)} = 4; k_{b(0)b(4)} = 2; k_{b(0)b(5)} = 1;$$
$$k_{b(1)b(2)} = 4; k_{b(1)b(3)} = 4; k_{b(1)b(4)} = 2; k_{b(1)b(5)} = 1;$$
$$k_{b(2)b(3)} = 3; k_{b(2)b(4)} = 2; k_{b(2)b(5)} = 1;$$
$$k_{b(3)b(4)} = 2; k_{b(3)b(5)} = 1;$$
$$k_{b(4)b(5)} = 1.$$

It is already clear from the examples given that with an increase in the dimension of the factors of the product of polytopes, the values of the incidence coefficients increase. With a further increase in the dimension of the factors of the product of polytopes and the number of elements of different dimensions in the factors, the values of the incidence coefficients are steadily increasing. They can be identified by the presented images of polytopes, which are works of polytopes.

Earlier, the author introduced the concept of the incidence coefficient of a polytope element in relation to an element, the dimension of which is one more than the dimension of the given element. It can be considered a local incidence coefficient it is denoted as follows $k_{b(i)}(n)$ (b is element, i is its dimension, n is dimension of the polytope). It was proved (Zhizhin, 2019 a) that equality

$$k_{b(i)}(n) = n - i, \tag{6.33}$$

holds for polytopes of simplex type and a cube of any dimension. This equality is very important, since it shows that the local incidence coefficient of elements increases linearly with the growth of the polytope dimension. In particular, with an increase in the dimension of the polytope, flat (two-dimensional) faces should be common to an increasing number of three-dimensional faces. The number of three-dimensional faces that have a common two-dimensional face, the calculation is linear with increasing dimension. The same pattern is found in the works of polytopes. It can be noted that it is observed in the examples of the products of polytopes given in this section. Besides, for example, for a polytopic prismahedron that is a product of a square and a tetrahedron (Figure 13), the incidence values of elements of a boundary complex of different dimensions have the form:

$$k_{b(i=0)}(5) = 5, k_{b(i=1)}(5) = 4, k_{b(i=2)}(5) = 3, k_{b(i=3)}(5) = 2, k_{b(i=4)}(5) = 1.$$

This completely coincides with equality (6.33).

For the polytopic prismahedron, which is the product of a tetrahedron on a tetrahedron (Figure 14), the incidence values of elements of a boundary complex of different dimensions have the form:

$$k_{b(i=0)}(6) = 6, k_{b(i=1)}(6) = 5, k_{b(i=2)}(6) = 4, k_{b(i=3)}(6) = 3, k_{b(i=4)}(6) = 2, k_{b(i=5)}(6) = 1.$$

This also agrees with equality (6.33).

However, for polytopic prismahedron in which a cross-polytopes participates (any dimension)), there is a difference from equality (6.33). The incidence of vertices in this case is greater than the value determined equality (6.33). For example, for a polytopic prismahedron that is the product of an octahedron per triangle (Figure 10), the incidence values of the elements of the boundary complex have the form:

$$k_{b(i=0)}(5)=6, k_{b(i=1)}(5)=4, k_{b(i=2)}(5)=3, k_{b(i=3)}(5)=2, k_{b(i=4)}(5)=1.$$

The incidence of vertices in this case is greater than the dimension of the polytopic prismahedron. This difference was already observed when considering the product of the Egyptian pyramid on a one - dimensional segment (Figure 3).

The participation of a cross - polytopes as elements of the boundary complex of the polytopic prismahedrons significantly changes the incidence of the elements of the boundary complex, as this has already been noted when considering polytopic prisms. The influence of a cross - polytopes on the incidence of elements of the boundary complex in complicated polytopic prismahedrons has yet to be investigated.

Multi-Shell Systems of Polytopes from the Product of Polytopes with Increasing Dimension Inside Systems

When considering polytopes, which are works of polytopes of lesser dimension, we made sure that the products of polytopes are incorrect polytopes. Due to this, the boundary complexes (Grünbaum, 1967) of these polytopes have external continuous closed surfaces with a dimension less than the dimension of the product of polytopes by one. This opens up the possibility of the existence of coaxial structures of systems of unconnected shells of polytopes with an increase in dimension as we go deeper into the system. Indeed, since the surface of the boundary complex of the product of polytopes has a dimension one less than the dimension of the product of polytopes, then a polytope representing the product of polytopes can be located in a space of dimension one less than the dimension of the product of polytopes. At the interior points of the product of polytopes, the dimension is one greater than the dimension of the surface. Therefore, in principle, at the interior points of the product of polytopes, a polytope with a surface one greater than the dimension of the surface of the product of polytopes can be located. But in this case, a polytope with a surface can be located in this polytope, the dimension of which is equal to the dimension of the polytope. However, within this polytope, the dimension is one greater than the dimension of the original polytope. This process can be continued by locating polytopes with increasing dimension at the interior points between the shells. This, in principle, becomes possible if inside each polytope, which is a product of polytopes, there are internal points, i.e. points not belonging to the boundary complex of the given product of polytopes. In order to make sure of this, consider the simplest example of the product of a tetrahedron by a one-dimensional segment, i.e. tetrahedral prism (Figure 1).

We decide whether the tetrahedral prism has internal points, i. e. points not belonging to its boundary complex. For this purpose, in Figure 1, coordinate systems are introduced, as well as when the tetrahedron is considered. Taking into account that in this case we are dealing with a space of dimension 4, a fourth coordinate t is added to the coordinate system parallel to the generating lines of the tetrahedral prism. Assuming the length of all edges of the tetrahedral prism per unit, one can establish the coordinates of all the vertices of the tetrahedral prism (Figure 1) in the given coordinate system (x, y, z, t)

$$b_1\left(0,\frac{1}{\sqrt{3}},0,0\right),b_2\left(\frac{1}{2},-\frac{1}{2\sqrt{3}},0,0\right),b_3\left(-\frac{1}{2},-\frac{1}{2\sqrt{3}},0,0\right),b_4\left(0,0,\sqrt{\frac{2}{3}},0\right),$$
$$b_5\left(0,\frac{1}{\sqrt{3}},0,1\right),b_6\left(\frac{1}{2},-\frac{1}{2\sqrt{3}},0,1\right),b_7\left(-\frac{1}{2},-\frac{1}{2\sqrt{3}},0,1\right),b_8\left(0,0,\sqrt{\frac{2}{3}},1\right).$$

Then the set of interior points of the tetrahedral prism is determined by the system of inequalities $z>0$, $0<t<1$;

$$z<2\sqrt{\frac{2}{3}}x-\sqrt{2}y+\sqrt{\frac{2}{3}},-\frac{1}{2}<x<0,-\frac{1}{2\sqrt{3}}<y<\frac{1}{\sqrt{3}},0<\mathrm{t}<1;$$

$$z<2\sqrt{2}y+\sqrt{\frac{2}{3}},-\frac{1}{2}<x<\frac{1}{2};-\frac{1}{2\sqrt{3}}<y<0,0<t<1;$$

$$z<-\sqrt{6}x-\sqrt{2}y+\sqrt{\frac{2}{3}},0<x<\frac{1}{2},-\frac{1}{2\sqrt{3}}<y<\frac{1}{\sqrt{3}},0<t<1.$$

Similarly, in principle, one can define a set of internal points and other polytopic prisms.

CONCLUSION

In monographs of Zhizhin (Zhizhin, 2017 b, 2019 a, b, c; 2020) it showed, that many molecules have a dimension greater than three if they are described in the form of geometric convex figures. Many of them have tetrahedral and octahedral coordination. But any substance is not a separate molecule, but a collection of molecules. This raises the problem of filling spaces with molecules, i. e. convex geometric figures of different dimensions. The addition of other molecules (indifferent or not) leads to the formation of a cluster. There are various patterns in the formation of clusters (Lord, Mackay & Ranganathan, 2006). One can distinguish different levels of organization in clusters. From a geometric point of view, this amounts to a hierarchical filling of the space (Zhizhin, 2012). However, the hierarchical filling of space is always limited by the size of some area of space, primarily because when distance from the original molecule the distance between the particles (or points) increases significantly (Zhizhin, 2014 a). This inevitably leads to a weakening of the bonds between molecules and, as a consequence, to the detachment of molecules. Therefore, in order to further fill of the space, it is necessary to proceed to space translation a certain set of atoms or molecules. If a molecule in geometric relation is a multidimensional formation, then the problem arises of filling the space by means of translation by a multidimensional convex body. If a molecule or molecular formation is a multidimensional body, then this body must be translated from the coordinates of a multidimensional space. Moreover, in order for such a translation to be possible, the body itself must have a certain structure, i. e. to have selected directions along the coordinates in itself. Then translation of this body in the selected directions will not cause a violation of the body shape. The products of polytopes considered in this chapter just create forms in which there are selected directions along the coordinates of a multidimensional space. They are created by the very mechanism of the product polytopes. The mathematical apparatus developed in the chapter for determining the shape of a body that is a product of polytopes can be applied to determining the shape of bodies. The diffraction pattern of the quasicrystal can see on structures of different substances (Shechtman et al., 1984; Mukhopadhyay et al., 1993; Zhang & Kelton, 1993). The diffraction pattern clearly shows the system of parallel lines formed as result of the translation of molecules along different directions of coordinates (Shevchenko, Zhizhin & Mackay, 2013). At the same time, the lines formed as result of the hierarchical filling of space are visible. In fact, from each point one can see the formation of lines as a result of the hierarchical filling of space (if we neglect the influence of the experimental features on the diffraction pattern - the axial symmetry caused by the axial symmetry of the electron beam directed to the sample of matter). Just as there is an expansion of space at each point of space in the model of the expansion of the Universe (Silk, 1980).

In the monographs Zhizhin G.V. (Zhizhin, 2017 b, 2021) it was established that the fundamental area in the filling of the n - dimensional space is the polytopic prismahedron, which is a prism or a complex of prisms with bases in the form of polytopes of dimension n. It is polytopes provide the filling of the n - dimensional space with a face into the face without gaps. Thus, the polytopic prismahedron is the sterehedron, the existence of which was suggested by B.N. Delone (Delone, 1961; Delone & Sandakova, 1961). However, Delone never succeeded in constructing any particular stereohedron in a space with a dimension greater than three. It was shown that a special case of a polytopic prismahedron is a gold hyper - rhombohedron of dimension 4 and higher, which is filled by a quasicrystal, revealing the periodic structure of quasicrystals in a space of higher dimension (Zhizhin, 2014 b, 2015; Zhizhin, & Diudea, 2016). It was shown (Zhizhin, 2017b, 2019 a) that a special case of a polytopic prismahedron is a gold hyper - rhombohedron of dimension 4 and higher, which is filled by a quasi - crystal, revealing the periodic structure of quasi - crystals in a space of higher dimension. By its design, a polytopic prismahedron is the product of a polytope to another polytope, or in a particular case to a one - dimensional segment. Pontryagin mentioned the structures resulting from the product of a polyhedron by a one - dimensional segment the cylinder ones (Pontryagin, 1976). We can say that the product of a polytope by a segment is a prism (Robertson, 1984) with a base in form of a polytope. To distinguish it from a usual three - dimensional prism, can call it polytopic prism. Ziegler noted that the product of polytopes is not a simplex even if the factors are simplexes, so the polytopes are of considerable interest (Ziegler, 1995). Especially taking into account that a multi - dimensional world has its own peculiarities having no analogues in the three - dimensional world (Ziegler, 1995). In this regard, the theory of simplicial polytopes for the analysis of polytopes product becomes inapplicable, especially in the case of high - dimensional factors.

The product of two polytopes is result of the product of one of them by one - dimensional edges of another polytope. Thus, the product of polytopes is a complex of polytopic prism. Let`s call this complex a polytopic prismahedron. The polytopic prism is particular case of polytopic prismahedron. The interest to the study of polytopic prismahedrons is connected, in first place, with the novelty of this field and, secondly, with the fact that polytopic prismahedrons, due to their construction, can be "bricks" to fill the spaces of higher dimension face in face.

It is proved that for the product of polytopes to satisfy the Euler-Poincaré equation, it is necessary that the factors of the product satisfy the Euler-Poincaré equation.

One of the valuable properties of polytopes, which are products of polytopes of lesser dimension, as found in this chapter, is the incorrectness of these polytopes. Moreover, this property appears as a result of the product operation, regardless of the geometric shape of the factors. It turns out that this very property made it possible to solve the 18th problem of Hilbert (Gilbert, 1901; Alexandrov, 1975, 1979) 120 years after its formulation (Zhizhin, 2019 b, 2021), i. e. constructing n-dimensional spaces from congruent figures. In addition, due to the incorrectness of the polytopes representing the products of polytopes, as found in this chapter, these polytopes have a continuous closed boundary with a dimension less than the dimension of the product of polytopes by one. This opens up the possibility of creating multi-shell systems of polytope products with no connection between shells, which are characterized by a continuous increase in dimension when penetrating to the center of the system. This unique property still needs to be understood in order to use it.

In general, I would like to note that the combination of irregularity and the highest dimension of polytopes, as works of polytopes, gives rise to a new mathematical world, distinguished by novelty, beauty of forms, and closeness to nature. I would like to call this new mathematics polytypictops mathematics, noting its uniqueness, brightness and natural closeness to living forms.

REFERENCES

Alexandrov, P. S. (1975). *Introduction to homological dimension theory and general combinatorial topology*. Science.

Alexandrov, P. S. (1979). *Selected work. The general theory of homology*. Science.

Delone, B. (1961). Proof of the main theorem of the theory of stereohedrons. *Reports of the Academy of Sciences of the USSR, 138*(6), 1270–1272.

Delone, B., & Sandakova, N. (1961). Theory of stereohedrons. *Proceedings of the Mathematical Institute. V.A. Steklov, 64*, 28 – 51.

Grünbaum, B. (1967). *Convex Polytopes*. Springer.

Hilbert, D. (1901). Gesammelte Abhandlangen. *Archiv f. Mathematics and Physics, 3*(1), 44–63, 213–237.

Lord, E. A., Mackay, A. L., & Ranganathan, S. (2006). *New geometry for new materials*. Cambridge University Press.

Mukhopadhyay, N.K. (1993). Diffraction studies of icosahedral phases in $Al_{70}Fe_{20}W_{10}$. *Journal of Non-Crystalline Solids, 153 – 154*, 1193 – 1197.

Poincare, A. (1895). Analysis situs. *J. de é Ecole Polytechnique, 1*.

Pontryagin, L. S. (1976). *The foundations of combinatorial topology*. Science.

Robertson, S. A. (1984). *Polytopes and Symmetry*. Cambridge University Press.

Shechtman, D., Blech, I., Gratias, D., & Cahn, J. W. (1984). Metallic phase with longerange orientational order and no translational symmetry. *Physical Review Letters, 53*(20), 1951–1953. doi:10.1103/PhysRevLett.53.1951

Shevchenko, V. Ya., Zhizhin, G. V., & Mackay, A. (2013). On the structure of the quasi - crystals in the high dimension space. In M. V. Diudea (Ed.), *Diamonds and related nanostructures* (pp. 311–320). Springer. doi:10.1007/978-94-007-6371-5_17

Silk, J. (1980). *The big bang. The creation and evolution of the Universe*. W.H. Freeman and Company.

Snegirev, I. L., & Frantsev, Yu. P. (1938). *Ancient Egypt*. OGIZ.

Zang, X., & Kelton, K. (1993). High-order crystal approximant alloys $Ti_{54}Zr_{26}Ni_{20}$. *Journal of Non-Crystalline Solids, 153-154*, 114–118. doi:10.1016/0022-3093(93)90325-R

Zhizhin, G. V. (2012, October). *Hierarchical filling of spaces with polytopes.* Paper presented at "St. Petersburg Scientific Forum: Science and Human Progress". 7th St.- Petersburg meeting of Nobel Prize laureates, St. Petersburg, Russia.

Zhizhin, G. V. (2014a). *World – 4D*. Polytechnic Service.

Zhizhin, G. V. (2014b). Fractal nature of disproportionate phases. In *10th All-Russian Scientific School "Mathematical research in the natural sciences."* Geological Institute KSC RAS.

Zhizhin, G. V. (2015, November). *Polytopic prismahedrons – fundamental regions of the n-dimension nanostructures.* Paper presented at the International conference "Nanoscience in Chemistry, Physics, Biology and Mathematics", Cluj-Napoca, Romania.

Zhizhin, G.V. (2017a). N–prismahedrons and their dual polytopes. *International Journal Chemical Modeling, 5*(4).

Zhizhin, G. V. (2017b). *Chemical Compounds Structures and the Higher Dimension of Molecules: Emerging Research and Opportunities*. IGI Global.

Zhizhin, G. V. (2019a). *The Geometry of Higher - Dimensional polytopes*. IGI Global. doi:10.4018/978-1-5225-6968-8

Zhizhin, G. V. (2019b). On the Structure of Parallelohedrons of Higher Dimension: Hilbert's 18th Problem. *International Journal of Chemoinformatics and Chemical Engineering, 8*(2), 69–96. doi:10.4018/IJCCE.20190701.oa2

Zhizhin, G. V. (2019c). *Attractors and Higher Dimensions in Population and Molecular Biology: Emerging Research and Opportunities*. IGI Global. doi:10.4018/978-1-5225-9651-6

Zhizhin, G. V. (2020). *Nanotechnologies and Clusters in the Spaces of Higher Dimension: Emerging Research and Opportunities*. IGI Global.

Zhizhin, G. V. (2021). *Normal Partitions and Hierarchical Fillings of N-Dimensional Spaces*. IGI Global. doi:10.4018/978-1-7998-6768-5

Zhizhin, G. V., & Diudea, M. V. (2016). Space of nanoworld. In M. V. Putz & C. M. Marius (Eds.), *Sustainable nanosystems, development, properties, and applications* (pp. 221–245). IGI Global.

Ziegler, G. (1995). *Lectures on polytopes*. Springer. doi:10.1007/978-1-4613-8431-1

KEY TERMS AND DEFINITIONS

5-Cross-Prism: The product of 4-cross-polytope by one dimension segment.

5-Simplex-Prism: The product of 4-simplex by one dimension segment.

6-Complex-Polytopic Prismahedrons: The product of tetrahedron by the tetrahedron.

6-Complex-Tetrahedral Prismahedrons and Octahedral Prismahedrons: The product octahedral by the tetrahedral.

6-Simplex-Polytopic Prismahedron: The product of 4-simplex by the triangle.

Cube-Polytopic Prismahedron: The product of cube by the triangle.

Dodeca-Polytopic Prismahedron: The product of dodecahedron by the triangle.

Dodecahedral Prism: The product of dodecahedral by one dimension segment.

Hierarchical and Translation Filling Spaces: At the same time hierarchical and translation filling.

Icosa-Polytopic Prismahedron: The product of icosahedron by the triangle.

Icosahedral Prism: The product of icosahedral by one dimension segment.

N*3-Angular Prismahedron: The product of n-angle by the triangle.

Octahedral Prism: The product of octahedral by one dimension segment.

Octahedral Prismahedron: The product of octahedral by the triangle.

Polytopic Prismahedron: The product of polytope by one dimension segment.

Tetrahedral Prism: The product of tetrahedral by one dimension segment.

Tetrahedral Prismahedron: The product of tetrahedral by the triangle.

Triangular Prismahedron: The product of triangle by the triangle.

Chapter 7
Higher-Dimensional Polytopes With Regular-Faceted Polyhedrons Are Different From Archimedes Bodies

ABSTRACT

The question of the existence of regular-faceted polytopes of higher dimension is investigated. To construct regular-faceted polytopes of the highest dimension, the results of the study of regular-faceted three-dimensional polyhedrons, carried out by the author, were used, in particular the regular-faceted polyhedrons discovered by the author with a large number of gonohedrons of various types. These polyhedrons, due to their sharp difference from the Archimedes bodies and the diversity of the type of vertices, are called polytypichedrons, and the polyhedron containing the maximum possible number of different types of vertices (six) is called the polytypichedron-max. The method of obtaining polytopes of the highest dimension from them by multiplying them by geometric elements of various dimensions (including the highest dimension) is applied to the polytypichedrons. This is how, in principle, an infinite set of polytopes of the highest dimension with regular two-dimensional edges and the participation of vertices of various types was obtained.

INTRODUCTION

In the second chapter of this work, the existence of three-dimensional polyhedrons with flat faces in the form of various regular polygons simultaneously present in one polyhedron with the participation of several types of gonohedrons (Fedorov, 1885, 1889, 1891) as vertices was proved. Moreover, it was found that the maximum number of different types of gonohedrons that can be simultaneously present in one polyhedron is 6. This single polyhedron PG_{14} can name polytypichedron - max. It represents the maximum possible difference between a polyhedron and Archimedes bodies, while maintaining flat faces in the form of regular polygons. The whole family of regular-faceted polyhedrons, in which three

DOI: 10.4018/978-1-7998-8374-6.ch007

or more of different types of genohedrons can be present shell named polytypichedrons. The question about existence of regular-faceted polyhedron was considered on the monograph by V.A. Zalgaller (Zalgaller, 1966). This monograph contains a bibliography of works in which, to one degree or another, the question of the existence of convex regularly faceted polyhedrons is considered. It is proved that the assertion, expressed earlier (Johnson, 1960), that (apart from prisms and antiprisms) there are only 92 regularly faceted polytopes, which have 28 simple regularly faceted polytopes in their composition (Zalgaller, 1966 a). Accordingly, N. Johnson's assumption on the existence of only 92 total number regularly faceted polyhedrons is wrong too. In the proof Johnson's assumption Zalgaller used Cauchy's lemma (Cauchy, 1944; Hadamard,1938) and the theorem of A.D. Aleksandrov (Aleksandrov, 1950). In the author's work (Zhizhin, 2009), the list of regular polyhedrons proposed by Zalgaller was refined and corrected. In particular, the existence of nine previously unknown regular-faceted polyhedrons was proved. Moreover, It is these polyhedrons that have a fairly large number of types of different gonohedrons, the diversity of which has prompted to call these polyhedrons polytypichedrons.

In the sixth chapter of this work, the geometry of polytopes, which are products of polytopes, was studied in detail. A number of valuable properties of these polytopes were discovered, which made it possible, in particular, to solve the 18th Hilbert problem (Hilbert, 1901; Zhizhin, 2019 b, 2021) on the construction of n-dimensional spaces, as well as systems with a steady increase in dimension as we go deeper into the systems. Since polytopes, which are products of polytopes, form a very important class of polytopes, and polytypichedrons are interesting for the presence of different types of gonohedrons in them, the problem arises of finding polytopes of higher dimension with the simultaneous participation of different types of gonohedrons in them. This can be done by multiplying the polytypichedrons by polytopes. This chapter is devoted to this research. As objects for the study, polyhedrons were selected, the existence of which was established by the author PG_8, PG_9, PG_{10}, PG_{11}, PG_{14} (polytypichedron - max), PG_{35} (Chapter 2).

THE POPYTOPES WHICH ARE THE PRODUCTS OF POLYHEDRONS PG_9 BY ONE – DIMENSIONAL SEGMENT, TRIANGLE, TETRAHEDRON

The Polytope which is Product of the Polyhedron PG_9 by One-Dimensional Segment

A diagram of the PG_9 polyhedron is presented in Chapter 2 in Figure 2.19.9. However, the small scale of Figure 2.18 makes it difficult to analyze the structure of this polyhedron. Let us represent the PG_9 polyhedron in an enlarged form (Figure 1).

Figure 1. Polyhedron PG_9

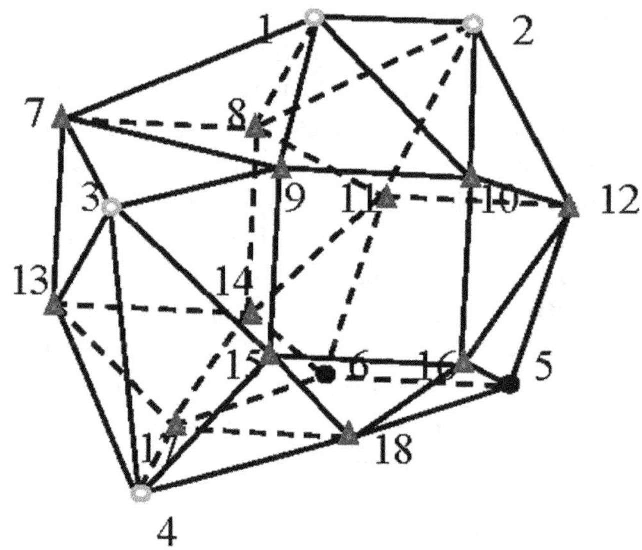

The set of vertices of the PG_9 polyhedron form gonohedrons 3, 10_4, 12_4, i. e. three type of gonohedrons. In all other of the listed polyhedrons PG_8, PG_{10}, PG_{11}, PG_{14}, PG_{35} the number of types of gonohedrons involved in the creation of each of these polyhedrons is more than three. The geometry of gonohedrons 3, 10_4, 12_4 is described in Chapter 2. The polyhedron in Figure 1 has a total the number of vertices 18, f_0=18. There are edges between the vertices

1 – 2, 1 – 10, 1 – 8, 1 – 9, 1 – 7, 2 – 8, 2 – 10, 2 – 11, 2 – 12, 3 – 7, 3 – 9, 3 – 13, 3 – 15, 3 – 4, 4 – 13, 4 – 17, 4 – 18, 4 – 15, 5 – 6, 5 – 16, 5 – 12, 5 – 18, 6 – 11, 6 – 14, 6 – 17, 7 – 8, 7 – 9, 7 – 13, 8 – 11, 8 – 14, 9 – 10, 9 – 15, 10 – 12, 10 – 16, 11 – 12, 11 – 14, 12 – 16, 13 – 14, 13 – 17, 14 – 17, 15 – 16, 15 – 18, 16 – 18, 17 - 18.

The total number of edges is 44, f_1=44. The edges on the surface of the PG_9 polyhedron form 24 triangles

1 – 2 – 10, 1 – 2 – 8, 1 – 7 – 9, 1 – 7 – 8, 1 – 9 – 10, 2 – 10 – 12, 2 – 11 – 12, 2 – 8 – 11, 3 – 7 – 9, 3 – 7 – 13, 3 – 9 – 15, 3 – 4 – 15, 3 – 13 – 4, 4 – 17 – 18, 4 – 13 – 17, 4 – 15 – 18, 5 – 12 – 16, 5 – 16 – 18, 6 – 14 – 11, 6 – 14 – 17, 8 – 11 – 14, 10 – 12 – 16, 13 – 14 – 17, 15 – 16 - 18;

and four squares 7 – 8 – 13 – 14, 9 – 10 – 15 – 16, 5 – 6 – 17 – 18, 6 – 11 – 12 - 5.

Thus, the total number of flat faces of the PG_9 polyhedron is 28, f_2= 28. Substitute these values into the Euler-Poincaré equation (Euler, 1736; Poincaré, 1895)

$$\sum_{i=0}^{n-1}(-1)^i f_i(n) = 1 + (-1)^n, \tag{7.1}$$

in (7.1) $f_i(n)$ is the number of elements of a polytope P of dimension n having dimension i.

Then we have $18 - 4 + 28 = 2$.

We are convinced that the polyhedron PG$_9$ satisfies the Euler - Poincaré equation for $n = 3$, i. e. it is three-dimensional. The structure of this polyhedron, taking into account the obtained values of the numbers of various geometric elements included in the polyhedron, can be written as follows $P_3^{18}(24P_2^3, 4P_2^4)$. Consider the product of the PG$_9$ polyhedron by a one-dimensional segment $P_3^{18}(24P_2^3, 4P_2^4) \times Q_1^2$.

According to theorem 6.1 and the ratios (6.5) there are:

$$f_0^{\times} = f_0^{P_3} f_0^{Q_1} = 18 \cdot 2 = 36,$$

$$f_1^{\times} = f_1^{P_3} f_0^{Q_1} + f_0^{P_3} f_1^{Q_1} = 44 \cdot 2 + 18 \cdot 1 = 106,$$

$$f_2^{\times} = f_2^{P_3} f_0^{Q_1} + f_1^{P_3} f_1^{Q_1} = 28 \cdot 2 + 44 \cdot 1 = 100, \tag{7.2}$$

For receiving of the composition of the faces in the dimension 2 polytopes product one of the faces, observing the same rules as at the definition of the product polytopes, i. e. dimensions are added, and the numbers of vertices are multiplied. At that, indexes P and Q are replaced by index ×. Then 100 faces obtained in (7.2), are deciphered so

$$100 = 28P_2^{\times,3} + 44P_2^{\times,4}. \tag{7.3}$$

This means that 100 faces of dimension 2 in the product include twenty and eight faces $P_2^{\times,3}$ (i.e., triangles) and forty and four faces $P_2^{\times,4}$ (i.e., quadrangles). In future, equations similar to (7.2), (7.3) will be recorded in a single line.

For the faces of dimension 3, there is:

$$f_3^{\times} = f_3^{P_3} f_0^{Q_1} + f_2^{P_3} f_1^{Q_1} = 1 \cdot 2 + 24 \cdot 1 + 4 \cdot 1 = 30 = 2P_3^{\times,18} + 24P_3^{\times,6} + 4P_3^{\times,8}.$$

Thus, 30 faces of dimension 3 in the product consist of two polyhedrons $P_3^{18}(24P_2^3, 4P_2^4)$, twenty and four triangular prisms $P_3^6(3P_2^4, 2P_2^3)$ and four quadrangular prisms $P_3^8(6P_2^4)$.

For the faces of dimension 4, there is:

$$f_4^{\times} = f_3^{P_3} f_1^{Q_1} = 1 \cdot 1 = 1 = P_4^{\times,36}.$$

According to the preceding equation, can obtain the structural formula of the product:

$$P_3^{18}(24P_2^3, 4P_2^4) \times Q_1^2 = P_4^{36}[2P_3^{18}(24P_2^3, 4P_2^4), 24P_3^6(3P_2^4, 2P_2^3), 4P_3^8(6P_2^4)]. \tag{7.4}$$

Therefore, the product of the polyhedron PG_9 by the segment is polytope 4*D*, which includes twenty and four triangular prisms, four quadrangular prisms and 2 polyhedron PG_9. You can call a polytope (7.4) PG_9 - prism. The picture of the polytope (7.4) is shown in Figure 2.

Figure 2. The polytope PG_9 - prism

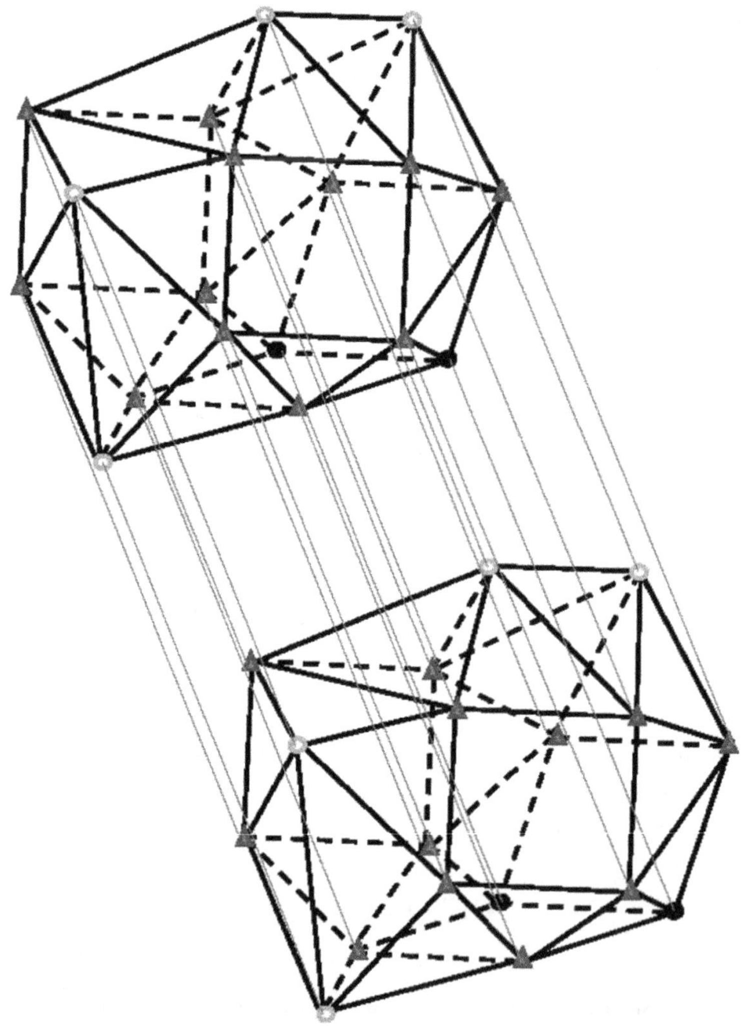

According to Theorem 6.3, the Euler-Poincaré equation for the product of polytopes is satisfied if this equation holds for the factors of the product of polytopes. Therefore, there is no need to check the fulfillment of the Euler-Poincaré equation for the product of polytopes, if it has already been proved that this equation holds for the factors.

When depicting the PG_9 - prism in Figure 2, the edges connecting two PG_9 polyhedrons have a length significantly longer than the lengths of the edges in the PG_9 polyhedrons themselves. This is done on purpose so that the images of the two polyhedrons do not overlap. This would significantly complicate the perception of the drawing. For the same reason, images of the products of Tanbulan polyhedrons by geometric elements of dimension greater than one are not given, due to their complexity. The cited analytical expressions of all these products make it possible to fairly fully represent their structure. In any case, when a polyhedron is multiplied by a triangle, tetrahedron, or elements of higher dimension, the structure of the product is that each vertex of the polyhedron forms this element in space. Thus, the structure consists in the fact that each element, as a factor, is repeated in the structure as many times as there are vertices in the polyhedron. Moreover, these elements are connected by edges, as elements of unit dimension. Examples of spatial construction of the product of polytopes in simpler cases are given in Chapter 6.

The Polytope which is Product of the Polyhedron PG8 by Triangle

There we consider the product $P_3^{18}(24P_2^3,4P_2^4)\times Q_2^3$. According to theorem 6.1 and ratios (6.5) there are:

$$f_0^{\times}=f_0^{P_3}f_0^{Q_2}=18\cdot 3=54,$$

$$f_1^{\times}=f_1^{P_3}f_0^{Q_2}+f_0^{P_3}f_1^{Q_2}=44\cdot 3+18\cdot 3=186,.$$

$$f_2^{\times}=f_2^{P_3}f_0^{Q_2}+f_1^{P_3}f_1^{Q_2}+f_0^{P_3}f_2^{Q_2}=(24+4)\cdot 3+44\cdot 3+18\cdot 1=234=72P_2^{\times,3}+12P_2^{\times,4}+132P_2^{\times,4}+18P_2^{\times,3},$$

(thus, in set of the elements with dimension 2 include 90 triangles and 144 quadrats)

$$f_3^{\times}=f_3^{P_3}f_0^{Q_2}+f_2^{P_3}f_1^{Q_2}+f_1^{P_3}f_2^{Q_2}=1\cdot 3+(24+4)\cdot 3+44\cdot 1=131=3P_3^{\times,18}+72P_3^{\times,6}+12P_3^{\times,8}+44P_3^{\times,6},$$

(thus, in set of the elements with dimension 3 include 3 polyhedrons P_3^{18}, 116 triangles prisms and 12 cubes)

$$f_4^{\times}=f_3^{P_3}f_1^{Q_2}+f_2^{P_3}f_2^{Q_2}=1\cdot 3+(24+4)\cdot 1=31=3P_4^{\times,36}+24P_4^{\times,9}+4P_4^{\times,12},$$

(thus, in set of the elements with dimension 4 include 3 polytopes PG_9 – prisms, 24 triangular prismahedrons and four 4*3 – angular prismahedrons, see Chapter 6)

$$f_5^{\times}=f_3^{P_3}f_2^{Q_2}=1\cdot 1=1=P_5^{\times,54}(3P_4^{\times,36},24P_4^{\times,9},4P_4^{\times,12}).\qquad(7.5)$$

Thus, the product $P_3^{18}(24P_2^3,4P_2^4)\times Q_2^3$ is polytope with dimension 5 and hyperfaces in view of PG_9 – prisms, triangular and 4*3 – angular prismahedrons.

The Polytope which is Product of the Polyhedron PG9 by Tetrahedron

There we consider the product $P_3^{18}(24P_2^3, 4P_2^4) \times Q_3^4$. According to theorem 6.1 and ratios (6.5) there are:

$$f_0^\times = f_0^{P_3} f_0^{Q_3} = 18 \cdot 4 = 72,$$

$$f_1^\times = f_1^{P_3} f_0^{Q_3} + f_0^{P_3} f_1^{Q_3} = 44 \cdot 4 + 18 \cdot 6 = 284,.$$

$$f_2^\times = f_2^{P_3} f_0^{Q_3} + f_1^{P_3} f_1^{Q_3} + f_0^{P_3} f_2^{Q_3} = (24+4) \cdot 4 + 44 \cdot 6 + 18 \cdot 4 = 448 = 96P_2^{\times,3} + 16P_2^{\times,4} + 264P_2^{\times,4} + 72P_2^{\times,3},$$

(thus, in set of the elements with dimension 2 include 164 triangles and 280 quadrats)

$$f_3^\times = f_3^{P_3} f_0^{Q_3} + f_2^{P_3} f_1^{Q_3} + f_1^{P_3} f_2^{Q_3} + f_0^{P_3} f_3^{3} = 1 \cdot 4 + (24+4) \cdot 6 + 44 \cdot 4 + 18 \cdot 1$$
$$= 366 = 4P_3^{\times,18} + 144P_3^{\times,6} + 24P_3^{\times,8} + 176P_3^{\times,6} + 18P_3^{\times,4},$$

(thus, in set of the elements with dimension 3 include 4 polyhedrons P_3^{18}, 320 triangles prisms, 24 cubes and 18 tetrahedrons)

$$f_4^\times = f_3^{P_3} f_1^{Q_3} + f_2^{P_3} f_2^{Q_3} + f_1^{P_3} f_3^{Q_3} = 1 \cdot 6 + (24+4) \cdot 4 + 44 \cdot 1 = 162 = 6P_4^{\times,36} + 96P_4^{\times,9} + 16P_4^{\times,12} + 44P_4^8,$$

(thus, in set of the elements with dimension 4 include 6 polytopes PG_9 – prisms, 96 triangular prismahedrons, 16 4*3 – angular prismahedrons, and 44 tetrahedral prisms, see Chapter 6)

$$f_5^\times = f_3^{P_3} f_2^{Q_3} + f_2^{P_3} f_3^{Q_3} = 1 \cdot 4 + (24+4) \cdot 1 = 32 = 4P_5^{\times,54} + 24P_5^{\times,12} + 4P_5^{\times,16},$$
$$f_6^\times = f_3^{P_3} f_3^{Q_3} = 1 \cdot 1 = 1 = P_6^{\times,72}(4P_5^{\times,54}, 24P_5^{\times,12}, 4P_5^{\times,16}). \tag{7.6}$$

Thus, the product $P_3^{18}(24P_2^3, 4P_2^4) \times Q_3^4$ is polytope with dimension 6 and hyperfaces in view of $P_5^{54}, P_5^{12}, P_5^{16}$.

THE POPYTOPES WHICH ARE THE PRODUCTS OF POLYHEDRONS PG_8 BY ONE – DIMENSIONAL SEGMENT, TRIANGLE, TETRAHEDRON

The Polytope which is Product of the Polyhedron PG8 by One-Dimensional Segment

A diagram of the PG_8 polyhedron is presented in Chapter 2 in Figure 2.19.8. However, the small scale of Figure 2.19.8 makes it difficult to analyze the structure of this polyhedron. Let us represent the PG_8 polyhedron in an enlarged form (Figure 3).

Figure 3. Polyhedron PG_8

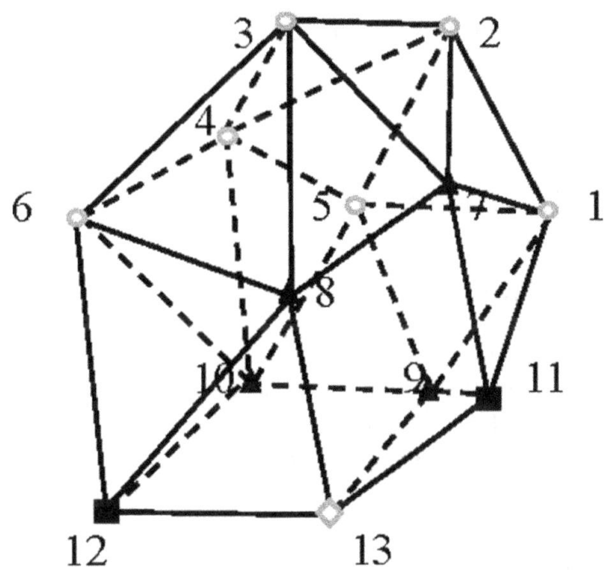

In the PG_8 polyhedron, the number of types of gonohedrons participating in the formation of the polyhedron is one more than in the PG_9 polyhedron. Oddly enough, this led to a decrease in the number of vertices in the polyhedron. The set of vertices of the PG_8 polyhedron form gonohedrons 3, 11_4, 12_4, 9_4, i. e. four type of gonohedrons. The geometry of gonohedrons 3, 11_4, 12_4, 9_4 is described in Chapter 2. The polyhedron in Figure 3 has a total the number of vertices 13, f_0=13. There are edges between the vertices

1 – 11, 1 – 7, 1 – 5, 1 – 9, 1 – 2, 2 – 7, 2 – 5, 2 – 4, 2 – 3, 3 – 7, 3 – 4, 3 – 6 3 – 8, 4 – 5, 4 – 6, 4 – 10, 5 – 9, 5 – 10, 6 – 8, 6 – 10, 6 – 12, 7 – 8, 7 – 11, 8 – 12, 8 – 13, 9 – 10, 9 – 11, 9 – 13, 10 – 12, 11 – 13, 12 – 13.

The total number of edges is 31, f_1= 31. The edges on the surface of the PG_8 polyhedron form 18 triangles

1 – 2 – 7, 1 – 7 – 11, 1 – 2 – 5, 1 – 5 – 9, 1 – 9 – 11, 2 – 3 – 7, 2 – 4 – 3, 2 – 4 – 5, 3 – 7 – 8, 3 – 6 – 8, 3 – 4 – 6, 4 – 5 – 10, 4 – 6 – 10, 5 – 9 – 10, 6 – 8 – 12, 6 – 10 – 12, 8 – 12 – 13, 9 – 11 – 13;

and two squares 7 – 8 – 13 – 11, 9 – 10 – 12 – 13. Thus, the total number of flat faces of the PG_8 polyhedron is 20, $f_2 = 20$.

Substitute these values into the Euler-Poincaré equation (Poincaré, 1895) (7.1).

Then we have 13 – 31 + 20 = 2.

We are convinced that the polyhedron PG_8 satisfies the Euler - Poincaré equation for $n = 3$, i. e. it is three-dimensional. The structure of this polyhedron, taking into account the obtained values of the numbers of various geometric elements included in the polyhedron, can be written as follows $P_3^{13}(18P_2^3, 2P_2^4)$. Consider the product of the PG_8 polyhedron by a one-dimensional segment $P_3^{13}(18P_2^3, 2P_2^4) \times Q_1^2$.
According to theorem 6.1 and the ratios (6.5) there are:

$$f_0^{\times} = f_0^{P_3} f_0^{Q_1} = 13 \cdot 2 = 26,$$

$$f_1^{\times} = f_1^{P_3} f_0^{Q_1} + f_0^{P_3} f_1^{Q_1} = 31 \cdot 2 + 13 \cdot 1 = 75,$$

$$f_2^{\times} = f_2^{P_3} f_0^{Q_1} + f_1^{P_3} f_1^{Q_1} = (18+2) \cdot 2 + 31 \cdot 1 = 71, \tag{7.7}$$

For receiving of the composition of the faces in the dimension 2 polytopes product one of the faces, observing the same rules as at the definition of the product polytopes, i. e. dimensions are added, and the numbers of vertices are multiplied. At that, indexes P and Q are replaced by index ×. Then 71 faces obtained in (7.7), are deciphered so

$$71 = 36P_2^{\times,3} + 4P_2^{\times,4} + 31P_2^{\times,4}. \tag{7.8}$$

This means that 71 faces of dimension 2 in the product include thirty and six faces $P_2^{\times,3}$ (i.e., triangles) and thirty and five faces $P_2^{\times,4}$ (i.e., quadrangles).
For the faces of dimension 3, there is:

$$f_3^{\times} = f_3^{P_3} f_0^{Q_1} + f_2^{P_3} f_1^{Q_1} = 1 \cdot 2 + 18 \cdot 1 + 2 \cdot 1 = 22 = 2P_3^{\times,13} + 18P_3^{\times,6} + 2P_3^{\times,8}.$$

Thus, 22 faces of dimension 3 in the product consist of two polyhedrons $P_3^{13}(18P_2^3, 2P_2^4)$, eighteen triangular prisms $P_3^6(3P_2^4, 2P_2^3)$ and two quadrangular prisms $P_3^8(6P_2^4)$.
For the faces of dimension 4, there is:

$$f_4^{\times} = f_3^{P_3} f_1^{Q_1} = 1 \cdot 1 = 1 = P_4^{\times,26}.$$

According to the preceding equation, can obtain the structural formula of the product:

$$P_3^{13}(18P_2^3,2P_2^4)\times Q_1^2 = P_4^{26}[2P_3^{13}(18P_2^3,2P_2^4),18P_3^6(3P_2^4,2P_2^3),2P_3^8(6P_2^4)]. \tag{7.9}$$

Therefore, the product of the polyhedron PG_8 by the segment is polytope 4*D*, which includes eighteen triangular prisms, two quadrangular prisms and 2 polyhedron PG_8. You can call a polytope (7.9) PG_8 - prism. The picture of the polytope (7.9) is shown in Figure 4.

Figure 4. The polytope PG_8 - prism

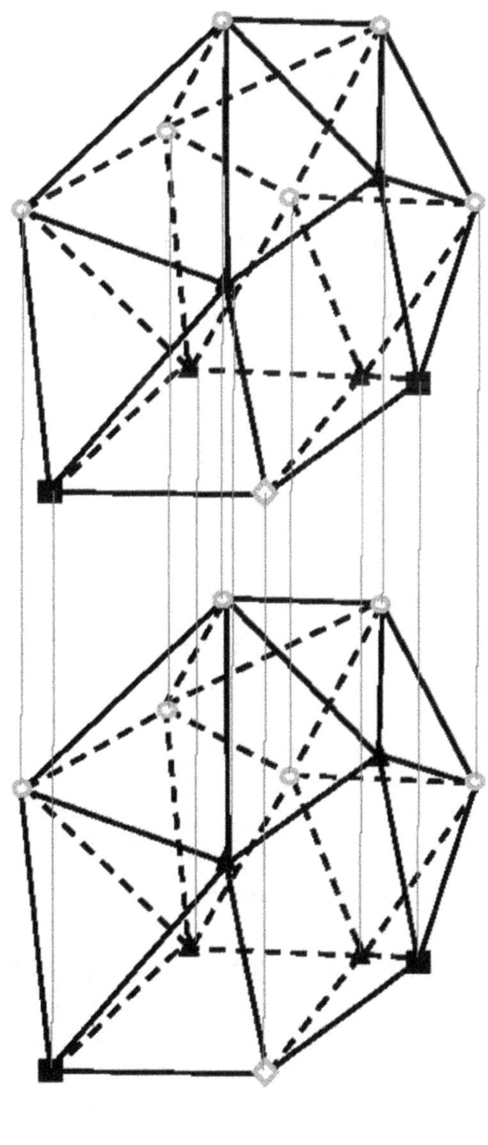

According to Theorem 6.3, the Euler-Poincaré equation for the product of polytopes is satisfied if this equation holds for the factors of the product of polytopes. Therefore, there is no need to check the fulfillment of the Euler-Poincaré equation for the product of polytopes, if it has already been proved that this equation holds for the factors.

The Polytope which is Product of the Polyhedron PG8 by Triangle

There we consider the product $P_3^{13}(18P_2^3, 2P_2^4) \times Q_2^3$. According to theorem 6.1 and ratios (6.5) there are:

$$f_0^{\times} = f_0^{P_3} f_0^{Q_2} = 13 \cdot 3 = 39,$$

$$f_1^{\times} = f_1^{P_3} f_0^{Q_2} + f_0^{P_3} f_1^{Q_2} = 31 \cdot 3 + 13 \cdot 3 = 132,.$$

$$f_2^{\times} = f_2^{P_3} f_0^{Q_2} + f_1^{P_3} f_1^{Q_2} + f_0^{P_3} f_2^{Q_2} = (18+2) \cdot 3 + 31 \cdot 3 + 13 \cdot 1 = 166 = 54P_2^{\times,3} + 6P_2^{\times,4} + 93P_2^{\times,4} + 13P_2^{\times,3},$$

(thus, in set of the elements with dimension 2 include 67 triangles and 99 quadrats)

$$f_3^{\times} = f_3^{P_3} f_0^{Q_2} + f_2^{P_3} f_1^{Q_2} + f_1^{P_3} f_2^{Q_2} = 1 \cdot 3 + (18+2) \cdot 3 + 31 \cdot 1 = 94 = 3P_3^{\times,13} + 54P_3^{\times,6} + 6P_3^{\times,8} + 31P_3^{\times,6},$$

(thus, in set of the elements with dimension 3 include 3 polyhedrons P_3^{13}, 85 triangles prisms and 6 cubes)

$$f_4^{\times} = f_3^{P_3} f_1^{Q_2} + f_2^{P_3} f_2^{Q_2} = 1 \cdot 3 + (18+2) \cdot 1 = 23 = 3P_4^{\times,26} + 18P_4^{\times,9} + 2P_4^{\times,12},$$

(thus, in set of the elements with dimension 4 include 3 polytopes PG_8 – prisms, 18 triangular prismahedrons and two 4*3 – angular prismahedrons, see Chapter 6)

$$f_5^{\times} = f_3^{P_3} f_2^{Q_2} = 1 \cdot 1 = 1 = P_5^{\times,39}(3P_4^{\times,26}, 18P_4^{\times,9}, 2P_4^{\times,12}). \tag{7.10}$$

Thus, the product $P_3^{13}(18P_2^3, 2P_2^4) \times Q_2^3$ is polytope with dimension 5 and hyperfaces in view of PG_8 – prisms, triangular and 4*3 – angular prismahedrons.

The Polytope which is Product of the Polyhedron PG8 by Tetrahedron

There we consider the product $P_3^{13}(18P_2^3, 2P_2^4) \times Q_3^4$. According to theorem 6.1 and ratios (6.5) there are:

$$f_0^{\times} = f_0^{P_3} f_0^{Q_3} = 13 \cdot 4 = 52,$$

$$f_1^{\times} = f_1^{P_3} f_0^{Q_3} + f_0^{P_3} f_1^{Q_3} = 31 \cdot 4 + 13 \cdot 6 = 202,.$$

$$f_2^{\times} = f_2^{P_3} f_0^{Q_3} + f_1^{P_3} f_1^{Q_3} + f_0^{P_3} f_2^{Q_3} = (18+2) \cdot 4 + 31 \cdot 6 + 13 \cdot 4 = 318 = 72P_2^{\times,3} + 8P_2^{\times,4} + 186P_2^{\times,4} + 52P_2^{\times,3},$$

(thus, in set of the elements with dimension 2 include 124 triangles and 194 quadrats)

$$f_3^{\times} = f_3^{P_3} f_0^{Q_3} + f_2^{P_3} f_1^{Q_3} + f_1^{P_3} f_2^{Q_3} + f_0^{P_3} f_3^{3} = 1 \cdot 4 + (18+2) \cdot 6 + 31 \cdot 4 + 13 \cdot 1$$
$$= 261 = 4P_3^{\times,13} + 106P_3^{\times,6} + 12P_3^{\times,8} + 124P_3^{\times,6} + 13P_3^{\times,4},$$

(thus, in set of the elements with dimension 3 include 4 polyhedrons P_3^{13}, 230 triangles prisms, 12 cubes and 13 tetrahedrons)

$$f_4^{\times} = f_3^{P_3} f_1^{Q_3} + f_2^{P_3} f_2^{Q_3} + f_1^{P_3} f_3^{Q_3} = 1 \cdot 6 + (18+2) \cdot 4 + 31 \cdot 1 = 117 = 6P_4^{\times,26} + 72P_4^{\times,9} + 8P_4^{\times,12} + 31P_4^{8},$$

(thus, in set of the elements with dimension 4 include 6 polytopes PG_8 – prisms, 72 triangular prismahedrons, either 4*3 – angular prismahedrons, and 31 tetrahedral prisms, see Chapter 6)

$$f_5^{\times} = f_3^{P_3} f_2^{Q_3} + f_2^{P_3} f_3^{Q_3} = 1 \cdot 4 + (18+2) \cdot 1 = 24 = 4P_5^{\times,39} + 18P_5^{\times,12} + 2P_5^{\times,16},$$
$$f_6^{\times} = f_3^{P_3} f_3^{Q_3} = 1 \cdot 1 = 1 = P_6^{\times,52}(4P_5^{\times,39}, 18P_5^{\times,12}, 2P_5^{\times,16}). \quad (7.11)$$

Thus, the product $P_3^{13}(18P_2^3, 2P_2^4) \times Q_3^4$ is polytope with dimension 6 and hyperfaces in view of $P_5^{39}, P_5^{12}, P_5^{16}$.

THE POPYTOPES WHICH ARE THE PRODUCTS OF POLYHEDRONS PG_{10} BY ONE – DIMENSIONAL SEGMENT, TRIANGLE, TETRAHEDRON

The Polytope which is Product of the Polyhedron PG10 by One-Dimensional Segment

A diagram of the PG_{10} polyhedron is presented in Chapter 2 in Figure 2.19.10. However, the small scale of Figure 2.19.10 makes it difficult to analyze the structure of this polyhedron. Let us represent the PG_{10} polyhedron in an enlarged form (Figure 5).

Figure 5. Polyhedron PG_{10}

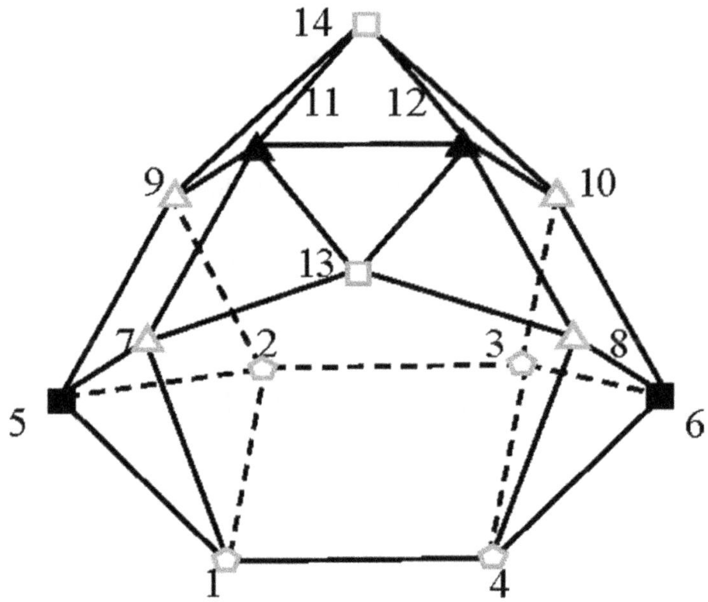

In the PG_{10} polyhedron, the number of types of gonohedrons participating in the formation of the polyhedron is two more than in the PG_9 polyhedron. Oddly enough, this led to a decrease in the number of vertices in the polyhedron. The set of vertices of the PG_{10} polyhedron form gonohedrons 11_5, 11_4, 12_4, $16_{4,5}$, $17_{4,5}$, i. e. five type of gonohedrons. The geometry of gonohedrons 11_5, 11_4, 12_4, $16_{4,5}$, $17_{4,5}$ is described in Chapter 2. The polyhedron in Figure 5 has a total the number of vertices 14, f_0=14. There are edges between the vertices

1 – 2, 3 – 4, 2 – 3, 1 – 4, 3 – 6, 2 – 5, 4 – 6, 1 – 5, 4 – 8, 1 – 7, 3 – 10, 2 – 9, 6 – 10, 9 – 5, 8 – 6, 5 – 7, 11 – 9, 12 – 10, 7 – 13, 8 – 13, 13 – 11, 13 – 12, 7 – 13, 8 – 12, 12 – 14, 11 – 14, 9 – 14, 10 – 14, 11 – 12.

The total number of edges is 29, f_1= 29. The edges on the surface of the PG_{10} polyhedron form 12 triangles

1 – 2 – 5, 1 – 7 – 5, 9 – 2 – 5, 4 – 6 – 3, 4 – 8 – 6, 6 – 3 – 10, 7 – 13 – 11, 8 – 12 – 13, 13 – 11 – 12, 11 – 12 – 14, 12 – 14 – 10, 9 – 11 – 14; and three squares 6 – 8 – 10 – 12, 9 – 5 – 7 – 11, 1 – 2 – 3 – 4,

and two pentagon 2 – 9 – 14 – 10 – 3, 1 – 7 – 13 – 8 – 4.

Thus, the total number of flat faces of the PG_{10} polyhedron is 17, f_2= 17. Substitute these values into the Euler-Poincaré equation (Poincaré, 1895) (7.1).

Then we have 14 – 29 + 17 = 2.

We are convinced that the polyhedron PG_{10} satisfies the Euler - Poincaré equation for n = 3, i. e. it is three-dimensional. The structure of this polyhedron, taking into account the obtained values of the numbers of various geometric elements included in the polyhedron, can be written as follows

$P_3^{14}(12P_2^3, 3P_2^4, 2P_2^5)$. Consider the product of the PG_{10} polyhedron by a one-dimensional segment $P_3^{14}(12P_2^3, 3P_2^4, 2P_2^5) \times Q_1^2$.
According to theorem 6.1 and the ratios (6.5) there are:

$$f_0^{\times} = f_0^{P_3} f_0^{Q_1} = 14 \cdot 2 = 28,$$

$$f_1^{\times} = f_1^{P_3} f_0^{Q_1} + f_0^{P_3} f_1^{Q_1} = 29 \cdot 2 + 14 \cdot 1 = 72,$$

$$f_2^{\times} = f_2^{P_3} f_0^{Q_1} + f_1^{P_3} f_1^{Q_1} = (12+3+2) \cdot 2 + 29 \cdot 1 = 63, \tag{7.12}$$

For receiving of the composition of the faces in the dimension 2 polytopes product one of the faces, observing the same rules as at the definition of the product polytopes, i. e. dimensions are added, and the numbers of vertices are multiplied. At that, indexes P and Q are replaced by index $\times$. Then 63 faces obtained in (7.12), are deciphered so

$$63 = 24P_2^{\times,3} + 6P_2^{\times,4} + 4P_2^{\times,5} + 31P_2^{\times,4}. \tag{7.13}$$

This means that 63 faces of dimension 2 in the product include twenty and four faces $P_2^{\times,3}$ (i.e., triangles), and thirty and seven faces $P_2^{\times,4}$ (i.e., quadrangles), and four faces $P_2^{\times,5}$.
For the faces of dimension 3, there is:

$$f_3^{\times} = f_3^{P_3} f_0^{Q_1} + f_2^{P_3} f_1^{Q_1} = 1 \cdot 2 + 12 \cdot 1 + 3 \cdot 1 + 2 \cdot 1 = 19 = 2P_3^{\times,14} + 12P_3^{\times,6} + 3P_3^{\times,8} + 2P_3^{\times,10}.$$

Thus, 19 faces of dimension 3 in the product consist of two polyhedrons $P_3^{14}(12P_2^3, 3P_2^4, 2P_2^5)$.,
eleven triangular prisms $P_3^6(3P_2^4, 2P_2^3)$, three quadrangular prisms $P_3^8(6P_2^4)$, and two pentangular prisms $P_3^{10}(5P_2^4, 2P_2^5)$.
For the faces of dimension 4, there is:

$$f_4^{\times} = f_3^{P_3} f_1^{Q_1} = 1 \cdot 1 = 1 = P_4^{\times,28}.$$

According to the preceding equation, can obtain the structural formula of the product:

$$\begin{aligned} &P_3^{14}(12P_2^3, 3P_2^4, 2P_2^5) \times Q_1^2 \\ &= P_4^{28}[2P_3^{14}(12P_2^3, 3P_2^4, 2P_2^5), 12P_3^6(3P_2^4, 2P_2^3), 3P_3^8(6P_2^4), 2P_3^{10}(5P_2^4, 2P_2^5)]. \end{aligned} \tag{7.14}$$

Therefore, the product of the polyhedron PG_{10} by the segment is polytope 4*D*, which includes eleven triangular prisms, three quadrangular prisms, two pentagonal prisms, and 2 polyhedron PG_{10}. You can call a polytope (7.14) PG_{10} - prism. The picture of the polytope (7.14) is shown in Figure 6.

Figure 6. The polytope PG_{10} - prism

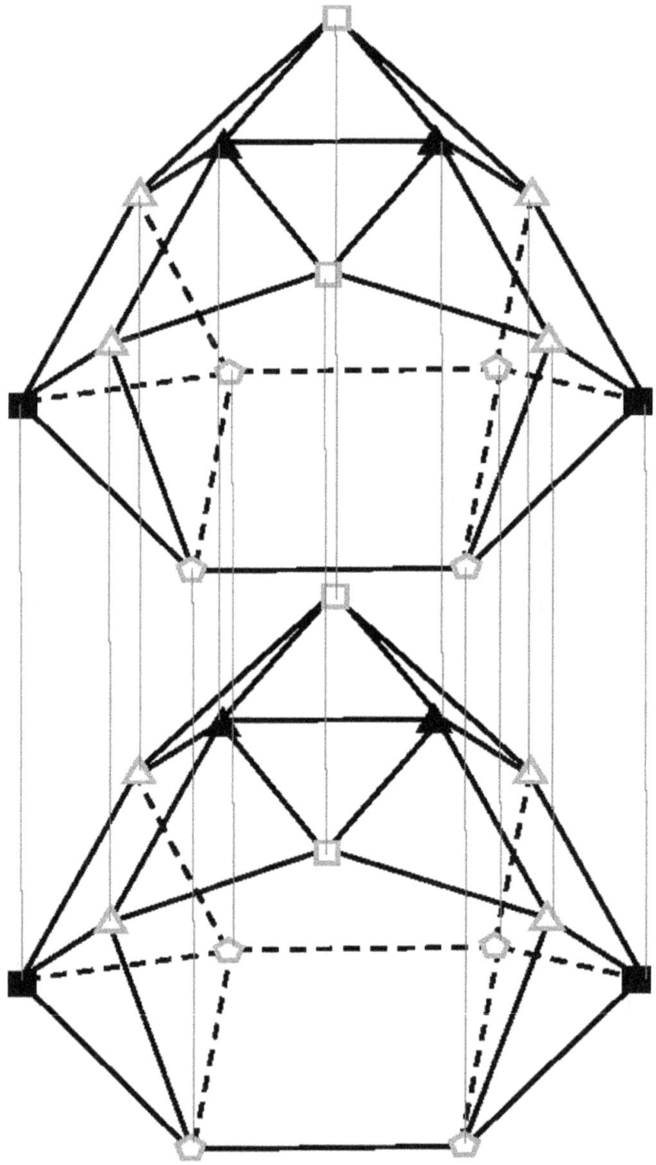

According to Theorem 6.3, the Euler-Poincaré equation for the product of polytopes is satisfied if this equation holds for the factors of the product of polytopes. Therefore, there is no need to check the fulfillment of the Euler-Poincaré equation for the product of polytopes, if it has already been proved that this equation holds for the factors.

The Polytope which is Product of the Polyhedron PG10 by Triangle

There we consider the product $P_3^{14}(12P_2^3, 3P_2^4, 2P_2^5) \times Q_2^3$. According to theorem 6.1 and ratios (6.5) there are:

$f_0^{\times} = f_0^{P_3} f_0^{Q_2} = 14 \cdot 3 = 42,$

$f_1^{\times} = f_1^{P_3} f_0^{Q_2} + f_0^{P_3} f_1^{Q_2} = 29 \cdot 3 + 14 \cdot 3 = 129,.$

$f_2^{\times} = f_2^{P_3} f_0^{Q_2} + f_1^{P_3} f_1^{Q_2} + f_0^{P_3} f_2^{Q_2} = (12+3+2) \cdot 3 + 29 \cdot 3 + 14 \cdot 1 = 152 =$
$36P_2^{\times,3} + 9P_2^{\times,4} + 6P_2^{\times,5} + 87P_2^{\times,4} + 14P_2^{\times,3},$

(thus, in set of the elements with dimension 2 include 50 triangles, and 96 quadrats, and 6 pentagon)

$f_3^{\times} = f_3^{P_3} f_0^{Q_2} + f_2^{P_3} f_1^{Q_2} + f_1^{P_3} f_2^{Q_2} = 1 \cdot 3 + (12+3+2) \cdot 3 + 29 \cdot 1 = 83 =$
$3P_3^{\times,14} + 36P_3^{\times,6} + 9P_3^{\times,8} + 6P_3^{10} + 29P_3^{\times,6},$

(thus, in set of the elements with dimension 3 include 3 polyhedrons P_3^{14}, 65 triangles prisms, and 9 cubes, and 6 pentangles prisms)

$f_4^{\times} = f_3^{P_3} f_1^{Q_2} + f_2^{P_3} f_2^{Q_2} = 1 \cdot 3 + (12+3+2) \cdot 1 = 20 = 3P_4^{\times,28} + 12P_4^{\times,9} + 3P_4^{\times,12} + 2P_4^{\times,15},$

(thus, in set of the elements with dimension 4 include 3 polytopes PG_{10} – prisms, 12 triangular prismahedrons, and three 4*3 – angular prismahedrons, and two 5*3 – angular prismahedrons, see Chapter 6)

$$f_5^{\times} = f_3^{P_3} f_2^{Q_2} = 1 \cdot 1 = 1 = P_5^{\times,42}(3P_4^{\times,28}, 12P_4^{\times,9}, 3P_4^{\times,12}, 2P_4^{15}). \tag{7.15}$$

Thus, the product $P_3^{14}(12P_2^3, 3P_2^4, 2P_2^5) \times Q_2^3$. is polytope with dimension 5 and hyperfaces in view of PG_{10} – prisms, triangular prismahedrons, and 4*3 – angular prismahedrons, and 5*3 – angular prismahedrons.

The Polytope which is Product of the Polyhedron PG10 by Tetrahedron

There we consider the product $P_3^{14}(12P_2^3, 3P_2^4, 2P_2^5) \times Q_3^4$. According to theorem 6.1 and ratios (6.5) there are:

$f_0^{\times} = f_0^{P_3} f_0^{Q_3} = 14 \cdot 4 = 56,$

$f_1^{\times} = f_1^{P_3} f_0^{Q_3} + f_0^{P_3} f_1^{Q_3} = 29 \cdot 4 + 14 \cdot 6 = 200,.$

$f_2^{\times} = f_2^{P_3} f_0^{Q_3} + f_1^{P_3} f_1^{Q_3} + f_0^{P_3} f_2^{Q_3} = (12+3+2) \cdot 4 + 29 \cdot 6 + 14 \cdot 4 = 298 =$
$48P_2^{\times,3} + 12P_2^{\times,4} + 8P_2^5 + 174P_2^{\times,4} + 56P_2^{\times,3},$

(thus, in set of the elements with dimension 2 include 104 triangles and 186 quadrats, and 8 pentagons)

$$f_3^{\times} = f_3^{P_3} f_0^{Q_3} + f_2^{P_3} f_1^{Q_3} + f_1^{P_3} f_2^{Q_3} + f_0^{P_3} f_3^{3} = 1\cdot 4 + (12+3+2)\cdot 6 + 29\cdot 4 + 14\cdot 1$$
$$= 236 = 4P_3^{\times,14} + 72P_3^{\times,6} + 18P_3^{\times,8} + 12P_3^{10} + 116P_3^{\times,6} + 14P_3^{\times,4},$$

(thus, in set of the elements with dimension 3 include 4 polyhedrons P_3^{14}, 188 triangles prisms, 18 cubes, and 12 pentagon prisms, and 14 tetrahedrons)

$$f_4^{\times} = f_3^{P_3} f_1^{Q_3} + f_2^{P_3} f_2^{Q_3} + f_1^{P_3} f_3^{Q_3} = 1\cdot 6 + (12+3+2)\cdot 4 + 29\cdot 1 = 103 =$$
$$6P_4^{\times,28} + 68P_4^{\times,9} + 12P_4^{\times,12} + 8P_4^{15} + 29P_4^{8},$$

(thus, in set of the elements with dimension 4 include 6 polytopes PG_{10} – prisms, 68 triangular prismahedrons, eleven 4*3 – angular prismahedrons, either 5*3 – angular prismahedrons, and 29 tetrahedral prisms, see Chapter 6)

$$f_5^{\times} = f_3^{P_3} f_2^{Q_3} + f_2^{P_3} f_3^{Q_3} = 1\cdot 4 + (12+3+2)\cdot 1 = 21 = 4P_5^{\times,42} + 12P_5^{\times,12} + 3P_5^{\times,16} + 2P_5^{\times,20},$$
$$f_6^{\times} = f_3^{P_3} f_3^{Q_3} = 1\cdot 1 = 1 = P_6^{\times,56}(4P_5^{\times,42}, 12P_5^{\times,12}, 3P_5^{\times,16}, 2P_5^{20}). \tag{7.16}$$

Thus, the product $P_3^{14}(12P_2^3, 3P_2^4, 2P_2^5)\times Q_3^4$ is polytope with dimension 6 and hyperfaces in view of $P_5^{42}, P_5^{12}, P_5^{16}, P_5^{20}$.

THE POPYTOPES WHICH ARE THE PRODUCTS OF POLYHEDRONS PG_{11} BY ONE – DIMENSIONAL SEGMENT, TRIANGLE, TETRAHEDRON

The Polytope which is Product of the Polyhedron PG11 by One-Dimensional Segment

A diagram of the PG_{11} polyhedron is presented in Chapter 2 in Figure 2.19.11. However, the small scale of Figure 2.19.11 makes it difficult to analyze the structure of this polyhedron. Let us represent the PG_{11} polyhedron in an enlarged form (Figure 7).

Figure 7. Polyhedron PG_{11}

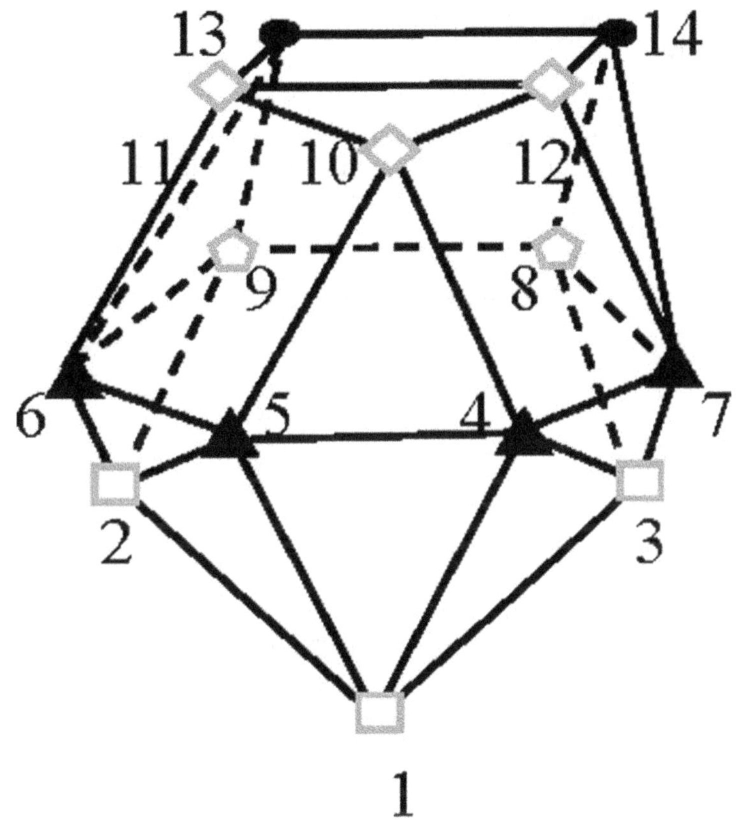

In the PG_{11} polyhedron, the number of types of gonohedrons participating in the formation of the polyhedron is two more than in the PG_9 polyhedron. Oddly enough, this no led to a decrease in the number of vertices in the polyhedron. The set of vertices of the PG_{11} polyhedron form gonohedrons 11_5, 10_4, 12_4, 9_4, $17_{4,5}$, i. e. five type of gonohedrons. The geometry of gonohedrons 11_5, 10_4, 12_4, 9_4, $17_{4,5}$ is described in Chapter 2. The polyhedron in Figure 7 has a total the number of vertices 14, f_0=14. There are edges between the vertices

1 – 2, 3 – 1, 1 – 5, 1 – 4, 2 – 5, 4 – 3, 2 – 9, 3 – 8, 2 – 6, 3 – 7, 5 – 10, 4 – 10, 6 – 5, 4 – 7, 11 – 6, 12 – 7, 11 – 10, 12 – 10, 6 – 13, 7 – 14, 13 – 11, 14 – 12, 9 – 13, 8 – 14, 6 – 9, 7 – 8, 5 – 4, 9 – 8, 11 – 12, 13 – 14.

The total number of edges is 30, f_1= 30. The edges on the surface of the PG_{11} polyhedron form 13 triangles

1 – 2 – 5, 1 – 4 – 5, 1 – 3 – 4, 2 – 6 – 5, 4 – 3 – 7, 5 – 4 – 10, 10 – 12 – 11, 2 – 6 – 9, 3 – 7 – 8, 7 – 8 – 14, 6 – 9 – 13, 6 – 11 – 13, 7 – 12 - 14; and fore squares 6 – 11 – 10 – 5, 9 – 13 – 8 – 14, 7 – 10 – 12 – 4, 11 – 13 – 12 – 14,

and one pentagon 2 – 1 – 8– 9 – 3. Thus, the total number of flat faces of the PG$_{11}$ polyhedron is 18, f_2= 18. Substitute these values into the Euler-Poincaré equation (Poincaré, 1895) (7.1).
Then we have 14 – 30 + 18 = 2.

We are convinced that the polyhedron PG$_{11}$ satisfies the Euler - Poincaré equation for $n = 3$, i. e. it is three-dimensional. The structure of this polyhedron, taking into account the obtained values of the numbers of various geometric elements included in the polyhedron, can be written as follows $P_3^{14}(13P_2^3, 4P_2^4, P_2^5)$. Consider the product of the PG$_{11}$ polyhedron by a one-dimensional segment $P_3^{14}(13P_2^3, 4P_2^4, P_2^5)\times Q_1^2$.
According to theorem 6.1 and the ratios (6.5) there are:

$$f_0^{\times} = f_0^{P_3} f_0^{Q_1} = 14 \cdot 2 = 28,$$

$$f_1^{\times} = f_1^{P_3} f_0^{Q_1} + f_0^{P_3} f_1^{Q_1} = 30\cdot 2 + 14\cdot 1 = 74,$$

$$f_2^{\times} = f_2^{P_3} f_0^{Q_1} + f_1^{P_3} f_1^{Q_1} = (13+4+1)\cdot 2 + 30\cdot 1 = 66, \tag{7.17}$$

For receiving of the composition of the faces in the dimension 2 polytopes product one of the faces, observing the same rules as at the definition of the product polytopes, i. e. dimensions are added, and the numbers of vertices are multiplied. At that, indexes P and Q are replaced by index ×. Then 66 faces obtained in (7.17), are deciphered so

$$66 = 26P_2^{\times,3} + 8P_2^{\times,4} + 2P_2^{\times,5} + 30P_2^{\times,4}. \tag{7.18}$$

This means that 66 faces of dimension 2 in the product include twenty and six faces $P_2^{\times,3}$ (i.e., triangles), and thirty and eight faces $P_2^{\times,4}$ (i.e., quadrangles), and two faces $P_2^{\times,5}$.
For the faces of dimension 3, there is:

$$f_3^{\times} = f_3^{P_3} f_0^{Q_1} + f_2^{P_3} f_1^{Q_1} = 1\cdot 2 + 13\cdot 1 + 4\cdot 1 + 1\cdot 1 = 20 = 2P_3^{\times,14} + 13P_3^{\times,6} + 4P_3^{\times,8} + P_3^{\times,10}.$$

Thus, 20 faces of dimension 3 in the product consist of two polyhedrons $P_3^{14}(13P_2^3, 4P_2^4, P_2^5)$., thirteen triangular prisms $P_3^6(3P_2^4, 2P_2^3)$, four quadrangular prisms $P_3^8(6P_2^4)$, and one pentangular prisms $P_3^{10}(5P_2^4, 2P_2^5)$.
For the faces of dimension 4, there is:

$$f_4^{\times} = f_3^{P_3} f_1^{Q_1} = 1\cdot 1 = 1 = P_4^{\times,28}.$$

According to the preceding equation, can obtain the structural formula of the product:

$$P_3^{14}(13P_2^3, 4P_2^4, P_2^5) \times Q_1^2 \\ = P_4^{28}[2P_3^{14}(13P_2^3, 4P_2^4, P_2^5), 13P_3^6(3P_2^4, 2P_2^3), 4P_3^8(6P_2^4), P_3^{10}(5P_2^4, 2P_2^5)]. \quad (7.19)$$

Therefore, the product of the polyhedron PG_{11} by the segment is polytope 4*D*, which includes eleven triangular prisms, three quadrangular prisms, two pentagonal prisms, and 2 polyhedron PG_{11}. You can call a polytope (7.19) PG_{11} - prism. The picture of the polytope (7.19) is shown in Figure 8.

Figure 8. The polytope PG_{11} - prism

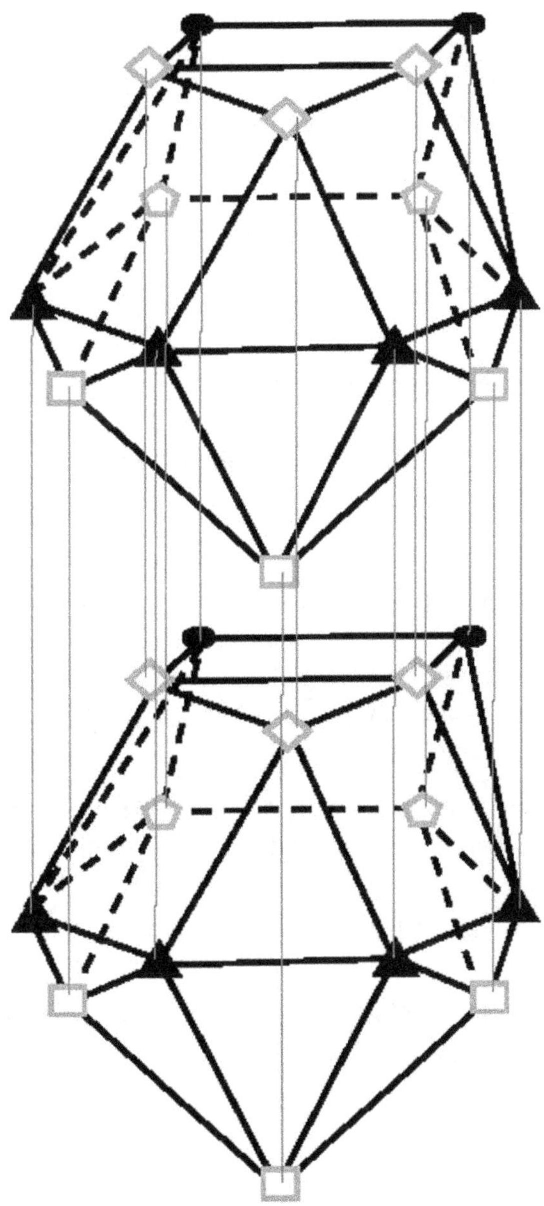

According to Theorem 6.3, the Euler-Poincaré equation for the product of polytopes is satisfied if this equation holds for the factors of the product of polytopes. Therefore, there is no need to check the fulfillment of the Euler-Poincaré equation for the product of polytopes, if it has already been proved that this equation holds for the factors.

The Polytope which is Product of the Polyhedron PG11 by Triangle

There we consider the product $P_3^{14}(13P_2^3, 4P_2^4, P_2^5) \times Q_2^3$. According to theorem 6.1 and ratios (6.5) there are:

$$f_0^{\times} = f_0^{P_3} f_0^{Q_2} = 14 \cdot 3 = 42,$$

$$f_1^{\times} = f_1^{P_3} f_0^{Q_2} + f_0^{P_3} f_1^{Q_2} = 30 \cdot 3 + 14 \cdot 3 = 132, .$$

$$f_2^{\times} = f_2^{P_3} f_0^{Q_2} + f_1^{P_3} f_1^{Q_2} + f_0^{P_3} f_2^{Q_2} = (13+4+1) \cdot 3 + 30 \cdot 3 + 14 \cdot 1 = 158 =$$
$$39P_2^{\times,3} + 12P_2^{\times,4} + 3P_2^{\times,5} + 90P_2^{\times,4} + 14P_2^{\times,3},$$

(thus, in set of the elements with dimension 2 include 53 triangles, and 102 quadrats, and 3 pentagon)

$$f_3^{\times} = f_3^{P_3} f_0^{Q_2} + f_2^{P_3} f_1^{Q_2} + f_1^{P_3} f_2^{Q_2} = 1 \cdot 3 + (13+4+1) \cdot 3 + 30 \cdot 1 = 87 =$$
$$3P_3^{\times,14} + 39P_3^{\times,6} + 12P_3^{\times,8} + 3P_3^{10} + 30P_3^{\times,6},$$

(thus, in set of the elements with dimension 3 include 3 polyhedrons P_3^{14}, 69 triangles prisms, and 12 cubes, and 3 pentangles prisms)

$$f_4^{\times} = f_3^{P_3} f_1^{Q_2} + f_2^{P_3} f_2^{Q_2} = 1 \cdot 3 + (13+4+1) \cdot 1 = 21 = 3P_4^{\times,28} + 13P_4^{\times,9} + 4P_4^{\times,12} + P_4^{\times,15},$$

(thus, in set of the elements with dimension 4 include 3 polytopes PG_{11} – prisms, 13 triangular prismahedrons, and four 4*3 – angular prismahedrons, and one 5*3 – angular prismahedrons, see Chapter 6)

$$f_5^{\times} = f_3^{P_3} f_2^{Q_2} = 1 \cdot 1 = 1 = P_5^{\times,42}(3P_4^{\times,28}, 13P_4^{\times,9}, 4P_4^{\times,12}, P_4^{\times,15}). \tag{7.20}$$

Thus, the product $P_3^{14}(13P_2^3, 4P_2^4, P_2^5) \times Q_2^3$. is polytope with dimension 5 and hyperfaces in view of PG_{11} – prisms, triangular prismahedrons, and 4*3 – angular prismahedrons, and 5*3 – angular prismahedrons.

The Polytope which is Product of the Polyhedron PG11 by Tetrahedron

There we consider the product $P_3^{14}(13P_2^3, 4P_2^4, P_2^5)\times Q_3^4$. According to theorem 6.1 and ratios (6.5) there are:

$$f_0^{\times} = f_0^{P_3} f_0^{Q_3} = 14 \cdot 4 = 56,$$

$$f_1^{\times} = f_1^{P_3} f_0^{Q_3} + f_0^{P_3} f_1^{Q_3} = 30 \cdot 4 + 14 \cdot 6 = 204,.$$

$$f_2^{\times} = f_2^{P_3} f_0^{Q_3} + f_1^{P_3} f_1^{Q_3} + f_0^{P_3} f_2^{Q_3} = (13+4+1)\cdot 4 + 30\cdot 6 + 14\cdot 4 = 308 = \\ 42P_2^{\times,3} + 16P_2^{\times,4} + 4P_2^5 + 180P_2^{\times,4} + 56P_2^{\times,3},$$

(thus, in set of the elements with dimension 2 include 98 triangles and 196 quadrats, and 4 pentagons)

$$f_3^{\times} = f_3^{P_3} f_0^{Q_3} + f_2^{P_3} f_1^{Q_3} + f_1^{P_3} f_2^{Q_3} + f_0^{P_3} f_3^{3} = 1\cdot 4 + (13+4+1)\cdot 6 + 30\cdot 4 + 14\cdot 1 \\ = 354 = 4P_3^{\times,14} + 78P_3^{\times,6} + 24P_3^{\times,8} + 6P_3^{10} + 120P_3^{\times,6} + 14P_3^{\times,4},$$

(thus, in set of the elements with dimension 3 include 4 polyhedrons P_3^{14}, 198 triangles prisms, 24 cubes, and 6 pentagon prisms, and 14 tetrahedrons)

$$f_4^{\times} = f_3^{P_3} f_1^{Q_3} + f_2^{P_3} f_2^{Q_3} + f_1^{P_3} f_3^{Q_3} = 1\cdot 6 + (13+4+1)\cdot 4 + 30\cdot 1 = 108 = \\ 6P_4^{\times,28} + 52P_4^{\times,9} + 16P_4^{\times,12} + 4P_4^{15} + 30P_4^{8},$$

(thus, in set of the elements with dimension 4 include 6 polytopes PG_{11} – prisms, 52 triangular prismahedrons, sixteen 4*3 – angular prismahedrons, four 5*3 – angular prismahedrons, and 30 tetrahedral prisms, see Chapter 6)

$$f_5^{\times} = f_3^{P_3} f_2^{Q_3} + f_2^{P_3} f_3^{Q_3} = 1\cdot 4 + (13+4+1)\cdot 1 = 22 = 4P_5^{\times,42} + 13P_5^{\times,12} + 4P_5^{\times,16} + P_5^{\times,20}, \\ f_6^{\times} = f_3^{P_3} f_3^{Q_3} = 1\cdot 1 = 1 = P_6^{\times,56}(4P_5^{\times,42}, 13P_5^{\times,12}, 4P_5^{\times,16}, P_5^{20}). \qquad (7.21)$$

Thus, the product $P_3^{14}(13P_2^3, 4P_2^4, P_2^5)\times Q_3^4$ is polytope with dimension 6 and hyperfaces in view of $P_5^{42}, P_5^{12}, P_5^{16}, P_5^{20}$.

THE POPYTOPES WHICH ARE THE PRODUCTS OF POLYHEDRONS PG_{35} BY ONE – DIMENSIONAL SEGMENT, TRIANGLE, TETRAHEDRON

The Polytope which is Product of the Polyhedron PG35 by One-Dimensional Segment

A diagram of the PG_{35} polyhedron is presented in Chapter 2 in Figure 2.19.35. However, the small scale of Figure 2.19.35 makes it difficult to analyze the structure of this polyhedron. Let us represent the PG_{35} polyhedron in an enlarged form (Figure 9).

Figure 9. Polyhedron PG_{35}

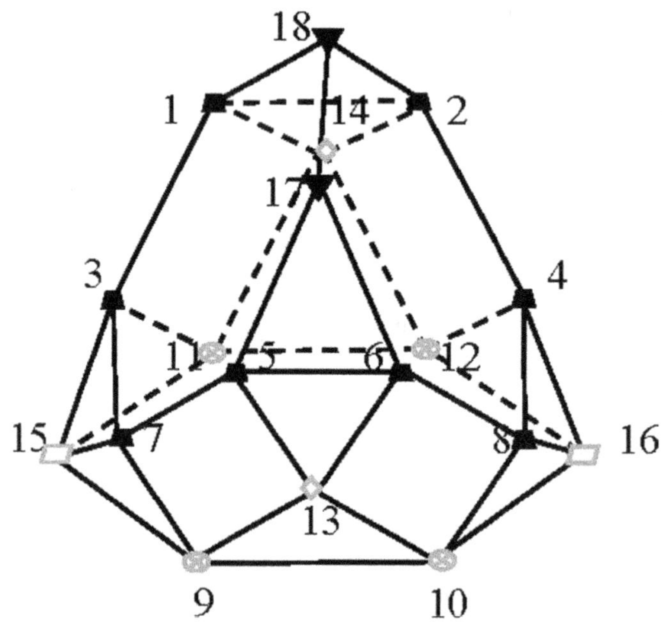

In the PG_{35} polyhedron, the number of types of gonohedrons participating in the formation of the polyhedron is two more than in the PG_9 polyhedron. Oddly enough, this no led to a decrease in the number of vertices in the polyhedron. The set of vertices of the PG_{35} polyhedron form gonohedrons 11_6, $6_{3,6}$, $16_{4,6}$, 9_4, $17_{4,6}$, i. e. five type of gonohedrons. The geometry of gonohedrons 11_6, $6_{3,6}$, $16_{4,6}$, 9_4, $17_{4,6}$ is described in Chapter 2. The polyhedron in Figure 9 has a total the number of vertices 18, f_0=18. There are edges between the vertices

1 – 2, 11 – 12, 6 – 5, 9 – 10, 18 – 17, 1 – 18, 2 – 18, 1 – 14, 2 – 14, 3 – 1, 2 – 4, 14 – 11, 14 – 12, 3 – 11, 4 – 12, 3 – 7, 4 – 8, 3 – 15, 16 – 4, 17 – 5, 6 – 17, 7 – 5, 6 – 8, 7 – 15, 16 – 8, 5 – 13, 6 – 13, 9 – 7, 8 – 10, 15 – 9, 10 – 16, 9 – 13, 10 – 13, 12 – 16, 15 – 11.

The total number of edges is 35, f_1= 35. The edges on the surface of the PG_{35} polyhedron form 12 triangles

1 – 2 – 18, 1 – 14 – 2, 11 – 12 – 14, 3 – 11 – 15, 4 – 12 – 16, 5 – 6 – 17, 5 – 6 – 13, 4 – 8 – 16, 10 – 16 – 8, 7 – 3 – 15, 7 – 3 – 15, 9 – 10 – 13; and fore squares 1 – 11 – 14 – 3, 2 – 12 – 4 – 14, 7 – 13 – 5 – 9, 6 – 13 – 10 – 8,

and three hexagons 18 – 1 – 17 – 5 – 7 – 3, 17 – 18 – 2 – 4 – 8 – 6, 9 – 10 – 16 – 12 – 11 - 15. Thus, the total number of flat faces of the PG_{35} polyhedron is 19, f_2= 19. Substitute these values into the Euler-Poincaré equation (Poincaré, 1895) (7.1).
Then we have 18 – 35 + 19 = 2.

We are convinced that the polyhedron PG_{35} satisfies the Euler - Poincaré equation for n = 3, i. e. it is three-dimensional. The structure of this polyhedron, taking into account the obtained values of the numbers of various geometric elements included in the polyhedron, can be written as follows $P_3^{18}(12P_2^3,4P_2^4,3P_2^6)$. Consider the product of the PG_{35} polyhedron by a one-dimensional segment $P_3^{18}(12P_2^3,4P_2^4,3P_2^6)\times Q_1^2$.
According to theorem 6.1 and the ratios (6.5) there are:

$$f_0^{\times} = f_0^{P_3} f_0^{Q_1} = 18 \cdot 2 = 36,$$

$$f_1^{\times} = f_1^{P_3} f_0^{Q_1} + f_0^{P_3} f_1^{Q_1} = 35 \cdot 2 + 18 \cdot 1 = 88,$$

$$f_2^{\times} = f_2^{P_3} f_0^{Q_1} + f_1^{P_3} f_1^{Q_1} = (12+4+3) \cdot 2 + 35 \cdot 1 = 73, \tag{7.22}$$

For receiving of the composition of the faces in the dimension 2 polytopes product one of the faces, observing the same rules as at the definition of the product polytopes, i. e. dimensions are added, and the numbers of vertices are multiplied. At that, indexes P and Q are replaced by index ×. Then 73 faces obtained in (7.22), are deciphered so

$$73 = 24P_2^{\times,3} + 8P_2^{\times,4} + 6P_2^{\times,6} + 35P_2^{\times,4}. \tag{7.23}$$

This means that 73 faces of dimension 2 in the product include twenty and four faces $P_2^{\times,3}$ (i.e., triangles), and forty and three faces $P_2^{\times,4}$ (i.e., quadrangles), and six faces $P_2^{\times,6}$.
For the faces of dimension 3, there is:

$$f_3^{\times} = f_3^{P_3} f_0^{Q_1} + f_2^{P_3} f_1^{Q_1} = 1 \cdot 2 + 12 \cdot 1 + 4 \cdot 1 + 3 \cdot 1 = 21 = 2P_3^{\times,18} + 12P_3^{\times,6} + 4P_3^{\times,8} + P_3^{\times,12}.$$

Thus, 21 faces of dimension 3 in the product consist of two polyhedrons $P_3^{18}(12P_2^3,4P_2^4,3P_2^6)$, eleven triangular prisms $P_3^6(3P_2^4,2P_2^3)$, four quadrangular prisms $P_3^8(6P_2^4)$, and three hexagonal prisms $P_3^{12}(6P_2^4,2P_2^6)$.
For the faces of dimension 4, there is:

$f_4^{\times} = f_3^{P_3} f_1^{Q_1} = 1 \cdot 1 = 1 = P_4^{\times,36}.$

According to the preceding equation, can obtain the structural formula of the product:

$$\begin{aligned} & P_3^{18}(12P_2^3, 4P_2^4, 3P_2^6) \times Q_1^2 \\ & = P_4^{36}[2P_3^{18}(12P_2^3, 4P_2^4, 3P_2^6), 12P_3^6(3P_2^4, 2P_2^3), 4P_3^8(6P_2^4), 3P_3^{12}(6P_2^4, 2P_2^6)]. \end{aligned} \tag{7.24}$$

Therefore, the product of the polyhedron PG_{35} by the segment is polytope 4*D*, which includes eleven triangular prisms, four quadrangular prisms, three hexogonal prisms, and 2 polyhedron PG_{35}. You can call a polytope (7.24) PG_{35} - prism. The picture of the polytope (7.24) is shown in Figure 10.

Figure 10. The polytope PG_{35} - prism

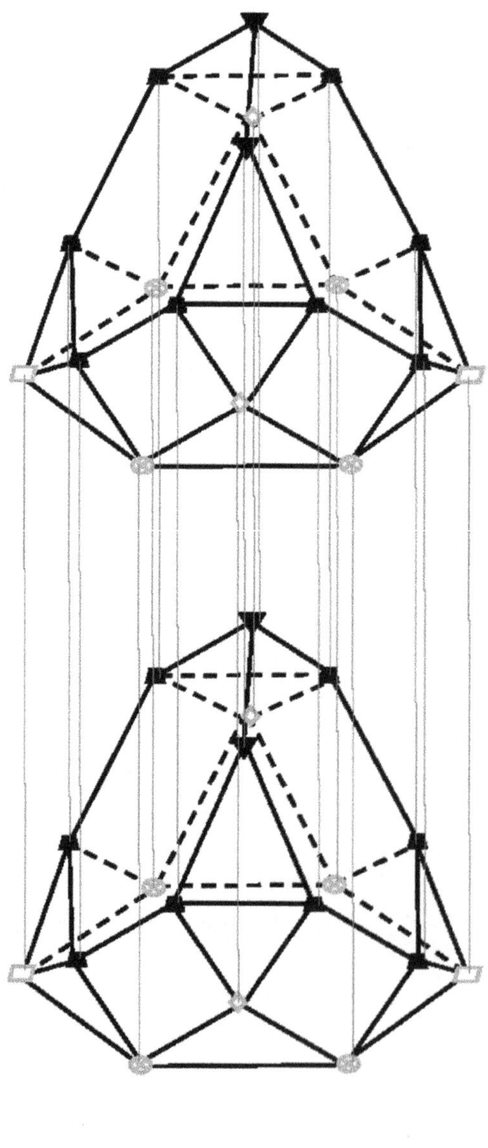

According to Theorem 6.3, the Euler-Poincaré equation for the product of polytopes is satisfied if this equation holds for the factors of the product of polytopes. Therefore, there is no need to check the fulfillment of the Euler-Poincaré equation for the product of polytopes, if it has already been proved that this equation holds for the factors.

The Polytope which is Product of the Polyhedron PG35 by Triangle

There we consider the product $P_3^{18}(12P_2^3, 4P_2^4, 3P_2^6) \times Q_2^3$. According to theorem 6.1 and ratios (6.5) there are:

$$f_0^{\times} = f_0^{P_3} f_0^{Q_2} = 18 \cdot 3 = 54,$$

$$f_1^{\times} = f_1^{P_3} f_0^{Q_2} + f_0^{P_3} f_1^{Q_2} = 35 \cdot 3 + 18 \cdot 3 = 159,$$

$$f_2^{\times} = f_2^{P_3} f_0^{Q_2} + f_1^{P_3} f_1^{Q_2} + f_0^{P_3} f_2^{Q_2} = (12+4+3) \cdot 3 + 35 \cdot 3 + 18 \cdot 1 = 180 =$$
$$36P_2^{\times,3} + 12P_2^{\times,4} + 9P_2^{\times,6} + 105P_2^{\times,4} + 18P_2^{\times,3},$$

(thus, in set of the elements with dimension 2 include 54 triangles, and 117 quadrats, and 9 hexagons)

$$f_3^{\times} = f_3^{P_3} f_0^{Q_2} + f_2^{P_3} f_1^{Q_2} + f_1^{P_3} f_2^{Q_2} = 1 \cdot 3 + (12+4+3) \cdot 3 + 35 \cdot 1 = 95 =$$
$$3P_3^{\times,18} + 36P_3^{\times,6} + 12P_3^{\times,8} + 9P_3^{12} + 35P_3^{\times,6},$$

(thus, in set of the elements with dimension 3 include 3 polyhedrons P_3^{18}, 71 triangles prisms, and 12 cubes, and 9 hexagons prisms)

$$f_4^{\times} = f_3^{P_3} f_1^{Q_2} + f_2^{P_3} f_2^{Q_2} = 1 \cdot 3 + (12+4+3) \cdot 1 = 22 = 3P_4^{\times,36} + 12P_4^{\times,9} + 4P_4^{\times,12} + 3P_4^{\times,18},$$

(thus, in set of the elements with dimension 4 include 3 polytopes PG_{35} – prisms, 12 triangular prismahedrons, and four 4*3 – angular prismahedrons, and three 6*3 – angular prismahedrons, see Chapter 6)

$$f_5^{\times} = f_3^{P_3} f_2^{Q_2} = 1 \cdot 1 = 1 = P_5^{\times,54}(3P_4^{\times,36}, 12P_4^{\times,9}, 4P_4^{\times,12}, 3P_4^{\times,18}). \quad (7.25)$$

Thus, the product $P_3^{18}(12P_2^3, 4P_2^4, 3P_2^6) \times Q_2^3$ is polytope with dimension 5 and hyperfaces in view of PG_{35} – prisms, triangular prismahedrons, and 4*3 – angular prismahedrons, and 6*3 – angular prismahedrons.

The Polytope which is Product of the Polyhedron PG35 by Tetrahedron

There we consider the product $P_3^{18}(12P_2^3, 4P_2^4, 3P_2^6)\times Q_3^4$. According to theorem 6.1 and ratios (6.5) there are:

$$f_0^{\times} = f_0^{P_3} f_0^{Q_3} = 18 \cdot 4 = 726,$$

$$f_1^{\times} = f_1^{P_3} f_0^{Q_3} + f_0^{P_3} f_1^{Q_3} = 35\cdot 4 + 18\cdot 6 = 248,.$$

$$f_2^{\times} = f_2^{P_3} f_0^{Q_3} + f_1^{P_3} f_1^{Q_3} + f_0^{P_3} f_2^{Q_3} = (12+4+3)\cdot 4 + 35\cdot 6 + 18\cdot 4 = 358 =$$
$$48P_2^{\times,3} + 16P_2^{\times,4} + 12P_2^6 + 210P_2^{\times,4} + 72P_2^{\times,3},$$

(thus, in set of the elements with dimension 2 include 120 triangles and 226 quadrats, and 12 hexagons)

$$f_3^{\times} = f_3^{P_3} f_0^{Q_3} + f_2^{P_3} f_1^{Q_3} + f_1^{P_3} f_2^{Q_3} + f_0^{P_3} f_3^{3} = 1\cdot 4 + (12+4+3)\cdot 6 + 35\cdot 4 + 18\cdot 1$$
$$= 276 = 4P_3^{\times,18} + 72P_3^{\times,6} + 24P_3^{\times,8} + 18P_3^{12} + 140P_3^{\times,6} + 18P_3^{\times,4},$$

(thus, in set of the elements with dimension 3 include 4 polyhedrons P_3^{18}, 212 triangles prisms, 24 cubes, and 18 hexagon prisms, and 18 tetrahedrons)

$$f_4^{\times} = f_3^{P_3} f_1^{Q_3} + f_2^{P_3} f_2^{Q_3} + f_1^{P_3} f_3^{Q_3} = 1\cdot 6 + (12+4+3)\cdot 4 + 35\cdot 1 = 117 =$$
$$6P_4^{\times,36} + 48P_4^{\times,9} + 16P_4^{\times,12} + 12P_4^{18} + 35P_4^{8},$$

(thus, in set of the elements with dimension 4 include 6 polytopes PG_{35} – prisms, 48 triangular prismahedrons, sixteen 4*3 – angular prismahedrons, eleven 6*3 – angular prismahedrons, and 35 tetrahedral prisms, see Chapter 6)

$$f_5^{\times} = f_3^{P_3} f_2^{Q_3} + f_2^{P_3} f_3^{Q_3} = 1\cdot 4 + (12+4+3)\cdot 1 = 23 = 4P_5^{\times,54} + 12P_5^{\times,12} + 4P_5^{\times,16} + 3P_5^{\times,24},$$
$$f_6^{\times} = f_3^{P_3} f_3^{Q_3} = 1\cdot 1 = 1 = P_6^{\times,72}(4P_5^{\times,54}, 12P_5^{\times,12}, 4P_5^{\times,16}, 3P_5^{24}). \quad (7.26)$$

Thus, the product $P_3^{18}(12P_2^3, 4P_2^4, 3P_2^6)\times Q_3^4$ is polytope with dimension 6 and hyperfaces in view of $P_5^{54}, P_5^{12}, P_5^{16}, P_5^{24}$.

THE POPYTOPES WHICH ARE THE PRODUCTS OF POLYHEDRONS PG_{14} BY ONE – DIMENSIONAL SEGMENT, TRIANGLE, TETRAHEDRON

The Polytope which is Product of the Polyhedron PG14 by One-Dimensional Segment

A diagram of the PG_{14} polyhedron is presented in Chapter 2 in Figure 2.19.14. However, the small scale of Figure 2.19.14 makes it difficult to analyze the structure of this polyhedron. Let us represent the PG_{14} polyhedron in an enlarged form (Figure 11).

Figure 11. Polyhedron PG_{14}

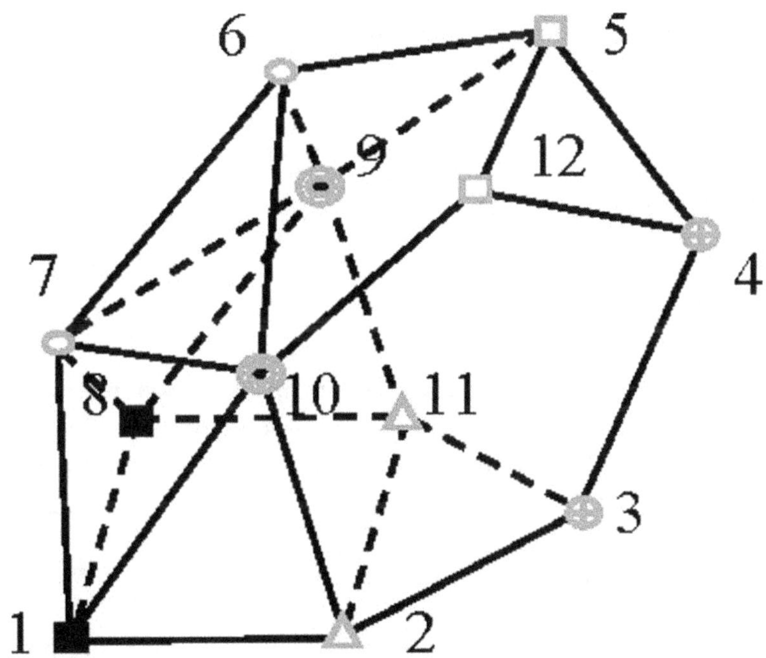

In the polytypichedron - max PG_{14} polyhedron, the number of types of gonohedrons participating in the formation of the polyhedron is three more than in the PG_9 polyhedron. Oddly enough, this led to a decrease in the number of vertices in the polyhedron. The set of vertices of the PG_{14} polyhedron form gonohedrons 11_4, $6_{3,5}$, $16_{4,5}$, 3, 12_5, 11_5, i. e. sex type of gonohedrons. The geometry of gonohedrons 11_4, $6_{3,5}$, $16_{4,5}$, 3, 12_5, 11_5 is described in Chapter 2. The polyhedron in Figure 11 has a total the number of vertices 12, f_0=12. There are edges between the vertices

5 – 12, 11 – 2, 1 – 8, 7 – 6, 3 – 4, 1 – 7, 7 – 8, 1 – 10, 8 – 9, 7 – 10, 7 – 9, 8 – 11, 1 – 2, 3 – 11, 3 – 2, 2 – 10, 6 – 10, 6 – 9, 10 – 12, 9 – 5, 6 – 12, 6 – 5, 12 – 4, 5 – 4.

The total number of edges is 25, f_1= 25. The edges on the surface of the PG_{14} polyhedron form 12 triangles

1 – 7 – 8, 5 – 4 – 12, 11 – 2 – 3, 6 – 12 – 5, 7 – 10 – 1, 7 – 9 – 8, 7 – 6 – 10, 7 – 6 – 9, 10 – 6 – 12, 6 – 9 – 5, 8 – 9 – 11, 1 – 10 – 2;

and one squares 1 – 11 – 8 – 2, and two pentagons 2 – 3 – 12 – 10 – 4, 3 – 11 – 9 – 4 – 5. Thus, the total number of flat faces of the PG_{14} polyhedron is 15, f_2= 15. Substitute these values into the Euler-Poincaré equation (Poincaré, 1895) (7.1).
Then we have 12 – 25 + 15 = 2.

We are convinced that the polyhedron PG_{14} satisfies the Euler - Poincaré equation for n = 3, i. e. it is three-dimensional. The structure of this polyhedron, taking into account the obtained values of the numbers of various geometric elements included in the polyhedron, can be written as follows $P_3^{12}(12P_2^3, P_2^4, 2P_2^5)$. Consider the product of the PG_{14} polyhedron by a one-dimensional segment $P_3^{12}(12P_2^3, P_2^4, 2P_2^5)\times Q_1^2$.
According to theorem 6.1 and the ratios (6.5) there are:

$$f_0^{\times} = f_0^{P_3} f_0^{Q_1} = 12 \cdot 2 = 24,$$

$$f_1^{\times} = f_1^{P_3} f_0^{Q_1} + f_0^{P_3} f_1^{Q_1} = 25 \cdot 2 + 12 \cdot 1 = 62,$$

$$f_2^{\times} = f_2^{P_3} f_0^{Q_1} + f_1^{P_3} f_1^{Q_1} = (12+1+2) \cdot 2 + 12 \cdot 1 = 62, \tag{7.27}$$

For receiving of the composition of the faces in the dimension 2 polytopes product one of the faces, observing the same rules as at the definition of the product polytopes, i. e. dimensions are added, and the numbers of vertices are multiplied. At that, indexes P and Q are replaced by index ×. Then 62 faces obtained in (7.27), are deciphered so

$$62 = 24P_2^{\times,3} + 2P_2^{\times,4} + 4P_2^{\times,5} + 12P_2^{\times,4}. \tag{7.28}$$

This means that 62 faces of dimension 2 in the product include twenty and four faces $P_2^{\times,3}$ (i.e., triangles), and fourteen faces $P_2^{\times,4}$ (i.e., quadrangles), and four faces $P_2^{\times,5}$.
For the faces of dimension 3, there is:

$$f_3^{\times} = f_3^{P_3} f_0^{Q_1} + f_2^{P_3} f_1^{Q_1} = 1 \cdot 2 + 12 \cdot 1 + 1 \cdot 1 + 2 \cdot 1 = 17 = 2P_3^{\times,12} + 12P_3^{\times,6} + P_3^{\times,8} + 2P_3^{\times,10}.$$

Thus, 17 faces of dimension 3 in the product consist of two polyhedrons $P_3^{12}(12P_2^3, P_2^4, 2P_2^5)$, eleven triangular prisms $P_3^6(3P_2^4, 2P_2^3)$, one quadrangular prisms $P_3^8(6P_2^4)$, and two pentagonal prisms $P_3^{10}(5P_2^4, 2P_2^5)$.
For the faces of dimension 4, there is:

$$f_4^{\times} = f_3^{P_3} f_1^{Q_1} = 1 \cdot 1 = 1 = P_4^{\times,24}.$$

According to the preceding equation, can obtain the structural formula of the product:

$$\begin{aligned} & P_3^{12}(12P_2^3, P_2^4, 2P_2^5)\times Q_1^2 \\ & = P_4^{24}[2P_3^{12}(12P_2^3, P_2^4, 2P_2^6), 12P_3^6(3P_2^4, 2P_2^3), P_3^8(6P_2^4), 2P_3^{10}(5P_2^4, 2P_2^5)]. \end{aligned} \tag{7.29}$$

Therefore, the product of the polyhedron PG_{14} by the segment is polytope 4*D*, which includes eleven triangular prisms, one quadrangular prism, two pentagonal prisms, and 2 polyhedron PG_{14}. You can call a polytope (7.29) PG_{14} - prism. The picture of the polytope (7.29) is shown in Figure 12.

Figure 12. The polytope PG_{14} - prism

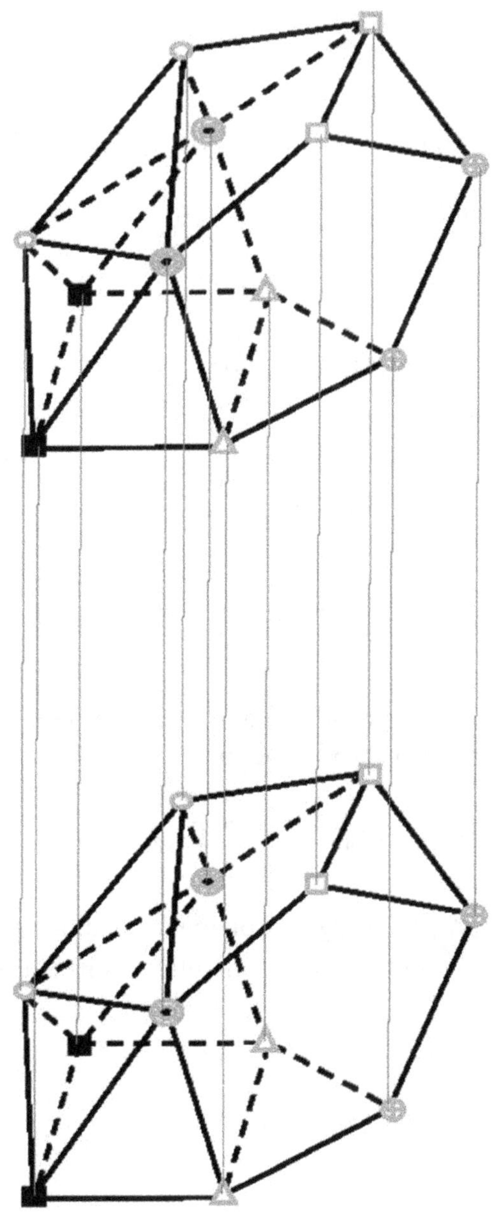

According to Theorem 6.3, the Euler-Poincaré equation for the product of polytopes is satisfied if this equation holds for the factors of the product of polytopes. Therefore, there is no need to check the fulfillment of the Euler-Poincaré equation for the product of polytopes, if it has already been proved that this equation holds for the factors.

The Polytope which is Product of the Polyhedron PG14 by Triangle

There we consider the product $P_3^{12}(12P_2^3, P_2^4, 2P_2^5)\times Q_2^3$ According to theorem 6.1 and ratios (6.5) there are:

$$f_0^{\times} = f_0^{P_3} f_0^{Q_2} = 12 \cdot 3 = 36,$$

$$f_1^{\times} = f_1^{P_3} f_0^{Q_2} + f_0^{P_3} f_1^{Q_2} = 25\cdot 3 + 12\cdot 3 = 111,$$

$$f_2^{\times} = f_2^{P_3} f_0^{Q_2} + f_1^{P_3} f_1^{Q_2} + f_0^{P_3} f_2^{Q_2} = (12+1+2)\cdot 3 + 25\cdot 3 + 12\cdot 1 = 132 = 36P_2^{\times,3} + 3P_2^{\times,4} + 6P_2^{\times,5} + 75P_2^{\times,4} + 12P_2^{\times,3},$$

(thus, in set of the elements with dimension 2 include 48 triangles, and 78 quadrats, and 6 pentagons)

$$f_3^{\times} = f_3^{P_3} f_0^{Q_2} + f_2^{P_3} f_1^{Q_2} + f_1^{P_3} f_2^{Q_2} = 1\cdot 3 + (12+1+2)\cdot 3 + 25\cdot 1 = 73 = 3P_3^{\times,12} + 36P_3^{\times,6} + 3P_3^{\times,8} + 6P_3^{10} + 25P_3^{\times,6},$$

(thus, in set of the elements with dimension 3 include 3 polyhedrons P_3^{12}, 61 triangles prisms, and 3 cubes, and 6 pentagons prisms)

$$f_4^{\times} = f_3^{P_3} f_1^{Q_2} + f_2^{P_3} f_2^{Q_2} = 1\cdot 3 + (12+1+2)\cdot 1 = 18 = 3P_4^{\times,36} + 12P_4^{\times,9} + P_4^{\times,12} + 2P_4^{\times,15},$$

(thus, in set of the elements with dimension 4 include 3 polytopes PG_{12} – prisms, 12 triangular prismahedrons, and one 4*3 – angular prismahedrons, and two 5*3 – angular prismahedrons, see Chapter 6)

$$f_5^{\times} = f_3^{P_3} f_2^{Q_2} = 1\cdot 1 = 1 = P_5^{\times,36}(3P_4^{\times,36}, 12P_4^{\times,9}, P_4^{\times,12}, 2P_4^{\times,15}). \tag{7.30}$$

Thus, the product $P_3^{12}(12P_2^3, P_2^4, 2P_2^5)\times Q_2^3$ is polytope with dimension 5 and hyperfaces in view of PG_{14} – prisms, triangular prismahedrons, and 4*3 – angular prismahedrons, and 5*3 – angular prismahedrons.

The Polytope which is Product of the Polyhedron PG14 by Tetrahedron

There we consider the product $P_3^{12}(12P_2^3, P_2^4, 2P_2^5)\times Q_3^4$. According to theorem 6.1 and ratios (6.5) there are:

$$f_0^\times = f_0^{P_3} f_0^{Q_3} = 12\cdot 4 = 48,$$

$$f_1^\times = f_1^{P_3} f_0^{Q_3} + f_0^{P_3} f_1^{Q_3} = 25\cdot 4 + 12\cdot 6 = 172,.$$

$$f_2^\times = f_2^{P_3} f_0^{Q_3} + f_1^{P_3} f_1^{Q_3} + f_0^{P_3} f_2^{Q_3} = (12+1+2)\cdot 4 + 25\cdot 6 + 12\cdot 4 = 258 =$$
$$48P_2^{\times,3} + 4P_2^{\times,4} + 8P_2^5 + 150P_2^{\times,4} + 48P_2^{\times,3},$$

(thus, in set of the elements with dimension 2 include 96 triangles and 154 quadrats, and 8 pentagons)

$$f_3^\times = f_3^{P_3} f_0^{Q_3} + f_2^{P_3} f_1^{Q_3} + f_1^{P_3} f_2^{Q_3} + f_0^{P_3} f_3^{3} = 1\cdot 4 + (12+1+2)\cdot 6 + 25\cdot 4 + 12\cdot 1$$
$$= 206 = 4P_3^{\times,12} + 72P_3^{\times,6} + 6P_3^{\times,8} + 12P_3^{10} + 100P_3^{\times,6} + 12P_3^{\times,4},$$

(thus, in set of the elements with dimension 3 include 4 polyhedrons P_3^{12}, 172 triangles prisms, 6 cubes, and 12 pentagon prisms, and 12 tetrahedrons)

$$f_4^\times = f_3^{P_3} f_1^{Q_3} + f_2^{P_3} f_2^{Q_3} + f_1^{P_3} f_3^{Q_3} = 1\cdot 6 + (12+1+2)\cdot 4 + 25\cdot 1 = 91 =$$
$$6P_4^{\times,24} + 48P_4^{\times,9} + 4P_4^{\times,12} + 8P_4^{15} + 25P_4^8,$$

(thus, in set of the elements with dimension 4 include 6 polytopes PG_{14} – prisms, 48 triangular prismahedrons, four 4*3 – angular prismahedrons, eight 5*3 – angular prismahedrons, and 25 tetrahedral prisms, see Chapter 6)

$$\begin{aligned} f_5^\times &= f_3^{P_3} f_2^{Q_3} + f_2^{P_3} f_3^{Q_3} = 1\cdot 4 + (12+1+2)\cdot 1 = 19 = 4P_5^{\times,36} + 12P_5^{\times,12} + P_5^{\times,16} + 2P_5^{20}, \\ f_6^\times &= f_3^{P_3} f_3^{Q_3} = 1\cdot 1 = 1 = P_6^{\times,48}(4P_5^{\times,36}, 12P_5^{\times,12}, P_5^{\times,16}, 2P_5^{20}). \end{aligned} \tag{7.31}$$

Thus, the product $P_3^{12}(12P_2^3, P_2^4, 2P_2^5)\times Q_3^4$ is polytope with dimension 6 and hyperfaces in view of $P_5^{36}, P_5^{12}, P_5^{16}, P_5^{20}$.

The Polytope which is Product of the Polyhedron PG14 by N-Simplex

The PG_{14} polyhedron contains the largest number of types of gonohedrons in its composition and, at the same time, a rather small total number of vertices of the polyhedron. This is ensured by the compact symmetric arrangement of the vertices in the polyhedron. If we talk about other polyhedrons with

regular-faceted flat faces, remote from the Archimedes bodies due to the presence of several types of gonohedrons, then in addition to the polyhedrons considered in this chapter, we can indicate several more polyhedrons, given in Chapter 2 (Table 2.6). For example, polyhedrons PG_7, PG_{20}, PG_{32}, PG_{33}. Each of them can lead to the formation of polytopes of the highest dimension with regular-faceted polyhedrons and gonohedrons of various types distributed in space after multiplying them by geometric elements of various dimensions. The construction of these polytopes can be carried out in the manner described in this chapter. However, the greatest interest in this regard is the polytypichedron-max PG_{14}, as the most distant from the bodies of Archimedes. Therefore, let us consider in general the polytope formed from the product of the polytypichedron-max PG_{14} the n - simplex. This will allow in a more general form and at the same time to present in more detail the mechanism for obtaining such polytopes.
Thus, we consider the polytope which is the product

$$P_3^{12}(12P_2^3, P_2^4, 2P_2^5) \times Q_n^{n+1}(C_{n+1}^n P_{n-1}^n, C_{n+1}^{n-1} P_{n-2}^{n-1}, ..., C_{n+1}^2 P_1^2, C_{n+1}^1 P_0^1). \qquad (7.32)$$

The right part of equality (7.32) is structure form of a polytope n – simplex rite with account the formular for the number of faces with dimension i in polytope n - simplex - $f_i(n) = C_{n+1}^{i+1}$ the number of combinations of n + 1 elements by i + 1 element. (Zhizhin, 2019). On the right side of the equality (7.32) near the polytope symbol P, the subscript still indicates the dimension of the polytope, and the superscript indicates the number of vertices of the polytope.

According to theorem 6.1 and ratios (6.5) calculate the number of elements of different dimensions in the desired product, noting also the main characteristics of these elements, i. e. the dimension of the polytopes and the number of vertices. When calculating these characteristics in the product of polytopes, the superscripts (the number of vertices) are multiplied, and the subscripts (dimensions) are added:

$$f_0^{\times} = f_0^{P_3} f_0^{Q_n} = 12 \cdot C_{n+1}^1 P_o^{\times,1},$$

$$f_1^{\times} = f_1^{P_3} f_0^{Q_n} + f_0^{P_3} f_1^{Q_n} = 25 \cdot C_{n+1}^1 P_1^{\times,2} + 12 \cdot C_{n+1}^2 P_1^{\times,2} = (25 \cdot C_{n+1}^1 + 12 \cdot C_{n+1}^2) P_1^{\times,2},$$

$$f_2^{\times} = f_2^{P_3} f_0^{Q_n} + f_1^{P_3} f_1^{Q_n} + f_0^{P_3} f_2^{Q_n} = (12P_2^{\times,3} + P_2^{\times,4} + 2P_2^{\times,5}) \cdot C_{n+1}^1 + 25 \cdot C_{n+1}^2 P_2^{\times,4}$$
$$+12 \cdot C_{n+1}^3 P_2^{\times,3} = (12C_{n+1}^1 + 12C_{n+1}^3) P_2^{\times,3} + (C_{n+1}^1 + 25C_{n+1}^2) P_2^{\times,4} + 2C_{n+1}^1 P_2^{\times,5},$$

(thus, in set of the elements with dimension 2 include $12C_{n+1}^1 + 12C_{n+1}^3$ triangles and $C_{n+1}^1 + 25C_{n+1}^2$ quadrats, and $2C_{n+1}^1$ pentagons)

$$f_3^{\times} = f_3^{P_3} f_0^{Q_n} + f_2^{P_3} f_1^{Q_n} + f_1^{P_3} f_2^{Q_n} + f_0^{P_3} f_3^{Q_n} = C_{n+1}^1 \cdot P_3^{\times,12} + (12P_2^{\times,3} + P_2^{\times,4} + 2P_2^{\times,5}) \cdot C_{n+1}^2 \cdot P_1^{\times,2} +$$
$$25 \cdot C_{n+1}^3 P_2^3 P_1^2 + 12C_{n+1}^4 P_3^4 = C_{n+1}^1 \cdot P_3^{\times,12} + (12C_{n+1}^2 + 25C_{n+1}^3) P_3^{\times,6} + C_{n+1}^2 P_3^{\times,8} + 2C_{n+1}^3 P_3^{10} + 12C_{n+1}^4 P_3^{\times,4},$$

(thus, in set of the elements with dimension 3 include C_{n+1}^1 polyhedrons P_3^{12}, $12C_{n+1}^2 + 25C_{n+1}^3$ triangles prisms, C_{n+1}^2 cubes, and $2C_{n+1}^3$ pentagon prisms, and $12C_{n+1}^4$ tetrahedrons),

$$
\begin{aligned}
f_4^{\times} &= f_3^{P_3} f_1^{Q_n} + f_2^{P_3} f_2^{Q_n} + f_1^{P_3} f_3^{Q_n} + f_o^{P_3} f_4^{Q_n} = P_1^{\times,2} C_{n+1}^2 \cdot P_3^{\times,12} \\
&+(12P_2^{\times,3} + P_2^{\times,4} + 2P_2^{\times,5}) \cdot C_{n+1}^3 \cdot P_2^{\times,3} + 25 \cdot P_1^{\times,2} C_{n+1}^4 P_3^{\times,4} + 12C_{n+1}^5 P_4^{\times,5} \\
&= C_{n+1}^2 P_4^{\times,24} + 12C_{n+1}^3 P_4^{\times,9} + C_{n+1}^3 P_4^{\times,12} + 2C_{n+1}^3 P_4^{\times,15} + 25C_{n+1}^4 P_4^{\times,8} + 12C_{n+1}^5 P_4^{\times,5},
\end{aligned}
$$

(thus, in set of the elements with dimension 4 include C_{n+1}^2 polytopes PG_{14} – prisms, $12C_{n+1}^3$ triangular prismahedrons, C_{n+1}^3 4*3 – angular prismahedrons, $2C_{n+1}^3$ 5*3 – angular prismahedrons, and $25C_{n+1}^4$ tetrahedral prisms, and $12C_{n+1}^5$ four-dimensional simplices, see Chapter 6)

$$
\begin{aligned}
f_5^{\times} &= f_3^{P_3} f_2^{Q_n} + f_2^{P_3} f_3^{Q_n} + f_1^{P_3} f_4^{Q_n} + f_o^{P_3} f_5^{Q_n} = P_2^{\times,3} C_{n+1}^3 \cdot P_3^{\times,12} \\
&+(12P_2^{\times,3} + P_2^{\times,4} + 2P_2^{\times,5}) \cdot C_{n+1}^4 \cdot P_3^{\times,4} + 25 \cdot P_1^{\times,2} C_{n+1}^5 P_3^{\times,5} + 12C_{n+1}^6 P_4^{\times,6} \\
&= C_{n+1}^3 P_5^{\times,36} + 12C_{n+1}^4 P_5^{\times,12} + C_{n+1}^4 P_5^{\times,16} + 2C_{n+1}^4 P_5^{\times,20} + 25C_{n+1}^5 P_5^{\times,10} + 12C_{n+1}^6 P_5^{\times,6}.
\end{aligned}
$$

Continuing these calculations for polytopes of higher and higher dimensions, we come to calculating the characteristics of the facets of the sought product:

$$
f_{n+2}^{\times} = f_3^{P_3} f_{n-1}^{Q_n} + f_2^{P_3} f_n^{Q_n} = C_{n+1}^n P_{n+2}^{\times,12n} + 12P_{n+2}^{\times,3(n+1)} + P_{n+2}^{\times,4(n+1)} + 2P_{n+2}^{\times,5(n+1)}.
$$

Thus, the final product of the work has the structure:

$$
f_{n+3}^{\times} = f_3^{P_3} f_n^{Q_n} = 1 \cdot 1 = P_{n+3}^{\times,12(n+1)}(!\ {}_{B+1}^{B} P_{n+2}^{\times,12n}, 12P_{n+2}^{\times,3(n+1)}, P_{n+2}^{\times,4(n+1)}, 2P_{n+2}^{\times,5(n+1)}).
$$

In the same way, we can calculate the final product of the product of the polytypichedron-max PG_{14} by polytopes of higher dimensions of another type, for example, n-cube or n-cross-polytopes, using formulas for the numbers of elements of different dimensions in these polytopes obtained earlier (Zhizhin, 2019).

It should be noted that due to the product of polytopes and the presence of the polytypichedron as a factor in this product, the space of the polytope resulting from this product is filled simultaneously with copies of the polytypichedron and copies of the second factor of the product. Moreover, in each copy of the second factor, the vertices are constructed from some vertices of the Tanbulan polyhedron. Since the polytypichedron includes vertices of different types, each copy of the second factor consists of vertices of one type. Thus, the space of the product result is filled with copies of the second factor, each of which contains vertices of the same type (the same gonohedron). This is done due to the higher dimensionality of the product of polytopes and the variety of types of vertices in polytypichedron-max. This is most clearly manifested in the works with the polytypichedron-max PG_{14}, which has the largest number of types of gonohedrons.

The considered process of constructing higher-dimensional spaces is a fundamentally new process that differs from the known processes of constructing higher-dimensional spaces from congruent figures (Zhizhin, 2021) by the variety of vertices, which is typical for the formation of higher-dimensional spaces in natural chemical compounds.

CONCLUSION

The question of the existence of convex regular polyhedrons in which different regular polygons can simultaneously be flat faces has been of interest to people for a long time. Since such polyhedrons are found in the structures of biological tissues (Thompson A`Arcy, 1942), in the forms of cells (Lewis, 1923), in various natural structures (Petrov, 2007), in particular, the large-scale structure of the Universe (Zhizhin, 2008). The shapes of regular polyhedrons are of interest in construction, nanotechnology. Research in this direction has been carried out so far only for the case of three-dimensional polyhedrons. The question of the existence of regular-faceted polytopes of the highest dimension has not yet been raised. However, the author's recent studies in various fields of knowledge have shown that polytopes of the highest dimension are widespread in nature and their research is of fundamental importance for all natural science disciplines: mathematics, physics, chemistry, biology (Zhizhin, 2018, 2019 a, b; 2021 a, b). Therefore, it is very important to raise the question of the existence of regular-faceted polytopes of the highest dimension. It should be noted here that since polytopes of higher dimension have many faces of different dimensions, it is necessary to clarify which faces, i. e. what dimensions are we talking about when we talk about regular-faceted polytopes of higher dimension. But if we keep in mind that all two-dimensional faces are regular faces, then faces of any dimension become regular-faceted. To construct regular-faceted polytopes of the highest dimension, in this chapter, we used the results of the study of regular-faceted three-dimensional polyhedrons carried out by the author. In particular, the regular polyhedrons discovered by the author with a large number of gonohedrons of various types. These polyhedrons, due to their sharp difference from the Archimedes bodies and the diversity of the type of vertices, are called polytypichedrons, and the polyhedron containing the maximum possible number of different types of vertices (six) is called the Tanbulan - max polyhedron. The method of obtaining polytopes of the highest dimension from them by multiplying them by geometric elements of various dimensions (including the highest dimension) is applied to the polytypichedrons. This is how, in principle, an infinite set of polytopes of the highest dimension with regular two-dimensional edges and the participation of vertices of various types was obtained. A method has been developed that allows calculating the number of faces of different dimensions in these polytopes with an indication of the shape of the faces (i.e., the number of vertices and dimensions of faces).

It is interesting that due to the product of polytopes and the presence of the polytypichedron as a factor in this product, the space of the polytope resulting from this product is filled simultaneously with copies of the polytypichedron and copies of the second factor of the product. Moreover, in each copy of the second factor, the vertices are constructed from some vertices of the polytypichedron. Since the polytypichedron includes vertices of different types, each copy of the second factor consists of vertices of one type. Thus, the space of the product result is filled with copies of the second factor, each of which contains vertices of the same type. Higher-dimensional polytopes obtained on the basis of three-dimensional polytypichedrons can be called polytypictopes of higher dimension, since they have manifolds of different types of vertices distributed in the polytope space.

The considered process of constructing higher-dimensional spaces is a fundamentally new process that differs from the known processes of constructing higher-dimensional spaces from congruent figures (Zhizhin, 2021) by the variety of vertices, which is typical for the formation of higher-dimensional spaces in natural chemical compounds.

REFERENCES

Alexandrov, A. D. (1950). *The convex polyhedrons*. Gostechizdat.

Cauchy, O. (1944). Research on polyhedrons. *Advances in Mathematical Sciences*, *10*, 5–17.

D'Arsy, T. (1942). *On Growth and Form*. Cambridge University Press.

Euler, L. (1736). Solutio problematisad geometriam situs pertinentis. *Comment Acad Sci I Petropolitanae*, *8*, 128–140.

Fedorov, E.S. (1885). Beginning of the teachings on the figures. *Notes of the St. Petersburg Mineralogical Society, 21*.

Fedorov, E.S. (1889). Symmetry of finite figures. *Notes of the St. Petersburg Mineralogical Society, 25*.

Fedorov, E.S. (1891). Symmetry of the correct systems of figures. *Notes of the St. Petersburg Mineralogical Society, 28*.

Hadamard, J. (1938). *Elementary geometry. Part 2*. Uchpedgiz.

Hilbert, D. (1901). Gesammelte Abhandlangen. *Archiv f. Mathematics and Physics*, *3*(1), 44–63, 213–237.

Johnson, W. W. (1960). Convex polyhedrons with regular faces. *Notices of the American Mathematical Society*, *7*, 952.

Petrov, O. V. (2007). *Dissipative of structure of the Earth as display of fundamental wave properties of a matter*. A.P. Karpinsky Russian Geological Research Institute.

Poincare, A. (1895). Analysis situs *.J. de é Ecole Polytechnique, 1*.

Zalgaller, V. A. (1966). *Convex polyhedrons with correct sides*. Science.

Zalgaller, V.A. (1966). List of all convex polyhedrons with correct sides. *The International Congress of Mathematicians*, 28 – 29.

Zhihin, G.V. (2005a). Correct both semiregular polyhedrons and filling of space with them. *Life and Safety*, *1-2*, 551-559.

Zhizhin, G. V. (2005b). *Dissipative of structures in chemical, geological and ecological systems*. Science.

Zhizhin, G. V. (2006). Mathematical of a basis economically and ecologically rational use building and finishing materials. Architecture, Construction, Ecology, 215-262.

Zhizhin, G. V. (2008). To a question on possible mesh large-scale structure of the universe. In *Practical cosmology*. St Petersburg: Russian Geographical Society.

Zhizhin, G. V. (2009). *Convex right-faceted polyhedrons*. Polytechnika-service.

Zhizhin, G. V. (2014). *4D world*. St. Petersburg: Polytechnica-Service.

Zhizhin, G. V. (2019). *The Geometry of Higher - Dimensional polytopes*. IGI Global. doi:10.4018/978-1-5225-6968-8

Zhizhin, G. V. (2019 b). On the Structure of Parallelohedrons of Higher Dimension: Hilbert's 18th Problem. *International Journal of Chemoinformatics and Chemical Engineering*, *8*(2), 69–96. doi:10.4018/IJCCE.20190701.oa2

Zhizhin, G. V. (2021). *Nanotechnologies and Clusters in the Spaces of Higher Dimension: Emerging Research and Opportunities*. IGI Global. doi:10.4018/978-1-7998-3784-8

Zhizhin, G. V. (2021). *Normal Partitions and Hierarchical Fillings of N-Dimensional Spaces*. IGI Global. doi:10.4018/978-1-7998-6768-5

KEY TERMS AND DEFINITIONS

Convex Regularly Faceted Polyhedrons: Convex three-dimensional polytopes whose faces are regular polygons.

Gonohedron: A polyhedral corner whose faces are regular polygons.

N*3-Angular Prismahedron: The product of n-angle by the triangle.

Polytopic Prismahedron: The product of polytope by one dimension segment.

Polytypichedron (Max): The regular-faceted polyhedron with max number of different types of genohedrons.

Polytypichedrons: The regular-faceted polyhedrons, in which three or more of different types of genohedrons can be present.

Regular Polyhedron: Convex polyhedron with faces of regular polygons of one type.

Semiregular Polyhedrons: Convex polyhedrons with faces of regular polygons of two or more types, if all gonohedrons of the polyhedron have the same composition and arrangement of faces.

Tetrahedral Prism: The product of tetrahedral by one dimension segment.

Tetrahedral Prismahedron: The product of tetrahedral by the triangle.

Triangular Prismahedron: The product of triangle by the triangle.

Chapter 8
Polytopes Dual to Higher-Dimensional Polytopes

ABSTRACT

The transitions from regular and semi-regular polytopes of various dimensions to dual polytopes are considered in detail. It is shown that any deviation of polytopes from the correct forms leads, in their dual modifications, to new classes of polytopes. Polytopes dual to the products of polytopes form the class of polyincidental polytopes. Their images are given, and their composition is precisely indicated. For example, the polytope dual to the product of a decagon and a decagon contains 100 tetrahedra. All of them are listed by vertices in the dual polytope. Regularly faceted polytopes with a large number of different types of gonohedrons (polytypichedron and polytypictops) for dual transitions form a new class of polytopes with faces of irregular shape and, accordingly, a new type of gonohedron. This can be important in the analysis of chemical structures, often characterized by deviations from regular geometric shapes.

INTRODUCTION

From the previous chapters it follows that there are different classes of polytopes of the highest dimension: regular, semi-regular, regular - faceted, quasi-regular with edges of different incidence, polytopic prismahedrons, polytypictopes. Polytopes of each of these classes can have dual polytopes. For a general definition of the dual polytope to some polytope of higher dimension P_n, we introduce the concept of a tower of a polytope Φ. A tower Φ of a polytope P_n is a set $\{F_{j_0},\dots,F_{j_{r-1}}\}$ of faces of P_n such that, i. e. their sequence is subject to strict inclusion

$$F_{j_0} \subset F_{j_1} \subset \dots \subset F_{j_{r-1}}. \tag{8.1}$$

Thus, the dimensions of the faces in the tower Φ satisfy the inequalities:

DOI: 10.4018/978-1-7998-8374-6.ch008

$$0 \le j_0 < j_1 < \ldots < j_{r-1} \le n-1. \tag{8.2}$$

If $P = n$, the tower Φ is said to be maximal.

Let there be a polytope P. Then they say that a polytope P * is dual to a polytope P if there is a one-to-one correspondence between the set of faces of the polytope P and the set of faces of the polytope P *, which reverses the inclusion relation (Grünbaum, 1967). Let us apply this definition to some three-dimensional polytope. In it f_0 is the number of vertices, f_1 is the number of edges, f_2 is the number of flat faces. These numbers satisfy the Euler equation (Euler, 1736):

$$f_0 - f_1 + f_2 = 2.$$

According to the definition of a dual polytope for it, the faces that have dimension zero in the original polytope become faces that have dimension 2 in the dual polytope. This can be done by associating a vertex in the dual polytope with each planar face in the original polytope (for example, choosing the center of the planar face of the original polytope beyond the vertex of the dual polytope). The resulting vertices are connected by edges that form flat faces of the dual polyhedron in an amount equal to the number of vertices in the original polyhedron. In this case, the number of edges in the dual polytope must be equal to the number of edges in the original polytope so that the Euler equation for the dual polyhedron is satisfied:

$$f_{0d} - f_{1d} + f_{2d} = 2,$$
$$f_{0d} = f_2, f_{1d} = f_1, f_{2d} = f_0.$$

For the well-known regular polyhedrons this is indeed the case. For example, an octahedron inscribed in a cube (Figure 1), a tetrahedron inscribed in a tetrahedron (Figure 2).

Figure 1. The dual polyhedrons an octahedron and a cube

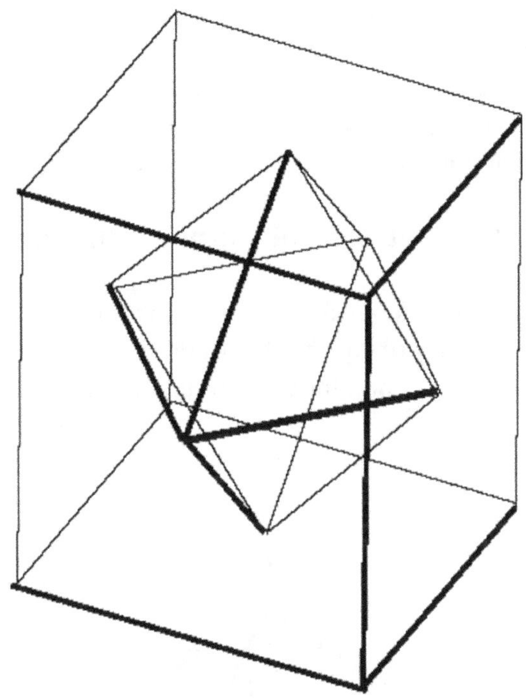

Figure 2. The dual two the tetrahedrons

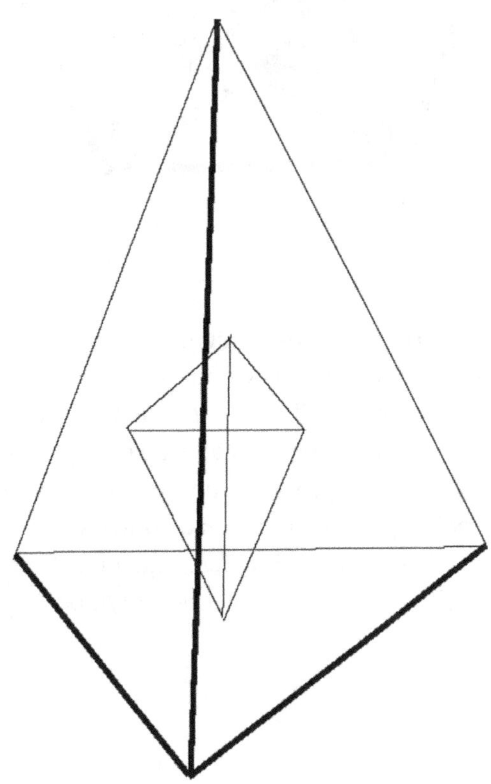

The same situation is observed when the dodecahedron is inscribed in the icosahedron or when the icosahedron is inscribed in the dodecahedron (Steinhaus, 1950; Coxeter, 1963). Thus, the tetrahedron is dual to itself, the octahedron is dual to the cube, the icosahedron is dual to the dodecahedron. Dual polytopes are related to each other by the inclusion relation and together characterize the structure of the polytope. Therefore, it is of interest to construct and study dual polytopes included in the already defined classes of polytopes. This chapter is devoted to this analysis.

Polytopes Dual to Regular Polytopes of Higher Dimension

In Chapter 4, an image of a four-dimensional simplex-type polytope was given (Figure 4.3). Obviously, another 4-simplex polytope can be inscribed into this figure. Let us place an arbitrary point v_1 on the two-dimensional face *ebd*, which is common for two tetrahedrons *abed* and *bcde* (Figure 3).

Figure 3. Two dual polytopes 4-simplex

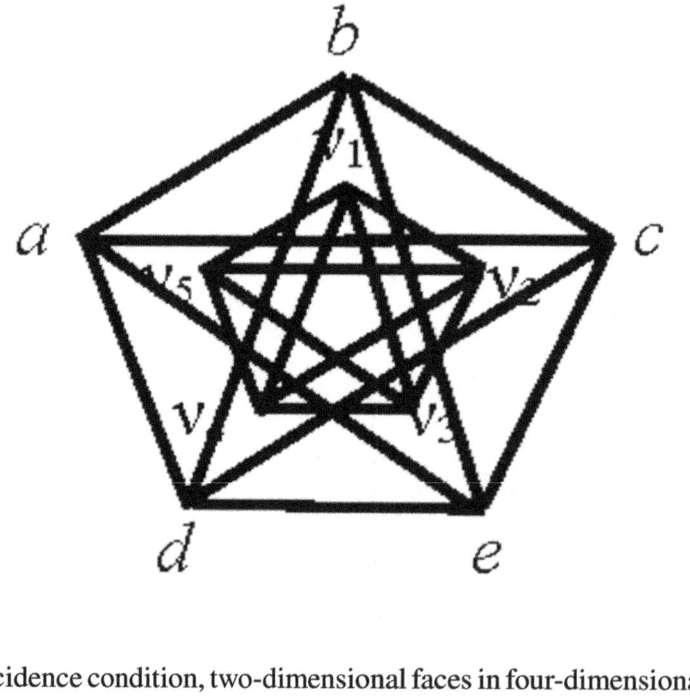

According to the incidence condition, two-dimensional faces in four-dimensional polytopes belong to two hyperfaces (Chapter 3). Let us place an arbitrary point v_2 on the two-dimensional face *acd*, which is common for two tetrahedrons *abcd* and *acde*. Place the points v_3, v_4, v_5 in the same way (Figure 3). We connect the points $v_1 \div v_5$ with edges to each other so that each of these five points is connected by edges with all other points by edges. We get the second 4-simplex polytope inscribed in the first 4-simplex polytope. It is obvious that these polytopes are dual to each other. Both the first 4-simplex polytope and the second 4-simplex polytope satisfy the Euler-Poincaré equation (Poincaré, 1895).

In the first polytope is $f_0 - f_1 + f_2 - f_3 = 0$, in the second polytope is $f_{0d} - f_{1d} + f_{2d} - f_{3d} = 0$, moreover $f_{0d} = f_3, f_{1d} = f_2, f_{2d} = f_1, f_{3d} = f_0$.

The exclusion relation (8.1) is strictly fulfilled between the faces in both polytopes, and inequalities (8.2) are satisfied between the dimensions of the faces in both cases. An increase in the dimension of the n-simplex polytope does not violate the possibility of inscribing the n-simplex polytopes into each other for any n.

Consider two polytopes 4-cube and 4-cross-polytope. Let's use the 4-cross-polytope image shown in chapter 4 (Figure 4.14). To compare the 4-cube and 4-cross-polytopes, it is convenient to use the image of the 4-cube polytope obtained earlier (Zhizhin, 2014, 2019). This image is shown in Figure 4.

Figure 4. Polytope 4 - cube in the form of a figure with central symmetry

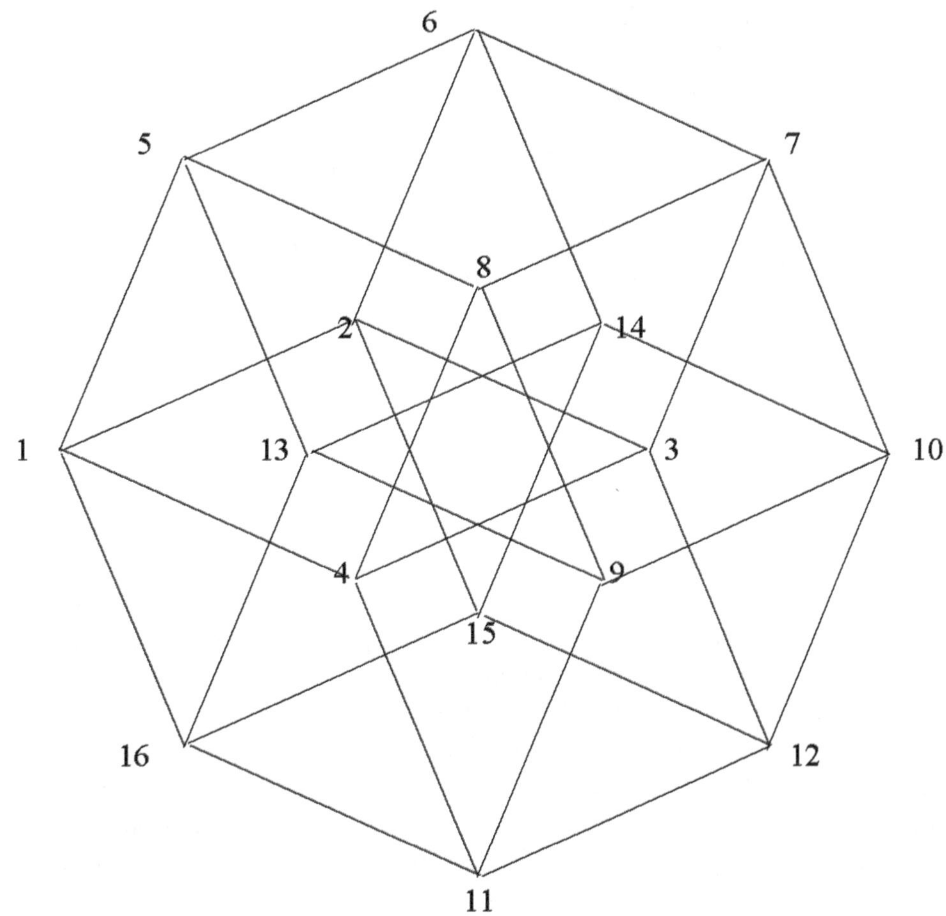

Comparing the images of polytopes in Figure 4.14 and Figure 4, it is easy to see that one of the images fits into the other and vice versa. For example, let's select in the center of the two-dimensional faces of the cubes, bordering on the outer contour of the polytope 4-cube, points $\mu 1 \div \mu 8$ (Figure 5).

Figure 5. The dual polytopes 4-cube and 4-cross-polytope

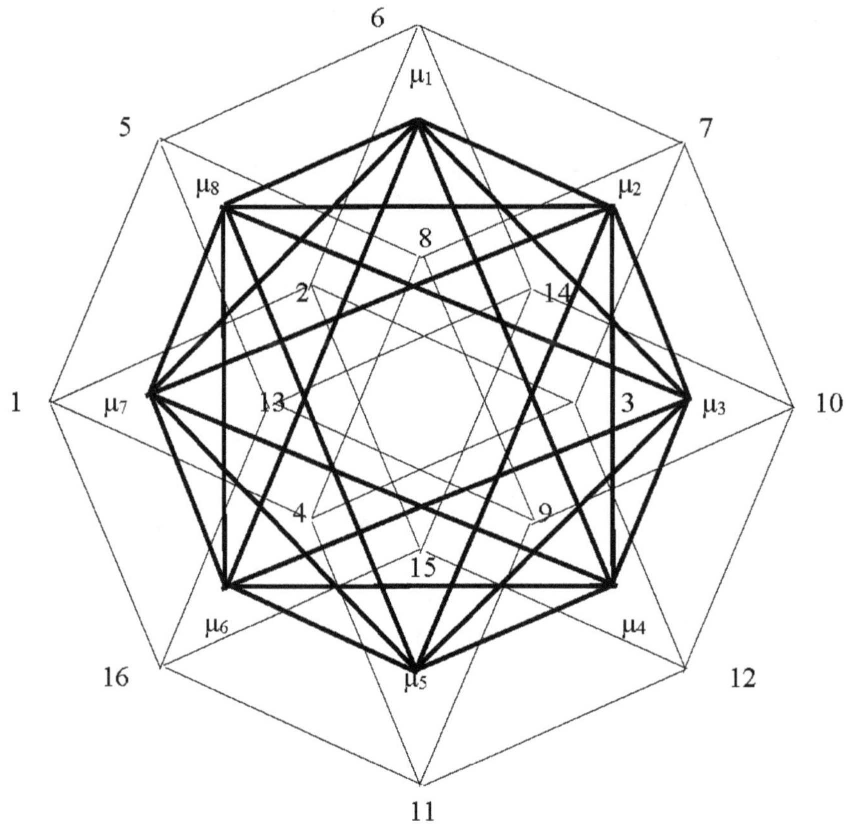

The two-dimensional faces of the cubes, according to the incidence conditions, belong simultaneously to two polyhedrons of the 3-cube. We connect the points with $\mu 1 \div \mu 8$ edges so that only the points located opposite to the center of symmetry remain unconnected. Then we get a 4-cross-polytope inscribed in a 4-cube polytope. In this case, the strict inclusion relations (8.1) and the relations between the dimensions of the faces of both polytopes (8.2) are satisfied. This proves that 4-cube and 4-cross-polytopes are dual to each other. An increase in the dimension of polytopes, taking into account the fundamental preservation of the structures of these polytopes with an increase in the dimension, does not change the possibility of inscribing the polytope n-*c*ross-polytope and n-*c*ube into each other at the same value of n. This preserves the duality of n-*c*ube and n-*c*ross-polytopes for any n.

Consider, for example, 5-cube and 5-cross-polytopes. The symmetrical image of a 5-cube polytope was obtained in the author's monograph (Zhizhin, 2019). It is shown in Figure 6.

Figure 6. 5-cube polytope in symmetrical view.

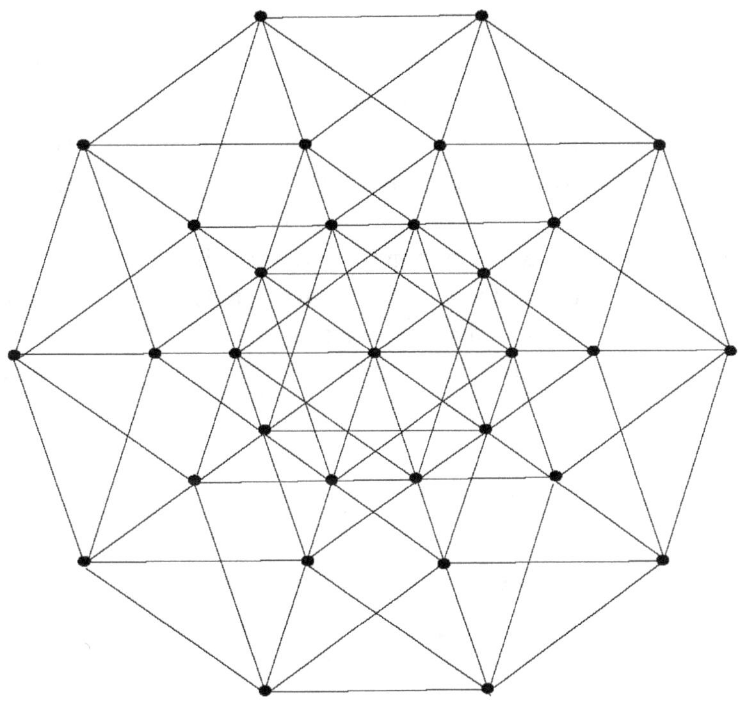

In the same monograph, there is an image of a 5-cross-polytope (Figure 7).

Figure 7. Image of 5-cross-polytope

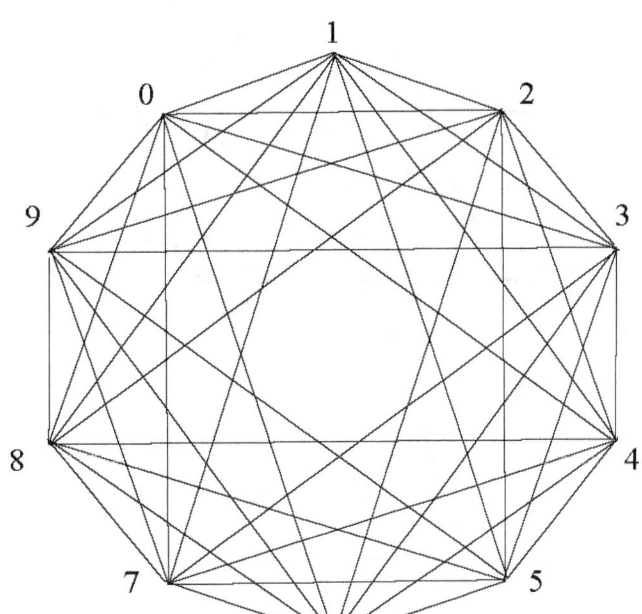

Let's show that these images are dual to each other. The number of faces of different dimensions in the *n*-cube polytope is determined by the formula $f_i(n) = 2^{n-i} C_n^i (i = 0 \div 5)$. According to this formula, a 5-cube polytope has 32 vertices, f_0=32; 80 edges, f_1=80; 80 flat faces (squares), f_2=80; 40 three-dimensional faces (3-cube), f_3=40; 10 four-dimensional faces (4-cube), f_4=10. The number of faces of different dimensions in a 5-cross-polytope is determined by the formula $f_i(n) = 2^{1+i} C_n^{n-1-i} (i = 0 \div 5)$. According to this formula, a 5-cross-polytope has 10 vertices, f_0=10; 40 edges, f_1=40; 80 flat faces (triangles), f_2=80; 80 three-dimensional faces (tetrahedrons), f_3=80; 32 four-dimensional faces (4-simplices), f_4=32. Each of these facets has a common tetrahedron with some other facet of this list. Let's imagine the connection structure of these facets in the form of a polytope whose vertices will correspond to the facets, and the edges will correspond to the common between the pair of facet tetrahedrons. By being, one will build a polytope dual to the 5 – cross - polytope. This will lead us to the symmetric structure in Figure 8.

Figure 8. The polytope dual to the 5 – cross - polytope

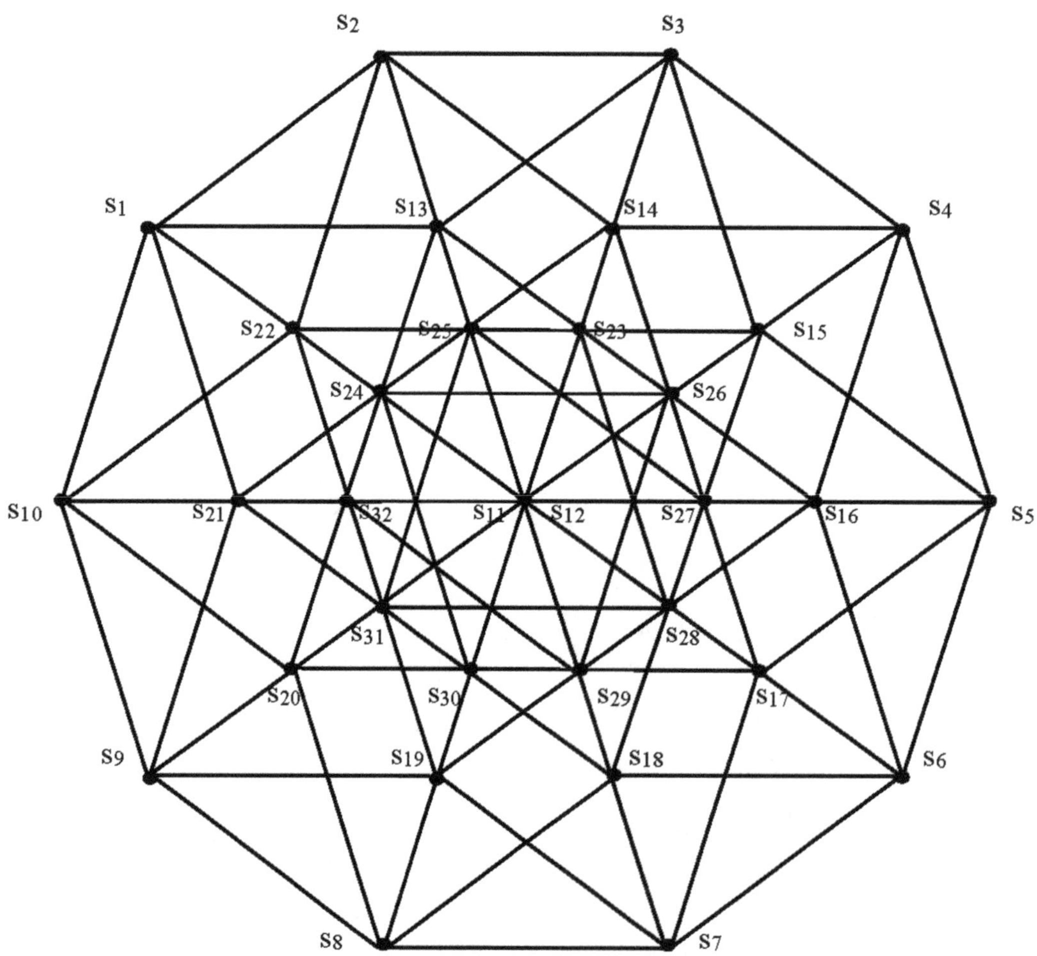

One can regard this structure as the factor - structure of the boundary complex of the 5 – cross - polytope, since each vertex of this structure is a polytope of dimension 4. In this structure, some edges overlap one another and two vertices in the center coincide. In order for the structure in Figure 8 to be more understandable, one slightly expand the coinciding edges and vertices (Figure 9).

Figure 9. The polytope dual to the 5 – cross – polytope with slightly expand the coinciding edges and vertices

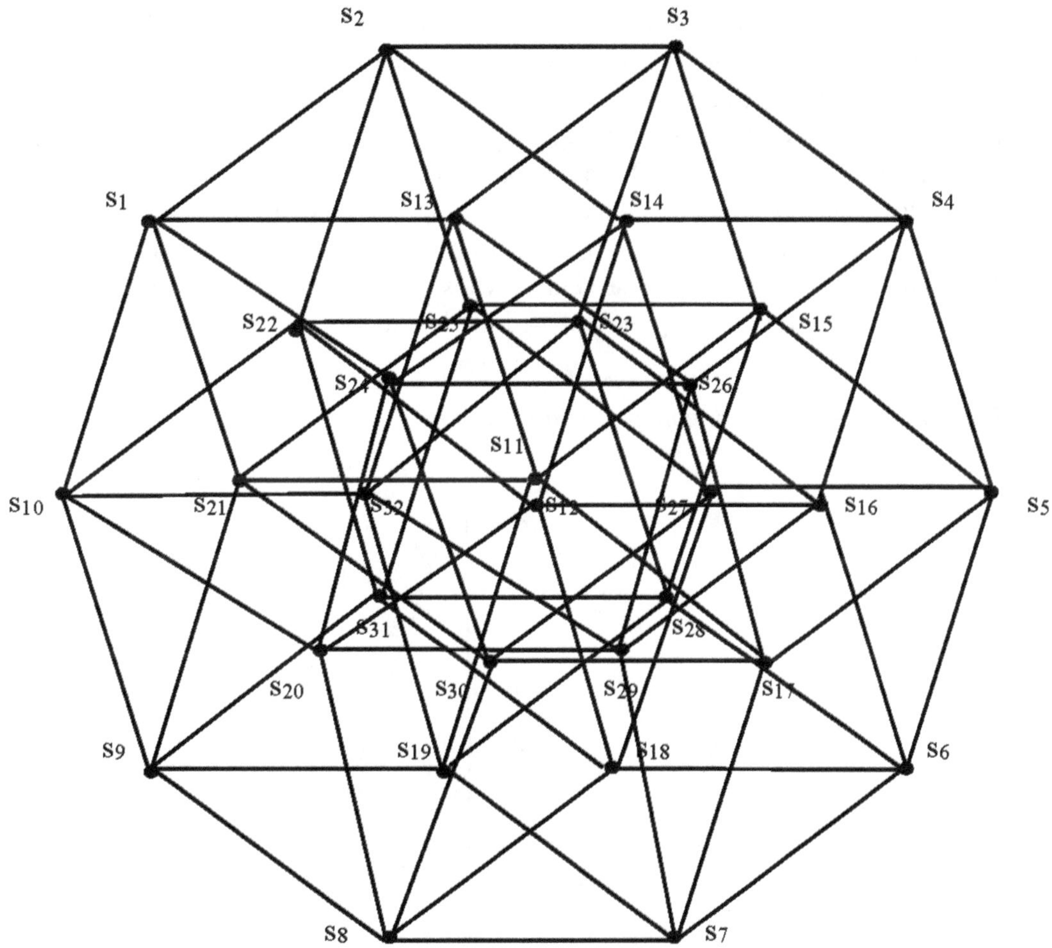

It follows from Figure 9 that the obtained structure is the image of the projection of the polytope of the 5 - cube, since it contains images of 40 cubes adjacent to each other along entire sides. The total number of vertices (32), edges (80), squares (80), cubes (40), and 4 - cubes (10) in Figure 9 and Figure 8 exactly coincide with the required quantities of elements of different dimensions for the polytope 5 – cube. Thus, the 5-cross-polytope is dual to the 5-cube. Obviously, similar images of n - dimensional cubes and n-cross-polytopes can be obtained even if $n > 5$.

Consideration of the duality of the 600-cell and 120-cell polytopes mentioned in the literature does not make sense in connection with the proof in Chapter 4 of the impossibility of the existence of these polytopes. It can be assumed that a 24-cell, as well as a 4-simplex, is dual to itself, since this four-dimensional polytope, as well as a 4-simplex, has the same number of vertices and facets, and also the numbers of edges and two-dimensional faces are equal. However, there is currently no rigorous proof of this fact (Zhizhin, 2005).

Polytopes Dual to Semiregular Polytopes and to Polytopes with Different Incidence of Edges

The analysis of duality, as an analysis of the unity of opposites, is used in various fields of science: in mathematics, logic, philosophy, physics and chemistry (Bogdanov, 1989; Whitehead, 1990; Feyman, 1968; Hegel; 1998; Bucur and Delanu, 1972).

One will consider the polytope as dual to a given polytope of dimension n if the facets of a given polytope are replaced on the vertices of the polytope. Moreover, the vertices of a new polytope are connected by an edge if the corresponding facets have a common face of dimension n - 2 in the given polytope. Consider a rectangular prism with triangular bases. Can introduce the centers of the planar faces of this prism and join the edges of those centers whose faces have a common edge. One will get a double pyramid (Figure 10). It is an irregular polyhedron dual to this prism.

Figure 10. The triangle prism and double pyramid dual to it

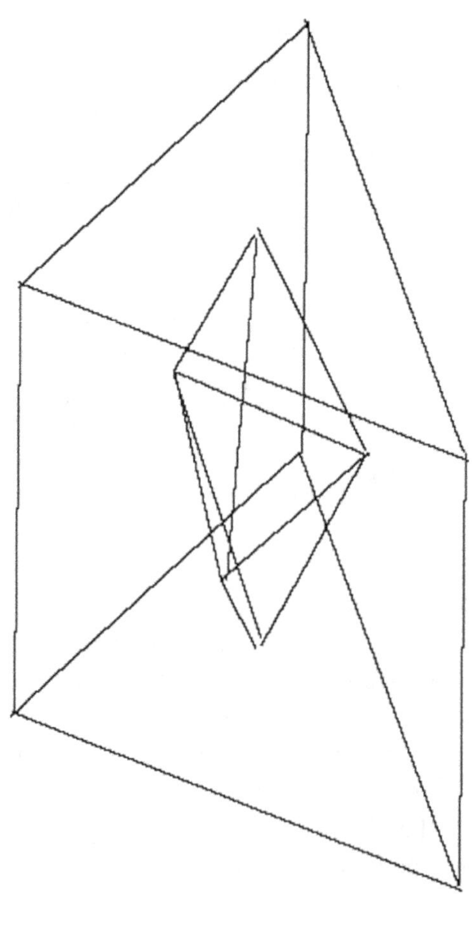

In a double pyramid two vertices are incident to three edges, and another two vertices are incident to four edges. The same picture will be if we take the pentagonal prism.

Two vertices of double pentagonal pyramid are incident to 5 edges, and 5 of the remaining vertices are incident to 4 edges (Figure 11).

Figure 11. The pentagonal prism and a double pyramid dual to it

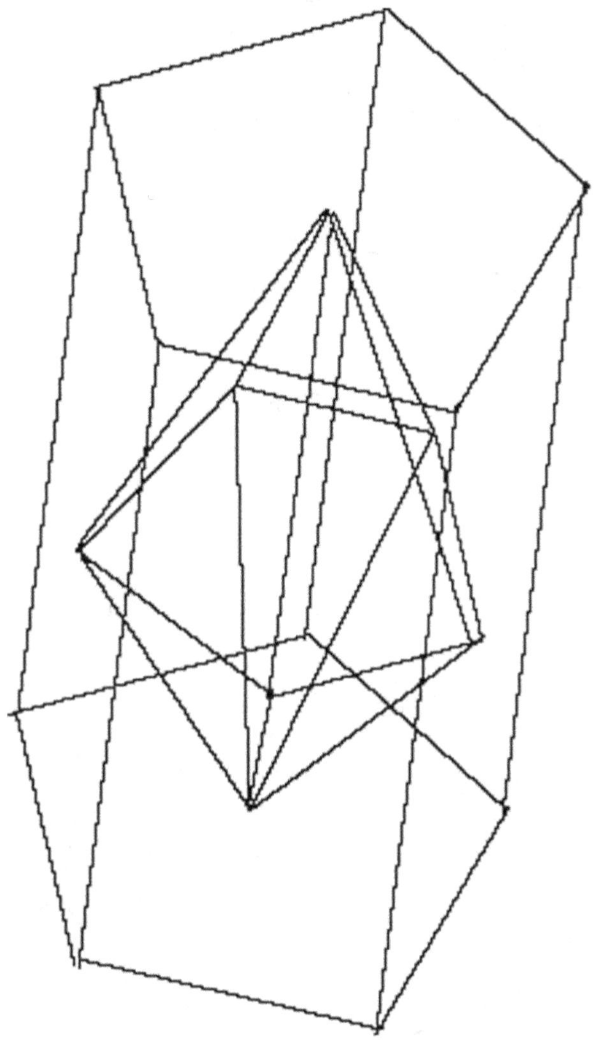

Further it will be shown that the polytopes dual to polytopes products have not only vertices with different values of incidence to the edges, but the edges with different values of incidence to three - dimensional figures. This new type of polytopes is called poly - incident polytopes.

To construct polytopes dual to polytopes from tetrahedra with varying degrees of edge use (with different edge incidence values) (Chapter 4), it is necessary to take the vertices of the corresponding graph as the vertices of the polytope, and the edges of the graph to be taken as the edges of the polytope. Then the polytope of the dual polytope in Figure 4.26 will be the triangular prismahedron in Figure 12.

Figure 12. Triangular prismahedron

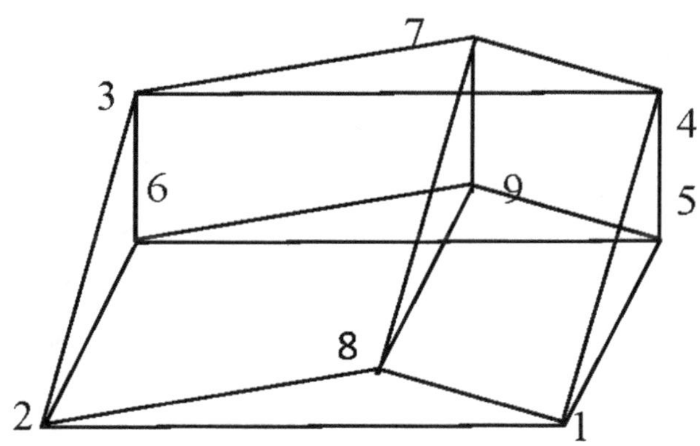

The polytope dual to the polytope whose graph is shown in Figure 4.27 is the pentagonal prismahedron in Figure 13.

Figure 13. Pentagonal prismahedron

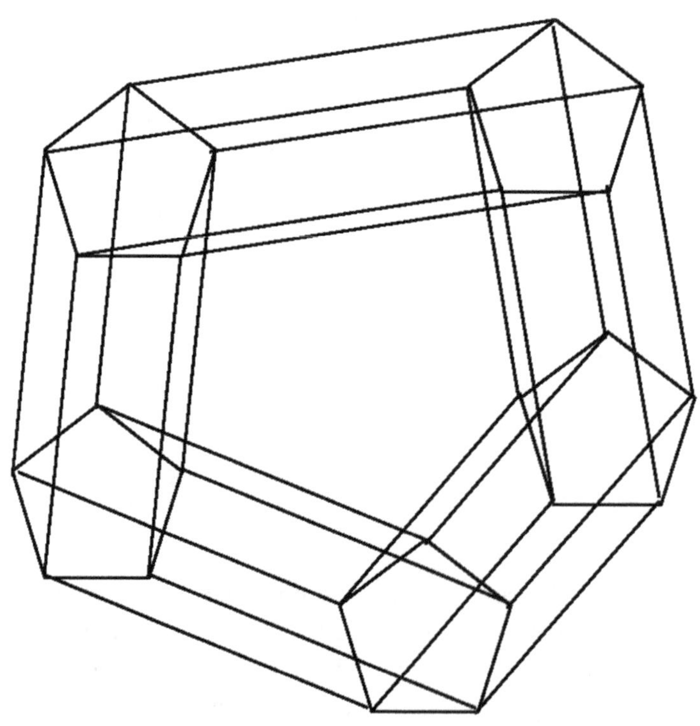

From the previous it follows that similarly it is possible to construct polytopes from tetrahedra (not necessarily regular) with different degrees of edge use (with different incidence of edges), one of the degrees of which is 4, and the corresponding dual polytopes. For example, the hexagonal prismahedron in Figure 14.

Figure 14. Hexagonal prismahedron

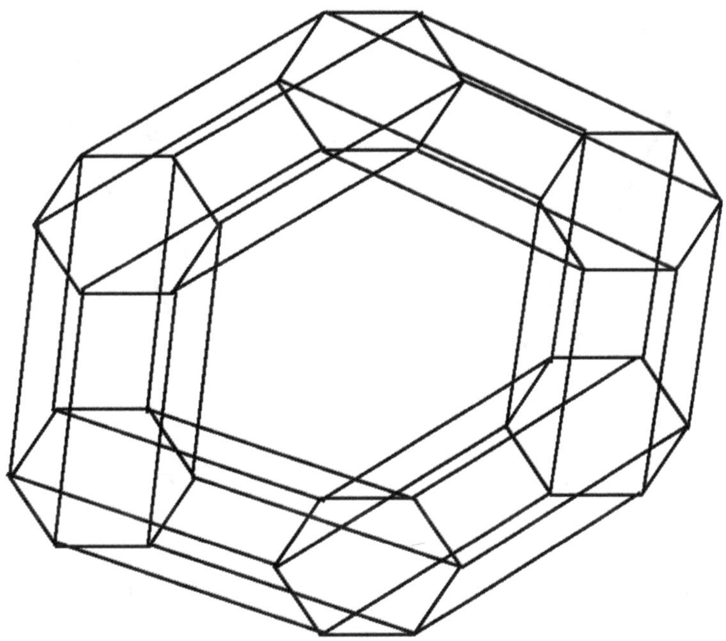

Let's denote vertices of the polytope in Figure 12 by symbols $a_1,\ldots,a_9$. Let's enumerate triangular prismahedrons (facets) forming this polytope:

1) $a_1a_2a_3a_7a_8a_9$; 2) $a_2a_3a_4a_5a_8a_9$; 3) $a_1a_2a_5a_6a_7a_8$;
4) $a_1a_2a_3a_5a_6a_4$; 5) $a_1a_6a_3a_4a_7a_9$; 6) $a_4a_6a_5a_7a_8a_9$.

Since the edges of a dual polytope shall connect the centers of polytope facets adjoining each other, so, by introduced designations of facets, we define neighboring facets having common flat faces by introduced designations of facets. Therefore, just these facets shall be connected by edge in a dual polytope (Zhizhin, 2017). Executing this analysis, we'll come to the projection of a dual polytope on a two-dimensional plane (Figure 15).

Figure 15. Polytope dual to triangular prismahedron

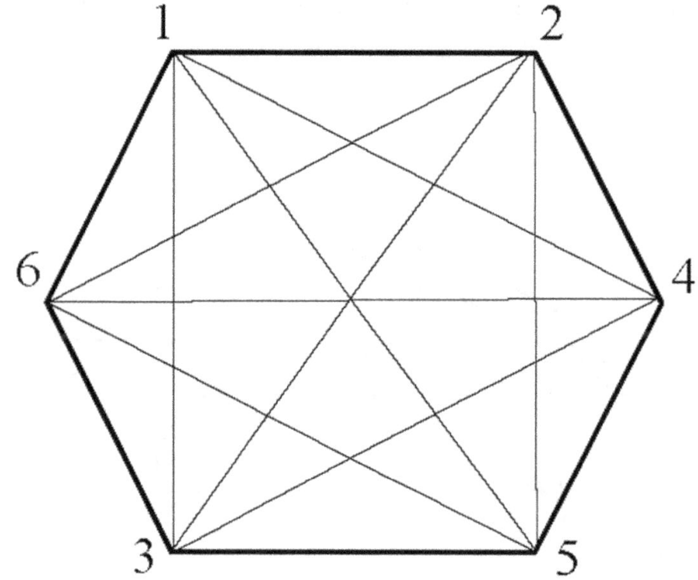

In Figure 15 using the condition that each triangle shall be overlapped twice, we really find 9 tetrahedrons:

$$T_1 = 1246, T_2 = 2345, T_3 = 1245, T_4 = 6435, T_5 = 1635, T_6 = 1362, T_7 = 1253, T_8 = 1645, T_9 = 6243.$$

Let's enumerate the edges and define their belonging to the listed tetrahedrons.
The edges: 61, 12, 24, 45, 35, 63, 13, 14, 34, 62, 25, 65, 15, 32, 64;

$$61 \in T_1, T_6, T_5, T_8; 12 \in T_1, T_3, T_6, T_7; 24 \in T_1, T_2, T_3, T_9; 45 \in T_2, T_4, T_8, T_3; 35 \in T_2, T_4, T_5, T_7;$$
$$63 \in T_6, T_4, T_5, T_9; 13 \in T_7, T_5, T_6; 14 \in T_1, T_3, T_8; 34 \in T_2, T_4, T_9; 62 \in T_1, T_6, T_9; 25 \in T_2, T_3, T_7;$$
$$65 \in T_4, T_5, T_8; 15 \in T_3, T_5, T_7, T_9; 32 \in T_2, T_6, T_7, T_9; 64 \in T_1, T_4, T_8, T_9.$$

Thus, there is a wonderful fact: the edges have different values of incidence to edges. The edges 13, 34, 14, 62, 25, 65 have value of incidence 3; the edges 61, 12, 24, 45, 35, 63, 15, 64, 32 have value of incidence 4. It occurs when the triangular prismahedron has all edges with equal values of incidence.

GENERAL ANALYSIS OF THE PRODUCT OF ANY TWO CONVEX POLYGONS

As a consequence of theorem 6.6 of Chapter 6 was received the structural formula (6.32) of the product of polytope P_n^m by a polytope Q_π^l . If both factors in (6.32) of Chapter 6 are polygons, the formula (6.32) take the form of

$$P_2^m(mP_1^2)\times Q_2^l(lP_1^2)=P_{2+2}^{lm}(mP_3^{2l},lP_3^{2m}). \tag{8.3}$$

According to Chapter 6, we'll call polytopes (8.3) ml-angular prismahedrons.

Theorem 8.1. Ml-angular prismahedron is a 4-polytope consisting of m l-angular prisms and l m-angular prisms.

***Proof*.** As a result of the product of m-angular polygon by l–angular polygon, according to the definition of polytopes product (Ziegler, 1995), is a polytope with facets forms. These facets are the products of each side of m- angular polygon by l-angular polygon and, accordingly, the products of each sides of l-angular polygon by m-angular polygon. The first from the products give m l- angular prisms, the second ones – give l m-angular prisms. Let's check by Euler-Pioncaré equation (Pioncaré, 1895)

$$\sum_{i=0}^{n-1}(-1)^i f_i=1-(-1)^n, \tag{8.4}$$

that the polytopes formed as a result of the product of polygons satisfy this equation for polytope dimension $n=4$. In this case the number of elements of dimension 0 (vertices) owing to constructing polytopes is equal to ml. The number of one-dimension elements (edges) is composed of the number of edges issuing from all vertices with account of each edge passing through two vertices. Since 4 edges issue from each vertex owing to polygons product, the total number of edges is $\frac{ml4}{2}=2ml$. The number of two-dimensional faces is composed of numbers m and l, correspondingly l- angular and m – angular bases of prisms and 4- angular faces of prisms ml.

Substituting the received values of numbers of elements of different dimension in Euler-Pioncaré equation

ml – 2ml +(m+l+ml) – (m +l)=0.

Thus, Euler – Pioncaré equation is satisfied, which proves theorem 8.1.

Consequence. Owing to duality in a polytope dual to ml-angular prismahedron, the number of vertices is equal to $m+l$, the number of edges is equal to $m+l+ml$, the number of two-dimensional faces is equal to $2ml$, the number of three-dimensional faces is equal to ml.

Each three-dimensional face has the number of two-dimensional faces equal to double ratio $\frac{2ml}{ml}$, i.e. 4. Thus, in the general case of the product of any two convex polytopes the tree-dimensional figure is a tetrahedron, since only a tetrahedron, as a three-dimensional figure, has 4 two-dimensional faces.

4*3-ANGULAR PRISMAHEDRON AND ITS DUAL POLYTOPE

In Chapter 6 the image of 4*3-angular prismahedron (Figure 6.6) was received. Let's denote vertices of this polytope $a_1,\ldots,a_{12}$ (Figure 16).

*Figure 16. 4*3-angular prismahedron*

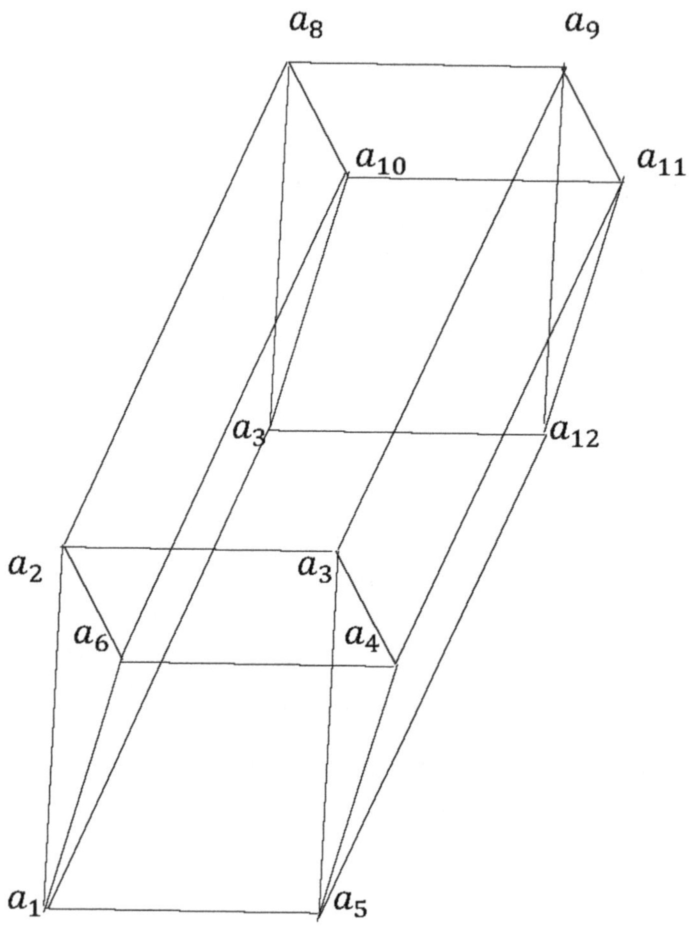

4*3-angular prismahedron includes three-dimensional figures: triangular prisms

1) $a_1a_2a_6a_7a_8a_{10}$, 2) $a_3a_4a_5a_9a_{11}a_{12}$, 3) $a_1a_2a_6a_3a_4a_5$, 4) $a_7a_8a_{10}a_{11}a_{12}$;

tetragonal prisms

5) $a_2a_6a_3a_4a_8a_{10}a_9a_{11}$, 6) $a_1a_6a_4a_5a_7a_{10}a_{11}a_{12}$, 7) $a_1a_2a_3a_5a_7a_8a_9a_{12}$.

For constructing dual polytope we define contacts of the prisms by equal sets of symbols a_i. In the result we receive a projection of dual polytope (Figure 17).

*Figure 17. Polytope dual to 4*3-angular prismahedron*

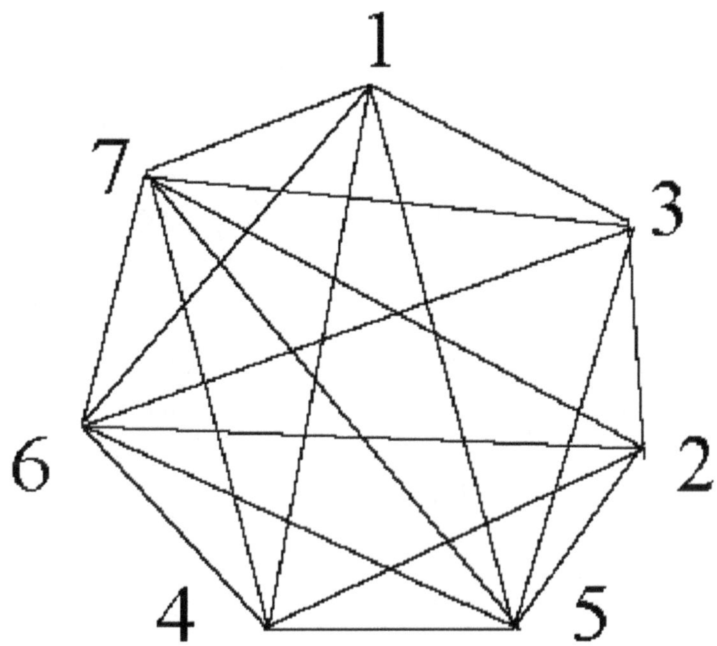

In polytope in Figure 17 one can single out 12 tetrahedrons with flat faces overlapped two times: 2564, 4527, 2537, 2563, 4561, 4571, 5731, 2746, 1536, 3627, 6714, 7136. This polytope have 19 edges: 17, 16, 15, 14, 13, 37, 36, 35, 32, 27, 26, 24, 25, 57, 56, 54, 47, 46, 67. From them edges 17, 14, 32, 24 have incidence value 3 with respect to tetrahedrons and the remaining 15 edges have incidence value 4. The vertices of dual polytope 3, 2, 1, 4 are incident to 5 edges and vertices 5, 6, 7 are incident to 6 edges.

3*6-ANGULAR PRISMSHEDRON AND A POLYTOPE DUAL TO IT

In Figure 18 a projection of 3*6-angular prismahedron with marked 18 vertices is shown.

*Figure 18. 3*6-angular prismahedron*

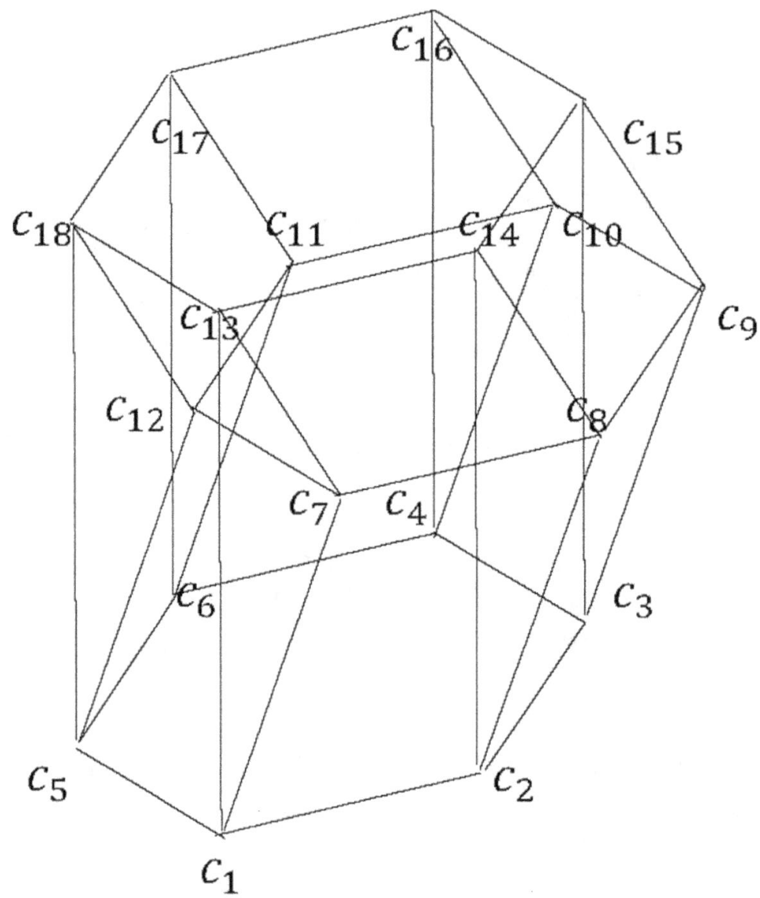

It has three 6-angular prisms

1) $c_1c_2c_3c_4c_5c_6c_7c_8c_9c_{10}c_{11}c_{12}$, 2) $c_7c_8c_9c_{10}c_{11}c_{12}c_{13}c_{14}c_{15}c_{16}c_{17}c_{18}$, 3) $c_1c_2c_3c_4c_5c_6c_{13}c_{14}c_{15}c_{16}c_{17}c_{18}$,

and six 3-angular prisms

4) $c_1c_7c_6c_{13}c_{12}c_{18}$, 5) $c_2c_3c_8c_9c_{14}c_{15}$, 6) $c_3c_4c_9c_{10}c_{15}c_{16}$, 7) $c_4c_5c_{16}c_{10}c_{11}c_{17}$, 8) $c_5c_6c_{11}c_{12}c_{17}c_{18}$.

Analyzing contacts of the prisms on coinciding sets of symbols, we construct a projection of the polytope dual to 6 * 3-angular prismahedron (Figure 19).

*Figure 19. Polytope dual to 3*6-angular prismahedron*

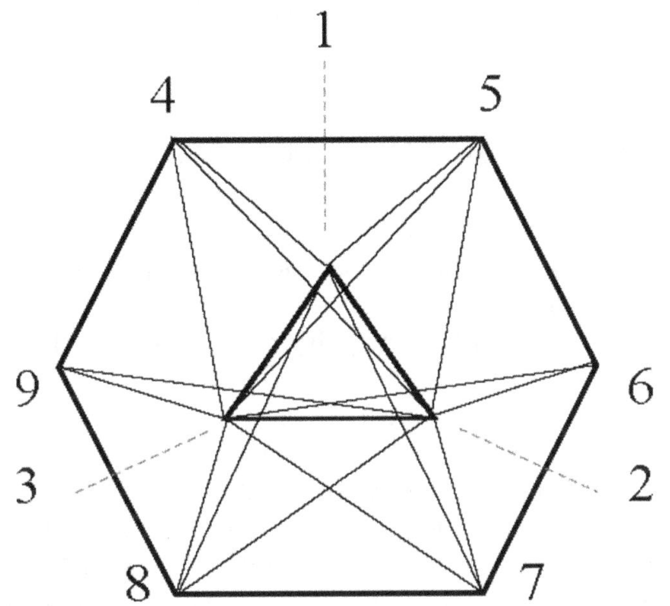

In polytope in Figure 19 we can single out 18 tetrahedrons:

$T_1 = 1234,\ T_2 = 1235,\ T_3 = 1236,\ T_4 = 1237,\ T_5 = 1238,\ T_6 = 1239,$
$T_7 = 1265,\ T_8 = 1267,\ T_9 = 2378,\ T_{10} = 3289,\ T_{11} = 1394,\ T_{12} = 1398,$
$T_{13} = 4325,\ T_{14} = 4315,\ T_{15} = 4215,\ T_{16} = 3267,\ T_{17} = 1378,\ T_{18} = 1278.$

The polytope has 27 edges with different values of incidence to tetrahedrons: edge

$45 \in T_{13}, T_{14}, T_{15};\ 12 \in T_1, T_8, T_{15}, T_{18};\ 15 \in T_2, T_7, T_{14}, T_{15};\ 14 \in T_1, T_{11}, T_{14}, T_{15};$
$16 \in T_3, T_7, T_8;\ 56 \in T_7;\ 19 \in T_6, T_{11}, T_{12};\ 94 \in T_{11};\ 13 \in T_1, T_6, T_{11}, T_{12}, T_{14}, T_{17};$
$42 \in T_1, T_{13}, T_{15};\ 35 \in T_2, T_{13}, T_{14};\ 43 \in T_1, T_{11}, T_{13}, T_{14};\ 32 \in T_1, T_6, T_9, T_{10}, T_{13}, T_{16};$
$93 \in T_6, T_{10}, T_{12};\ 26 \in T_3, T_7, T_8, T_{16};\ 92 \in T_6, T_{10};\ 36 \in T_3, T_{16};\ 17 \in T_4, T_8, T_{17}, T_{18};$
$52 \in T_2, T_7, T_{13}, T_{14};\ 87 \in T_9, T_{17}, T_{18};\ 18 \in T_5, T_{12}, T_{17};\ 37 \in T_4, T_9, T_{16}, T_{17};\ 82 \in T_5, T_9, T_{10}, T_{18};$
$27 \in T_4, T_8, T_9, T_{16}, T_{18};\ 83 \in T_5, T_9, T_{10}, T_{12}, T_{17};\ 76 \in T_8, T_{16};\ 98 \in T_{10}, T_{12}.$

Thus, edges 56, 94 have incidence value equal to 1; edges 92, 36, 76 have incidence value equal to 2; edges 45, 16, 19, 42, 35, 87 have incidence value equal to 3; edges 15, 14, 43, 93, 26, 17, 52, 82 have incidence value equal to 4; edges 27, 83 have incidence value equal to 5; edges 12, 13, 32 have incidence value equal to 10.

In addition, we note that vertices 1, 2, 3 are incident to 14 tetrahedrons; vertices 4, 5 are incident to 5 tetrahedrons; vertices 6, 9 are incident to 4 tetrahedrons; vertices 7, 8 are incident to 6 tetrahedrons.

This method can be used for construction of dual polytopes to many discussed in this paper, the polytopes with simultaneous use of two-dimensional faces of different shapes. All these polytopes are poly-incident, i. e. they have edges (and vertices) with different values of incidence.

N –PRISMAHEDRONS AND POLYTOPES DUAL TO THEM

Let's call n-prismahedron a polytope of dimension 4, with facets in the form of prisms with n - angular bases. In item 4*3-angular prismahedron and its dual polytope a 3- prismahedron was considered. Now let'us turn to the study of n - prismahedrons for any n.

Theorem 8.2. The product of two n-angular polygons is an n-prismahedron with the number of facets $2n$.

Theorem 2 follows immediately from theorem 8.1 when $m = l = n$.

Corollary. In a polytope dual to n - prismahedron the number of vertices is equal to $2n$, the number of edges is equal to $2n + n^2$, the number of two-dimensional faces is equal to $2\,n^2$, the number of three-dimensional faces is equal to n^2. This follows directly from the corollary to theorem 8.1.

Theorem 3. Polytopes dual to n - prismahedrons are those composed of tetrahedrons. For $n = 4$ the dual polytope is a 4-cross-polytope, for $n = 10$ the dual polytope is a previously unknown polytope consisting of 100 tetrahedrons (100-cell one) with a degree of incidence equal to 5, for other values of n ³ 3 the dual are the poly-incident polytopes with a medium degree of edge incidence tending to 6 with an infinite increase of n (Zhizhin, 2016).

Proof. As polytopes dual to n-prismhedrons have the number of vertices equal to $2n$, owing to the inverse of inclusion relation, the number of edges is equal to $2n + n^2$, the number of 2-faces is equal to $2n^2$, the number of 3-faces is equal to n^2. Then each 3-figure has the number of 2-faces $2\frac{2n^2}{n^2} = 4$ (multiplication by 2 is due to the fact that each 2-face belongs at once to two 3-shapes). Thus, a 3-figure is a tetrahedron, since only tetrahedron, as a 3-figure, has 4 2-faces. As everybody knows, the number of edges in a tetrahedron is 6. We can write this number on the basis of set numbers of different dimension in a dual polytope: $6 = k\frac{2n+n^2}{n^2}$, k is the degree of edge incidence (the edge belongs to k-tetrahedrons) under the assumption that all the edges have this degree of incidence, the same for all. Hence, we find the degree of incidence:

$$k = \frac{6n}{n+2}, n \geq 3. \tag{8.5}$$

Equation (8.5) in integers has two solutions: $k = 4$ when $n = 4$ and $k = 5$ when $n = 10$. For the other possible values of n, according to (8.5) the degree of incidence of edges is a fractional value. This means that in these cases in a dual polytope edges with varying degrees of incidence present simultaneously, i.e. it is a poly-incident polytope.

Figure 12, Figure 13, Figure 14 give examples of 3-prismahedron, 5-prismahedron, 6-prismahedron respectively. The known hypercube is 4 - prismahedron.

For $n = 4$ in a dual polytope the number of vertices is $2n = 8$, the number of edges is $2n+n^2=24$, the number of $2D$-faces is $2n^2=32$, the number of $3D$-faces is $n^2=16$. This is a 4 - cross-polytope dual to a hypercube. Each edge of this polytope has a degree of incidence 4.

For $n = 5$ in a dual polytope the number of vertices is $2n = 10$, the number of edges is $2n+n^2=35$, the number of $2D$-faces is $2n^2=50$, the number of $3D$-faces is $n^2=25$. In the projection on the plane this polytope is presented in Figure 20.

Figure 20. Polytope dual to 5-prismahedron

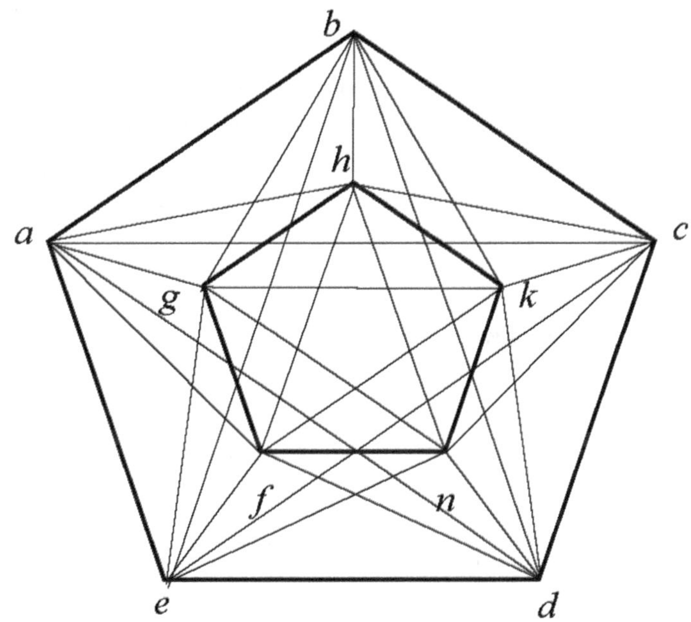

In Figure 20, tetrahedrons are:

abgh, 2)dckh, 3)kcnd, 4)efnd, 5)agfe, 6)hkgn, 7)knfh, 8)fngk, 9)gfhn, 10)ghfk, 11)ahfg, 12)gbkh, 13) hcnk, 14)fkdn, 15)egnf, 16)abec, 17)bcad, 18)cdeb, 19)edac, 20)aebd, 21)agbe, 22)bhac, 23)kcbd, 24) ndec, 25)efad. 1)

Every vertice of the polytope belongs to 10 tetrahedrons. Every $2D$-face belongs to two tetrahedrons (also as for $n = 3; 4$). From 35 edges 10 edges are incident to 3 tetrahedrons

$gb \in$ 1), 12), 21); $bk \in$ 2), 12), 23);$hc \in$ 2), 13), 22); $nc \in$ 3), 13), 24); $kd \in$ 3), 14), 23); $fd \in$ 4), 14), 25);$en \in$ 4), 15), 24); $eg \in$ 5), 15), 21);$af \in$ 5), 11), 25); $ah \in$ 1), 11), 22).

From 35 edges 15 edges are incident to 4 tetrahedrons

$ag \in$ 1), 5), 11), 21); $bh \in$ 1), 2), 12), 22); $kc \in$ 2), 3), 13), 23); $nd \in$ 3), 4), 14), 24); $ef \in$ 4), 5), 15), 25); $eb \in$ 16), 18), 20), 21); $bd \in$ 17), 18), 20), 23); $ac \in$ 16), 17), 19), 22); $ec \in$ 16), 18), 19), 24); *ad*

$\in$ 17), 19), 20), 25); *hf* $\in$ 7), 9), 10), 11); *hn* $\in$ 6), 7), 9), 13); *kg* $\in$ 6), 8), 10), 12); *kf* $\in$ 7), 8), 10), 14); *gn* $\in$ *6), 8), 9), 15).*

From 35 edges 10 edges are incident to 6 tetrahedrons:

ab $\in$ 1), 16), 17), 20), 21), 22); *bc* $\in$ 2), 16), 17), 18), 22),23); *cd* $\in$ 3), 17), 18), 19), 23), 24); *ed* $\in$ 4), 8), 19), 20), 24), 25); *ae* $\in$ 5), 16), 19), 20), 21), 25); *gh* $\in$ 1), 6), 9), 10), 11), 12); *hk* $\in$ 2), 6), 7), 10), 12), 13); *kn* $\in$ 3), 6), 7), 8), 13), 14); *fn* $\in$ 4), 7), 8), 13), 14); *gf* $\in$ 5), 8), 9), 10), 11), 15).

Thus, the average value of incidence degree is equal to: $\bar{k} = \frac{10 \cdot 6 + 15 \cdot 4 + 10 \cdot 3}{35} = \frac{30}{7}$ and coincides exactly with the degree of incidence according to equation (8.5).

By increasing n, the structure of a dual polytope gets greatly more complicated and there are difficulties in its constructing. For $n = 6$ in a dual polytope the number of vertices is $2n = 12$, the number of edges is $2n+n^2=48$, the number of $2D$-faces is $2n^2=72$, the number of $3D$-faces is $n^2=36$. This polytope is presented in Figure 21 in the projection on the plane.

Figure 21. Polytope dual to 6-prismahedron

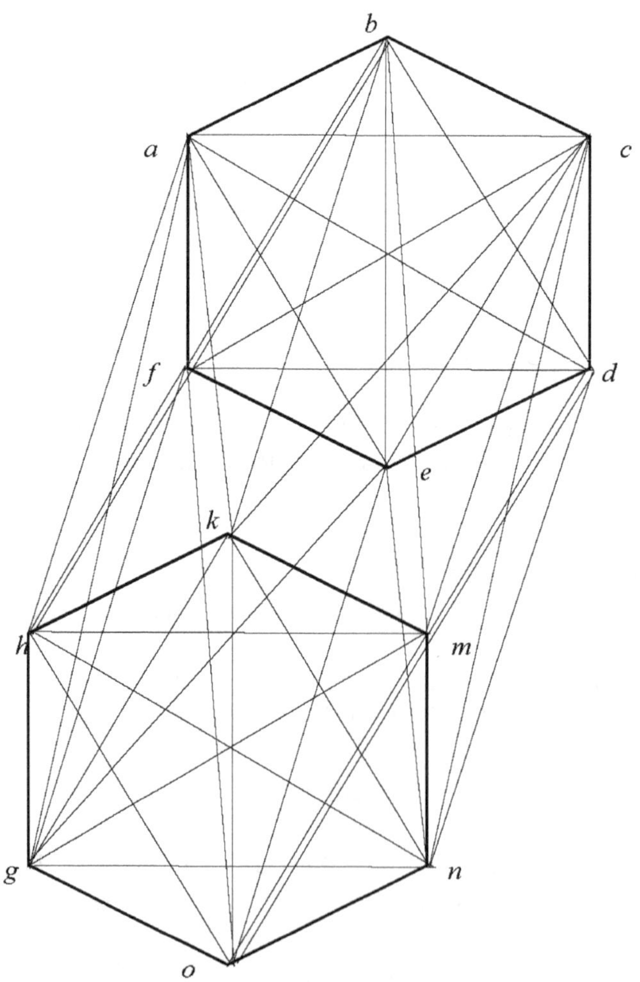

In Figure 21 tetrahedrons are:

1) *hmng*, 2) *acdf*, 3) *hkmg*, 4) *abcf*, 5) *hkmn*, 6) *abcd*, 7) *hkno*, 8) *abde*, 9) *gmno*, 10) *fcde*, 11) *kmno*, 12) *bcde*, 13) *gkmo*, 14) *fbce*, 15) *ghko*, 16) *fabe*, 17) *ghno*, 18) *fade*, 19) *hkmb*, 20) *abck*, 21) *kmnc*, 22) *bdcm*, 23) *mnod*, 24) *cden*, 25) *gone*, 26) *fedo*, 27) *hgof*, 28) *afeg*, 29) *ghka*, 30) *fabh*, 31) *kmbc*, 32) *mncd*, 33) *oned*, 34) *gofe*, 35) *hgaf*, 36) *hkab*.

From 48 edges each of 12 edges is incident to 7 tetrahedrons from the following list:

ab, bc, cd, de, fe, af, hk, km, mn, no, og, hg; each of 24

edges is incident to 4 tetrahedrons:

ko, gm,hn, ho, kg, mo, kb, oe, fc, be, ad, ac, fd, bd, mc, gf, hm, gn, kn, ae, bf, ce, nd, ha;

each of 12 edges is incident to 3 tetrahedrons:

bm, kc, ge, fo, md, cn, ga, hf, ne,od, hb, ak.

Thus, the average value of edges incidence degree is equal to $\bar{k} = \dfrac{12\cdot 7 + 24\cdot 4 + 3\cdot 12}{48} = 4{,}5$, which fits exactly with the degree of incidence k according to equation (8.5).

Every vertex of this polytope is incident to 12 tetrahedrons. From 72 triangles, which are faces of tetrahedrons, some have degree of incidence 3 and the rest have degree of incidence 2.

The construction of a dual polytope with $n = 10$ is particularly interesting, since according to equation (8.5) in this case the degree of incidence of edges is the same for the whole polytope and equals to 5. For $n = 10$ in a dual polytope, the number of vertices is $2n = 20$, the number of edges is $2n+n^2=120$, the number of $2D$-faces is $2n^2=200$, the number of $3D$-faces is $n^2=100$. This polytope is presented in Figure 22 in the projection on the plane.

Figure 22. Poytope dual to 10-prismahedron

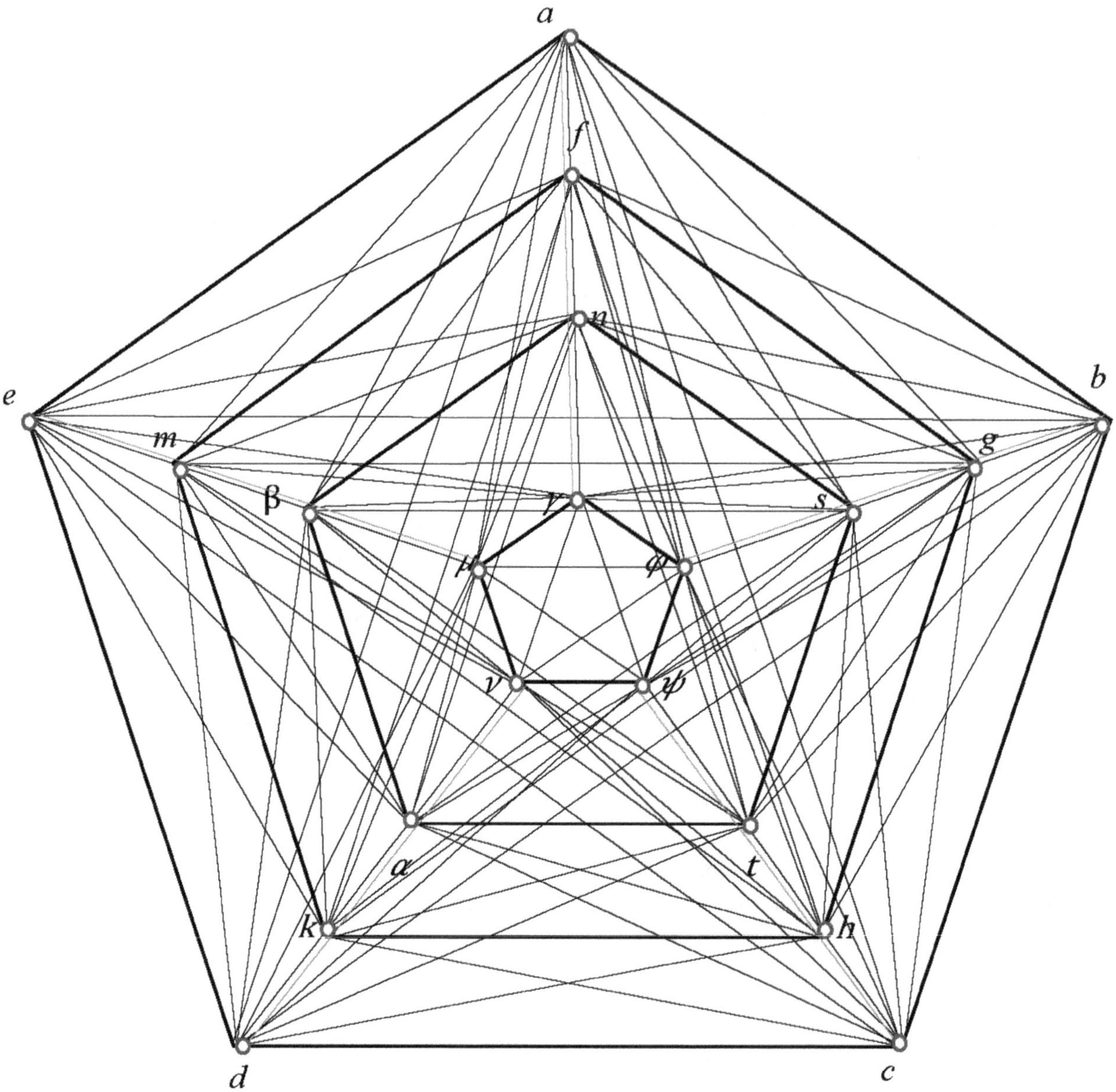

Tetrahedrons in Figure 22 are:

1) *dkhc*, 2) *cgbh*, 3) *fagb*, 4) *aefm*, 5) *demk*, 6) *kthα*, 7) t*hsg*, 8) n*fsg*, 9) m*fnβ*, 10) *β*mk*α*,
11) αtψν, 12) φψst, 13) γφns, 14) μγnβ, 15) βμαν, 16) φψγμ, 17) φψμγ, 18) φψγν,
19) γψνφ, 20) φγνμ, 21) dhbc, 22) cgab, 23) bfae, 24) daem, 25) ekcd, 26) ktgh, 27) hsfg,
28) mgnf, 29) fkmβ, 30) mhkα, 31) dcνψ, 32) φcbψ, 33) aγbφ, 34) μaeγ, 35) νdeμ, 36) νdψμ,
37) cνφψ, 38) νφψb, 39) aφγμ, 40) eγνμ, 41) φγνμ, 42) φνtψ, 43) φγψs, 44) μnφγ, 45) βγνμ,
46) kdth, 47) hcsg, 48) gbnf, 49) afmβ, 50) αemk, 51) αkhc, 52) thgb, 53) asgf, 54) fnem, 55) mdkβ,

56) kαtψ, 57) thφs, 58) sgγn, 59) nfμ β, 60) αvmβ, 61) thαv, 62) ψsgt, 63) snfφ, 64) nβγμ, 65) kβμα, 66) ktαβ, 67) thsα, 68) sgnt, 69) fnmg, 70) mnβα, 71) tαvβ, 72) tsαψ, 73) sntφ, 74) nsβγ, 75) βαμn, 76) mhαv, 77) tkgψ, 78) sfhφ, 79) γnmg, 80) kf βμ, 81) decv, 82) cdbψ, 83) tacφ, 84) ebγa, 85) daμe, 86) dmhv, 87) ckgψ, 88) bfhφ, 89) amgγ, 90) efkμ, 91) kcαψ, 92) thbφ, 93) sgγa, 94) enfμ, 95) mvβd, 96) kμeα, 97) maβγ, 98) fnbφ, 99) fgψc, 100) thdv.

In the given above list of tetrahedrons each line corresponds to a specific type of tetrahedron projection, which differ only by angle of rotation relative to the polytope projection center divisible by 72 degrees. The structure of a 100-cell polytope is quite complicated, but it is defined by enumerating all the tetrahedrons. If all the edges of a polytope have the same degree of incidence, the degree of incidence of vertices and triangles, which are faces of tetrahedrons, is variable.

By increasing the value of n in equation (8.5), the average degree of edges incidence tends to its maximum value of 6.

Theorem 8.3 is proved.

In this Chapter polytopes with different incidence values of edges to three-dimensional figures included in the polytope were considered. Graphs of these polytopes were constructed, with each three-dimensional body being taken for the top of the graph. By construction, these graphs are topologically equivalent to polytopes, dual to poly-incident polytopes. It is now clear that polytopic prizmahedrons, which play an important role in filling the space of higher dimension (Zhizhin, 2021), are polytopes dual to poly-incident polytopes.

Polytopes Dual to Polytypictops

In Chapter 7, the existence of higher-dimensional polytopes with regular-faceted polyhedrons (Johnson, 1960; Zalgaller, 1966; Zhizhin, 2009) was first established, which are very different from Archimedes bodies (polytypictops) due to the presence in them of a large number of types of different vertices (gonohedrons). Within the framework of this chapter, devoted to the consideration of polytopes dual to polytopes of various classes, it is of interest to determine the polyhedrons dual to polytypichedrons and the possibility of obtaining polytopes of higher dimension with the participation of polytopes dual to polytypichedrons Consider three polyhedrons from the family of polytipichedrons PG10, PG11 and PG14 as the polyhedron containing the largest number of different types of vertices (polytypichedrons – max).

The image of PG11 was presented in Figure 2.19.11 of Chapter 2. To construct the polyhedron dual to the polyhedron PG11, it is necessary to select a point in each two-dimensional face of the PG11 polyhedron as a vertex of the dual polyhedron and then connect these vertices with each other by edges. In this case, only the nearest vertices located on opposite sides of the edge of the flat face PG11 are connected by edges. Figure 23 shows the PG11 polyhedron, and the edges of the polyhedron dual to the PG11 polyhedron are indicated by dash-dotted lines in it.

Figure 23. The polyhedron PG_{11} and dual to it polyhedron PG^*_{11}

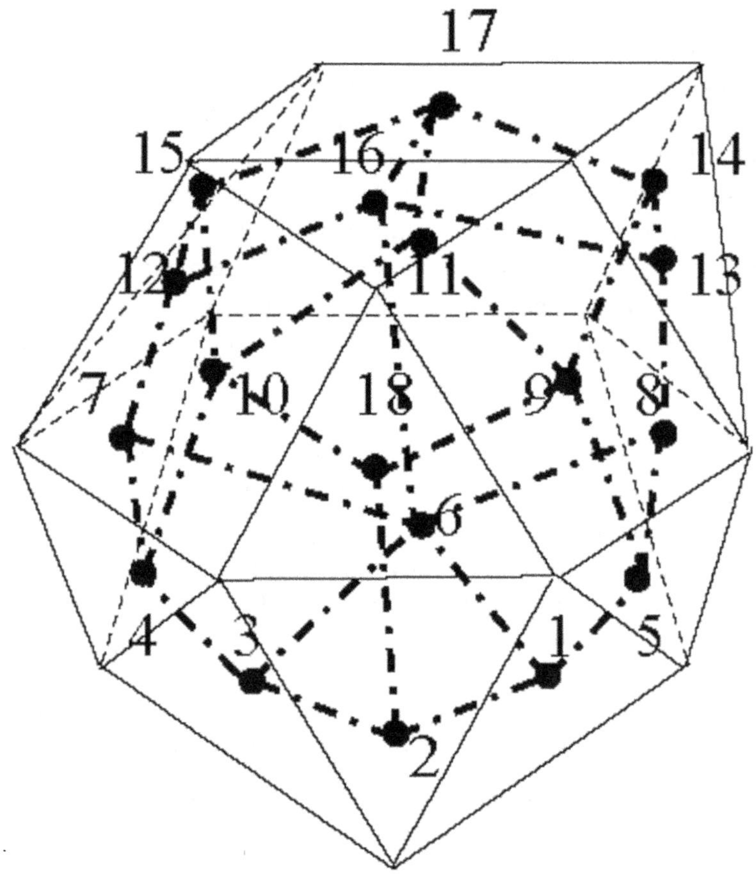

The bold points denote the vertices of the dual polyhedron. The number of these vertices is equal to the number of flat faces of the PG11 polyhedron, i. e. 18. Let us introduce the usual notation for the number of faces of different dimensions of the dual polyhedron. In this case f0= 18. Let us list the edges of the dual polyhedron:

1 – 5, 1 – 6, 1 – 2, 2 – 18, 2 – 3, 3 – 6, 3 – 4, 4 – 10, 4 – 7, 5 – 8, 5 – 9, 6 – 7, 6 – 8, 6 – 16, 7 – 12, 8 – 13, 9 – 11, 9 – 14, 9 – 18, 10 – 11, 10 – 18, 10 – 15, 11 – 17, 12 – 15, 12 – 16, 13 – 14, 13 – 16, 14 – 17, 15 – 17, 16 – 17.

Thus, the number of the edges for the dual polyhedron is equal 30, f1=30. Let us list two-dimensional faces of the dual polyhedron. In it include 10 faces with four vertices:

17 – 14 – 13 – 16, 15 – 12 – 16 – 17, 15 – 10 – 11 – 17, 7 – 12 – 16 – 6, 17 – 14 – 9 – 11, 16 – 13 – 8 – 6, 11 – 10 – 18 – 9, 7 – 6 – 3 – 4, 6 – 8 – 5 – 1, 1 – 2 – 3 – 6.

In it include 4 faces with five vertices:

4 – 10 – 18 – 2 – 3, 2 – 18 – 9 – 5 – 1, 4 – 7 – 12 – 15 – 10, 5 – 8 – 13 – 14 – 9.

Thus, the common number of two-dimensional faces for the dual polyhedron is 14, f2=14. Let us check the fulfillment of the Euler equation for the polyhedron dual to the PG polyhedron. Substitute the obtained values fi, i=0,1,2 into the equation (8.4)

18 – 30 + 14 = 2.

This confirms that the dual polyhedron satisfies the Euler equation. *The closedness* of the dual polyhedron is proved by checking that each of the listed edges of the polyhedron belongs to some two flat faces of the polyhedron. Now let's move on to the analysis of the gonohedrons of the dual polyhedron, denote it as PG_{11}^{*}. It should be said that if the PG11 polyhedron has two-dimensional faces with regular polygons, then the dual polyhedron PG_{11}^{*} has two-dimensional faces, due to the fundamental difference in the structure of a regular polyhedron from a regular polyhedron, two-dimensional faces in the general case are not regular polygons. In this regard, when listing the gonohedrons of the dual polyhedron, we indicate the number of the vertex of the dual polyhedron PG_{11}^{*} according to Figure 23 and the number of this particular type of polygons that make up the gonohedron:

1 (2 quads, 1 pentagon), 2 (1 quad, 2 pentagons), 3 (2 quads, 1 pentagon), 4 (2 quads, 1 pentagon),

5 (1 quad, 2 pentagons), 6 (5 quads), 7 (2 quads, 1 pentagon), 8 (2 quads, 1 pentagon),
9 (2 quads, 1 pentagon), 10 (2 quads, 2 pentagons), 11 (3 quads), 12 (2 quads, 1 pentagon),
13 (2 quads, 1 pentagon), 14 (2 quads, 1 pentagon), 15 (2 quads, 1 pentagon), 16 (3 quads),
17 (3 quads), 18 (1 quad, 2 pentagons).

It follows from the list of gonohedrons that the number of different types of gonohedrons in the dual polyhedron PG_{11}^{*} is 5: (2 quads, 1 pentagon), (1 quad, 2 pentagons), (2 quads, 2 pentagons), (5 quads), (3 quads). This number is equal to the number of different types of gonohedrons in the PG11 polyhedron, i. e. the PG11 polyhedron and its dual polyhedron PG_{11}^{*} have the same number of different types of gonohedrons - 5.

The image of PG_{10} was presented in Figure 2.19.10 of Chapter 2. To construct the polyhedron dual to the polyhedron PG10, it is necessary to select a point in each two-dimensional face of the PG10 polyhedron as a vertex of the dual polyhedron and then connect these vertices with each other by edges. In this case, only the nearest vertices located on opposite sides of the edge of the flat face PG10 are connected by edges. Figure 24 shows the PG10 polyhedron, and the edges of the polyhedron dual to the PG10 polyhedron are indicated by dash-dotted lines in it.

Figure 24. The polyhedron PG_{10} and dual to it polyhedron PG^{*}_{10}

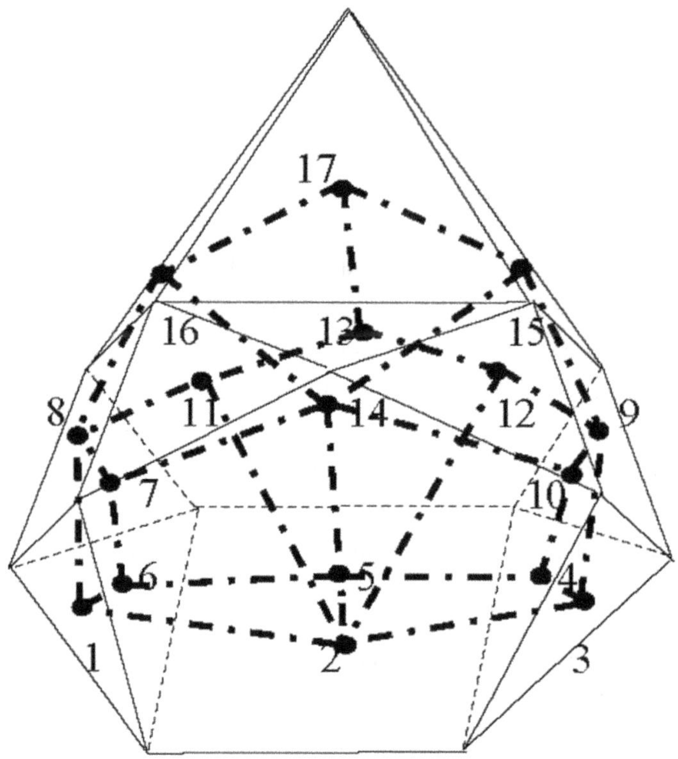

The bold points denote the vertices of the dual polyhed$_{\text{ro}}$n. The number of these vertices is equal to the number of flat faces of the PG10 polyhedron, i. e. 17. Let us introduce the usual notation for the number of faces of different dimensions of$_{\text{t}}$he dual polyhedron. In this case f0= 17. Let us list the edges of the dual polyhedron:

1 – 2, 1 – 6, 1 – 8, 2 – 5, 2 – 3, 2 – 11, 2 – 12, 4 – 3, 3 – 9, 4 – 10, 5 – 4, 6 – 5, 5 – 14, 6 – 7, 7 – 8, 7 – 14, 8 – 11, 8 – 16, 9 – 10, 9 – 12, 9 – 15, 10 – 14, 11 – 13, 12 – 13, 13 – 17, 15 – 14, 14 – 16, 15 – 17, 16 – 17.

Thus, the number of the edges for the dual polyhedron is equal 29, f1=29. Let us list two-dimensional faces of the dual polyhedron. In it include 12 faces with four vertices:

1 – 6 – 5 – 2, 1 – 6 – 8 – 7, 2 – 5 – 4 – 3, 4 – 3 – 10 – 9, 10 – 14 – 5 – 4, 6 – 14 – 5 – 7, 11 – 12 – 13 – 2, 2 – 12 – 3 – 9, 2 – 8 –11 – 1, 7 – 8 – 16 – 14, 9 – 10 – 14 – 15, 16 – 17 – 14 - 15.

In it include 2 faces with five vertices:

17 – 13 – 8 – 11 – 16, 17 – 13 – 9 – 12 – 15.

Thus, the common number of two-dimensional faces for the dual polyhedron is 14, f2=14. Let us check the fulfillment of the Euler equation for the polyhedron dual to the PG10 polyhedron. Substitute the obtained values fi, i=0,1,2 into the equation (8.4)

17 – 290 + 14 = 2.

This confirms that the dual polyhedron satisfies the Euler equation. The closedness of the dual polyhedron is proved by checking that each of the listed edges of the polyhedron belongs to some two flat faces of the polyhedron. Now let's move on to the analysis of the gonohedrons of the dual polyhedron, denote it as PG_{10}^*. It should be said that if the PG10 polyhedron has two-dimensional faces with regular polygons, then the dual polyhedron PG_{10}^* has two-dimensional faces, due to the fundamental difference in the structure of a regular polyhedron from a regular polyhedron, two-dimensional faces in the general case are not regular polygons. In this regard, when listing the gonohedrons of the dual polyhedron, we indicate the number of the vertex of the dual polyhedron PG_{10}^* according to Figure 24 and the number of this particular type of polygons that make up the gonohedron:

1 (3 quads), 2 (5 quads), 3 (3 quads), 4 (3 quads), 5 (4 quads), 6 (3 quads), 7 (3 quads),

8 (3 quads, 1 pentagon), 9 (3 quads, 1 pentagon), 10 (3 quads), 11 (2 quads, 1 pentagon),
12 (2 quads, 1 pentagon), 13 (1 quads, 2 pentagon), 14 (5 quads), 15 (2 quads, 1 pentagon),
16 (2 quads, 1 pentagon), 17 (1 quads, 2 pentagons).

It follows from the list of gonohedrons that the number of different types of gonohedrons in the dual polyhedron PG_{10}^* is 6: (2 quads, 1 pentagon), (1 quad, 2 pentagons), (3 quads, 1 pentagons), (5 quads), (3 quads), (4 quads). This number is on one more of the number of different types of gonohedrons in the PG10 polyhedron which have 5 of different types of gonohedrons.

The image of PG14 was presented in Figure 2.19.14 of Chapter 2. To construct the polyhedron dual to the polyhedron PG14, it is necessary to select a point in each two-dimensional face of the PG14 polyhedron as a vertex of the dual polyhedron and then connect these vertices with each other by edges. In this case, only the nearest vertices located on opposite sides of the edge of the flat face PG14 are connected by edges. Figure 25 shows the PG14 polyhedron, and the edges of the polyhedron dual to the PG14 polyhedron are indicated by dash-dotted lines in it.

Figure 25. The polyhedron PG_{14} and dual to it polyhedron PG^{*}_{14}

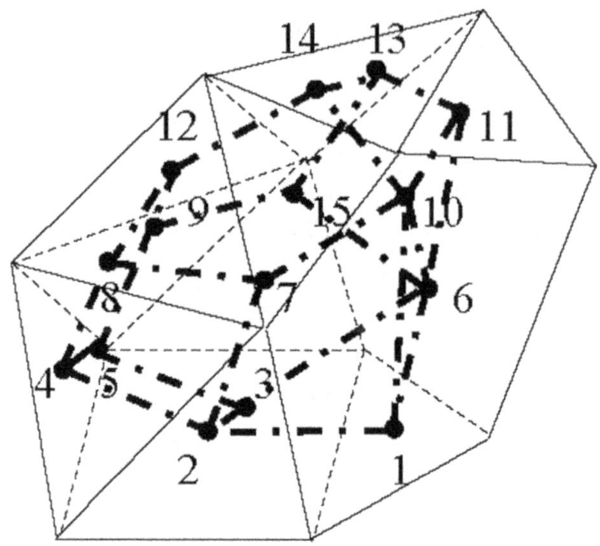

The bold points denote the vertices $_{o}$f the dual polyhedron. The number of these vertices is equal to the number of flat faces of the PG14 polyhedron, i. e. 15. Let us introduce the usual notation for the number of faces of dif$_{fe}$rent dimensions of the dual polyhedron. In this case f0= 15. Let us list the edges of the dual polyhedron:

1 – 10, 1 – 6, 1 – 2, 2 – 4, 2 – 3, 2 – 7, 3 – 6, 3 – 5, 4 – 5, 4 – 8, 5 – 9, 6 – 10, 6 – 11, 6 – 15, 7 – 8, 7 – 10, 8 – 12, 9 – 12, 9 – 15, 10 – 11, 10 – 14, 12 – 14, 11 – 13, 13 – 15, 13 – 14.

Thus, the number of the edges for the dual polyhedron is equal 25, f1=25. Let us list two-dimensional faces of the dual polyhedron. In it include 6 faces wit_{h} four vertices:

1 – 6 – 2 – 3, 1 – 2 – 10 – 7, 2 – 3 – 4 – 5, 7 – 2 – 8 – 4, 6 – 13 – 15 – 11, 14 – 13 – 11 – 10.

In it include 4 faces with five vertices:

3 – 5 – 15 – 9 – 6, 4 – 8 – 12 – 5 – 9, 8 – 13 – 12 – 15 – 14, 8 – 7 – 12 – 14 – 10.

In it include 2 faces triangle. Thus, the common number of two-dimensional faces for the dual polyhedron is 12, f2=12. Let us check the fulfillment of the Euler equation for the polyhedron dual to t_{h}e PG14 polyhedron. Substitute the obtained values fi, i=0,1,2 into the equation (8.4)

15 _ 25 + 12 = 2.

This confirms that the dual p_{o}lyhedron satisfies the Euler equation. The closedness of the dual polyhedron is proved by checking that each of the listed edges of the polyhedron belongs to some two flat faces of the polyhedron. Now let's move on to the analysis of the gonohedrons of the dual polyhedron, denote it as PG^{*}_{14}. It should be said that if the PG14 polyhedron has two-dimensional faces with regular polygons, then the dual polyh$_{ed}$ron PG^{*}_{14} has two-dimensional faces, due to the fundamental difference in the structure of a regular polyhedron from a regular polyhedron, two-dimensional faces in the general case are not regular polygons. In this regard, when listing the gonohedrons of the dual polyhedron, we indicate the number of the vertex of the dual polyhedron PG_{14}^{*} according to Figure 25 and the number of this particular type of polygons that make up the gonohedron:

1 (2 quads, 1 triangle), 2 (4 quads), 3 (2 quads, 1 pentagon), 4 (2 quads, 1 pentagon),
5 (1 quad, 2 pentagons), 6 (2 triangles, 2 quads, 1 pentagon), 7 (2 quads, 1 pentagon),
8 (1 quad, 2 pentagons), 9 (3 pentagons), 10 (2 triangles, 2 quads, 1 pentagon),
11 (1 triangle, 2 quads), 12 (3 pentagons), 13 (2 quads, 1 pentagon), 14 (1 quads, 2 pentagons),
15 (1 quad, 2 pentagons).

It follows from the list of gonohedrons that the number of different types of gonohedrons in the dual polyhedron PG_{14}^{*} is 6: (2 quads, 1 triangle), (1 quad, 2 pentagons), (2 quads, 2 triangles, 1 pentagon), (2 quads, 1 pentagon), (4 quads), (3 quads). This number is equal to the number of different types of gonohedrons in the PG14 polyhedron, i. e. the PG14 polyhedron and its dual polyhedron PG_{14}^{*} have the same number of different types of gonohedrons - 6.

Thus, in the examples considered, the maximum number of possible different types of vertices (gonohedrons), equal to six, which is characteristic of polytypichedrons, did not change for polyhedrons dual to polytypichedrons. In this case, the number of different types of gonohedrons in the transition to dual polyhedrons can increase, as follows from the consideration of the PG10 polyhedron. Multiplying the polyhedrons dual to the polytypichedrons by geometric elements of various dimensions (segments, triangles, polyhedrons, polytopes), one can obtain an infinite family of polytopes of higher dimension with polyhedrons of a new type, i. e. polyhedrons with irregular faces and a large number of different types of vertices. These polytopes also belong to polytypictops, as well as polytopes with regular -faced polyhedrons and with a large number of different types of vertices.

CONCLUSION

It is known that the study of dual structures enriches knowledge about the properties of these structures and their relationship with each other. If we consider regular dual polyhedrons or polytopes, then direct and dual objects remain in the same class of regular figures, i. e. new objects are not born at the same time. A dual transition only maps some elements of the class of regular shapes to other elements of this class or maps an element to itself. However, as shown in this chapter, any deviations in the correctness of figures of any dimension (polyhedrons or polytopes) lead to the fact that the dual elements form a new class of figures. This is an important finding of this study. In order to be convinced of this, at the beginning of the chapter, the process of the formation of dual figures in the class of regular three-dimensional figures and figures of the highest dimension is considered in detail. It is proved in detail how the dual transition from polytopes of higher dimensions such as simplex, cross-polytope and cube leads to dual figures, not being limited, as usual, with a brief statement that the n-simplex is dual to itself, the n-cross-polytope is dual to the n-cube. Consideration of polytopes dual to semiregular polytopes leads to surprises.

The dual to a four-dimensional semi-regular polytope, which is the product of two triangles (a trangular prismahedron), turned out to be a four-dimensional polytope with different incidence of edges in relation to the polytope facets. This property (polyincidence) was also observed in other polytopes, dual to the product of polygons to each other and to themselves. A whole class of polyincidental polytopes of the highest dimension has been formed. Their images are given, and their composition is precisely indicated. For example, the polytope dual to the product of a decagon and a decagon contains 100 tetrahedra. All of them are listed by vertices in the dual polytope. The most interesting results were obtained when constructing polytopes dual to regular-faceted polytopes and, in particular, to polytopes that are very different from Archimedes bodies. This difference is due to the large number of different types of vertices (gonohedrons) in the polyhedrons included in the polytope. These polytopes are called polytypictops. And the corresponding polyhedrons are named polytypicheadrons. It was found that the figures dual to these polytopes (polyhedrons) no longer have faces in the form of regular figures. They have a new class of irregular gonohedrons. Thus, the existence of an infinite class of polytopes of higher dimension was discovered, which include irregular polyhedrons. The latter will significantly enrich our knowledge about polytopes of higher dimension. It should be added that for all these polytopes, the fulfillment of the necessary conditions for their existence has been proven. It is to be expected that this knowledge will be applied in the analysis of chemical structures, in which the deviation of geometric objects from the correct figures is often observed (Zhizhin, 2017, 2021).

REFERENCES

Bogdanov, A. A. (1989). Tectology. In General Organizational Science. Moscow: Economics.

Buckour, I., *& Delanu, A. (1972). Introd*uction to the theory of categories. World.

Coxeter, *H.S.M. (1963). Regular Polytopes. John* Wiley & Sons, Inc.

Euler, L. (1*736). Solutio prob*lematisad geometriam situs pertinentis. Comment Acad Sci I Petropolitanae, 8, 128–140.

Feyma*n, R. (1968). The nature of physic*al laws. World.

Grunbaum, B. *(1967). Convex Polytopes. S*pringer.

Hegel, G. V. (1998*). The Science o*f Logic. Thought.

Johnson, W. *W. (1960). Convex po*lyhedrons with regular faces. Notices of the American Mathematical Societ*y, 7, 952.*

Poincare, A. (1895). Analysis si*t*us. J. de é Ecole Polytechnique, 1.

Stein*haus, H. (1950). Mathematical* Snapshots. Oxford Press.

*Whitehead, A. N. (*1990). Selected works on philosophy. Prog*ress.*

*Zalgaller, V. A. (1*966). Convex polyhedrons with correct *sides. Science.*

*Zhihin, G.V. (200*5). Correct both semiregular polyhedrons and filling of space with them. Life and Safety, 1-2, 551-55*9.*

Zhizhin, G. V. (2009). Convex right-faceted poly*hedrons. Polytechnika-service.*

Zhizhin, G. V. (2014). 4D world. St. Petersbu*rg: Poly*technica-Service.

Zhizhin, G.V. (2016). N–prismahedrons and their dual polytopes. International Journ*al Chemical Modeling, 5(4).*

Zhizhin, G. V. (2017). Chemical Compoun*d Structures and the Higher Dimension of Molecules: emerging Research and Opportunities. IGI Global.*

Zhizhin, G. V. (2019). The Geometry *of Higher-Dimensional polytopes. IGI Global.* doi:10.4018/978-1-5225-6968-8

Zhizhin, G. V. (2021). Nanotechno*logies and Clusters in the Spaces of Higher Dimension: Emerging Research and Opportunities. IGI Glob*al. doi:10.4018/978-1-7998-3784-8

Ziegler, G. (1995). Lecture*s on polytopes. Sprin*ger. doi:10.1007/978-1-4613-8431-1

KEY TERMS AND DEFINITIONS

Conve**x Regularly Faceted Polyhedrons: Conv**ex three-dimensional polytopes whose faces are regular polygons.

Dual**ity in Polytopes: If** the facets of one polytope correspond to the vertices of another polytope, and the edges of one polytope correspond to common elements of the facets of another polytope, then these polytopes are dual.

Gon**ohedron: A** polyhedral corner whose faces are regular polygons.

Incidence in Polytopes: Incidence in polytopes define the number of elements of higher dimension the given element of lower dimension belongs.

Mechanical Operations on a Polyhedron: Conceivable transformations of a polyhedron, including cutting a polyhedron (cutting off vertices), deforming a section of a cut to obtain a regular polygon in a section, rotating parts of a polyhedron after cutting relative to each other.

N*3-Angular Prismahedron: The product of n-angle by the triangle.

Poly-Incident Polytopes: Polytopes in which elements of lower dimension have different incidence values for elements of higher dimension. Polytopes that are dual to polytopes products are poly-incident polytopes.

Polytopic Prismahedron: The product of polytope by one dimension segment.

Polytypichedron (Max): The regular-faceted polyhedron with max number of different types of genohedrons.

Polytypichedrons: The regular-faceted polyhedrons, in which three or more of different types of genohedrons can be present.

Polytypictops: Polytopes of the highest dimension, which include regular-faceted three-dimensional polyhedra with a large number of types of vertices (gonohedrons).

Regular Polyhedron: Convex polyhedron with faces of regular polygons of one type.

Semiregular Polyhedrons: Convex polyhedrons with faces of regular polygons of two or more types, if all gonohedrons of the polyhedron have the same composition and arrangement of faces.

Tetrahedral Prism: The product of tetrahedral by one dimension segment.

Tetrahedral Prismahedron: The product of tetrahedral by the triangle.

Triangular Prismahedron: The product of triangle by the triangle.

Chapter 9
The Polytopes of the Higher Dimension in Physics and Chemistry and Construction of Spaces of the Higher Dimension

ABSTRACT

Polytopes of the highest dimension of molecules of compounds of inorganic and organic chemistry are considered on specific examples. It is shown that they all have dimensions greater than three. They may include polytopes of a known shape, discussed in previous chapters, but in general they differ significantly from the standard shapes. These include linear and nonlinear chains of metal atoms with ligands, closed chains of metal atoms with ligands, clusters with ligands and a metal polyhedral backbone. A class of polytopic prismahedrons (a special type of polytopes of higher dimension) is considered, from which parallelotopes of higher dimension are formed, which are necessary for constructing n-dimensional spaces, using them to create extended nanomaterials based on clusters of chemical compounds.

INTRODUCTION

The studies carried out in the monograph of G. Zhizhin "Chemical Compound Structures and Higher Dimension of Molecules: Emerging Research and Opportunities" (Zhizhin, 2018) convincingly showed that many elements of the periodic system of Mendeleev elements form molecules whose spatial structure can be geometrically represented as a convex polytope. Moreover, the dimension of this polytope, determined by the Euler-Poincaré relation between the geometric elements forming the polytope (Poincare, 1895), is usually higher than three. In addition, in this monograph it is noted that most elements of the periodic system exhibit valence in compounds higher than the valence determined by the location of the element in the periodic table. It is known that the calculation of the valence of a chemical element based on its location in the periodic table is based on the electronic structure of atoms (s -, p -, d -, f – orbitals), determined by solving the Schrodinger equation (Schrodinger, 1927) under the assumption of the three

DOI: 10.4018/978-1-7998-8374-6.ch009

- dimensionality of the space surrounding the nucleus of the atom. To explain the observed differences in the valence of atoms from valence, determined from the location of the element in the periodic table, researchers resorted to the construction of various models of the interaction of electrons in the vicinity of atomic nuclei. These models are quite diverse and still remain at the level of hypotheses.

Nevertheless, the concept of the electronic structure of atoms, based on the solution of the Schrödinger equation in a three - dimensional industrial space, is currently prevailing in the description of the structures of chemical compounds in both inorganic and organic chemistry. Moreover, these representations are supported by the opinion that there are no solutions to the Schrödinger equations in a space of dimension more than three (Büchel, 1963; Freeman, 1969; Gurevich & Mostepanenko, 1971). However, in the works of the author (Zhizhin, 2018, 2019 a), it was found that the proof of the impossibility of solving the Schrödinger equation in a space with a dimension greater than three contains an error associated with an incorrect writing of Coulomb's law in a space of higher dimension. When correcting this error, the author first obtained a solution to the Schrödinger equation in four - dimensional space (Zhizhin, 2018), and then obtained a solution to the Schrödinger equation in an n - dimensional space (Zhizhin, 2019 a). It is proved that an increase in the dimension of space in the vicinity of an atom leads to an increase in the number of quantum numbers that determine the structure of the electron cloud around the nucleus of the atom, as a consequence, an increase in the number of quantum cells in the electron orbitals of the atom. This may be the real reason for the experimentally observed increase in the valence of elements. As an illustration of the higher dimensionality of molecules, this chapter presents studies of molecules involving anomalous elements, binary compounds, and metal clusters with ligands. In addition, it is considered how spaces of higher dimension are constructed from polytopes of higher dimension.

CHEMICAL COMPOUNDS OF ANOMALOUS ELEMENTS

The desire of valence electrons to increase the number of d - electrons should lead to some features of the compounds of anomalous elements (Zhizhin, 2017 a). We consider which the features. We choose to consider the one s- element of the three mentioned groups of anomalous elements. The first anomalous element in the chromium group is Cr. It in consequence of the anomalies have one valence electron on the 4 s - orbital and five of the electrons on $3d$ - orbital. This allows have of chromium a valence equal 6 in many compounds. Since crystalline chromium oxide CrO_3 consists of chains of tetrahedrons CrO_4, united in two vertices. Each tetrahedron has located in the center of an atom of chromium associated by double bond with each of the four oxygen atoms at the vertices of a tetrahedron. All molecules have the form of a tetrahedron with the center, as shown in (Zhizhin & Diudea, 2016; Zhizhin, Khalaj & Diudea, 2016), have a dimension of 4, i.e., crystalline chromium oxide is a chain of polytopes of dimension 4, united in two vertices (Figure 1). If instead of double bonds are one-time connection with the chromium atom monovalent groups (such as hydroxyl groups), the molecule will be the center of the octahedron, which also would have dimension 4.

Figure 1. The chain of molecules CrO_3

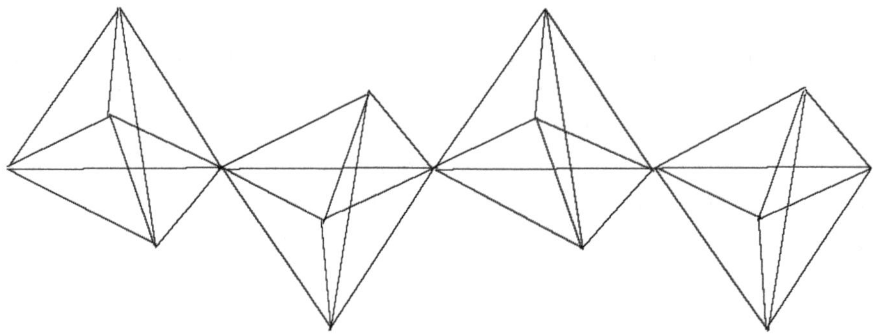

There are even more complex chromium compounds (see Gillespie, 1972; Gillespie & Hargittai, 1991), the dimension of molecules, which is more than four. It is clear that all the other anomalous elements with half (or nearly half) filled d - orbitals of the subshell will have similar compounds having a molecular of higher dimensions.

If the anomalous elements has one electron in the outer orbital s and subshell d completely (or almost completely) filled, then the element at the expense of s electron forms a linear molecule, such as a linear molecule oxide X_2 O, where X is the anomalous element (Cu, Pd, Ag, Pt, Au, Rg). However, due to the donor-acceptor bond linear molecule can form complex structures in the space. We choose element Cu from second group of anomalous elements. Figure 2 shows an exemplary structure formed by linear molecules Cu_2 O

Figure 2. The structure of the compound Cu_2O
A black small circle is oxygen atom. A brown circle is copper atom.

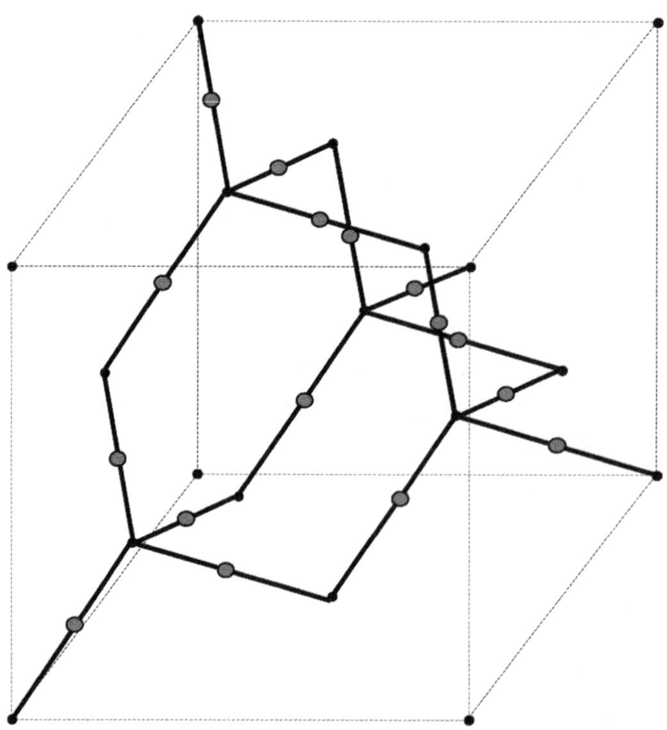

Each oxygen atom in the structure of Figure 2 bonded to four metal atoms (Cu). Two covalent bonds due to the formation of electron pairs divided: one *s* - electron metal atom and a *p* - electron atom of oxygen. In addition, there are two more donor-acceptor chemical bond due to the transfer of two electrons from the *s* - orbital and two electrons from the *p* - orbitals of the oxygen atom to vacant quantum cell of orbital metal. Thus, oxygen atom acquires valence equal four. In addition, each metal atom is linearly between two oxygen atoms.

In this structure, oxygen atoms (except oxygen atoms located at the vertices of a cube) form structure is topologically equivalent to the structure of carbon atoms in the molecule of adamantane. As shown in the article of Zhizhin (2014 a) on the basis the monograph Zhizhin (2014 b), the dimension of this molecule is 4. However, the two molecules comprising 10 oxygen atoms have free unallocated space. Therefore, if we set the task of finding the unit cell structure of copper oxide without filling cracks and gaps to help translation the entire space, we need to build polytopic prismahedron Zhizhin (2015 a), with bases in the form polytopes corresponding to these molecules. Taking a line segment equal to the length of the edge of the cube, in which is inscribed the structure including 10 oxygen atoms, multiply the polytope corresponding to this structure for this segment. We obtain polytopic prismahedron of dimension 5. With this polytopic prismahedron can fill space without gaps and clearances.

From the third group of anomalous elements we choose gold (Au). The outer shell of gold atom has one 6*s* - electron and a completely filled 5*d* - orbital. In the compound chlorine triphenylphosphine of gold $(Ph_3P)AuCl$ gold atom, giving one *s* - electron to chlorine atom, forms ion $Au(Ph_3P)_3^+$ with trigonal coordination (Gillespie, 1972; Perrin, Armarego & Perrin, 1980). Following Zhizhin (2017 b) we denoted phosphine molecule as a functional group of the compound.

Then ion $Au(Ph_3P)_3^+$ represented in the form of three tetrahedrons with the center, having one common vertex - a gold atom. In the center of each tetrahedron is located phosphorus atom and the remaining vertices of the tetrahedrons are occupied introduced functional groups Ph_3 (Figure 3).

Figure 3. Ion $Au(Ph_3P)_3^+$
A white circle is gold atom. A black small circle is phosphorus atom. A black big circle is functional groups Ph_3

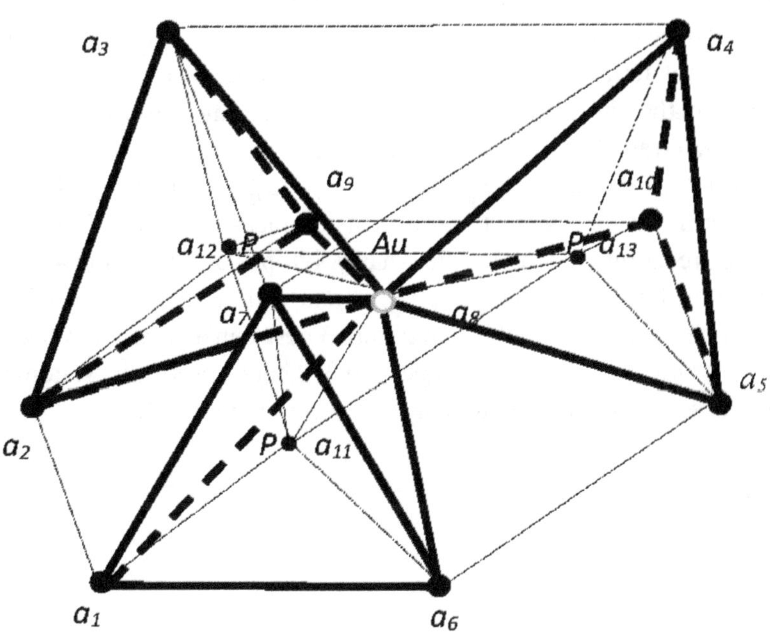

The functional dimension of each tetrahedron with center is still equal to 4. Thus, the ion $Au(Ph_3P)_3^+$ is a collection of three polytopes of dimension 4, having a common vertex. The assertion is proven

Theorem 9.1. The ion Au $(Ph_3P)_3^+$ has dimension 5.

Proof. To prove the necessity of the three tetrahedrons with a common vertex to form a convex shape. Connect the vertices of a_3, a_4, a_7 by line segments, forming a triangle $a_3\ a_4\ a_7$. Connect also the three centers of the tetrahedrons with each other, forming a triangle $a_7\ a_{11}a_{13}$. Connect the center of the tetrahedrons with vertices corresponding of the tetrahedrons and vertices in the grounds of the tetrahedrons, forming a hexagon $a_1\ a_2\ a_9\ a_{10}\ a_5\ a_6$ (thin lines on Figure 3). Define dimension polytope in Figure 3 on the Euler- Poincare equation (3.3) in Chapter 3 (Poincare, 1895)

To calculate the number of elements of large dimensions we turn first to a simple polytope, a part of a polytope in Figure 3. Temporarily excluded from Figure 3 the centers of the tetrahedrons – a_{11}, a_{12}, a_{13}, and all edges emanating from these vertices. Then, the polytope has 13 vertices, i. e.f_0=10. The number of edges of the polytope is sum the number edges of three tetrahedrons (6•3=18), the number of edges connecting tetrahedrons at the base figure (3), the number of the edges connecting vertices of the tetrahedrons at top of Figures 3. Thus, the number of edges polytope without centers of the tetrahedrons equal 24, i. e. f_1=24. The number of flat elements is sum of the flat faces of tetrahedrons (4•3=12), 1 hexagon, 3 triangles between tetrahedrons at base of figure, 3 lateral tetragons, 4 triangles of tetrahedron at top of figure. Thus, the number of flat elements is 23, i. e. f_2=23. The number of three-dimension elements is sum of 4 tetrahedrons, 3 figures between tetrahedrons, 1 hexagon at base and figure composed from boundary flat faces. Thus, the number of three-dimension elements is 9, i. e. f_3=9. Substituting the values f_i in equation (3.3), we see that it holds for $n = 4$

$$10 - 24 + 23 - 9 = 0.$$

Therefore, three tetrahedrons with common vertices is polytope with dimension 4.

For add centers in tetrahedrons the number of the vertices becomes equal 13, i. e. $(f_0)_c$=13. For this there add the number of the edges: 4•3=12 edges in tetrahedrons with centers, and 3 edges connecting centers. Thus, common number of edges on Figure 3 equal 39, i. e. $(f_1)_c$=39. The number flat faces there increases on 18 triangles in the tetrahedrons, 4 triangles in tetrahedron $a_{11}a_{12}a_{13}a_8$, 6 tetragons with vertices part which are centers of the tetrahedrons. Thus, common number of flat faces on Figure 3 equal 51, i. e. $(f_2)_c$=51. For add centers the number of three-dimensions faces increases on 4•3=12 tetrahedrons into tetrahedrons with centers, tetrahedron $a_{11}a_{12}a_{13}a_8$, figure $a_1\ a_2\ a_9\ a_{10}\ a_5\ a_6\ a_{11}a_{12}a_{13}a_8$, prism $a_{11}a_{12}a_{13}a_3a_4a_7$, 3 pyramids with vertex a_8 $(a_8a_1a_2a_{12}a_{11}$, $a_8a_5a_6a_{11}a_{13}$, $a_8a_9a_{10}a_{12}a_{13})$, 3 prism $(a_1a_2a_3a_7a_{11}a_{12}$, $a_5a_6a_4a_7a_{11}a_{13}$, $a_9a_{10}a_3a_4a_{12}a_{13})$. Thus, common number of three-dimension faces on Figure 3 equal 30, i. e. $(f_3)_c$=30.

It is known from the preceding that the Figure 3 has polytopes of dimension 4. Each tetrahedron with center there is polytope of dimension 4 and 3 tetrahedrons without center, but with a common vertex, there is a polytope of dimension 4. In addition, in Figure 3 between any two tetrahedrons with the center is polytope dimension 4. Obviously, such polytopes are 3. That to proof this statement we consider any polytope from them. For example, the polytope $a_4a_5a_6a_7a_{11}a_{13}$ (Figure 4).

Figure 4. The polytope $a_4a_5a_6a_7a_{11}a_{13}$

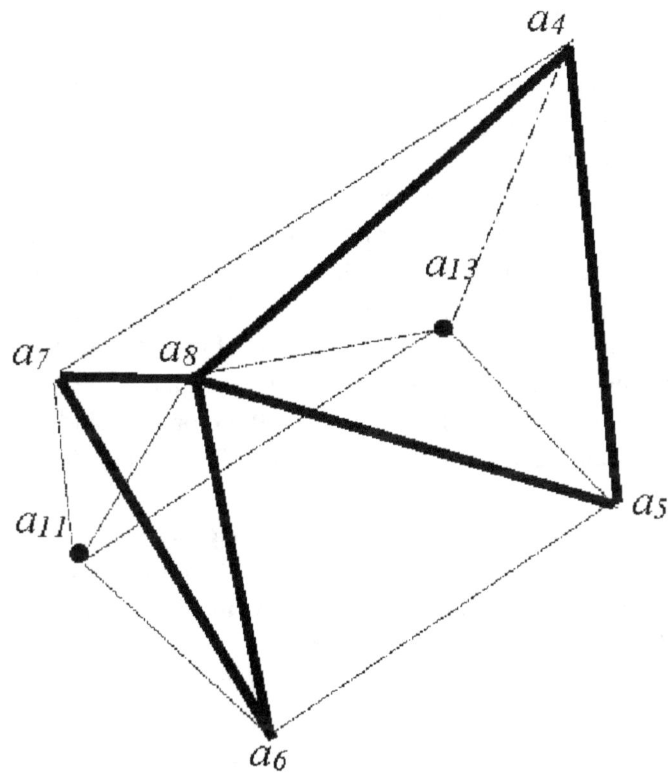

Figure 4 has 7 vertices (f_0=7); 15 edges $a_{11}a_7$, $a_{11}a_8$, $a_{11}a_6$, $a_{11}a_{13}$, a_7a_6, a_7a_8, a_7a_4, a_8a_4, a_8a_{13}, a_8a_5, a_6a_5, $a_{13}a_5$, a_4a_{13}, a_4a_5, a_8a_6 ; 14 flat faces $a_{11}a_8a_6$, $a_{11}a_7a_8$, $a_6a_7a_8$, $a_{11}a_7a_6$, $a_8a_5a_{13}$, $a_8a_{13}a_4$, $a_5a_{13}a_4$, $a_8a_5a_4$, $a_6a_8a_5$, $a_{11}a_8a_{13}$, $a_7a_8a_4$, $a_6a_{11}a_{13}a_5$, $a_6a_5a_7a_4$, $a_{11}a_7a_{13}a_4$; 6 three-dimension faces $a_6a_{11}a_8a_7$, $a_8a_5a_{13}a_4$, $a_7a_8a_{11}a_{13}a_4$, $a_6a_{11}a_8a_{13}a_5$, $a_{11}a_6a_7a_{13}a_5a_4$, $a_6a_7a_8a_4a_5$. Therefore, for Figure 4 are f_0=7, f_1=15, f_2=14, f_3=6 . Substituting the values f_i in equation (3.3), we see that it holds for $n = 4$

$$7 - 15 + 14 - 6 = 0.$$

This proof that Figure 4 has dimension 4.

As each figure in polytope on Figure 3 is 3, so common number polytopes with dimension 4 in Figure 3 equal 7, i. e. $(f_4)_c$=7. Substituting the values $(f_i)_c$ in equation (3.3), we see that it holds for $n = 5$

$$13 - 39 + 51 - 30 + 7 = 2.$$

Theorem 9.1 is proved.

A BINARY CHEMICAL COMPOUNDS WITH TRANSITION ELEMENTS AND THEIR STRUCTURE

The various binary chemical compounds have a limited number of typical structures (Fersman, 1937). In this section, the structures of binary compounds involving transition elements is considered. The simplest compound of transition elements with three - dimensional structure has a cubic unit cell, for example, oxides of transition elements. Since such structures form chlorides, bromides and iodides of alkali metals, we will refer to these structures rock salt structures. Many binary chemical compounds have the structure as adamantane molecule. In work of Zhizhin (2014 a) it was proved that the adamantane molecule consisting of 10 carbon atoms that make up the bulk of the unit cell of the diamond has of dimension 4. In compounds Ag_2O, Cu_2O at locations 10 of the carbon atoms are oxygen atoms, and atoms of copper and silver are arranged linearly between oxygen atoms. Many drugs are also a group of 10 carbon atoms as in the adamantane molecule. Among the inorganic and organometallic compounds have a number of structural analogues of adamantine (Table 9.1). All of these compounds have dimension 4 or even higher.

A series of binary compounds have a structure in the form of cube with centrum as in titanium chloride at which titanium ions are arranged in the centrum of the cube bat chlorine ions are arranged in vertices of the cube. We will call these structures titanium chloride structure. Haw it is shown in work of Zhizhin & Diudea (2016) this structure has dimension 4.

A series of binary compound have a structure of the mineral rutile TiO_2. In this compound each titanium atom is surrounded by six the oxygen atoms in the octahedral coordination. To compounds with such structure to concern for example fluorides of copper, zinc, magnesium, manganese, cobalt, nickel. We will call these structures rutile structure. In work of Zhizhin & Diudea (2016) it is shown that octahedron with centrum have dimension 4. Therefore, all these structures have dimension 4. A series of binary compounds have structure of wurtzite - mineral ZnS, in which from compound ZnS with the structure of the adamantane zinc atom and sulfur atom have the tetrahedron coordination. The centrum each tetrahedron is vertex of another tetrahedron. The wurtzite structure has compounds ZnO, CdS, ZnS. The dimension of this structure remains unknown. It will be defined in the next section. A series of binary compounds have a fluorite structure – mineral CaF_2 (fluorspar). Each calcium ion in this structure is in cube surrounded by fluorine ions, and each fluorine ion is in tetrahedron surrounded by calcium ions. The fluorite structure has for example chlorides of transition metals (Table 1).

Table 1. Binary compounds of the transition elements

N	Type of the structure	The compounds transition elements with this type of the structure
1	rock salt	MnO, FeO, CoO, NiO, CdO
2	Adamantane	Ag_2O, Cu_2O, ZnS, CuCl
3	titanium chloride	TiCl
4	Rutile	MnF_2, CoF_2, NiF_2, CuF_2, ZnF_2, MnO_2, MoO_2
5	Wurtzite	ZnO, CdO, ZnS
6	Fluorite	CdF_2, $MnCl_2$, $FeCl_2$, $CoCl_2$, $NiCl_2$, $ZnCl_2$, $CdCl_2$, Cu_2S

THE DIMENSION OF THE WURTZITE

In the structure of wurtzite every atom of one component has a tetrahedral environment of the atoms of the other component. This results to arrangement of tetrahedrons with center so that vertex one tetrahedron is the center of another tetrahedron (Figure 5).

Figure 5. The tetrahedral coordination atoms in wurtzite
A white circle is atom of one component. A black circle is atom of other component

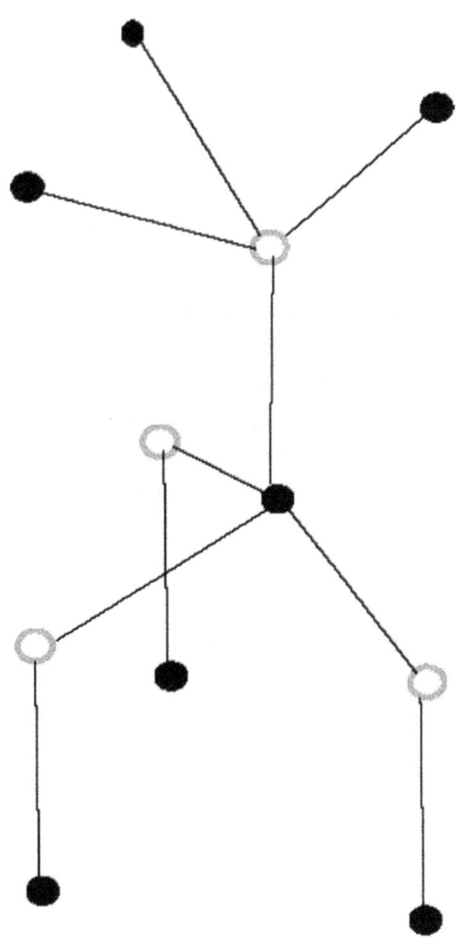

If to carry construction of atoms in Figure 5 on this principle, we obtain a spatial lattice, the unit cell the lattice is a convex shape it is shown in Figure 6.

Figure 6. The unit cell of the wurtzite

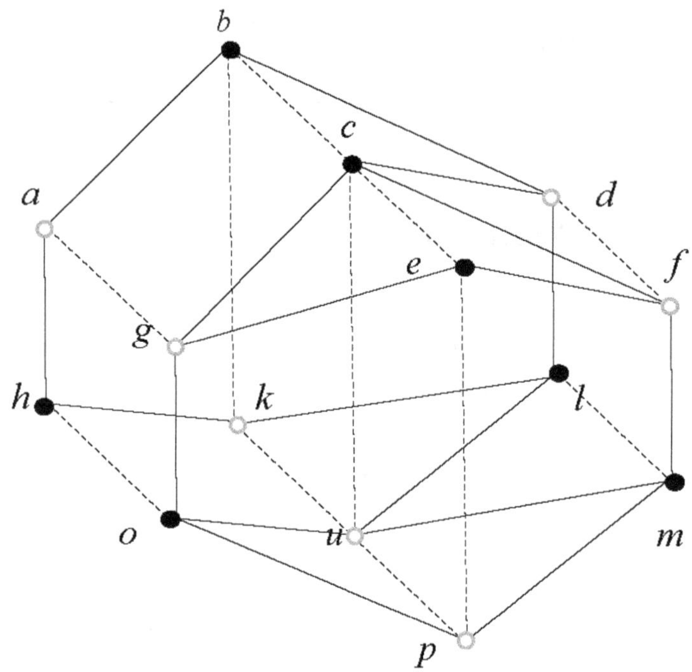

This figure is the unit cell structure of the wurtzite. In Figure 6, solid lines represent chemical bonds of the atoms, and the dotted lines are only geometric sense outlining contours of the figure. We define the dimension of this figure by the Euler - Poincare equation (3.3). The number of vertices of this figure is equal to 14, i. e. f_0=14. The number of edges is equal to 29, i. e. $f_1$29. The number of two - dimensional faces is the sum of the number of triangles (8) and number of quadrangles (13), i. e. f_2=21. The number of three - dimensional faces is equal to 6. This figures are *abcghkon, gceruo, cefump, cdfulm, bcdklu,* and all shape on Figure 6 without inner partitions, i.e. f_3=6. Substituting these values f_i, (i=0,1,2,3) in the Euler - Poincare equation (3.3), we find that it is satisfied for n = 4

$$14 - 29 + 21 - 6 = 0.$$

Thus, the dimension of polytope on Figure 6 is equal to 4, i. e. the unit cell structure of the wurtzite has dimension 4.

THE DIMENSION OF THE FLUORITE

On example of compound $MnCl_2$ we look at the structure of fluorite. Isolate magnesium atoms lying at the centers of the cube faces ( ), and chlorine atoms (), forming a smaller cube inside the bigger cube, which are located at the vertices of magnesium atoms (Figure 7).

Figure 7. The unit cell of the fluorite

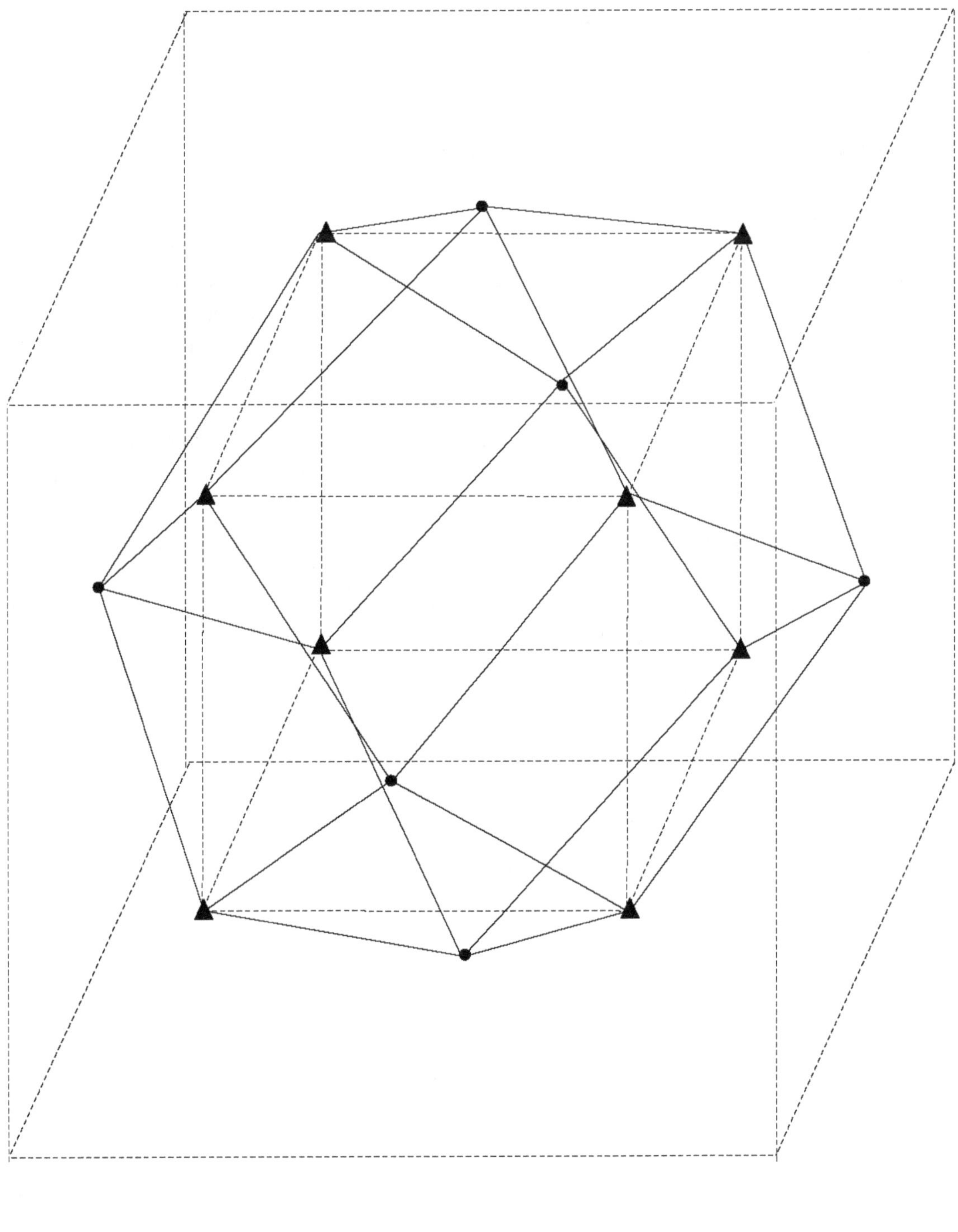

From Figure 7 it follows that the number of vertices is 14, i. e. f_0=14, the number of edges is 36, i.e. f_1=36, the number of flat faces is sum from number of triangles (24) and number of rectangles (6) (smaller cube faces), i.e. f_2=30. The number three - dimension shape to sum up from smaller cube (1), pyramids on its faces (6) and figure (22) without inner parts (1), i. e.f_3=8 8. Substituting these values f_i, (i=0,1,2,3) in the equation (3.3), we find that it is satisfied for $n = 4$

$$14 - 36 + 30 - 8 = 0.$$

Thus, the dimension of polytope on Figure 7 is equal to 4, i. e. the unit cell structure of the fluorite has dimension 4.

HYDROCARBONS AND CARBOHYDRATES

Hydrocarbons are the simplest class of organic compounds in terms of their elemental composition (they consist only of carbon and hydrogen). They are widespread on Earth: they are part of natural gas, oil and some solid fossil fuels. The simplest hydrocarbon is methane. Consider, for example, methane molecule CH_4. The carbon atom in this molecule binds around four hydrogen atoms. Geometrically, this molecule is a tetrahedron, whose vertices are located of the hydrogen atoms, and in the center is carbon atom. In Chapter 5 it is shown that the dimension of the molecules having the form of a tetrahedron with the center equals 4.

The carbon atom in the center of the methane molecule has the valence electrons $2s2p^3$. Valence electron orbitals of carbon atoms and hydrogen atoms $1s$ overlap and form four hybrid orbitals sp^3, directed from the carbon atom to the hydrogen atoms (Gray, 1965). If the distance from hydrogen atoms to carbon atoms is taken as unity, for the origin of coordinates to take the carbon atom, the directions hybrid orbitals send on four coordinates x, y, z, t, then the coordinates of the hydrogen atoms equal to (0, 0, 0, 1), (0, 0, 1, 0), (1, 0, 0, 0), (0, 1, 0, 0), and the carbon atom coordinates equal to (0, 0, 0, 0). So, we have the integer coordinates of vertex in the four-dimensional space (Figure 8).

Figure 8. The methane molecule (CH_4)

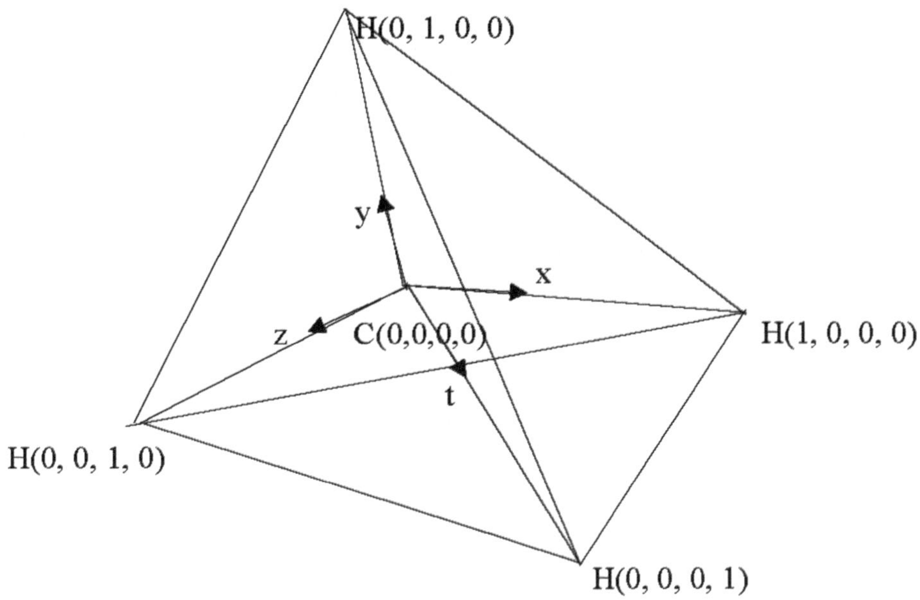

This is consistent with the evidence of four - dimension convex hull of the methane molecule on the Euler – Poincare equation (3.3) in Chapter 3. It is easy to see that the body in Figure 8, seen in the four -dimensional space, convex, because its edges belong to the body and enter into his boundary complex (Grunboum, 1967). Polytope in Figure 8 is a 4 - simplex, since each vertex of the polytope associated edges with all the other vertices of this polytope (Zhizhin, 2014 b).

If in the methane molecule a hydrogen atom replaced by a hydroxyl group - OH, then we get the simplest alcohol - methanol. If the hydroxyl group considered as the vertex of the polytope, then the dimension of this molecule will also be equal to 4. If each atom of the molecule of methanol is considered the vertex of the polytope, then connecting each vertex to all other vertices of edges, it turns out that it is equal to the dimension of the polytope to 5 and we have 5 - simplex. However, here we must remember that the accession of the hydroxyl group does not change the hybridization of the carbon atom, as the binding site as the place of one hydrogen atom took one oxygen atom of the hydroxyl group. Therefore, as a separate vertex in methanol molecule should take hydroxyl group entirely. Then the dimension of the methanol molecules is equal to 4 (Zhizhin, 2015 b, 2017 b, c).

Carbohydrates are a large group of organic substances that are widespread in nature. Representatives of carbohydrates are grape sugar (glucose), beet or cane sugar (sucrose), starch, cellulose. It can be considered that carbohydrates are the most abundant organic compounds. Carbohydrates can be divided into two main groups: simple carbohydrates and complex carbohydrates. Simple carbohydrates or monosaccharides are not hydrolyzed. Complex carbohydrates are capable of hydrolyzing and breaking down to form monosaccharides. Consider the simplest monosaccharide molecule containing three carbon atoms - the aldose monosaccharide molecule.

THE DIMENSION OF ALDOSE MONOSACCHARIDE

The combination of carbon atoms to form a chain of a certain length is a distinctive feature of biomolecules and organic chemistry compounds. Carbohydrates are the main source of energy for the body. All carbohydrates are made up of units that are saccharides. The simplest saccharide is an aldose monosaccharide, which contains three carbon atoms (Figure 9).

Figure 9. Shema of the molecule aldose monosaccharide

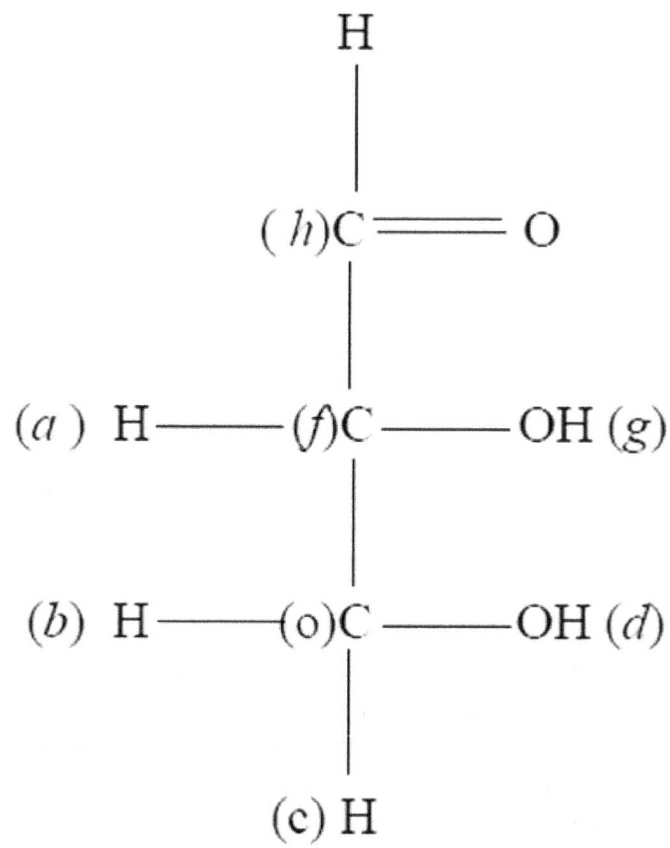

Spatial structure of the molecule aldose monosaccharide is shown in Figure 10 (Zhizhin, 2019 b).

Figure 10. Spatial structure of the molecule aldose monosaccharide

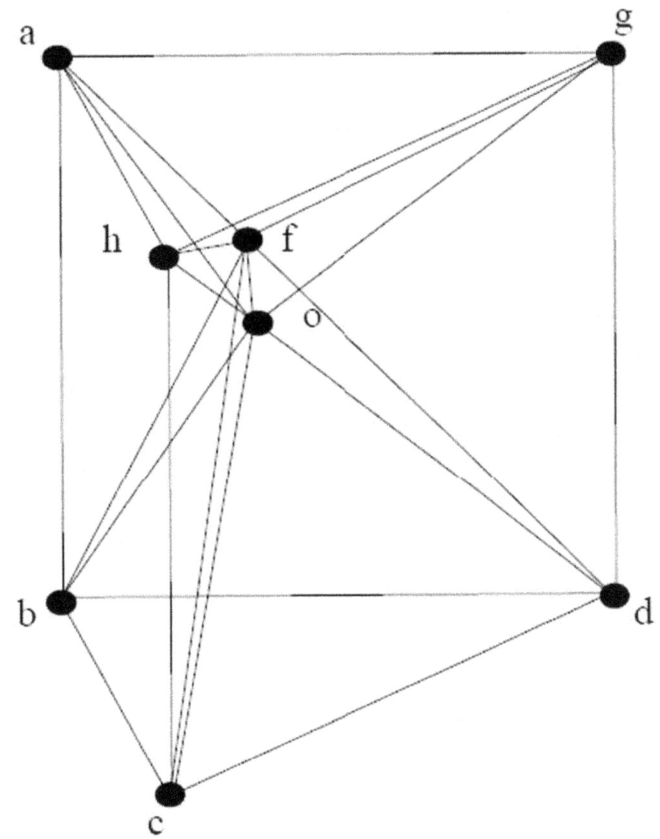

The dimension of a polyhedrons can be determined by the Euler-Poincaré equation (3.3) in Chapter 3.

In Figure 10 there is a tetrahedron *bcdf* with center *o* and a tetrahedron *oahg* with center *f*. Equation (3.3) implies that each tetrahedron with center has dimension 4 (Chapter 5). The vertex *f* of the first tetrahedron is the center of the second tetrahedron, and the vertex *o* of the second tetrahedron is the center of the first tetrahedron. Hydrogen (H) atoms are located at the vertices *a, b, c*, hydroxyl groups (OH) are located at the vertices *g, d*, carbon atoms (C) are located at the vertices *o, f, h*. The oxygen and hydrogen atoms following the carbon atom at the vertex *h* are not shown in Figure 14 for simplification. It is necessary to determine the dimension of the polytope *bcdfahgo*. The polytope in Figure 10 has 8 vertices ($f_0 = 8$), 22 edges (*ab, ag, af, ao, ah, gh, gf, go, gd, bf, bo, bd, bc, df, do, dc, hc, hf, ho, cf, co, fo*). Therefore, $f_1 = 22$. The polytope in Figure 10 has 29 planar faces, of which 26 are the triangles (*aho, afo, ahf, afg, ahg, aog, aob, afb, bfo, bco, bod, bfd, bfc, bcd, ghf, gho, gfo, god, gfd, dfo, dco, dfc, cof, chf, cho, hfo*) and 3 quadrangles (*abdg, hcgd, abhc*). Therefore, $f_2 = 29$. The polytope in Figure 10 has 20 three - dimensional figures, of which 13 are tetrahedrons (*bcfd, ahog, dcfo, bcdo, bfdo, cfdo, ahof, ahgf, hogf, aogf, fgod, fohc, foab*), 6 pyramids (*agbdo, ahbcf, ahbco, agbdf, Chgdf, chgdo*) and one (*ahgbcd)* prism. Therefore, $f_3 = 20$. It follows from the construction of the polytope in Figure 10 that it includes two tetrahedrons with the center *bcdfo* and *oahgf*. In addition to these two polytopes with dimension 4, five 4 - polytopes also appear in the polytope in Figure 10. Three of these polytopes have as their base three rectangular faces of the prism *ahgbcd*, whose vertices are connected with the vertices *f, o* located inside the prism. To prove their 4 - dimensionality, consider one of these polytopes *abhcfo*, since the proofs for the other two polytopes are similar. This polytope has 6 vertices ($f_0 = 6$); 13 edges (*ab, ah, hc, bc, af, hf, bf, cf, ho, ao, bo, co, fo*), $f_1 = 13$; 13 two - dimensional faces (*ahf, aho, abo, abf, afo, bfo, boc, ahbc*), $f_2 = 13$; 6 three -dimensional faces (*hfoc, abof, bfoc, afho, ahcbf, ahcbo*), $f_3 = 13$. Substituting the obtained values of the numbers of faces of different dimensions into equation (3.3), can find that equation (3.3) is satisfied for $n = 4$

$$6 - 13 + 13 - 6 = 0.$$

It is proved by the 4 - dimensionality of the polytope *abhcfo*.

The two polytopes of dimension 4 there are formed by the *ahgbcd* prism with the vertex *f* or *o* inside its. Consider the prism *ahgbcd* with the vertex *f* (the proof for the prism with vertex *o* is similar). The polytope *ahgbcdf* has 7 vertices, $f_0 = 7$; 15 edges (*ah, hg, ag, bd, bc, cd, ab, hc, gd, af, fh, fg, bf, fc, fd*), $f_1 = 15$; 14 two - dimensional faces (*ahg, bdc, ahf, hfg, afg, bfc, fcd, bfd, fhc, afb, fgd, ahbc, hcgd, agbd*), $f_2 = 14$; 6 three - dimensional faces (*ahgbcd, ahgf, bcdf, abdgf, hgcdf, ahbcf*), $f_3 = 6$. Substituting the values of the numbers of faces of various dimensions obtained for the polytope *ahgbcdf* into equation (3.3) can find that it is satisfied for $n = 4$

$$7 - 15 + 14 - 6 = 0.$$

This proves that the polytope *ahgbcdf* has a dimension of 4.

Thus, for the polytope in Figure 10 are $f_0 = 8, f_1 = 22, f_2 = 29, f_3 = 20, f_4 = 7$. Substituting these values into equation (3.3) can find that it is satisfied for $n = 5$

$$8 - 22 + 29 - 20 + 7 = 2.$$

This proves that the polytope in Figure 10 has dimension 5.

THE DIMENSION OF TARTARIC ACID MOLECULES

The phenomenon of optical activity (i. e. the ability of substances to rotate the plane of polarization of polarized light) is widespread among organic compounds. Optically active substances include many important natural substances, such as proteins, carbohydrates. This phenomenon has deep biological significance, as it is associated with the asymmetry of living matter and the phenomenon of life. In 1848, Louis Pasteur discovered the optical activity of liquid tartaric acid and proposed to explain this phenomenon by the different form of crystals of D and L tartaric acid in three-dimensional space. In this work, spatial images of D and L forms of tartaric acid molecules are construct and it is proved that the dimension of these molecules is five. Thus, the stereoisomerism of tartaric acid should be considered not in three-dimensional, but in five - dimensional space (Zhizhin, 2020).

The aldose monosaccharide has two an enantiomorphism configuration of D and *L*. The aldose monosaccharide molecules in both of these configurations have a dimension of 5 (Zhizhin, 2019 b). We will consider the difference in these configurations by the example of a closely related tartaric acid, which played a major role in the development of biology, starting with Pasteur's well - known works (Paster,

1960). However, instead of the known images of these molecules in the form of Fisher's projections (Figure 11, Figure 12), we will use images of space of higher dimension for their images.

Figure 11. Shema of the molecules D - tartaric acid

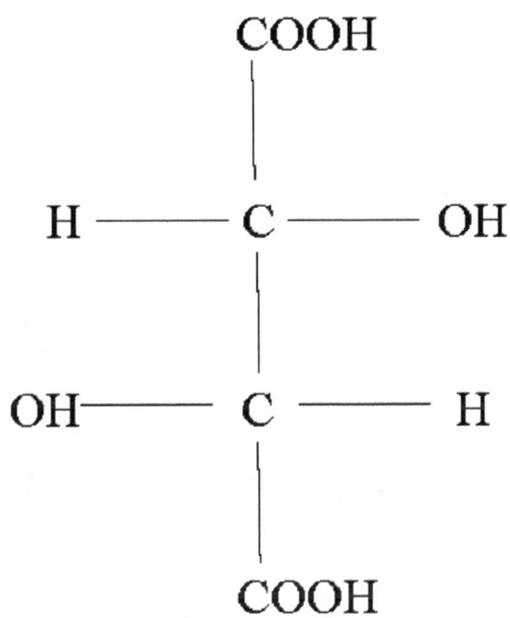

Figure 12. Shema of the molecules L – tartaric acid.

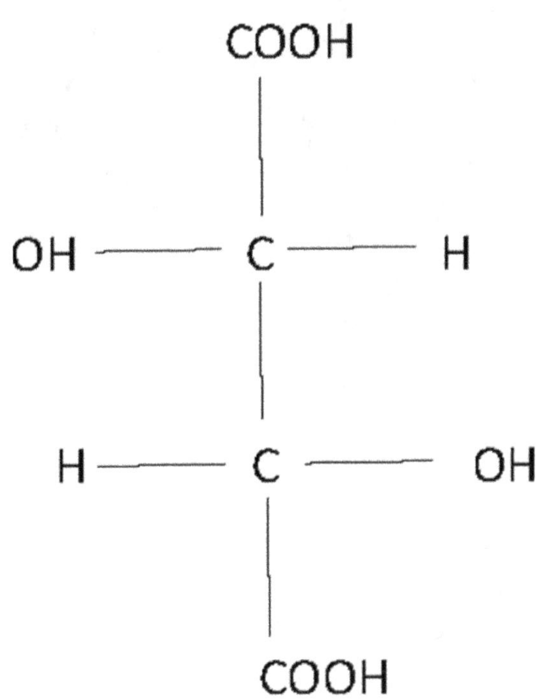

Comparing Fisher images of aldose monosaccharide (Figure 9) and tartaric acid (Figure 11, Figure 12) (Metzler, 1977; Lehninger, 1982) can see that these compounds have the same main part of the design. It has the form of two tetrahedrons with a center, and the center of each of them is simultaneously the vertex of another tetrahedron. There is some difference in functional groups of compounds. Enantiomorphism forms of tartaric acid differ in the mirror image of hydrogen ions and hydroxyl groups in the main part of the molecule's structure. The dimension of this construction how it is shown earlier equal 5. Thus, the dimension of the molecules tartaric acid in both forms is 5. Images of polytopes corresponding to a molecule of tartaric acid in the form *D* and form *L* are presented in Figures 13 and Figure 14.

Figure 13. Spatial structure of the D – tartaric acid

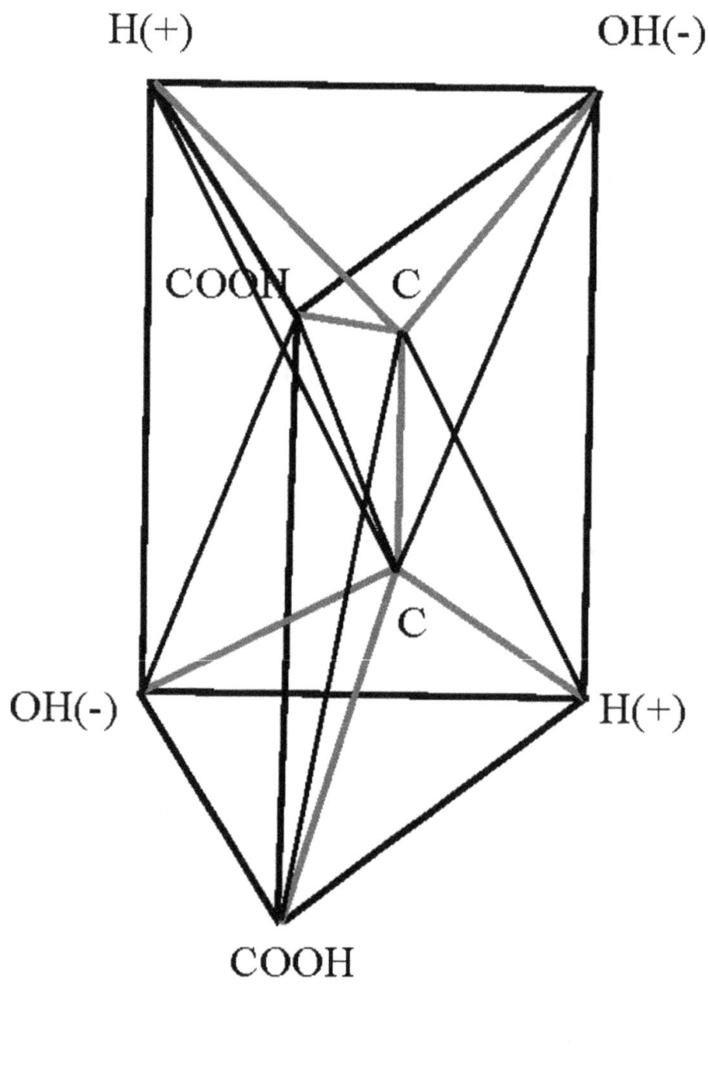

Figure 14. Spatial structure of the L – tartaric acid

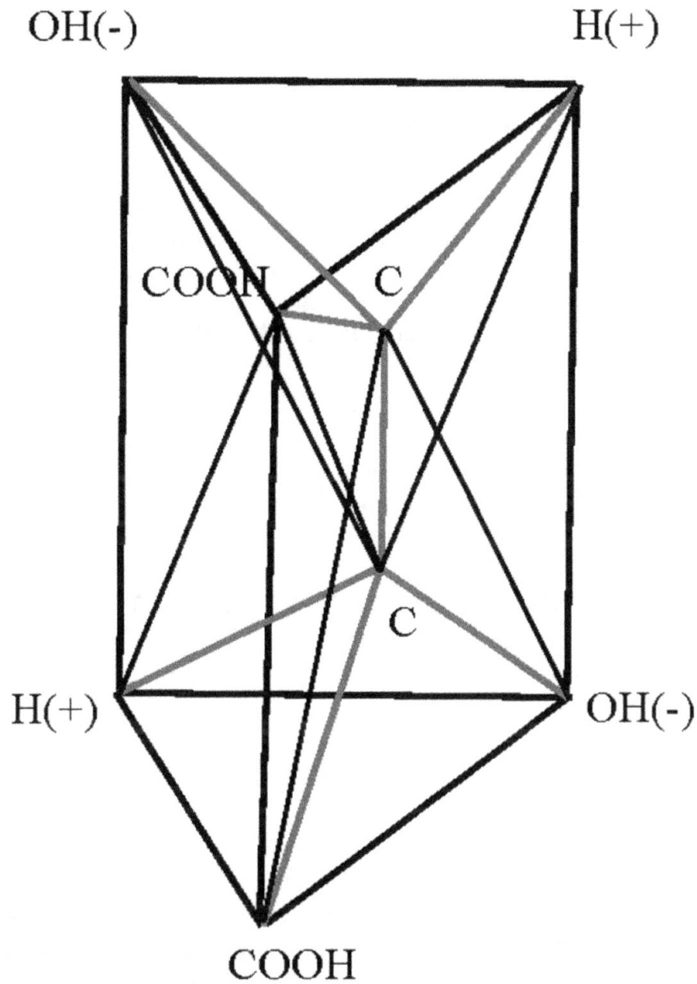

The brown color in Figures 13, 14 denotes the edges corresponding to the chemical bonds between the atoms. The black color in these figures denotes the edges that have values only as the edges of the convex body. The outer contour of both molecules in three-dimensional space is a triangular prism. There are two carbon atoms within these prisms. These two carbon atoms and lead to an increase in the dimension of the molecule to five. On the outer contour, two enantiomorphism forms have the opposite arrangement of hydrogen ions and a hydroxyl group. The images obtained make it possible to explain the main property of tartaric acid - rotation of the plane of polarization of the incident light in different directions: in the case of the *D* form to the right, in the case of the *L* form to the left. It is known devices for rotating the plane of polarization of light, having the appearance of two folded triangular prisms, the boundary between which serves to reflect light (Wood, 1936). Can say that the molecule of tartaric acid is a natural device for rotating the plane of light polarization. Two carbon atoms play the role of the reflecting partition in the molecule. The rotation occurs in the forms *D* and *L* in different directions because of the opposite arrangement of the charges of the hydrogen ions (+) and the hydroxyl group (-) in these forms. Thus, the reason for the different rotation of the plane of polarization of light lies not in the different forms of the crystals of *D* - tartaric acid and *L* - tartaric acid, as Pasteur suggested, but in different forms of molecules, clearly visible in the image in the space 5*D*.

LINEAR METAL CHAINS WITH LIGANDS

An example of a linear homo - element metal chain is shown in Figure 15 (Smart, Cook, & Woodward, 1977; Evans, Okrasinski, Pribula, & Norton, 1976; Gochin, & Moss, 1980).

Figure 15. Schematic structure of cluster connection $Os_3(CO)_{12}R_2$

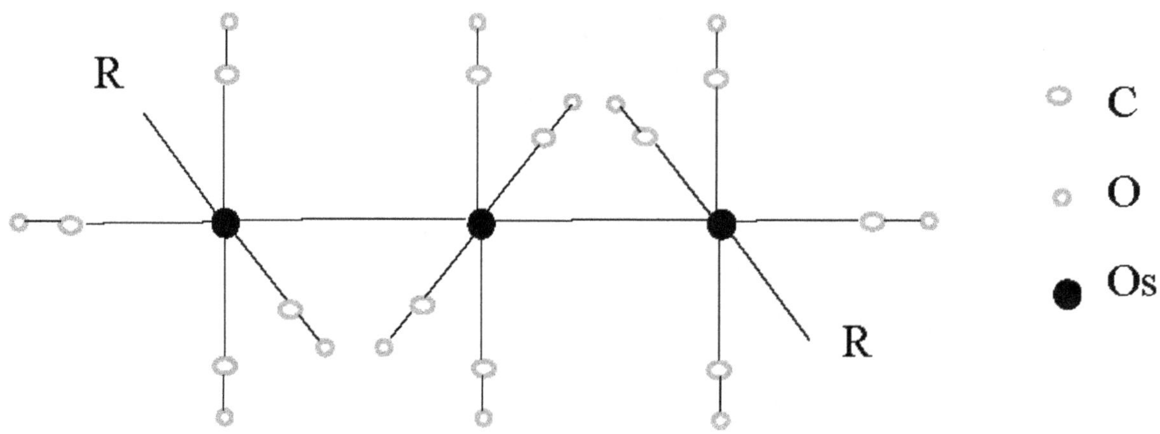

In this example, we illustrate the sequence of actions to calculate the dimension of the metal chain. When determining the dimension of metal chains, atoms that are linearly located and do not have branching of the chain at their location do not matter (Zhizhin, 2018). At the same time, not only metal atoms, but also ligand atoms are of importance for determining the dimension of a compound. Therefore, Figure 15 can be represented as Figure 16.

Figure 16. The geometric structure of cluster connection $Os_3(CO)_{12}R_2$

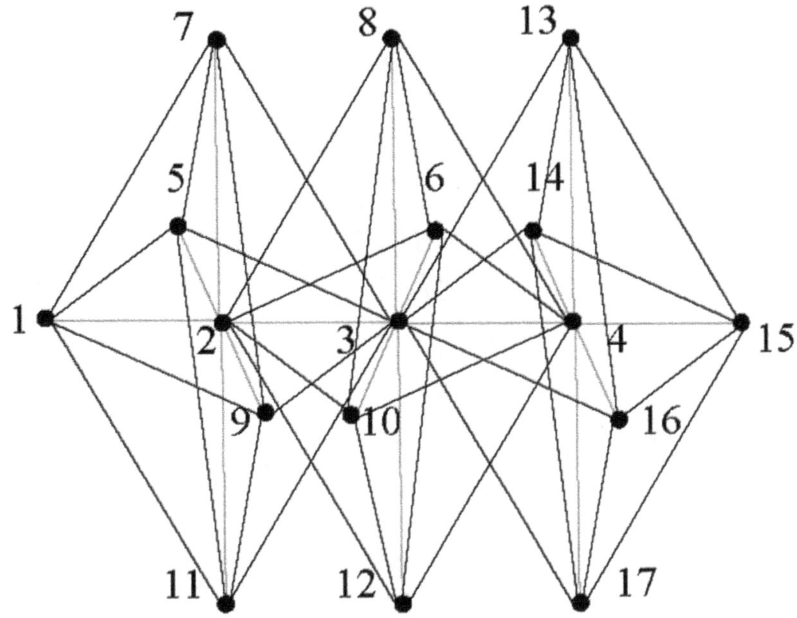

In Figure 16, the red edges correspond to chemical bonds, and the black edges serve only to create a convex closed figure. It can be seen that each link of the metal chain creates an octahedron with the center as a convex figure. Moreover, the center of each octahedron is simultaneously the vertex of another neighboring octahedron or two neighboring octahedrons. In Figure 16 at the centers of octahedrons are disposed Os atoms (vertices 2, 3, 4). At the vertices 1, 6, 7, 8, 9, 10, 11, 12, 13, 14, 15, 17 are disposed the functional group CO, at the vertices 5, 16 are disposed the functional group R (Zhizhin, 2016). It is easy to prove that an octahedron with a center has a dimension of 4. To do this, refer to Figure 17, which shows separately the first of the chain octahedron.

Figure 17. Octahedron with a center

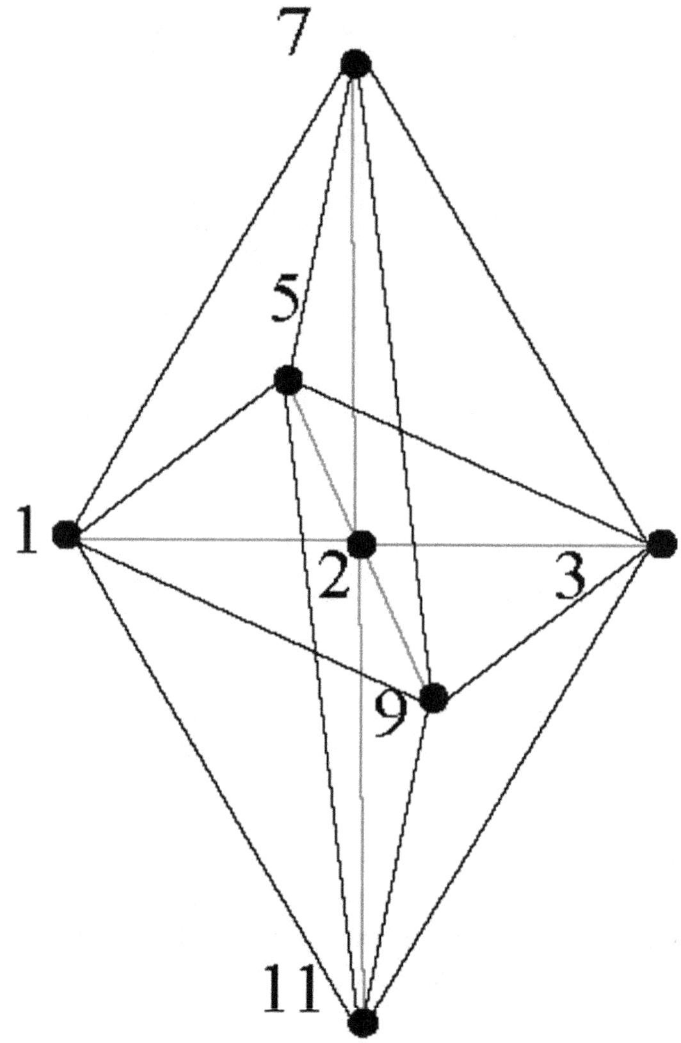

Polytope on Figure 17 has 7 vertices (f_0=7), 18 edges (f_1=18)

1 - 2, 2 - 3, 2 - 5, 2 - 9, 1 - 5, 5 - 3, 3 - 9, 1 - 9, 2 - 4, 2 - 11, 1 - 7, 5 - 7, 3 - 7, 7 - 9, 1 - 11, 5 - 11, 9 - 11, 3 - 11.

Polytope composed of eight tetrahedrons

1 - 2 - 9 - 11, 2 - 9 - 3 - 11, 2 - 5 - 3 - 11, 1 - 2 - 5 - 11, 1 - 2 - 5 - 7, 2 - 5 - 3 - 7, 2 - 9 - 3 - 7, 1 - 2 - 9 - 7.

To these three - dimensional figures you need to add an octahedron without a center 1 - 5 - 3 - 7 - 9 - 11. Thus, the total number of three - dimensional figures is equal to 9 (f_3=9). Each flat face of any of these nine polyhedrons is simultaneously one of the flat faces of some other of these nine polyhedrons. Therefore, the total number of two - dimensional faces in the polytope in Figure 17 is 16 (f_2=16) (Zhizhin, 2020).

Define dimension polytope in Figure 17 on the Euler - Poincaré equation (Poincaré, 1895).

Substituting the obtained values f_i(i=0,1,2,3) in the equation (3.3) in Chapter 3 you can see that the Euler – Poincaré equation is satisfied in this case with a value of n equal to 4. This proves that the dimension of an octahedron with center is 4.

In a linear homogeneous chain of metal atoms, tetrahedrons penetrate into each other (Figure 16). This greatly complicates the geometric analysis of the compound. Here one could use the same method of forming a convex figure, corresponding to a chain, as in the analysis of individual molecules. However, the construction of an additional system of edges, taking into account the possible growth of the chain, greatly complicates the convex figure. In this case, the dimension of the whole figure, the corresponding chain, will increase dramatically with the growth of the chain. This approach seems to be irrational. It would be logical to present these chains in the form of a convex figure consisting of a sequence of convex polytopes adjacent to each other in the geometric analysis of the chains of metal atoms.

Theorem 9.2. Linear homo - element metal chains with ligands are a sequence of convex polytopes of dimension 4 adjacent to each other along whole sections of octahedrons passing through the center and the vertices of the octahedrons.

Proof. Taking into account that metal chains with two chain links exist (Cotton, & Walton, 1982), consider two adjacent chain links in Figure 16. To create a closed shape from two octahedrons interconnected, it is necessary to fill the free space between the octahedrons. This can be done by connecting the vertices of two octahedrons, which were not yet connected to each other (Figure 18). These compounds (7 - 8, 5 - 6, 9 - 10, 11 - 12) in Figure 18 are indicated in green.

Figure 18. Convex figure built on a metal chain of two metal atoms with ligands

Metal chains: cluster compounds that have a skeleton in the form of metal chains, i.e. polymetallic chains formed by metal-metal localized covalent bonds.

Linear metal chains: metal chains in which metal atoms are located in a straight line.

Homo - element metal chains: clusters in which the core contains atoms of the same element.

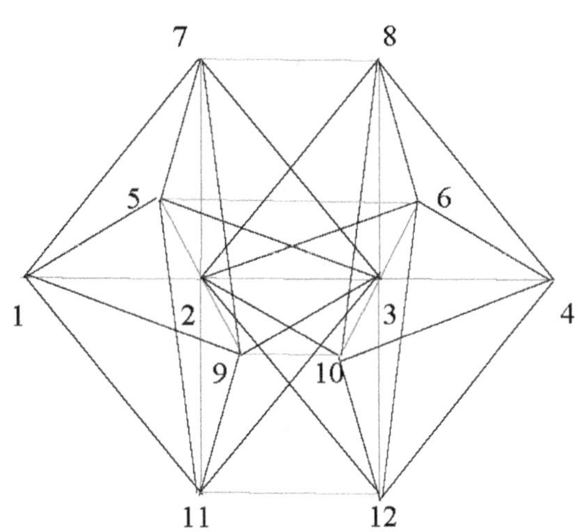

In Figure 18, the region in which the octahedrons enter each other 5 - 7 - 8 - 6 - 10 - 12 - 9 - 11 - 2 - 3 should be distinguished. As well as two regions in which only of the octahedron is present 1 - 5 - 7 - 9 - 11 - 2, 4 - 8 - 6 - 10 - 12 - 3 should be distinguished. An area in which only one of the octahedron is present, for example, 1 - 5 - 7 - 9 - 11 - 2 (Figure 19) has 6 vertices (f_0=6), 13 edges (f_1=13),12 flat two - dimensional faces (f_2=12), 5 three - dimensional areas (f_3=5)(Zhizhin, 2020).

Figure 19. An area in which only one the octahedron is present

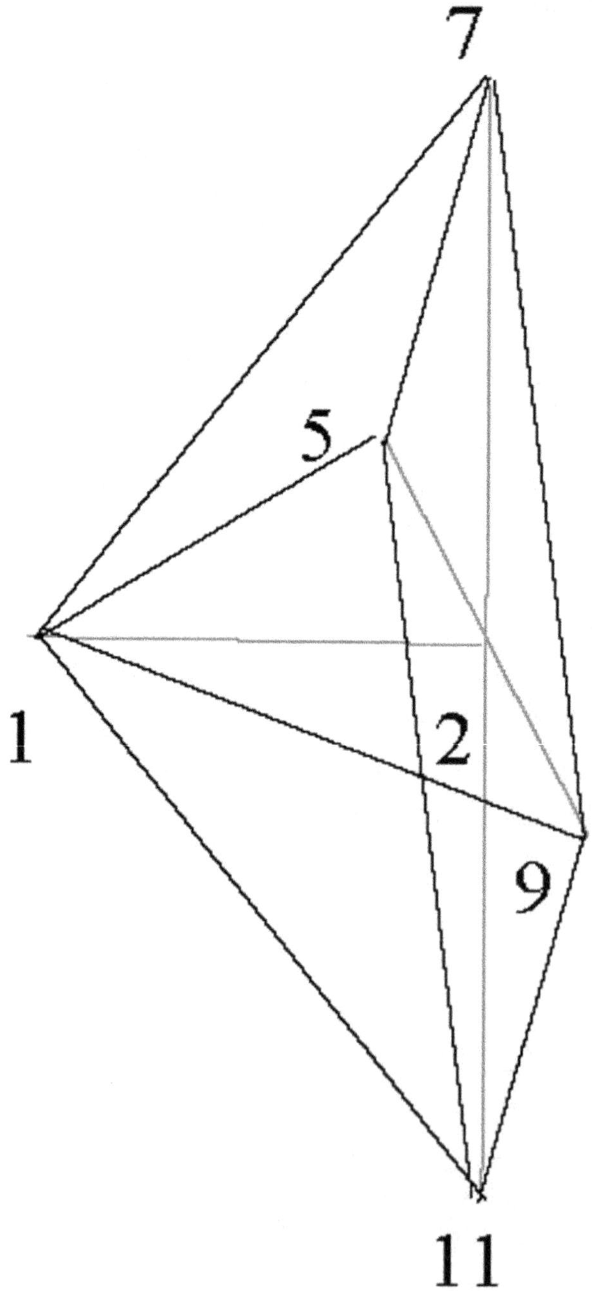

Substituting the obtained values $f_i(i=0,1,2,3)$ in the equation (3.3), you can see that the Euler - Poincaré equation is satisfied in this case with a value of n equal to 4

$6 - 13 + 12 - 5 = 0.$

This proves that each of these regions has dimension 4. It should be noted that the region 1 - 7 - 9 - 5 - 11 - 2 does not contain edge 1 - 2, therefore it has dimension 3. As soon as edge 1 - 2 is added, region has dimension 4. The base of this figure is constructed from four triangles that are faces of other three - dimensional figures in this polytope.

The region 5 - 7 - 8 - 6 - 10 - 12 - 9 - 11 - 2 - 3 in which the octahedrons penetrate into each other is quite complicated. However, here we must bear in mind that the edges of the octahedrons in this region, which are not valence bonds, were necessary for the creation of convex figures (octahedrons) when considering these figures separately. When the general region, which includes octahedrons, is considered, their purpose is lost, due to the introduction of other edges that create a unifying closed convex figure. Therefore, in this region one should leave only the edges corresponding to the valence bonds and the edges forming the boundary of the region. Other edges 7 - 3, 8 - 2, 5 - 3, 2 - 6, 2 - 10, 3 - 9 must be removed. In this case, the region takes the form shown in Figure 20.

Figure 20. Simplified region penetrated by adjacent octahedrons in chains

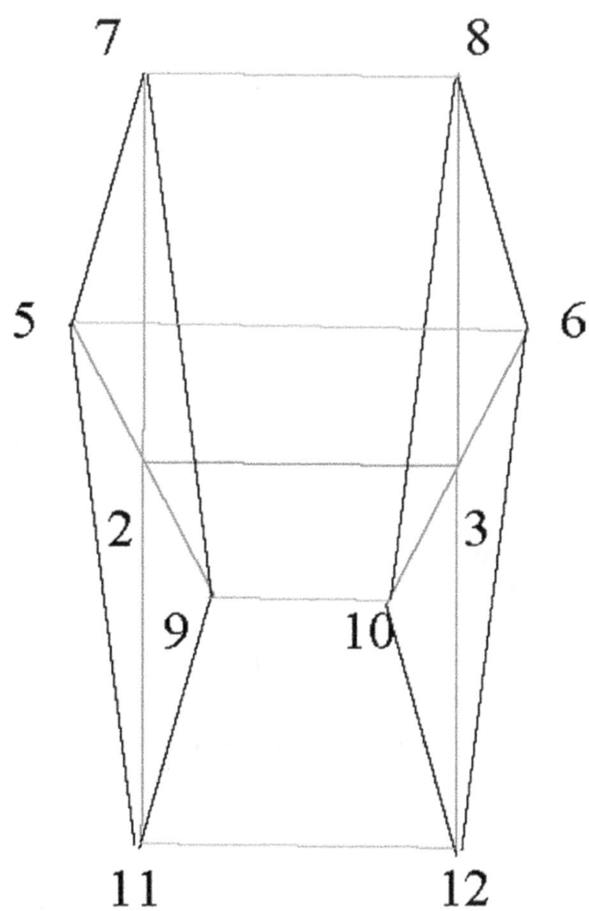

The polytope on Figure 20 has 10 vertices (f_0=10), 21 edges (f_1=21), 16 two - dimensional faces (f_2=16), 5 three - dimensional faces (f_3=5) (Zhizhin, 2020).

Substituting the obtained values f_i(i=0,1,2,3) in the equation (3.3) you can see that the Euler - Poincaré equation is satisfied in this case with a value of n equal to 4

$$10 - 21 + 16 - 5 = 0.$$

This proves that polytope on Figure 20 has dimension 4.

Let us return to the chain containing three metal atoms (Figure 16). Its image using spaces of higher dimension, according to the developed description, has the form shown in Figure 21.

Figure 21. Image of a chain with three metal atoms in a higher dimension space

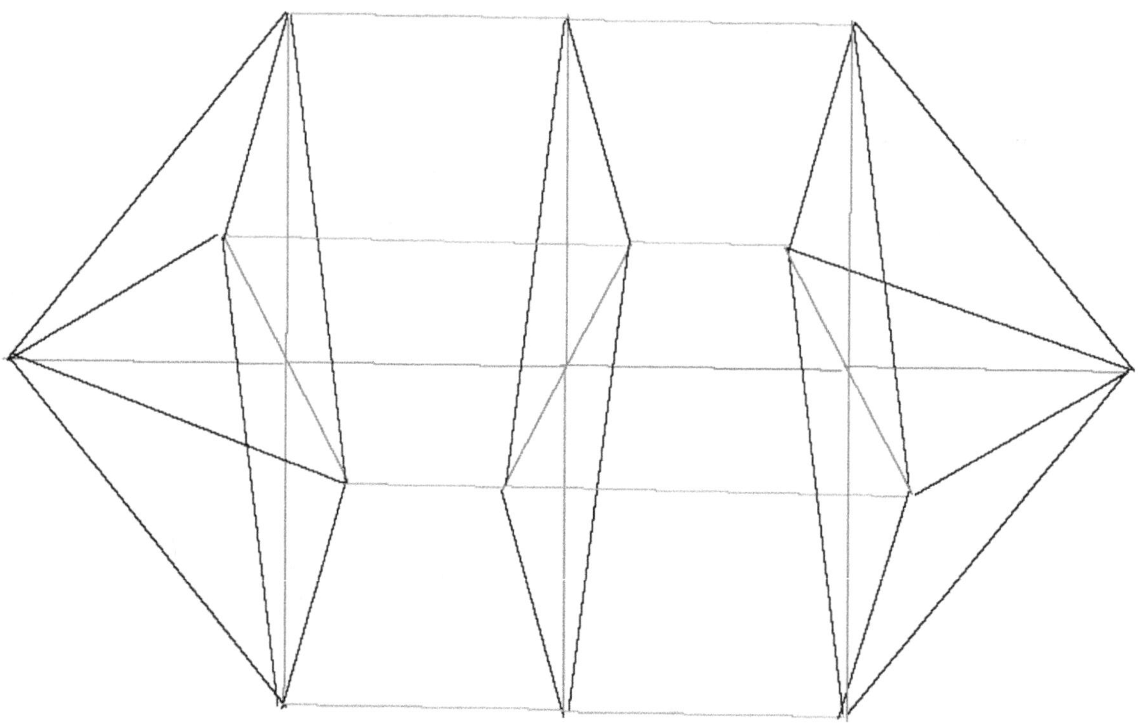

It is a chain of convex closed polytopes of dimension 4 adjoining each other along vertical sections of octahedrons passing through the center of the octahedrons and their vertices. It is obvious that the chain of polytopes can be continued. It will include a chain of polytopes in Figure 20, in alternating positions. At the ends of the chain are the polytopes in Figure 19, also in reversed positions. This is the spatial implementation of a polymetallic linear chain using a higher dimension space. Obviously, other linear metal chains with similar attachments of ligands will have the same appearance. For example, metal chains $[Mn_3(CO)_{14}]^-$ (Bau, Kirtley, Sorrel, & Winarko, 1974), in which the atoms of the metal are manganese atoms, and all ligands are functional groups CO.

Q.E.D.

CLUSTERS CONNECTIONS WITH A BACKBONE IN THE FORM OF NONLINEAR METALLOCHAIN

The existence of cluster compounds with a skeleton in the form of nonlinear (bent) homo - element metal chains was experimentally established. An example of such a compound can be hydride $H_2Os_3(CO)_{12}$ (Aleksandrov, Zolnikova, Kritskaya, & Pods, 1980). A diagram of it is presented in Figure 22.

Figure 22. One of the possible of hydride $H_2Os_3(CO)_{12}$structures

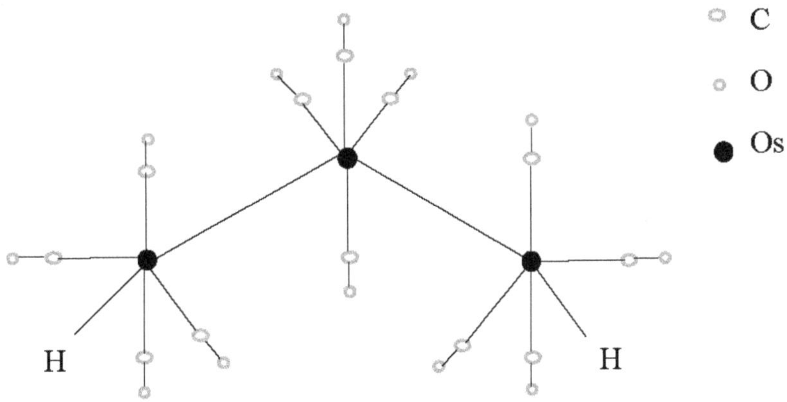

Theorem 9.3. The spatial figure correspondent to hydride $H_2Os_3(CO)_{12}$is a polytope consists of two parts, each of which is a polytope of dimension 4. Both polytopes of dimension 4 are adjacent to each other along a flat section that separates them and are symmetric with respect to this section.

Proof. The octahedral coordination of elements manifests itself around each of the three atoms of the osmium and the octahedrons (in the center of each octahedron, an osmium atom is located) penetrate each other. The spatial image of this compound is shown in Figure 23.

Figure 23. The spatial image of hydride $H_2Os_3(CO)_{12}$

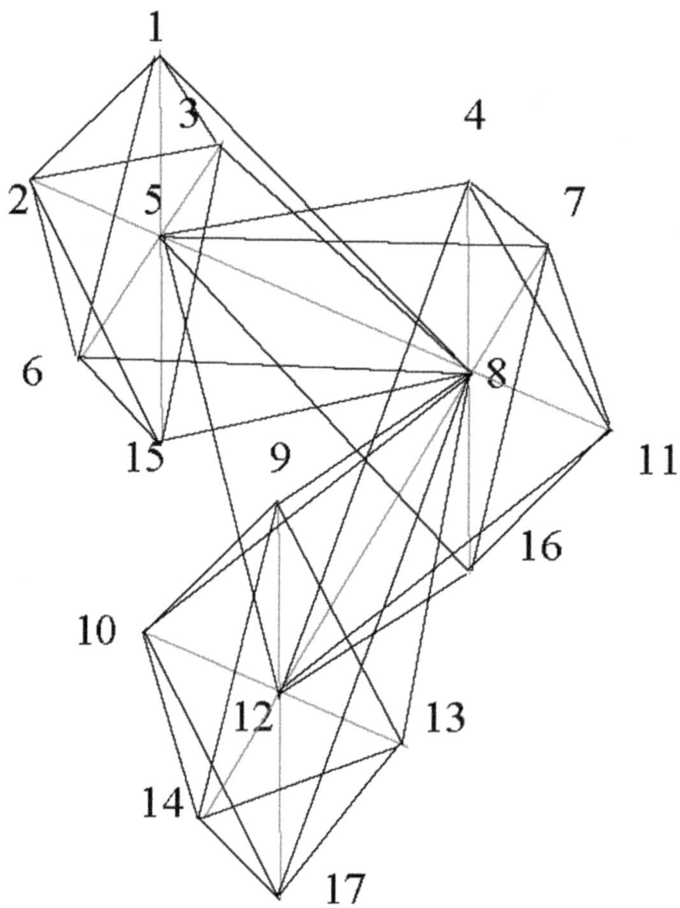

On Figure 23 the valence bonds are marked in red, it is taken into account that the metal - metal bond length is substantially greater than the bond length of the metal with the ligands. The edges of the octahedrons are marked in black. At the vertices 5, 8, 12 (centers of octahedrons), atoms of the osmium are located; hydrogen atoms are located at the vertices 2, 14; at vertices 3, 4, 6, 7, 9, 10, 11, 13, 14, 15, 16, 17 functional groups CO are located. Each octahedron with a center has a dimension of 4. However, the common connection of octahedrons is a non - convex and closed figure. To bring the image of this compound to the form of a figure closed and convex, one should fill the space between the octahedrons with additional figures. The result of such a construction will give a complex figure for analysis. However, here we must recall the procedure applied in the previous section. It is necessary to connect the vertices cut off from each other by the edges and remove the edges of the octahedrons, which have lost their significance for the creation of a convex closed figure around an Os atom and do not solve this problem in a combined figure. At same time, all the valence bonds are preserved and the edges delineating the outer contour of the figure are preserved. So, let is connect vertices of octahedrons 1, 4, 9 edges with each other (blue color), connect vertices of octahedrons 15, 16, 17 edges with each other (blue color). Connect vertex 3 with vertex 7, connect vertex 6 with vertex 10, connect vertex 11 with vertex 13 (all edges are black). Remove the edges 5 - 7, 1 - 8, 3 - 8, 12 - 11, 13 - 8, 5 - 16, 6 - 8, 15 - 8, 12 - 16. The number of remote edges turned out to be equal to the number of newly entered edges. Then all valence compounds will be inside the new closed convex figure. Moreover, the figure turned out to be a folded of two convex closed polytopes, applied to each other along a two - dimensional (horizontal) flat section 2 - 3 - 7 - 11 - 13 - 14 - 10 - 6 - 2 made up of several sections. The image of the upper polytope (above the horizontal section) is shown in Figure 24.

Figure 24. The top of the polytope corresponding to the compound $H_2Os_3(CO)_{12}$

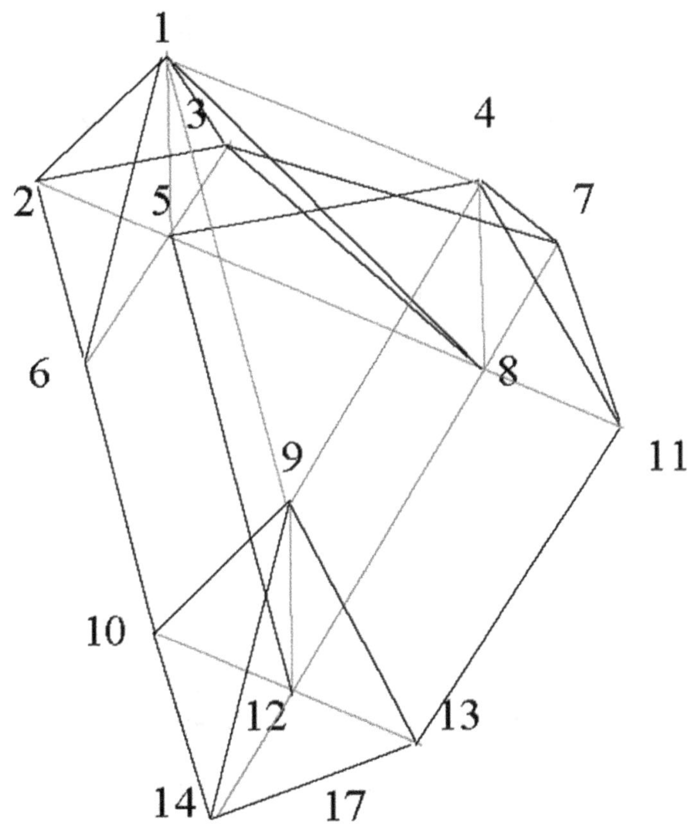

The image of the lower polytope (below the horizontal section) is shown in Figure 25.

Figure 25. The bottom of the polytope corresponding to the compound $H_2Os_3(CO)_{12}$

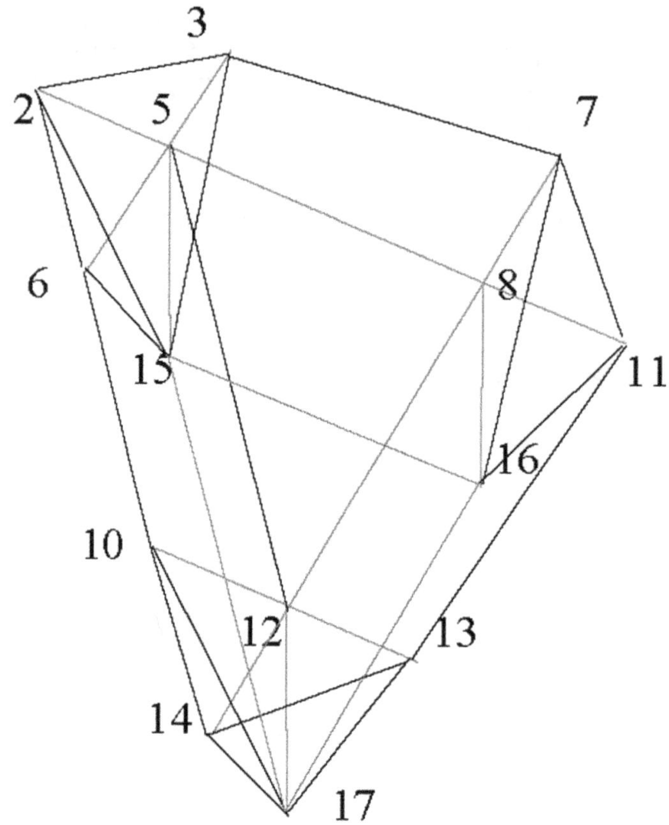

We define the dimension of the polytope in Figure 24. The polytope on Figure 24 has 14 vertices (f_0=14), 32 edges (f_1=32), 28 flat two - dimension sections (f_2=28), 10 three – dimension area (f_3=10) (Zhizhin, 2020).

Substituting the obtained values f_i(i=0,1,2,3) in the equation (3.3) you can see that the Euler - Poincaré equation is satisfied in this case with a value of n equal to 4

$$14 - 32 + 28 - 10 = 0.$$

This proves that polytope on Figure 24 has dimension 4.

We define the dimension of the polytope in Figure 25. The polytope on Figure 25 has 14 vertices (f_0=14), 32 edges (f_1=32), 28 flat two - dimension sections (f_2=28), 10 three - dimension area (f_3=10). Substituting the obtained values f_i(i=0,1,2,3) in the equation (3.3) you can see that the Euler - Poincaré equation is satisfied in this case with a value of n equal to 4

$$14 - 32 + 28 - 10 = 0.$$

This proves that polytope on Figure 25 has dimension 4.

It can be noted that the upper and lower parts of the polytope corresponding to the compound are symmetrical to each other with respect to the flat section separating them.

Q.E.D.

THE POLYTOPES OF THE HIGHER DIMENSION IN THE CONSTRUCTION OF SPACES OF THE HIGHER DIMENSION

Chapter 6 analytically examined the products of polytopes, which were called polytopic prismahedrons. It was noted that due to the presence of several families of parallel edges in them, they can serve as the basis for the creation of parallelohedrons of higher dimension (i.e., dimensions 4 and higher). This is important, since in monograph (Zhizhin, 2021) it was proved that the figures proposed by Delaunay (Delaunay, 1929, 1961) and Shtogrin (Shtogrin, 1973) as parallelohedrons of dimension 4 are not. The fact that polytopic prismahedrons can be the basis for creating parallelohedrons of higher dimension was discovered in the author's previous works (Zhizhin, 2012, 2013, 2014 a, b, 2015, 2018, 2019 a, d; 2021; Zhizhin & Duidea, 2016). One of the simplest types of the polytopic prismahedrons is the tetrahedral prism, i. e. a product tetrahedron by the segment (Chapter 6). Its structural formula has form (6.11).

Image tetrahedral prism is shown in Figure 26.

Figure 26. The tetrahedral prism

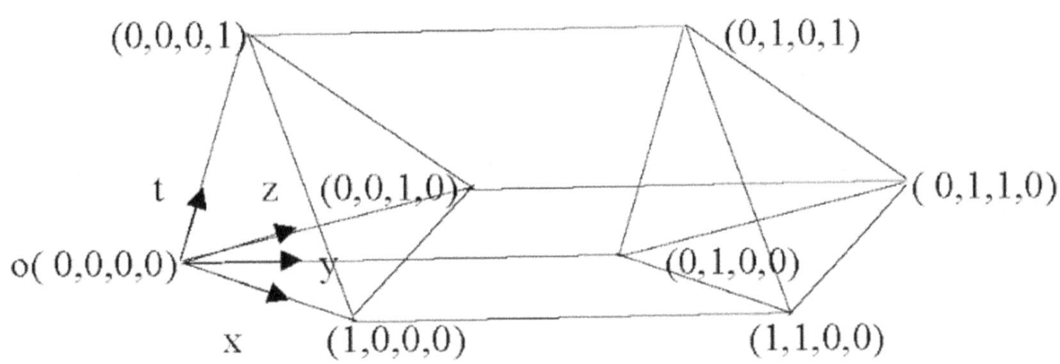

Can to introduce one of the vertices of the tetrahedral prism origin of the four - dimensional space (x, y, z, t). Orient the coordinates, such as indicated in Figure 26.

Assume that the length of each edge is equal to 1. Then, each node tetrahedral prism can be associated with a set of integers (Figure 26). Translating tetrahedral prism along the coordinates x, y, z, t, we obtain the lattice vertices. Let A_0 tetrahedral prism with the values of vertex coordinates in Figure 26. Then

$$A_0 = (0, 0, 0, 0), (1, 0, 0, 0), (0, 1, 0, 0), (0, 0, 1, 0), (0, 0, 0, 1), (1, 1, 0, 0), (0, 1, 1, 0), (0, 1, 0, 1), \quad (9.1)$$

A_1=A_0(x+1)= (1, 0, 0, 0), (2, 0, 0, 0), (1, 1, 0, 0), (1, 0, 1, 0), (1, 0, 0, 1), (2, 1, 0, 0), (1, 1, 1, 0), (1, 1, 0, 1),

A_2=A_0(z+1)= (0, 0, 1, 0), (1, 0, 1, 0), (0, 1, 1, 0), (0, 0, 2, 0), (0, 0, 1, 1), (1, 1, 1, 0), (0, 1, 2, 0), (0, 1, 1, 1),

A_3=A_0(x+1, z+1)=A_1(z+1)= (1, 0, 1, 0), (2, 0, 1, 0), (1, 1, 1, 0), (1, 0, 2, 0), (1, 0, 1, 1),(2, 1, 1, 0), (1, 1, 2, 0), (1, 1, 1, 1),

A_4=A_0(y+1, z+1)=A_2(y+1)= (0, 1, 1, 0), (1, 1, 1, 0), (0, 2, 1, 0), (0, 1, 2, 0), (0, 1, 1, 1), (1, 2, 1, 0), (0, 2, 2, 0), (0, 2, 1, 1),

A_5=A_0(x+1, y+1, z+1)=A_3(y+1)= (1, 1, 1, 0), (2, 1, 1, 0), (1, 2, 1, 0), (1, 1, 2, 0), (1, 1, 1, 1), (2, 2, 1, 0), (1, 2, 2, 0), (1, 2, 1, 1),

A_6=A_0(y+1)= (0, 1, 0, 0), (1, 1, 0, 0), (0, 2, 0, 0), (0, 1, 1, 0), (0, 1, 0, 1), (1, 2, 0, 0), (0, 2, 1, 0), (0, 2, 0, 1),

A_7=A_0(y+1, x+1)=A_1(y+1)= (1, 1, 0, 0), (2, 1, 0, 0), (1, 2, 0, 0), (1, 1, 1, 0), (1, 1, 0, 1), (2, 2, 0, 0), (1, 2, 1, 0), (1, 2, 0, 1),

A_8=A_0(t+1)= (0, 0, 0, 1), (1, 0, 0, 1), (0, 1, 0, 1), (0, 0, 1, 1), (0, 0, 0, 2), (1, 1, 0, 1), (0, 1, 1, 1), (0, 1, 0, 2),

A_9=A_0(y+1, t+1)=A_6(t+1)= (0, 1, 0, 1), (1, 1, 0, 1), (0, 2, 0, 1), (0, 1, 1, 1), (0, 1, 0, 2), (1, 2, 0, 1), (0, 2, 1, 1), (0, 2, 0, 2),

A_{10}=A_0(x+1, y=1, t+1)=A_7(t+1)= (1, 1, 0, 1), (2, 1, 0, 1), (1, 2, 0, 1), (1, 1, 1, 1), (1, 1, 0, 2), (2, 2, 0, 1), (1, 2, 1, 1), (1, 2, 0, 2),

A_{11}=A_0(x+1, t+1)=A_1(t+1)= (1, 0, 0, 1), (2, 0, 0, 1), (1, 1, 0, 1), (1, 0, 1, 1), (1, 0, 0, 2), (2, 1, 0, 1), (1, 1, 1, 1), (1, 1, 0, 2),

A_{12}=A_0(y+1, z+1)=A_2(t+1)= (0, 0, 1, 1), (1, 0, 1, 1), (0, 1, 1, 1), (0, 0, 2, 1), (0, 0, 1, 2), (1, 1, 1, 1), (0, 1, 2, 1), (0, 1, 1, 2),

A_{13}=A_0(z+1,y+1, t+1)=A_4(t+1)= (0, 1, 1, 1), (1, 1, 1, 1), (0, 2, 1, 1), (0, 1, 2, 1), (0, 2, 2, 1), (0, 1, 1, 2), (1, 2, 1, 1), (0, 2, 1, 2),

A_{14}=A_0(x+1,y+1, z+1, t+1)=A_5(t+1)= (1, 1, 1, 1), (2, 1, 1, 1), (1, 2, 1, 1), (1, 1, 2, 1), (1, 1, 1, 2), (2, 2, 1, 1), (1, 2, 2, 1), (1, 2, 1, 2),

A_{15}=A_0(x+1,t+1, z+1)=A_3(t+1)= (1, 0, 1, 1), (2, 0, 1, 1), (1, 1, 1, 1), (1, 0, 2, 1), (1, 0, 1, 2), (2, 1, 1, 1), (1, 1, 2, 1), (1, 1, 1, 2).

Between A_i type tetrahedral prisms arranged tetrahedral prism type B_i (see Figure 27) having a closest to them tetrahedral prisms A_i general quadrangular two - dimensional face.

Figure 27. The complex of the tetrahedral prisms

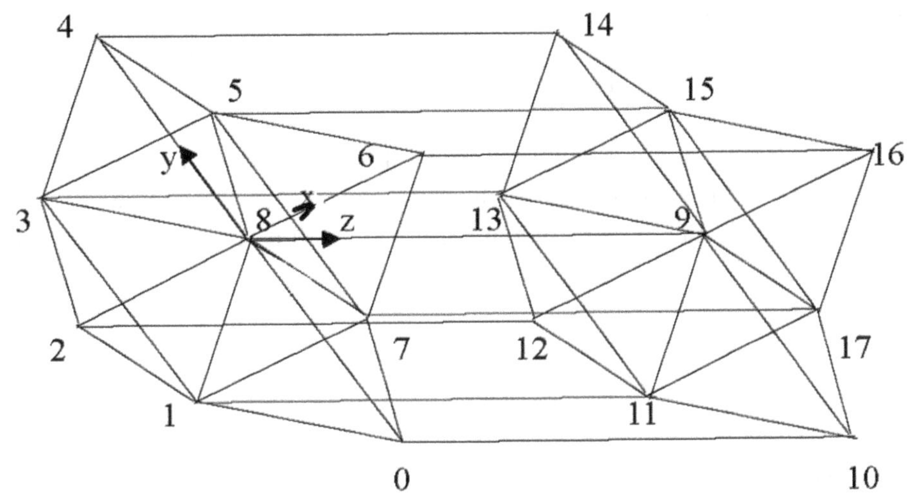

The tetrahedral prisms A_i and B_i have common edges, which are parties to the general quadrilateral faces. The tetrahedral prisms A_i and B_i are connected symmetry transformation - turning the 180^0 around a common edge of the tetrahedron. Forms A and B are two enantiomorphic forms of a tetrahedral prism. Let us prove that the complex of two pairs of enantiomorphic forms of tetrahedral prisms (Figure 27) has dimension 5. As follows from Figure 27, this complex is not a convex polytope. However, for its analysis, the Euler – Poincaré equation can be used, since Poincare showed that this equation is applicable to all connected forms (Poincaré, 1895).
The polytope in Figure 27 has 18 vertices, f_0=18. The polytope in Figure 27 has 47 edges:

0 – 1, 0 – 8, 0 – 7, 0 – 10, 1 – 7, 1 – 8, 1 – 3, 1 – 2, 1 – 11, 2 – 3, 2 – 8, 2 – 12, 3 – 8, 3 – 5, 3 – 4, 3 – 13, 4 – 8, 4 – 5, 4 – 14, 5 – 8, 5 – 6, 5 – 15, 6 – 8, 6 – 7, 6 – 16, 7 – 8, 7 – 17, 8 – 9, 9 – 10, 9 – 11, 9 – 12, 9 – 13, 9 – 14, 9 – 15, 9 – 16, 9 – 17, 10 – 11, 10 – 17, 11 – 12, 11 – 13, 12 – 13, 13 – 14, 13 – 15, 14 – 15, 15 – 16, 15 – 17, 16 – 17, i.e. f_1=47.

The polytope in Figure 27 has 50 two – dimensional elements:

0 – 8 – 9 – 10, 0 – 7 – 17 – 10, 0 – 1 – 11 – 10, 0 – 8 – 7, 0 – 1 – 8, 0 – 1 – 7, 1 – 8 – 7, 1 – 2 – 8, 1 – 2 – 3, 1 – 3 – 8, 1 – 2 – 12 – 11, 1 – 8 – 9 – 11, 1 – 17 – 7 – 11, 1 – 3 – 11 – 13, 2 – 3 – 13 – 12, 2 – 3 – 9 – 12, 2 – 3 – 8, 3 – 5 – 15 – 13, 4 – 3 – 14 – 13, 3 – 4 – 5, 3 – 5 – 8, 3 – 4 – 8, 4 – 5 – 8, 4 – 14 – 8 – 9, 4 – 5 – 15 – 14, 5 – 6 – 7, 5 – 8 – 7, 5 – 6 – 8, 5 – 15 – 8 – 9, 5 – 17 – 7 – 15, 5 – 6 – 15 – 16, 6 – 7 – 8, 6 – 16 – 7 – 17, 6 – 16 – 8 – 9, 7 – 8 – 9 – 17, 9 – 13 – 14, 9 – 13 – 15, 9 – 14 – 15, 9 – 15 – 16, 9 – 15 – 17, 9 – 16 – 17, 9 – 17 – 10, 9 – 10 – 11, 9 – 11- 17, 10 – 17 – 11, 11 – 12 – 13, 13 – 14 – 15, 15 – 10 – 17, 3 – 5 – 7 – 1,11 – 13 – 15 – 17, i.e. f_2=50.

The polytope in Figure 27 has 25 three – dimensional elements:

3 – 4 – 5 – 8, 14 – 15 – 13 – 9, 5 – 6 – 7 – 8,15 – 16 – 17 – 9, 1 – 2 – 3 – 8, 11 – 12 – 13 – 9, 8 – 7 – 0 – 1, 10 – 11 – 17 – 9, 0 – 8 – 7 – 10 – 17 – 9, 0 – 1 – 7 – 10 – 11 – 17, 0 – 1 – 8 – 10 – 11 – 9, 1 – 8 – 7 – 11 – 9 – 17, 1 – 2 8 – 11 – 12 – 9, 1 – 2 – 3 – 13 – 12 – 11, 2 – 3 – 8 – 12 – 13 – 9, 3 – 5 – 8 – 13 – 15 – 9, 3 – 4 – 5 – 13 – 14 – 15, 4 – 5 – 8 – 14 – 15 – 9, 5 – 8 – 6 – 15 – 16 – 9, 5 – 6 – 7 – 15 – 16 – 17, 5 – 8 – 7 – 15 – 9 – 17, 3 – 5 – 1 – 7 – 8, 13 – 15 – 17 – 11 – 9, 3 – 5 – 7 – 1 – 13 – 15 – 17 – 11, 1 – 2 – 3 – 4 – 5 – 6 – 7 – 8 – 9 – 10 – 11 – 12 – 13 – 14 – 15 – 16 – 17 – 18

(three - dimensional body bounded by the outer surface of the complex), i.e. f_3=25.

The polytope in Figure 27 has 6 four – dimensional elements: four of the tetrahedral prisms and two bases of the complex, each of which consists of four tetrahedrons. Consider one of these bases (Figure 28).

Figure 28. The base of the tetrahedral prisms complex

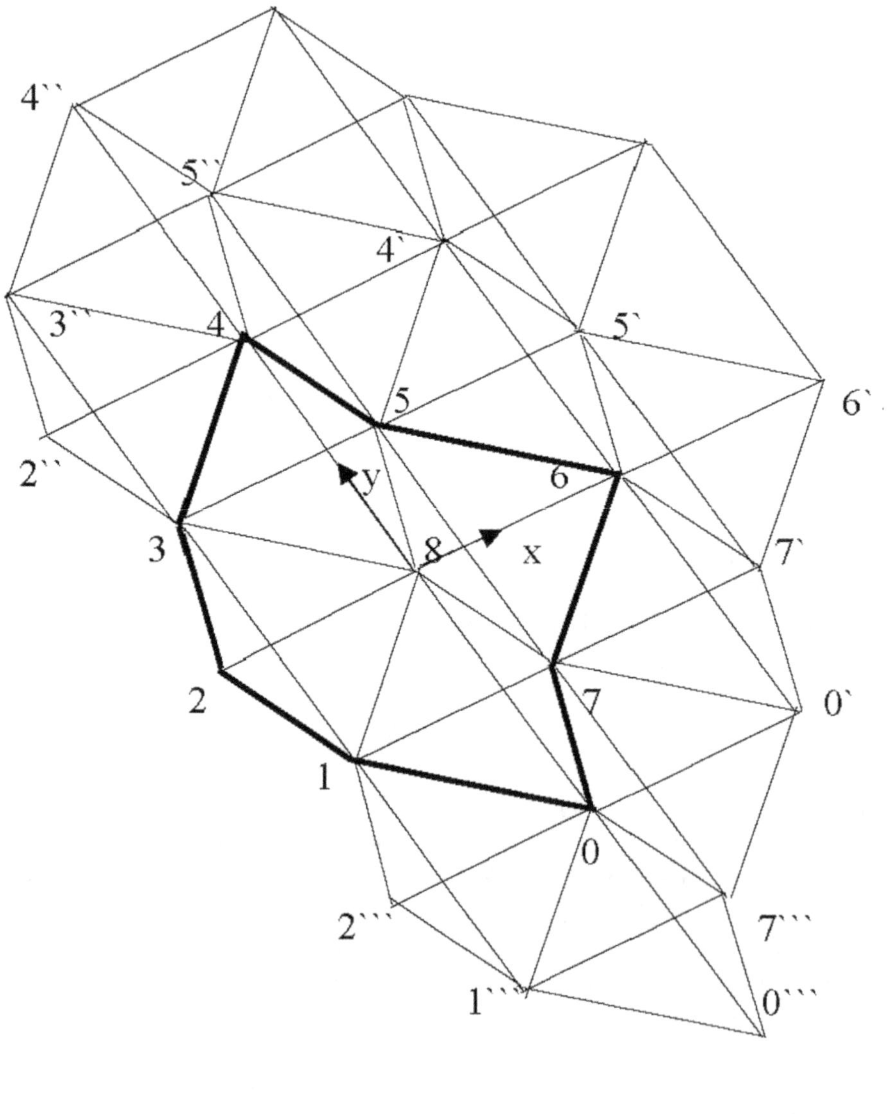

It has 9 vertices, i. e. f_0=9, 20 edges:

0 – 1, 1 – 2, 2 – 3, 3 – 4, 4 – 5, 5 – 6, 6 – 7, 0 – 7, 8 – 3, 8 – 4, 8 – 5, 8 – 6, 8 – 7, 8 – 0, 8 – 2, 8 – 1, 3 – 5, 1 – 7, 3 – 1, 5 – 7, f_1=20.

The complex from four tetrahedrons has 17 two – dimensional faces:

3 – 4 – 5, 3 – 5 – 8, 3 – 4 – 8, 3 – 8 – 2, 3 – 8 – 1, 3 – 1 – 2, 4 – 5 – 8, 5 – 6 – 8, 5 – 6 – 7, 5 – 8 – 7, 6 – 7 – 8, 1 – 7 – 8, 1 – 7 – 0, 1 – 8 – 0, 1 – 8 – 2, 1 – 3 – 5 – 7, i.e. f_2=17.

The complex from four tetrahedrons has 6 three – dimensional faces:

four tetrahedrons, one pyramid 3 – 5 – 7 – 1 – 8 and three - dimensional body bounded by the outer surface of the complex, i. e. f_3=6. Substituting the obtained values into the Euler - Poincaré equation (3.3), we obtain 9 – 20 + 17 – 6 = 0. The Euler - Poincaré equation for the base of the tetrahedral prism complex is satisfied for a dimension equal to 4, therefore this base has dimension 4. Then for the entire complex of tetrahedral prisms in Figure 27, the number of elements with dimension 4 is 6, i. e. f_4=6. Substituting the obtained values for the complex of tetrahedral prisms in the Euler - Poincaré equation (3.3), we obtain

$$18 - 47 - 50 - 25 + 6 = 2.$$

This means that in this case the Euler - Poincaré equation is satisfied for a dimension equal to 5, i. e. the tetrahedral prism complex in Figure 27 has dimension 5. It should be noted that not all edges in the complex of tetrahedral prisms form continuous lines, taking into account their possible continuation in neighboring tetrahedral prisms. Therefore, in this case, when choosing a coordinate system, it is necessary to select as the directions of the coordinate variables those edges that form continuous straight lines. There are three such directions, as can be seen from Figure 27. This is the direction along the generators of the complex of tetrahedral prisms (z) and two directions perpendicular to these generators (x, y) (Figure 27). Assuming that the length of the edges of the complex of tetrahedral prisms is equal to unity, we assign coordinates (x, y, z) to its vertices

0(0, -1, 0), 1(-1/2, -1/2, 0), 2(-1, 0, 0), 3(-1/2, 1/2, 0), 4(0, 1, 0), 5(1/2, 1/2, 0), 6(1, 0, 0), 7(1/2, -1/2, 0), 8(0, 0, 0), 9(0, 0, 1), (9.2)

10(0, -1, 1), 11(-1/2, -1/2, 1),12(-1, 0, 1), 13(-1/2, 1/2, 1),

14(0, 1, 1), 15(1/2, 1/2, 1), 16(1, 0, 1), 17(1/2, -1/2, 1).

The distribution of tetrahedral prisms B_i can be determined in the same way as the distribution of tetrahedral prisms A_i (see Zhizhin, 2019). However, in this case, of great interest is the question of the spatial motion of the entire complex of tetrahedral prisms, consisting of two pairs of enantiomorphic tetrahedral prisms.

Theorem 9.4. The complex, consisting of two pairs of enantiomorphic tetrahedral prisms adjacent to each other over whole quadrangular plane sections, is a parallelohedron of dimension 5.

Proof. The fact that the complex consisting of two pairs of enantiomorphic tetrahedral prisms has dimension 5 has already been proved. It remains to prove that it is a parallelohedron. For this, we assume that the initial state of the complex is that which is determined by the values of its vertices. Let us depict one of the bases (for example, the left) of the tetrahedral prism complex in a separate figure in bold line segments (Figure 28). We will not depict the generators of the tetrahedral prism complex so as not to clutter up the pattern. We remember that the generators come from each vertex of the base and remain parallel to themselves and the axis *z*. We will not depict the right base either, since it is parallel to the left base. We increase the values of the coordinate *x* by one step (one rib length) at all the vertices of the left base of the tetrahedral prism complex. This will correspond to the translation of the complex of tetrahedral prisms by one step along the *x* axis. Then the following mapping of the vertices of the left base of the complex

0(0, -1, 0) 0`(1, -1, 0), 1(-1/2, -1/2, 0) 7(1/2, -1/2, 0), 2(-1, 0, 0) 8(0, 0, 0),

3(-1/2, 1/2, 0) 5(1/2, 1/2, 0), 4(0, 1, 0) 4`(1, 1, 0), 5(1/2, 1/2, 0) 5`(3/2, 1/2, 0),

6(1, 0, 0) 6`(2, 0, 0), 7(1/2, -1/2, 0) 7`(3/2,-1/2, 0), 8(0, 0, 0) 6(1, 0, 0).

As a result of the mapping, a new base of the complex of tetrahedral prisms was formed 0` - 7 – 8 – 5 – 4` - 5` - 6` - 7` - 6. This base has dimension 4 and common tetrahedron 7 – 8 – 5 – 6 with initial left base.

Now we increase the values of the coordinate *y* by one step (one rib length) at all the vertices of the left base of the tetrahedral prism complex. This will correspond to the translation of the complex of tetrahedral prisms by one step along the *y* axis. Then the following mapping of the vertices of the left base of the complex

0(0, -1, 0) 8(0, 0, 0), 1(-1/2, -1/2, 0) 3(-1/2, 1/2, 0), 2(-1, 0, 0) 2``(-1, 1, 0),

3(-1/2, 1/2, 0) 3``(-1/2, 3/2, 0), 4(0, 1, 0) 4``(0, 2, 0), 5(1/2, 1/2, 0) 5``(1/2, 3/2, 0),

6(1, 0, 0) 4`(1, 1, 0), 7(1/2, -1/2, 0) 5(1/2, 1/2, 0), 8(0, 0, 0) 4(0, 1, 0).

As a result of the mapping, a new base of the complex of tetrahedral prisms was formed 8 – 3 – 2`` – 3`` – 4`` - 5`` - 4` - 5 - 4. This base has dimension 4 and common tetrahedron 3 – 8 – 5 – 4 with initial left base.

Now we decrease the values of the coordinate *y* by one step (one rib length) at all the vertices of the initial left base of the tetrahedral prism complex. This will correspond to the translation of the complex of tetrahedral prisms by one step opposite the *y* axis. Then the following mapping of the vertices of the left base of the complex

0(0, -1, 0) 0```(0, -2, 0), 1(-1/2, -1/2, 0) 1```(-1/2,-3/2, 0), 2(-1, 0, 0) 2```(-1, -1, 0),

3(-1/2, 1/2, 0) 1(-1/2, -1/2, 0), 4(0, 1, 0) 8(0, 0, 0), 5(1/2, 1/2, 0) 7(1/2, -1/2, 0),

6(1, 0, 0) 0`(1, -1, 0), 7(1/2, -1/2, 0) 7```(1/2,-3/2, 0), 8(0, 0, 0) 0(0, -1, 0).

As a result of the mapping, a new base of the complex of tetrahedral prisms was formed 0``` – 1``` – 2``` – 1 – 8 - 7 - 0` - 7``` - 0. This base has dimension 4 and common tetrahedron 1 – 8 – 7 – 0 with initial left base.

Similar broadcasts in the coordinate x and coordinate y can be continued. In all cases, we will receive new instances of the left bases of the tetrahedral prism complex. Moreover, each of these bases will have a common tetrahedron with any of the neighboring bases. Now recall that from each vertex of the base of the tetrahedral prism complex, the generators of the tetrahedral prism complex emanate. Therefore, the translation of the left base of the complex along the coordinates x and y corresponds to the translation of the entire complex of tetrahedral prisms along these coordinates. In these translations of the tetrahedral prism complex, the common between neighboring complexes will no longer be tetrahedrons, but tetrahedral prisms of dimension 4. In addition, there is still the possibility of translating the complex of tetrahedral prisms along the coordinate z. With this translation (one step), the left base of the complex takes the place of the right base of the initial position of the complex. Each of them has dimension 4. From the analysis it follows that the translation of the five - dimensional complex of tetrahedral prisms leads to a complete filling of the space without gaps with these complexes, which are in a parallel position relative to each other. This proves that the five - dimensional complex of tetrahedral prisms is a parallelohedron. Q.E.D.

CONCLUSION

The examples of the analysis of spatial forms of molecules of compounds of inorganic and organic chemistry given in this chapter have shown that the dimension of these molecules is more than three. In this case, the valence of chemical compounds is significantly different from the valency determined by their location in the periodic table of chemical elements. Transitional elements also form chemical compounds of large dimensions, for example, copper, gold, chromium. The simplest molecules of organic chemistry also have molecules of dimensions greater than three, for example, four or five. It is interesting that the well-known tartaric acid, which was studied by Pasteur, has a molecule of dimension 5. It is shown that it is precisely the high dimension of the tartaric acid molecule that ensures its ability to rotate the plane of polarized light, and not in a crystal, as Pasteur suggested, but in solution.

It is proved that the spatial images of chains of metal atoms with ligands are polytopes of the highest dimension, adjacent to each other in two-dimensional sections. This applies to linear, non-linear and closed circuits. An important class of polytopes of the highest dimension of chemical compounds are polytopic prismahedrons (products of polytopes). It is they, and not the Delaunay stereohedrons, that provide the possibility of solving the problem of constructing an n-dimensional space from congruent figures (Hilbert's 18th problem). This is necessary when creating extended nanomaterials that include clusters of chemical compounds. It is essential that polytopic prismahedrons form parallelotope of the highest dimension. In particular, it was shown that two pairs of enantiomorphic tetrahedral prisms, applied to each other along integral square faces, form a parallelotope of dimension 5.

REFERENCES

Aleksandrov, G.G., Zolnikova, G.P., Kritskaya, I.I., & Struchkov, Yu.T. (1980). Molecular structure $Os_3(\mu\text{–}Br)2(CO)1_0$ of the reaction product OS3(CO)2 with (3-bromomethylnaphthalene). *Coordination Chemistry, 6*(4), 626 - 628.

Bau, R., Kirtley, S. W., Sorrell, T. N., & Winarko, S. (1974). Preparation and structure determination of the manganese carbonyl [Mn3(CO)14]-. *Journal of the American Chemical Society*, *96*(4), 988–993. doi:10.1021/ja00811a006

Büchel, W. (1963). Warum hat ger Raume drei Dimensionen? *Physilalische Blatter*, *19*(12), 547–549. doi:10.1002/phbl.19630191204

Delaunay, B.N. (1929). Sur la partition ré gulière de l`espace A4 dimensions. *Bulletin de L`Académie des Sciences de L`Urss., 1*, 79 - 110, 147 - 164.

Delaunay, B. N. (1961). Proof of the main theorem of the theory of stereohedrons. *Reports. Academy of Sciences of the USSR*, *138*(6), 1270–1272.

Evans, J., Okrasunski, S. J., Pribula, A. J., & Norton, J. K. (1976). Synthesis of di- and trinuclear methyl osmium complexes via cis-hydridomethyltetracarbonylosmium. *Journal of the American Chemical Society*, *98*(13), 4000–4001. doi:10.1021/ja00429a045

Fersman, A. E. (1937). *Geochemistry. V.3.* Leningrad: ONTI. Freeman, I.M. (1069). Why is Space Three-Dimensional? *American Journal of Physics*, *37*, 1222–1224.

Gillespie, R. J. (1972). *Molecular Geometry*. Van Nostrand Reinhold Company.

Gillespie, R. J., & Hargittai, I. (1991). *The VSEPR Model of Molecular Geometry*. Allyn & Bacon.

Gochin, M., & Moss, Y. R. (1980). On the structure of H2Os3(CO)12. *Journal of Organometallic Chemistry*, *192*(3), 409–419. doi:10.1016/S0022-328X(00)81233-4

Gray, H. B. (1965). *Electrons and chemical bonding*. W.A. Benjamin INC.

Grunbaum, B. (1967). *Convex Polytopes*. Springer.

Gurevich, L., & Mostepanenko, V. (1971). On the Existence of Atoms in *n* - Dimensional Space. *Physics Letters. [Part A]*, *35*(3), 201–202. doi:10.1016/0375-9601(71)90148-4

Metzler, D. E. (1980). Biochemistry. The Chemical Reactions of Living Cells (vols. 1-3). Academic Press.

Pasteur, L. (1960). *Selected works*. Publishing House of the Academy of Sciences of the USSR.

Perrin, D. D., Armarego, W. L. F., & Perrin, D. R. (1980). *Purification of Laboratory Chemical*. Pergamon.

Poincare A. (1895). Analysis situs. *J. de é Ecole Polytechnique, 1.*

Schrodinger. (1928). *Abhandlungen zur Wellenmechanik*. Academic Press.

Shtogrin, M. I. (1973). Correct Dirichlet - Voronoi partitions for the second triclinic group. *Proceedings of the Mathematical Institute. V.A. Steklova, 123.*

Smart, L., Cook, N., & Woodward, P. (1977). Article. *Journal of the Chemical Society, Dalton Transactions: Inorganic Chemistry*, 4312.

Wood, R. (1936). *Researches in physical optics*. Wentworth Press.

Zhizhin, G. V. (2012, October). *Hierarchical filling of spaces with polytopes.* Paper presented at "St. Petersburg Scientific Forum: Science and Human Progress". 7th St. - Petersburg meeting of Nobel Prize laureates, St. Petersburg, Russia.

Zhizhin, G. V. (2013). Images of convex regular and semiregular n - dimensional polytopes. In *Proceedings of the 9th All-Russian Scientific School "Mathematical Research in the Natural Sciences".* Geological Institute KSC RAS.

Zhizhin, G. V. (2014a). On the higher dimension in nature. *Biosphere*, *6*(4), 313–318.

Zhizhin, G. V. (2014b). *World – 4D*. Polytechnic Service.

Zhizhin, G. V. (2015a, November). *Polytopic prismahedrons – Fundamental regions of the n-dimension nanostructures.* Paper presented at The International conference "Nanoscience in Chemistry, Physics, Biology and Mathematics", Cluj-Napoca, Romania.

Zhizhin, G. V. (2015b). The dimensions of supramolecular compounds. *Biosphere*, *7*(2), 149–154.

Zhizhin, G. V. (2017a). Systematics of Anomalies in the Filling of Electron Orbitals of Chemical Elements. *J. Chemoinformatics and Chemical Engineering*, *6*(1), 40–55. doi:10.4018/IJCCE.2017010104

Zhizhin, G. V. (2017b). The structure, topological and functional dimension of biomolecules. *J. Chemoinformatics and Chemical Engineering*, *5*(2), 44–58.

Zhizhin, G.V. (2017c). Dimensions of compounds in supramolecular chemistry. *International Journal Chemical Modeling, 5*(2).

Zhizhin, G. V. (2018). *Chemical Compound Structures and the Higher Dimension of Molecules: Emerging Research and Opportunities*. IGI Global. doi:10.4018/978-1-5225-4108-0

Zhizhin, G. V. (2019a). *The Geometry of Higher-Dimensional Polytopes*. IGI Global. doi:10.4018/978-1-5225-6968-8

Zhizhin, G. V. (2019b). *Attractors and Higher Dimensions in Population and Molecular Biology: Emerging Research and Opportunities*. IGI Global. doi:10.4018/978-1-5225-9651-6

Zhizhin, G. V. (2019 d). On the Structure of Parallelohedrons of Higher Dimension: Hilbert's 18th Problem. *International Journal of Chemoinformatics and Chemical Engineering*, *8*(2), 69–96. doi:10.4018/IJCCE.20190701.oa2

Zhizhin, G.V. (2020). Enantiomorphism of Tartaric Acid as a Property of Five-Dimensional Polypopes. *Acta Scientific Biotechnology, 1*(12), 2 – 7.

Zhizhin, G. V. (2020). *Nanotechnologies and Clusters in the Spaces of Higher Dimension: Emerging Research and Opportunities*. IGI Global.

Zhizhin, G. V. (2021). *Normal Partitions and Hierarchical Fillings of N-Dimensional Spaces*. IGI Global. doi:10.4018/978-1-7998-6768-5

Zhizhin, G. V., & Diudea, M. V. (2016). Space of Nanoworld. In M. V. Putz & M. C. Mirica (Eds.), *Sustainable Nanosystems, Development, Properties, and Applications* (pp. 214–236). IGI Global.

Zhizhin, G. V., Khalaj, Z., & Diudea, M. V. (2016). Geometrical and topology dimensions of the diamond. In A. R. Ashrafi & M. V. Diudea (Eds.), *Distance, symmetry and topology in carbon nanomaterials* (pp. 167–188). Springer. doi:10.1007/978-3-319-31584-3_12

KEY TERMS AND DEFINITIONS

Divided Electron Pair: The binding electron pair, which simultaneously belongs to two atoms in the molecule.

Homo-Element Metal Chains: Clusters in which the core contains atoms of the same element.

Linear Metal Chains: Metal chains in which metal atoms are located in a straight line.

Metal Chains: Cluster compounds that have a skeleton in the form of metal chains, i.e., polymetallic chains formed by metal-metal localized covalent bonds.

N-Cross-Polytope: The convex polytope of dimension n in which opposite related of centrum edges not have connection of edge.

N–Simplex: The convex polytope of dimension n in which each vertex is joined by edges with all remain vertices of polytope.

S- and P-Elements: The chemical elements in which is filling with electrons s- and p-orbitals of atoms.

Tetrahedral Coordination of Electron Pairs: The location of the electronic pairs of the outer and the pre-outer electron layer at the vertices of the tetrahedron.

Undivided Electron Pair: A non-bonding electron pair belonging to one atom in a molecule.

Chapter 10
Polytopes of the Highest Dimension in Biology

ABSTRACT

Polytopes of the highest dimension corresponding to sugar molecules with five and six carbon atoms have been constructed. Based on the study of the geometry of these polytopes of dimensions 12 and 15, simplified three-dimensional models of sugar with five and six carbon atoms were obtained, and the chains of these molecules were analyzed. The geometry of protein molecules in various conformations was investigated, taking into account their highest dimension. It was found that Pauling's quasi-flat models do not reflect the spatial structures of the polypeptide chains of protein molecules. An opinion was expressed that the highest dimension of biomolecules is a consequence of the conditions for the emergence of life from the LUCA protocells more than 4.5 million years ago.

INTRODUCTION

Life began in water and continues in water. Water is matter and the matrix of life. Water is not only outside of us, but also inside us. Every living organism is 80% water. Recent studies (Russel et al., 2005; Qian et al., 2010; Simakov et al., 2020) show that as a result of the interaction of water with pyrite FS_2, there is a possibility of the appearance of protocells LUCA in water layers on the surface of minerals, including on the surface of nanodiamonds, which can form under terrestrial conditions at moderate temperatures and pressures (Simakov, 2018). All biomolecules ultimately evolved from a protocell that emerged over 4.5 billion years ago. In this Chapter proves that various biomolecules in nature have the highest dimension and the water molecules themselves, from which biomolecules originated in contact with minerals, have the highest dimension. Thus, the highest dimension of biomolecules is a consequence of the highest dimension of water molecules. Water has the ability to adjust its structure to the structure of substances that are in it or border it. As an illustration of this property, the process of restructuring of water to match the structure of cyanide compounds dissolved in water is considered.

DOI: 10.4018/978-1-7998-8374-6.ch010

THE STRUCTURE OF WATER

The water molecule has a peculiar shape. Oxygen in the second energy level *p* orbital has one quantum cell with a pair of electrons with opposite spins and two quantum cells with one electron in each of them. These two electrons combine with two electrons of the hydrogen atoms to form a covalent chemical bond. In addition, the oxygen atom has one more pair of electrons at the second energy level in the *s* orbital. The hydrogen atoms, which donated their two electrons to form a bond, remain sufficiently distant from the oxygen atom and have a positive charge +. Unshared electron pairs of the oxygen atom of the outer energy level naturally have a negative charge -. Conventionally, the shape of a water molecule can be represented as shown in Figure 1.

Figure 1. The shape of a water molecule

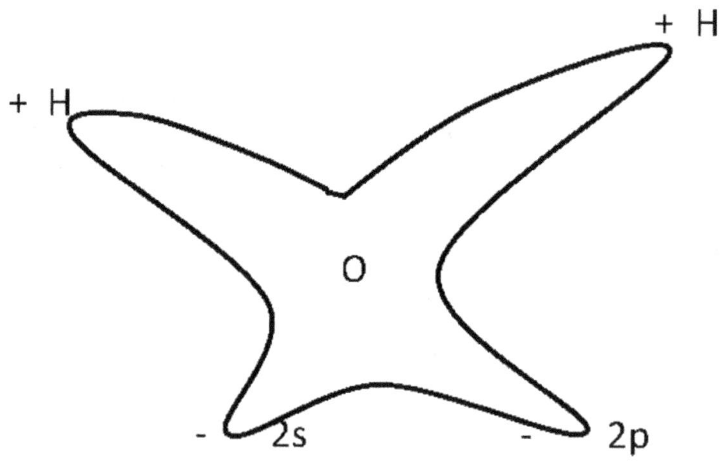

It is obvious that positively charged protons of other water molecules can join the orbitals with negative charges of the second energy level of the oxygen atom. The result is a tetrahedral structure with a center (Figure 2).

Figure 2. The tetrahedral structure with a center of molecules water with joined two protons

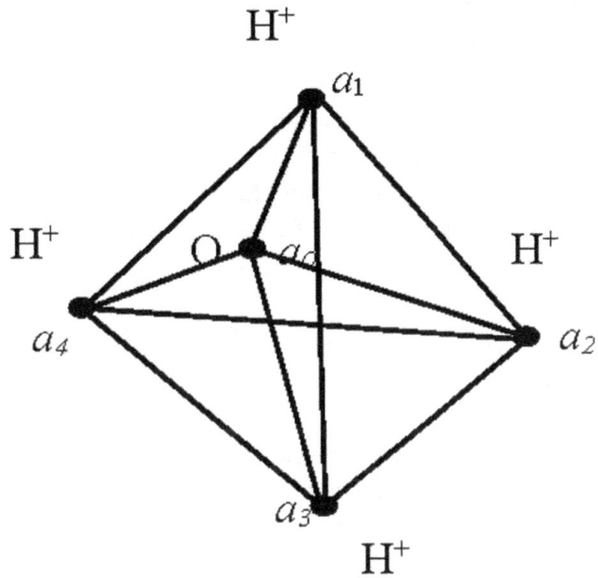

The tetrahedron with a center on Figure 2 has 5 vertices $a_0 \div a_4, f_0 = 5$;10 edges

$a_0a_1, a_0a_2, a_0a_3, a_0a_4, a_1a_2, a_3a_1, a_1a_4, a_4a_2, a_2a_3, a_4a_3, f_1 = 10$;

10 two - dimensional faces

$a_3a_4a_1, a_1a_3a_2, a_1a_4a_2, a_2a_3a_4, a_1a_0a_3, a_1a_0a_2, a_0a_4a_1, a_0a_4a_2, a_0a_3a_2, a_0a_3a_4, f_2 = 10$;

5 three-dimensional faces

$a_3a_4a_1a_2, a_0a_1a_3a_2, a_1a_4a_2a_0, a_3a_4a_1a_0, a_3a_4a_0a_2, f_2 = 5$.

To determine the dimension of a tetrahedron with a center, it is sufficient to use the Euler-Poincaré equation (3.3) in Chapter 3

$$\sum_{i=0}^{n-1}(-1)^i f_i(P) = 1 + (-1)^{n-1}. \tag{10.1}$$

There n is dimension of a polytope P, f_i is the number of elements with dimension i in the polytope P. In this case $f_0 = 5, f_1 = 10, f_2 = 10, f_3 = 5$.

Substituting the obtained values into equation (10.1), we obtain, i. e. Euler-Poincaré's equation for a tetrahedron with a center holds for $n = 4$. This proves that a molecule of water on Figure 2 has dimension 4.

The existence of a tetrahedral distribution of hydrogen atoms in the network of hydrogen bonds of water molecules is proved by the methods of computer experiment (Rahman & Stillinger, 1973; Eisenberg & Kautsman, 1975; Lyashchenko & Dunyashev, 2003). However, the possibility of the existence of non-tetrahedral dimers is noted. This is quite natural since hydrogen bonds do not differ in strength and chaotic rearrangements of the network of hydrogen bonds are possible in water. However, the tetrahedral distribution of hydrogen atoms in the structure of water is predominant. It should be remembered that oxygen atoms are in the center of the tetrahedron, and the tetrahedron with the center has dimension 4 (this is not mentioned in all works on the structure of water). Note that this is significant if we remember the statement that life originated in water.

THE MEMORY OF WATER

There is no doubt about the existence of a certain relaxation time required for the water to return to its original state after the cessation of external influences, for example, a magnetic or electric field (Klassen, 1973; Martynova, et al., 1969; Tatarsky, 1989; Lalalyans, 1990). Let us try to explain the existence of the memory of water about the chemical compounds dissolved in it, which are essential for the life of biological organisms. Let such compounds be cyanide compounds. It is well known that hydrocyanic acid HCN and its salts are strong poisons. The toxic effect of these compounds is that the ions CN^- formed in their solution inhibit the reduction of oxygen by the most important respiratory enzyme, cytochrodoxidation, in the tissue cells of living organisms (Lehninger, 1982; Ershov, 1989). Ions CN^- react with the oxidized form of cytochrome oxidase and form a complex compound with a trivalent iron atom of enzyme $[Fe(CN)_6]^{3-}$. As a result of the overlapping of the orbitals of the complexing agent Fe^{3+} by the ligands CN^- and the hybridization of the orbitals, the ion $[Fe(CN)_6]^{3-}$ has octahedral coordination (Figure 3). Geometrically, this ion is an octahedron with a center.

Figure 3. Geometric structure of the ion $\left[Fe(CN)_6\right]^{3-}$

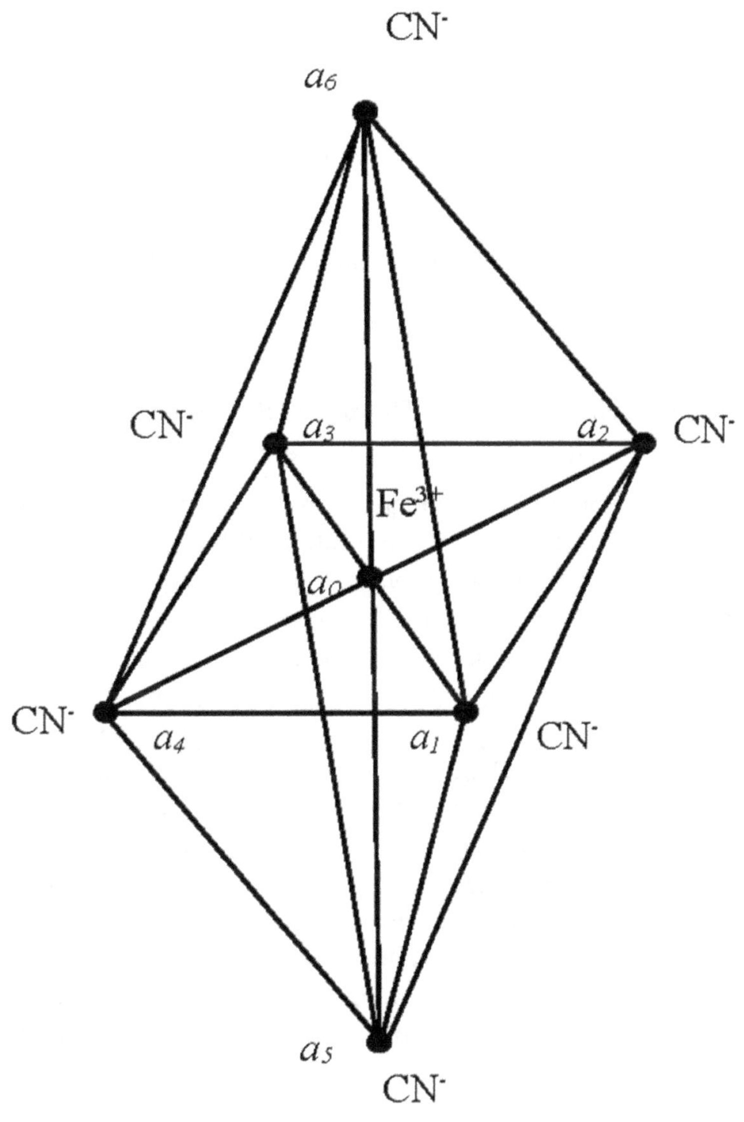

The functional dimension of this ion is 4. The octahedron with a center on Figure 3 has 7 vertices $a_0 \div a_6$, (f_0=7);18 edges

$$a_0a_1, a_0a_2, a_0a_3, a_0a_4, a_0a_5, a_0a_6, a_1a_2, a_2a_3, a_3a_4,$$
$$a_1a_4, a_6a_1, a_6a_2, a_6a_3, a_6a_4, a_1a_5, a_5a_2, a_5a_3, a_4a_5, (f_1 = 18);$$

20 two - dimensional faces

$a_0a_6a_1, a_0a_6a_2, a_0a_3a_6, a_0a_6a_4, a_1a_0a_5, a_5a_0a_2, a_3a_0a_5, a_0a_4a_5, a_1a_0a_2, a_2a_3a_0,$
$a_4a_0a_3, a_1a_0a_4, a_5a_2a_1, a_3a_5a_2, a_5a_3a_4, a_5a_1a_4, a_6a_2a_1, a_6a_3a_2, a_6a_3a_4, a_6a_1a_4, (f_2 = 20);$

9 three-dimensional faces

$a_0a_4a_1a_5, a_0a_1a_5a_2, a_3a_5a_2a_0, a_3a_4a_5a_0, a_1a_4a_0a_6, a_0a_1a_2a_6, a_3a_6a_0a_2, a_0a_4a_3a_6, a_1a_4a_2a_5a_3a_6, (f_3 = 9).$

Substituting the obtained numbers $f_i(i=0\div3)$ into equation (10.1), we obtain,

$7 - 18 + 20 - 9 = 0,$

i. e. the Euler-Poincaré equation holds for $n = 4$. This proves that the octahedron is centered, that is, ion $[Fe(CN)_6]^{3-}$ has dimension 4. In this compound, each pair of electrons 2s of the nitrogen atom in each of the six ions CN^- occupies one of the six vacant quantum cells of the ion Fe^{3+}. The ion Fe^{3+} has two vacant quantum cells in orbit 3d, one vacant quantum cell in orbit 4s and three vacant quantum cells in orbit 4p. Hybridization of the orbitals sp^3d^2 takes place. It has been experimentally established (Kostromina, et al., 1990) that anions CN^- bind most strongly with cations Fe^{3+} in comparison with other ions or molecules in the stoichiometric series in terms of the strength of interaction with the complexing agent. The strong bond of ions CN^- with cations Fe^{3+} disrupts the redox processes in cytochrome oxidase, consisting in the transfer of electrons to copper atoms and then to oxygen. As a result, the ions H^+ formed in the body during the decomposition of proteins, carbohydrates and amino acids do not react with oxygen to form water. The concentration of ions H^+ in the body increases sharply, which leads to poisoning of the body. Obviously, in this way, the degree of poisoning will be determined by the concentration of ions CN^- in the solution. They are formed during the dissociation of hydrocyanic acid or its salts. Moreover, hydrocyanic acid dissociates weakly, and the dissociation of salts goes almost to the end. The dissociation of hydrocyanic acid as a result of the formation of ions H^+ slows down the dissociation of water. An increase in the concentration of hydrocyanic acid leads to a decrease in the degree of dissociation of both hydrocyanic acid and water. Indeed, the concentration of ions and molecules in solution is determined by the relations

$$\begin{gathered}[H^+] = c_{H_2O}\alpha_{H_2O} + c_{HCN}\alpha_{HCN}, [OH^-] = c_{H_2O}\alpha_{H_2O}, [CN^-] = c_{HCN}\alpha_{HCN},\\ [H_2O] = c_{H_2O}(1-\alpha_{H_2O}), [HCN] = c_{HCN}(1-\alpha_{HCN}).\end{gathered} \tag{10.2}$$

There are α_{H_2O} – degree of dissociation of water, αH_{CN-} degree of dissociation of hydrocyanic acid, c_{H_2O} – concentration of water before dissociation, c_{HCN}- concentration of hydrocyanic acid before dissociation. Substituting relations (10.2) into the equations for the dissociation constants

$$K_{H_2O} = \frac{[H^+][OH^-]}{[H_2O]}, K_{HCN} = \frac{[H^+][CN^-]}{[HCN]}, \tag{10.3}$$

taking into account that the degrees of dissociation are small compared to unity, we obtain

$$K_{H_2O} = c_{H_2O}\alpha_{H_2O}^2 + c_{HCN}\alpha_{HCN}\alpha_{H_2O},$$
$$K_{HCN} = c_{H_2O}\alpha_{H_2O}\alpha_{HCN} + c_{HCN}\alpha_{HCN}^2. \tag{10.4}$$

Solving system (10.4) with respect to $\alpha H_{CN,}$ α_{H_2O}, taking into account that the mass fractions of water x_{H_2O} and acid x_{HCN} satisfy the conditions $x_{H_2O} + x_{HCN} = 1, x_{HCN} << 1$, we obtain (Zhizhin, 2020 d)

$$\alpha_{H_2O} = \frac{K_{H_2O}}{\sqrt{\rho(K_{H_2O} / \mu_{H_2O} + x_{HCN}K_{HCN} / \mu_{HCN})}}, \frac{\alpha_{HCN}}{\alpha_{H_2O}} = \frac{K_{HCN}}{K_{H_2O}}. \tag{10.5}$$

There are ρ- solution density, μ- molecular weight.

Relations (10.2), (10.5) make it possible to calculate the degree of dissociation and ion concentration for a given initial concentration of hydrocyanic acid. In particular, taking the values of the dissociation constants $K_{H_2O} = 10^{-14}$ mol/l, K_{HCN}=4.79•10^{-10} mol/l at very low concentrations of hydrocyanic acid (x_{HCN}®0), we find $\alpha_{H_2O} = 10^{-8}, \alpha_{HCN} = 4.79 \cdot 10^{-4}$. At a lethal concentration of hydrocyanic acid (Ershov, 1989) x_{HCN}1.4•10^{-6} these values are equal $\alpha_{H_2O} = 10^{-10}, \alpha_{HCN} = 4.6 \cdot 10^{-4}$. Thus, the dissociation of water in solutions of hydrocyanic acid and its salts is insignificant, and the degree of dissociation of hydrocyanic acid weakly depends on its concentration. From (10.2), (10.5) it follows that despite a decrease in the degree of dissociation of hydrocyanic acid with an increase in its concentration, the concentration of ions CN^- increases proportionally x_{HCN}.

When any substances are dissolved, the solvent molecules are strongly influenced by the molecules of the dissolved substance. In this case, the solvent molecules are modified. It was shown (Semenchenko, 1932) that the water molecules surrounding the ions of the dissolved substance are under the influence of forces equivalent to a pressure of 50,000 atmospheres. In the process of hydration, molecules of hydrated water (Zundel, 1972) pass into an excited state with vibrations of nuclei relative to new acquired positions that differ from unexcited positions. The nature of the fluctuations is determined by the nature of the hydrated ion. This is how the nature of the hydrated ion is transferred to the water molecules in its environment. Due to the excited vibrations, water molecules become more deformable and polarizable. It is these properties that put the ion CN^- on the first place in the spectrochemical series of the activities of ions and the forces of their interaction with the complexing agent. The greater the polarizability of the particle, the stronger its bond with the complexing agent. The greater the deformability of the molecule, the easier it is for the molecule to "fit" into the particular relief of the enzyme (in this case, cytochrome oxidase). The network of hydrogen bonds in water has the ability to transfer perturbations external to the network to the entire volume of the solution, while maintaining the stability of the structure of the H-network (Rodnikova, 1993). The phenomenon of negative solvation is associated with this property (Samoilov, 1957), i. e. an increase in the mobility of solvent molecules near ions. All this indicates that the excitation created by ions CN^- with an increase in concentration is transferred to almost all water molecules in solution. After dilution of this solution, the acquired properties of water are retained for some time, which is necessary for the relaxation of the H-bond network. The characteristic time for a change in the properties of water in the absence of external influences on it is, according to experiments (Miller, 1969), several tens of minutes. While the characteristic time for the absorption of ions and water molecules into the tissue cells of living organisms is fractions of a second (Ivanov, 1993). It follows from this that water molecules, having changed under the influence of ions CN^-, even after dilution of water, can retain their activity and, instead of ions CN^-, take their place as ligands around the complexing agent Fe^{3+} in the cytochrome oxidase enzyme, firmly binding with it. In this case, a pair of electrons 2s of the oxygen atoms in the water molecule is transferred to the vacant orbitals 3d, 4s, 4p of the iron atom of the complexing agent. Here, too, an octahedron with a center of dimension 4 is formed around the iron atom, and the violation of redox processes in cytochrome oxidase is preserved.

Since it is believed that biomolecules were formed in the process of evolution from the LUKA protocell more than 4.5 billion years ago in layers of water on the surface of minerals (Russel, et al., 2005; Qian, et al., 2010; Simakov, 2018; Simakov, et al., 2020), apparently the higher dimension of biomolecules may be a consequence of the higher dimension of water molecules. The highest dimension of water molecules is a consequence of the electronic structure of the atoms of the water molecule and the processes of hydration around the ions of mineral substances in water and on the surface of minerals. In this case, it should be taken into account that the elementary cells of minerals also have a higher dimension (Chapter 9, Zhizhin, 2018) and water during hydration changes its structure complementary to the vicinity of mineral ions.

THE DIMENSIONS AND SPATIAL IMAGES OF BIOMOLECULES

The Dimensions and Spatial Images of Molecules of Glucose

One of the most important monosaccharides is glucose. It makes up a significant part of all biomass (cellulose, starch), found in the blood of humans and animals. The most important aldohexose is D-glucose. The formula of Fischer for molecule D-glucose is presented on Figure 4

Figure 4. The molecule D-glucose

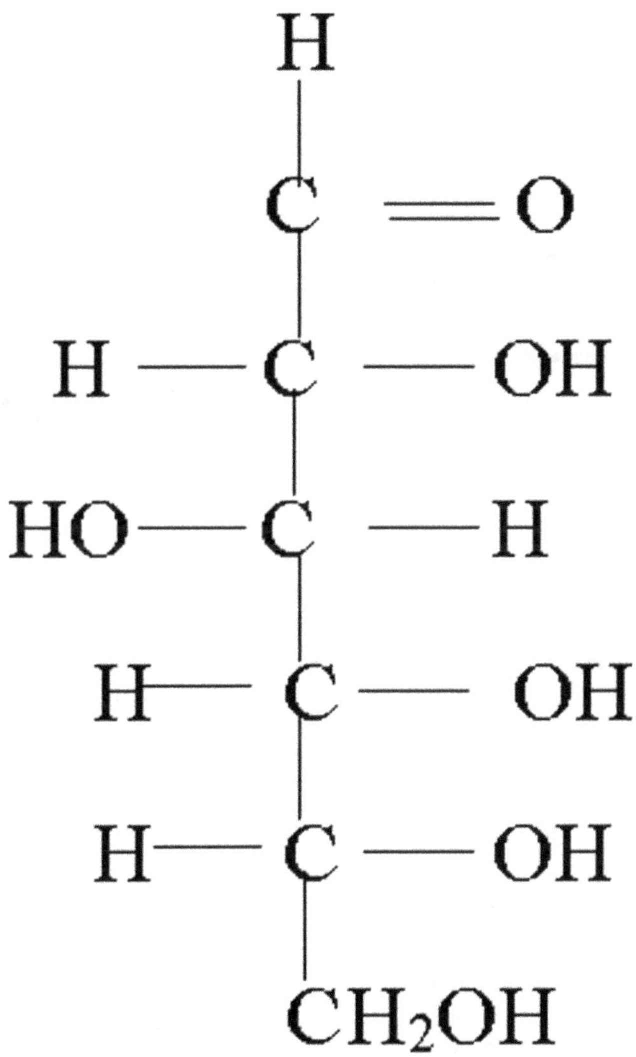

In solution, glucose molecules are present in the form of closed chains in two forms α - D-glucose and β-D-glucose. Haworth's projection formulas of these forms are presented in Figure 5 a, b

Figure 5. Pyranose forms of D-glucose
a) α -D- glucose, b) β -D- glucose

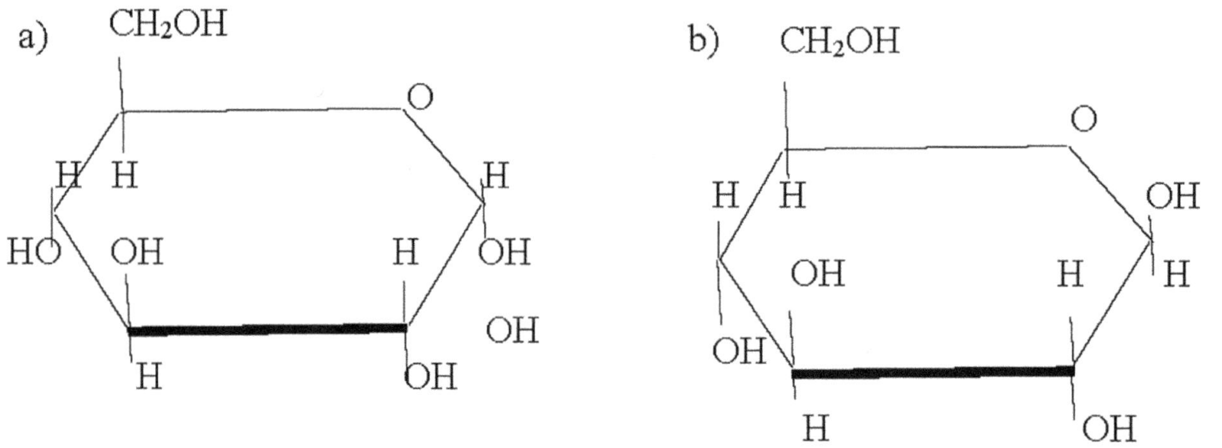

However, neither the Fischer formula or formula Haworth and their modifications (e.g., conformation as a “chair”) may not reflect the spatial structure of the monosaccharides. For this target the constructs described in the form of convex polytopes with boundary elements which form boundary complex (Grunbaum, 1967). Only when such a representation will be to determine the dimension of these molecules. Consider a molecule of α - D - glucose. Closing unbranched chain of carbon atoms of monosaccharides through an oxygen atom, considering functional groups vertices, connect the each vertex by edges with each other vertices, get polytope, depicted in Figure 6.

Figure 6. The molecule of α - D – glucose

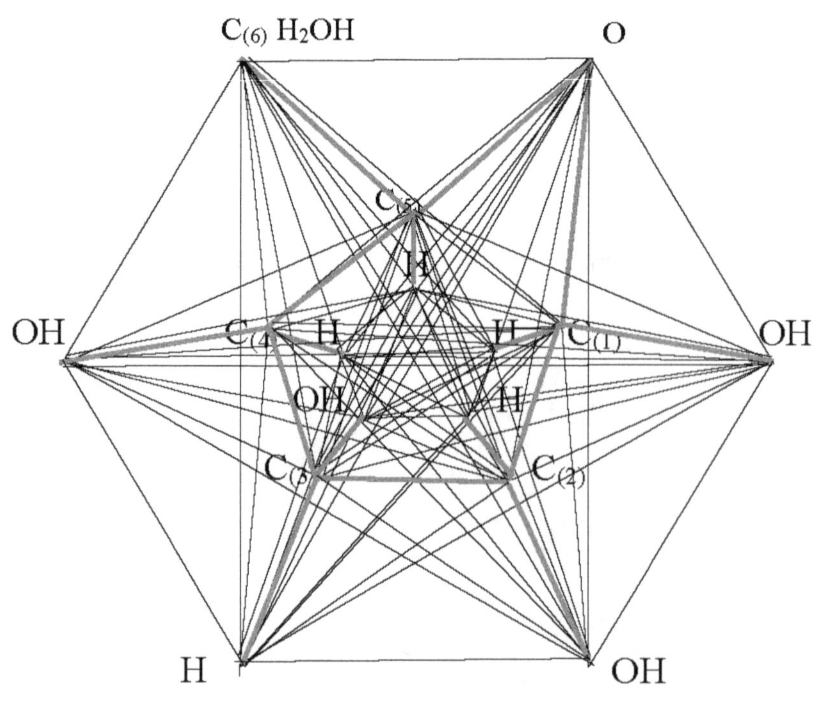

Edges marked in red in Figure 6 correspond to the chemical bonds in the molecule. The rest of the edges are only geometric sense, as the edges of the polytope.

Theorem 10.1. A molecule of α - D - glucose is a polytope type simplex with dimension 15 (Zhizhin, 2016).

Proof

The polytope in Figure 6 has 16 vertices, i. e. f_0=16=n+1; 120 edges ($f_1 = C_{n+1}^2 = 120$);

560 triangles ($f_2 = C_{n+1}^3 = 560$);

1820 tetrahedrons ($f_3 = C_{n+1}^4 = 1820$);

4368 4D – simplexes ($f_4 = C_{n+1}^5 = 4368$);

8008 5D – simplexes ($f_5 = C_{n+1}^6 = 8008$);

11440 6D – simplexes ($f_6 = C_{n+1}^7 = 11440$);

12870 7D – simplexes ($f_7 = C_{n+1}^8 = 12870$);

11440 8D – simplexes ($f_8 = C_{n+1}^9 = 11440$);

8008 9D – simplexes ($f_9 = C_{n+1}^{10} = 8008$);

4368 10D – simplexes ($f_{10} = C_{n+1}^{11} = 4368$);

1820 11D – simplexes ($f_{11} = C_{n+1}^{12} = 1820$);

560 12D – simplexes ($f_{12} = C_{n+1}^{13} = 560$);

120 13D – simplexes ($f_{13} = C_{n+1}^{14} = 120$);

16 14D – simplexes ($f_{14} = C_{n+1}^{15} = 16$).

Substituting the values f_i(0£i£15) in Euler's – Poincaré (Poincaré, 1895) equation (10.1), we see that it holds for n = 15

$$\sum_{i=0}^{14=n-1} f_i(-1)^i = 2.$$

This confirms that the polytope in Figure 6 is simplex of the dimension $n = 15$.
Theorem 10.1 it is proved.

High dimension of the molecule α - D – glucose is due to the fact that it contains a large number of differently oriented electronic atomic orbitals and, consequently, a large amount of energy. This is consistent with the established notions of large energy reserves in glucose, necessary for living organisms. Such an increase in energy and dimension occurs and other saccharides in the formation of closed loops.

In particular, the conformation of the β - D - glucose are interchanged only a hydroxyl group and a hydrogen atom bound to a carbon atom of the C (1) in Figure 6. Changes in the number of carbon atoms does not fundamentally change the picture of the molecule. The dimension of the polytope corresponds to the number of vertices of the polytope (not one less than the number of vertices).

When depicting the polytope in Figure 6, the assumption that the atoms of the carbon cycle are located in the plane was not used. Therefore, this image is suitable for any conformation of molecule α -D - glucose. In the model in Figure 6, the edges corresponding to chemical covalent bonds are indicated in red. The remaining edges have only the geometric meaning of creating a convex polytope. They are marked in black. In the aggregate, each vertex of a polytope (atom or functional group) is connected with all other vertices by edges. This means that the image in question is a simplex type polytopic. It is proved that the dimension of this polytope is 15 (Zhizhin, 2018).

To obtain a three - dimensional simplified model of the molecule α -D - glucose, in Figure 6 we leave only the edges corresponding to chemical covalent bonds, the edges of the external contour, and the edge connecting the $C_{(1)}$, $C_{(5)}$ atoms. We assume as a first approximation that the ring of carbon atoms $C_{(1)} \div C_{(5)}$ is a regular pentagon with a side length of 0.15 nm (the standard length of the carbon - carbon chemical bond). Since carbon atoms exhibit tetrahedral coordination, other valence bonds of carbon atoms must be located either above the plane of the ring or below the plane of the ring. It is convenient to assume for further geometric analysis that the atoms of hydrogen and the hydroxyl group located in Figure 6 inside the carbon ring are in space above the plane of the carbon ring. While the oxygen atom and the functional groups of the outer contour (larger) are under the plane of the carbon ring. A part of this spatial structure located under the plane of the carbon ring is shown in Figure 7 (Zhizhin, 2020 c).

Figure 7. The part of the glucose molecule located under the carbon ring

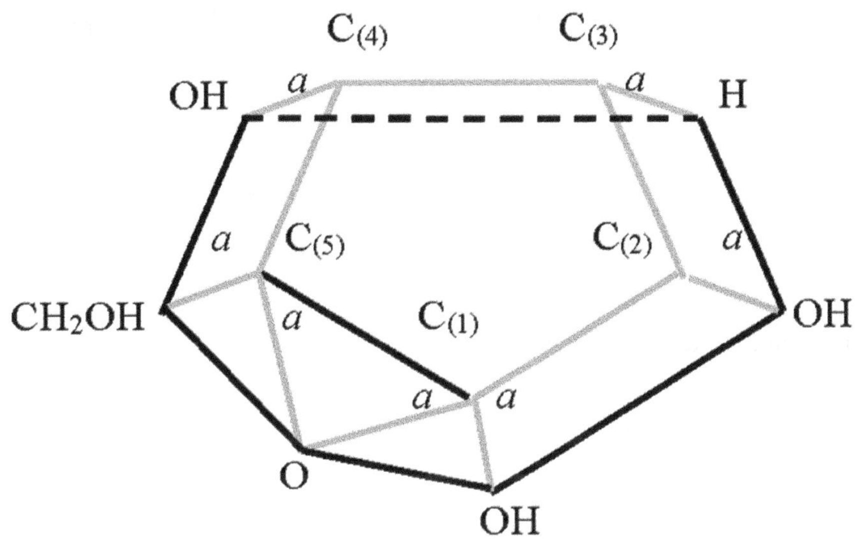

The length of the carbon – hydrogen and carbon – oxygen bonds is denoted by the letter $a = 0.1$ nm. Then the carbon - carbon bond length takes the form $1.5a$. Thus, the part of the construction of the α -D - glucose molecule under the carbon ring is a pyramid with a lower base in the form of an irregular hexagon and an upper base in the form of a regular pentagon.

To determine the yet unknown geometric characteristics of the figure in Figure 7, it is first necessary to find the value of the angle of inclination of the side edges of the pyramid to its base. For this, we consider separately the flat lateral faces of the pyramid in the vicinity of the oxygen atom. We double these faces and designate the vertices and some characteristic points of this structure in separate letters (Figure 8).

Figure 8. The enlarged part of the side surface of the pyramid (Figure 7) in the vicinity of the oxygen atom

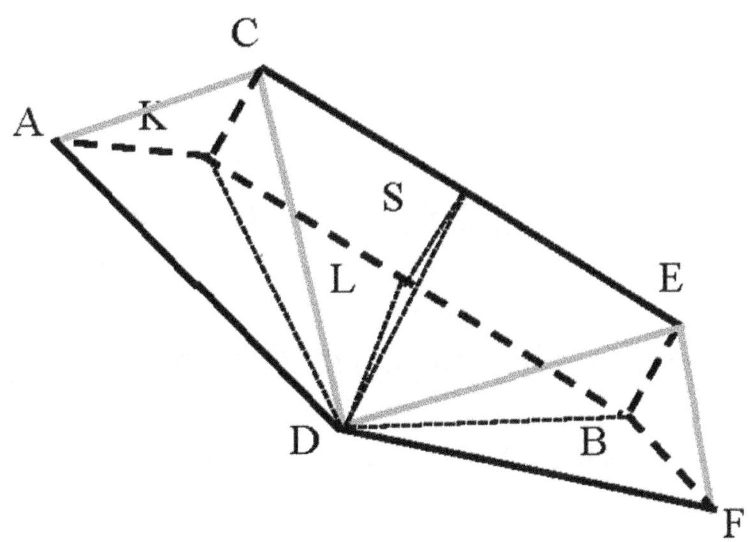

We denote the angle CAK by the letter φ. Given that Ð*AKC*=$\pi/2$ we have CK=*LS*=*BE*=*a* sin φ, so AC=*CD*=*DE*=*EF*=*a, AK*=*a* cosφ. Since the carbon cycle is a regular pentagon, then ÐAKL=$\pi/2$ – 54=126°. Since CE=1.5*a and* point S is *m*iddle of segment CE th*en*

$$DS = \sqrt{a^2 - (3/4)^2 a^2} = a\sqrt{7}/4.$$

From triangle DS*L we* have

$$DL = \sqrt{DS^2 - SL^2} = \frac{a}{4}\sqrt{7 - 16\sin^2\varphi}.$$

Therefore, from the triangle *KLD* we obtain

$$KD = \sqrt{DK^2 + KL^2} = a\cos\varphi = AK.$$

Then Ð*KDC*=φ. Besides,

$$\sin\angle DKL = \frac{DL}{KD} = \frac{\sqrt{7 - 16\sin^2\varphi}}{4\cos\varphi}.$$

Therefore,

$$\angle AKD = \angle AKL - \angle DKL = 126^o - \arcsin\frac{\sqrt{7 - 16\sin^2\varphi}}{4\cos\varphi}.$$

From the triangle *KAD* we obtain

$$\frac{AD}{2a\cos\varphi} = \sin\frac{\angle AKD}{2}.$$

From here we find

$$AD = 2\cos\varphi\sin(63^o - \frac{1}{2}\arcsin\frac{\sqrt{7 - 16\sin^2\varphi}}{4\cos\varphi}). \qquad (10.6)$$

Besides, from four-angle *AKLD* we have *KN/AK*=*cos (126º -90º).* Then,

$$KN = a\cos\varphi\cos 36^o = LM.AM = AN + NM = a\cos\varphi\sin 36^o + \frac{3}{4}a.$$

Therefore, $MD = DL - LM = \frac{a}{4}\sqrt{7-16\sin^2\varphi} - a\cos\varphi\cos 36^o$. Thus, we find also one express for segment AD

$$AD = \sqrt{MD^2 + AM^2} = \left[(\frac{a}{4}\sqrt{7-16\sin^2\varphi} - a\cos\varphi\cos 36^o)^2 + (a\cos\varphi\sin 36^o + \frac{3}{4}a)^2 \right]^{1/2}. \quad (10.7)$$

Comparing (10.6) and (10.7) after transformations, we obtain the equation for finding the angle φ for any a

$$\sin(63^o - \frac{1}{2}\arcsin\frac{\sqrt{7-16\sin^2\varphi}}{4\cos\varphi}) = \left[\frac{1}{2} + \frac{3}{8}\frac{\sin 36^o}{\cos\varphi} - \cos 36^o \frac{\sqrt{7-16\sin^2\varphi}}{8\cos\varphi} \right]^{1/2}. \quad (10.8)$$

Solving equation (10.8) numerically, we find that the angle $\varphi = 36o$. Thus, the height of the lower part of the α -*D* - glucose molecule is a sinφ= 0.058778 nm, the characteristic size of the upper base is 0.243 nm, and the characteristic size of the lower base is 0.243+2a*cos*φ= 0.405 nm.

The angle of inclination of the side edges of the lower part of the model, as follows from the solution, is different from 90. This reflects in averaged form the presence of atoms in the conformations of the α -*D* - glucose molecule located quite close to the equatorial plane of the molecule (Metzler, 1980; Lehninger, 1982; Koolman & Roem, 2013).

The upper part of the α -*D* - glucose molecule above the carbon cycle should be depicted in the form of a pentagonal prism, directing the volatile bonds from the carbon atoms of the cycle vertically upward. This is consistent with the tetrahedral coordination of atoms around a carbon atom. Thus, the α -*D* - glucose molecule in a simplified three - dimensional form is a pyramid in contact with the carbon cycle with hexagonal and pentagonal bases and a straight pentagonal prism. If the hydroxyl group bonded to the atom $C(_{1)}$ lies on the pentagonal base of the prism, then a simplified image of the α -*D* - glucose molecule (α- anomer) shown in Figure 9

Figure 9. The spatial simplified images of the α -D - glucose molecule (α- anomer)

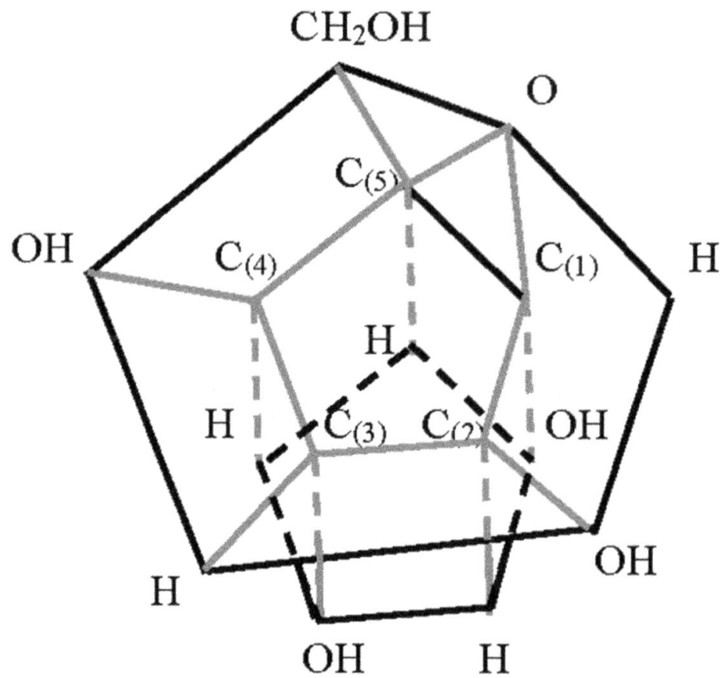

In the molecules of other pyranoses, the atoms or functional groups partially connected to the carbon atoms of the carbon cycle partially change. In this case, the structure of the molecules, up to functional groups, remains the same for glucose molecules in the pyranose cycle carbon atoms. Therefore, the dimension of molecules such as *D* - mannose, *D* - galactose, *L* - fucose is also equal to 15. Three - dimensional models of these molecules will retain their appearance, images are shown in Figure 9.

Monosaccharide Chains

Monosaccharide molecules combine with each other thanks to the combination of two hydroxyl groups with the release of a water molecule. The remaining oxygen atom connects the remaining part of the monosaccharides. The most common chains of *α -D* - glucose monosaccharide residues. In some cases, the carbon atom $C_{(2)}$ in the cycle will be involved in the compound. Branching chains of glucose molecules occurs at the sixth carbon atom in the functional group CH_2OH. Using the obtained simplified three - dimensional models of *α -D* - glucose molecules (*α*- anomer), we consider in more detail these compounds in a metric image, taking into account the angles between the valence bonds.

Let two *α* - anomers of the *α -D* - glucose molecule join together, linking the carbon atom $C_{(1)}$ of the first *α -D* - molecule to the carbon atom $C_{(4)}$ of the second *α -D* - molecule (maltose). Based on the geometric image in Figure 9, you can imagine a top view of this connection (Figure 10).

Figure 10. Projections of α - anomer disaccharides

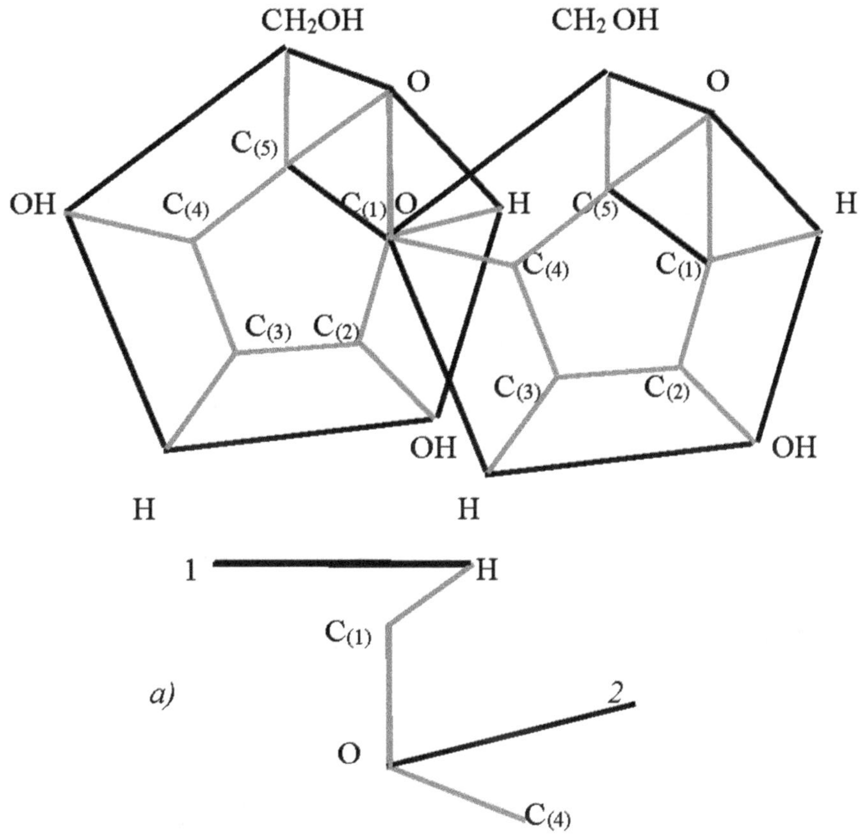

It is easy to see that the same attachment of the third *α -D* - glucose molecule leads to a nonlinear arrangement of molecules when viewed from above on this compound. Nonlinearity will save with the following similar connections. Thus, the chain of molecules turns to the right. However, given the three - dimensionality of the image, you should look at this connection from a different view. Figure 10 a) shows an image of the junction of two *α -D* - glucose molecules when viewed from the front. The black segments in Figure 10 a) are the traces of the intersection of the upper bases of the prism with a plane passing through the valence bonds of the oxygen atom and carbon atoms. According to the obtained solution of equation (10.8), the angle between the base of the prism and the valence bond is 36 degrees both the first *α -D* - glucose molecule and the second *α -D* - glucose molecule. Thus, on the Figure 10 a the angle 1H $C_{(1)}$ is 36 degrees and the angle $C_{(1)}OC_{(4)}$ is 112 degrees. Since, the angle $C_{(4)}O2$ is 36 degrees too, so the angle between bases 1H and O2 is 14 degrees. This means that the base of the second *α -D* - glucose molecule has a consequently slope to the base of the first *α -D* - glucose molecule to the left. It is clear that such an addition of a third *α -D* - glucose molecule will increase the angle of inclination of this molecule relative to the first molecule. Thus, chains of *α* - anomers of *α -D* - glucose molecules rotate simultaneously in opposite directions in perpendicular planes.

In the case of a chain linking of two β-D-glucose molecules to carbon C_1, in accordance with the Hewors' formulas (Figure 5), the hydroxyl group OH will be attached to the upper part of the spatial image of the molecule. This hydroxyl group can combine with a hydroxyl group attached to carbon atom C_4 of another molecule β - D-glucose (with the elimination of a water molecule), forming a linear chain.

The Dimensions and Spatial Images of Molecules of Deoxyribose

The deoxyribose (or the *D* – ribose) is molecule of carbohydrate with five atoms carbon. There are two enantiomer forms of this molecule. Its images in the form of Fisher's formula (Metzler, 1980) are shown in Figure 11 A, B.

Figure 11. A;B- Two enantiomer forms of the molecule D – ribose (formulas of Fisher`s)

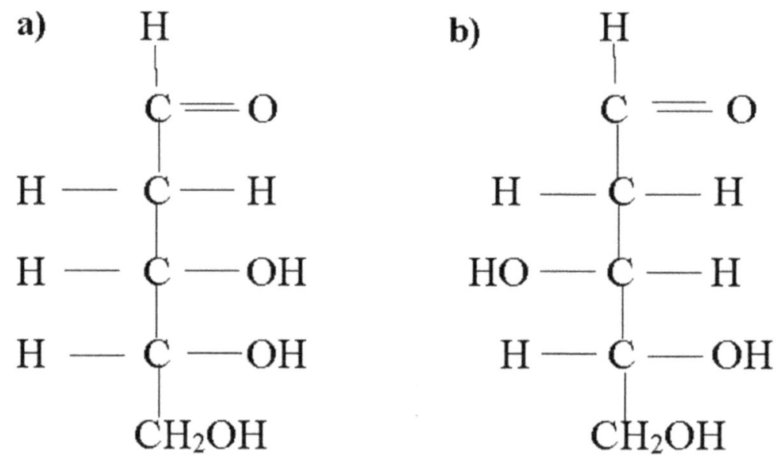

In the DNA molecule, *D* - ribose as a component enters with a closed carbon chain. The closed chain of the molecule in Figure 11 is formed due to the rupture of the double bond between oxygen and carbon (considering it the first in the chain), the liberation of the water molecule, and the addition of this oxygen atom to the penultimate (fourth) carbon atom in the carbon chain. The Fischer formula or formula Haworth and their modifications (Metzler, 1980; Lehninger, 1982) may not reflect the spatial structure of the *D* – ribose molecule. For this target the constructs described in the form of convex polytopes with boundary elements which form boundary complex (Grunbaum, 1967). Only when such a representation will be to determine the dimension of these molecules (Zhizhin, 2016; 2018). Figure 12 shows the result of the closure of a chain of four carbon atoms through an oxygen atom correspondent Figure 11 A. Chemical bonds are marked as before with red color.

Figure 12. The molecule D – ribose with closed a chain of carbon atoms (the form A)

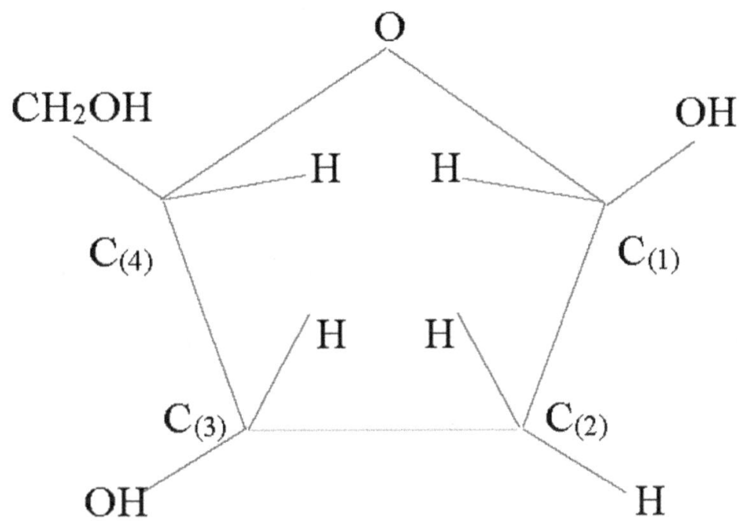

It is easy to see that, in addition to the form of the *D* - ribose molecule in Figure 12 (a form of *A*), there is another enantiomer form of this molecule. It connected by a symmetry transformation with respect to a straight line passing through an oxygen atom separating the projection of the form of *A* in half (we shall assume that it is a *B* shape). Form *B* of the *D* - ribose molecule is shown in Figure 13.

Figure 13. The molecule D – ribose with closed a chain of carbon atoms (the form B)

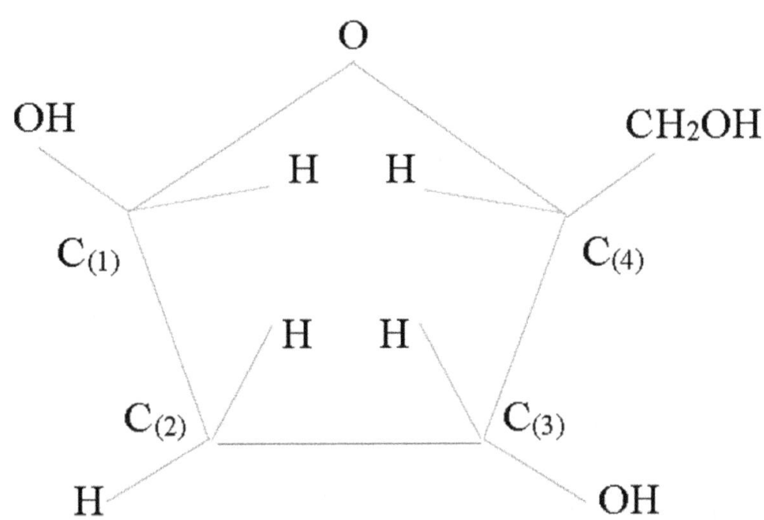

The ribose is molecule of carbohydrate with five atoms carbon. There are two enantiomer forms of this molecule. Its images in the form of Fisher's formula (Metzler, 1980) are shown in Figure 14 A, B.

Figure 14. A;B- Two enantiomer forms of the molecule ribose (formulas of Fisher`s)

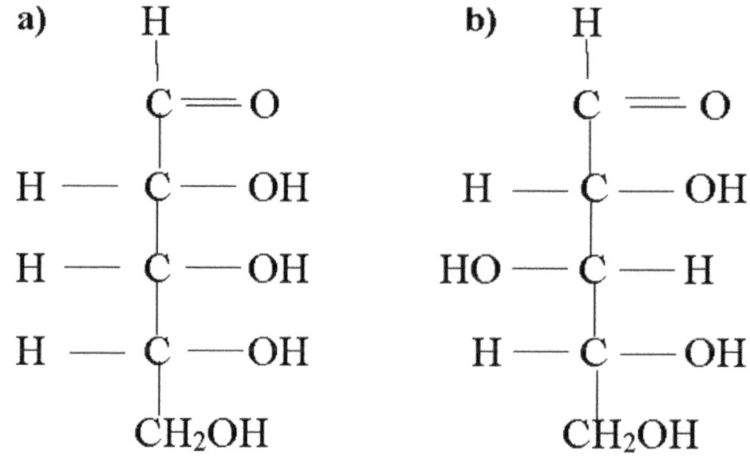

In the RNA molecule, ribose as a component enters with a closed carbon chain. The closed chain of the molecule in Figure 14 is formed due to the rupture of the double bond between oxygen and carbon (considering it the first in the chain), the liberation of the water molecule, and the addition of this oxygen atom to the penultimate (fourth) carbon atom in the carbon chain. Figure 15 shows the result of the closure of a chain of four carbon atoms through an oxygen atom correspondent Figure 14 A. Chemical bonds are marked as before with red color.

Figure 15. The molecule ribose with closed a chain of carbon atoms (the form A)

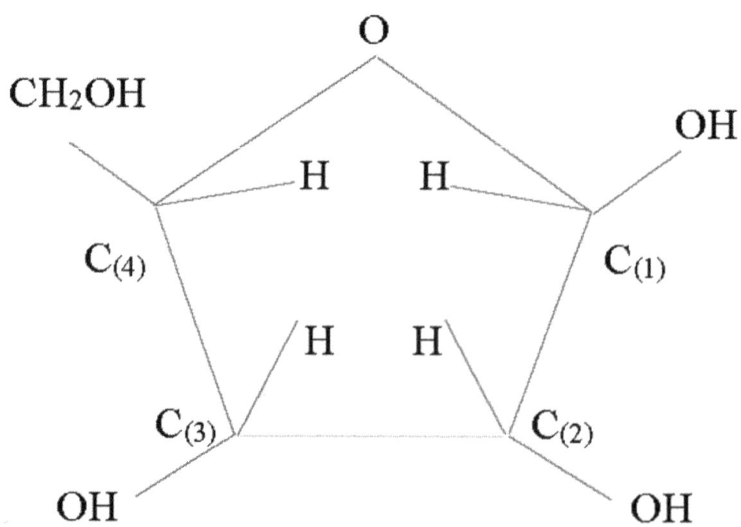

It is easy to see that, in addition to the form of the ribose molecule in Figure 14 (a form of *A*), there is another enantiomer form of this molecule. It connected by a symmetry transformation with respect to a straight line passing through an oxygen atom separating the projection of the form of *A* in half (we shall assume that it is a *B* shape). Form *B* of the ribose molecule is shown in Figure 16.

Figure 16. The ribose molecule with closed a chain of carbon atoms (the form B)

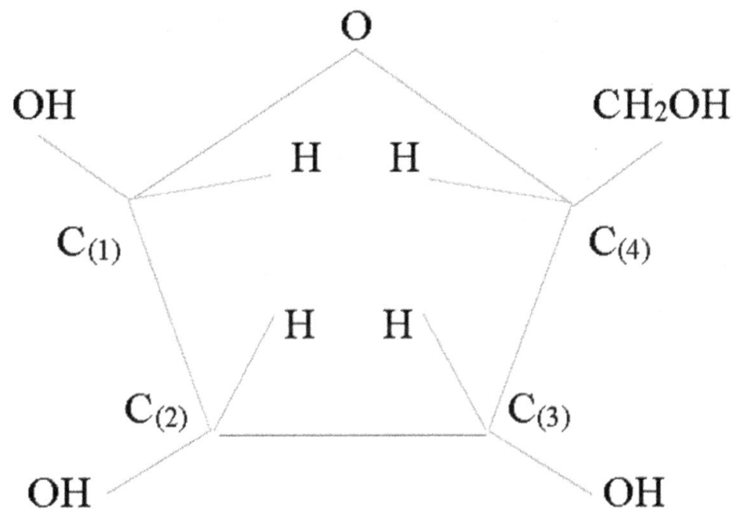

To determine the dimension of the molecules *D* - ribose and ribose of both enantimorphic forms (A, B), it is necessary to connect each vertex (each atom or functional group) with all other vertices of a given molecule. Thus, we received the polytope on Figure 17.

Figure 17. The five -carbon molecule sugar

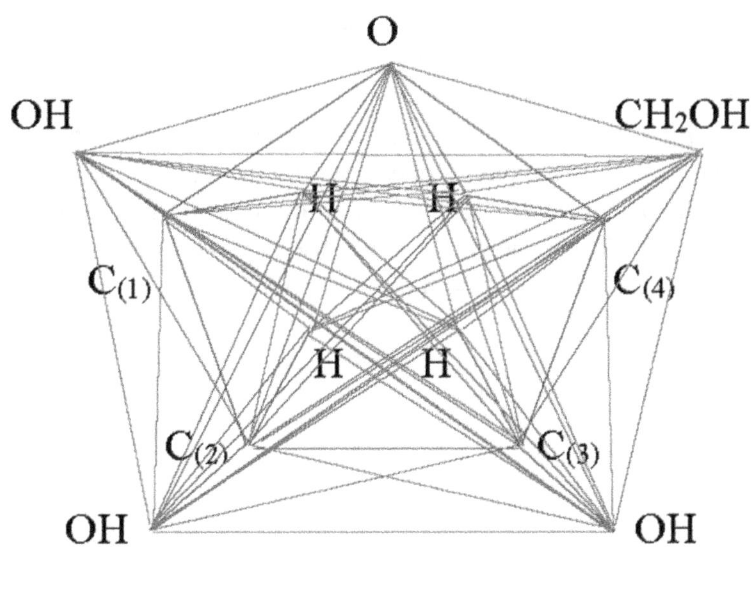

The polytope at Figure 17 has 13 vertices in which there are one oxygen atom, four hydrogen atoms, three hydroxyl groups, four carbon atoms, compound CH_2OH. It is known that the dimension of a simplex is one less than the number of its vertices (Grünbaum, 1967). Therefore, the dimension of the polytope in Figure 17, if the Euler - Poincaré equation (10.1) holds for it, equals $n=f_0-1=12$. The simplex is a simplicial polytope, and all its faces therefore are simplexes. Since all the vertices of a simplex are related to each other, the number of edges is determined by the number of combinations of the number of vertices of two. In this case, the number of edges of the polytope in Figure 17 is $f_1 = C_{n+1}^2 = 78$. The number of two - dimensional faces (triangles) of this polytope is $f_2 = C_{n+1}^3 = 286$.

The number of three - dimensional faces (tetrahedrons) of this polytope is $f_3 = C_{n+1}^4 = 715$.
The number of four – dimensional faces (4 – simplexes) of this polytope is $f_4 = C_{n+1}^5 = 1287$.
The number of five – dimensional faces (5 – simplexes) of this polytope is $f_5 = C_{n+1}^6 = 1716$.
The number of six – dimensional faces (6 – simplexes) of this polytope is $f_6 = C_{n+1}^7 = 1716$.
The number of seven – dimensional faces (7 – simplexes) of this polytope is $f_7 = C_{n+1}^8 = 1287$.
The number of eight – dimensional faces (8 – simplexes) of this polytope is $f_8 = C_{n+1}^9 = 715$.
The number of nine –dimensional faces (9 – simplexes) of this polytope is $f_9 = C_{n+1}^{10} = 286$.
The number of ten – dimensional faces (10 – simplexes) of this polytope is $f_{10} = C_{n+1}^{11} = 78$.
The number of eleven – dimensional faces (11 – simplexes) of this polytope is $f_{11} = C_{n+1}^{12} = 13$.

Substituting the values f_i(o£*i*£11) in the Euler - Poincaré equation (10.1), can see that it holds for n = 12 for polytope on Figure 17

$$13 - 78 + 286 - 715 + 1287 - 1716 + 1716 - 1287 + 715 - 286 + 78 - 13 = 0.$$

This finally proves that the polytope *D* - ribose (*A*) has dimension 12 and a simplex type. Obviously, the polytope *D* - ribose (*B*) has the same dimension and type.

The image in Figure 17 is a topological projection of polytope 12 - simplex on a two - dimensional plane. If a hydroxyl group OH is attached to the carbon atom $C_{(2)}$ on the outer contour of the image, this corresponds to a five - carbon sugar molecule (ribose B) in the RNA molecule. If a hydrogen atom H is attached to the carbon atom $C_{(2)}$ on the outer contour of the image, this corresponds to a five - carbon sugar molecule (D – ribose B) in the DNA molecule. If to simplify the image in Figure 17, discard all the blue lines that have only the geometric meaning of the edges of the convex polytope, with the exception of the outer contour, the image in Figure 18 will be obtained.

Figure 18. Simplified image of a molecule of five - carbon sugar (ribose B)

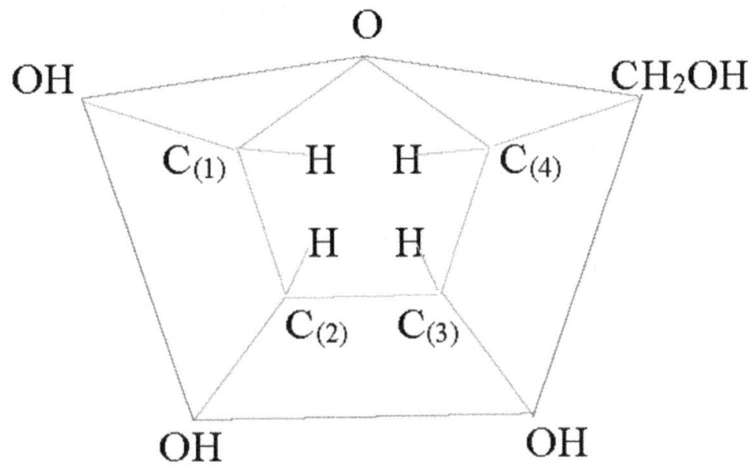

Since Figure 18 is a topological projection of a spatial figure, then by this projection it is possible to restore qualitatively the corresponding three - dimensional figure, the projection of which is Figure 18. If we take the carbon – carbon bond length and carbon – oxygen bond $a = 0.15$ nm, then the length of the edges in the carbon – oxygen pentagonal cycle is equal to a. One can assume that the chemical bonds emanating from the four carbon atoms of this cycle are spaced symmetrically relative to the chemical bonds of this cycle and have a certain angle φ with the plane of the cycle. Since the number of valence bonds emanating into the space of four carbon atoms of the oxygen - carbon cycle is eight, it should be assumed from the symmetry condition that four valence bonds extend into a half space above the cycle plane, and four valence bonds proceed into the half space under the cycle plane. Consider a half - space above the plane of the oxygen - carbon cycle (Figure 19).

Figure 19. The half - space above the plane of the oxygen - carbon cycle

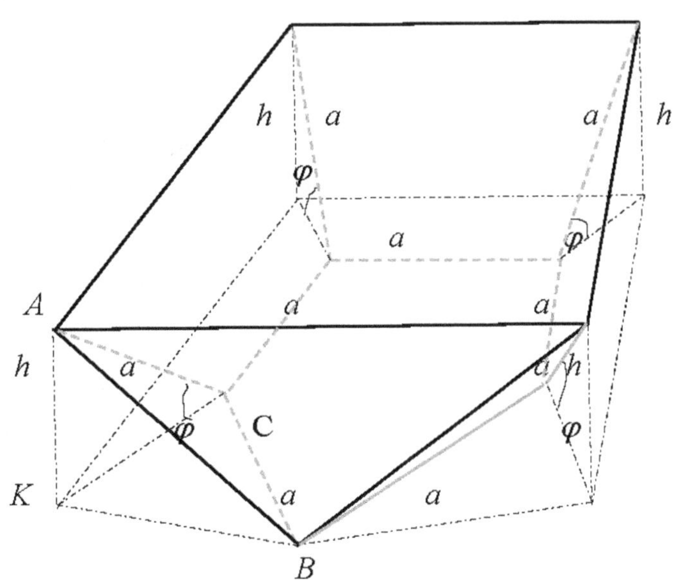

Then the length of the projection of these links on the plane of the cycle is equal to a cosφ, and the height of the figure is equal to h=a sinφ. In Figure 19, the orange segments are (solid and dashed) valence bonds, their length is assumed to be a. Fat solid black segments together with orange segments (dotted and solid) define a part of the three - dimensional shape of the sugar molecule.
In this figure, the lower pentagonal face and the upper quadrilateral face are parallel to each other.

In Figure 19, the angle φ is unknown. We define it from two triangles A*CB* and A*KB,* equating the length of the segment A*B* from each triangle. Note that the angle A*CB* linearly depends on the angleφ: ĐAC*B* = $\alpha\varphi+\beta$. One take into account that the angle KCB = *126°* due to the correctness of the lower pentagonal base of the polyhedron and the division of the external angle at the vertex C by *t*he segment KC in *h*alf. Since ĐACB = *126°* for φ=0 and ĐACB = *90°* for φ=90°, so ĐACB = $126^0 - \frac{2}{5}\varphi$. *Since* ACB tr*ian*gle is isosceles, so

$$AB = 2a\sin(63^0 - \frac{1}{5}\varphi). \qquad (10.9)$$

From triangle KCB *we* have $KB = a\sqrt{1+\cos^2\varphi - 2\cos\varphi\cdot\cos126^0}$. Since the AK*B t*riangle is rectangular, so

$$AB2^{=}a2^2(1 - \cos\varphi \bullet \cos126°). \qquad (10.10)$$

From (10.9) and (10.10) we have

$$1-\cos\varphi\cdot\cos126^0 = 2\sin^2(63^0 - \frac{1}{5}\varphi). \qquad (10.11)$$

This is the equation for finding the φ angle. The equation (10.11) we have two roots (Zhizhin, 2019 b): φ=0 and φ=90°. The first root corresponds to the flat shape of the sugar molecule. This does not correspond to the earlier proof of the multidimensionality of this molecule. One can want to find a three - dimensional view of the outer contour of the sugar molecule. This goal corresponds to the second root of equation (10.11), i. e. φ=90°. The second part of the three – dimensional figure is under flat of oxygen – carbon cycle. Obviously, here also the valence bonds emanating from carbon atoms are located at an angle φ=90° to the cycle plane. The parts below the cycle plane and above the cycle plane together form a three - dimensional model of the sugar molecule ribose B (Figure 20).

Figure 20. The three – dimensional model of the sugar molecule ribose B

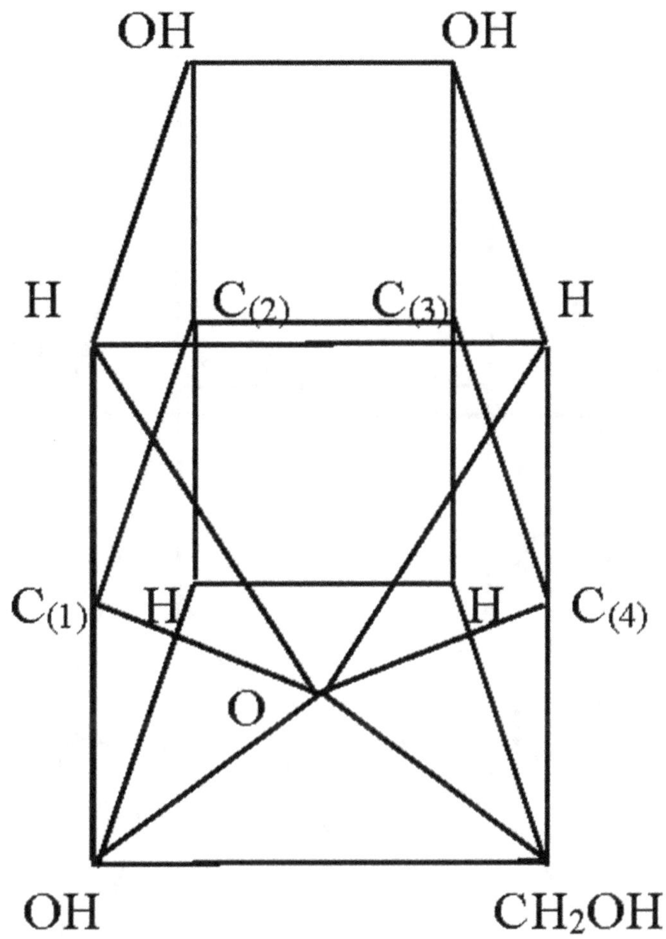

The resulting model of a five - carbon sugar molecule can be viewed as a development of the Haworth formula, since in the Haworth formula and the resulting sugar model, the valence bonds are perpendicular to the plane of the pentagonal cycle. The essential difference from Haworth's formula, however, is that the resulting model of a five - carbon sugar molecule is a convex closed polyhedron.

Similarly, three - dimensional models of five - carbon sugar molecules can be obtained ribose A (Figure 21), D - ribose A (Figure 22), D - ribose B (Figure 23).

Figure 21. The three – dimensional model of the sugar molecule ribose A

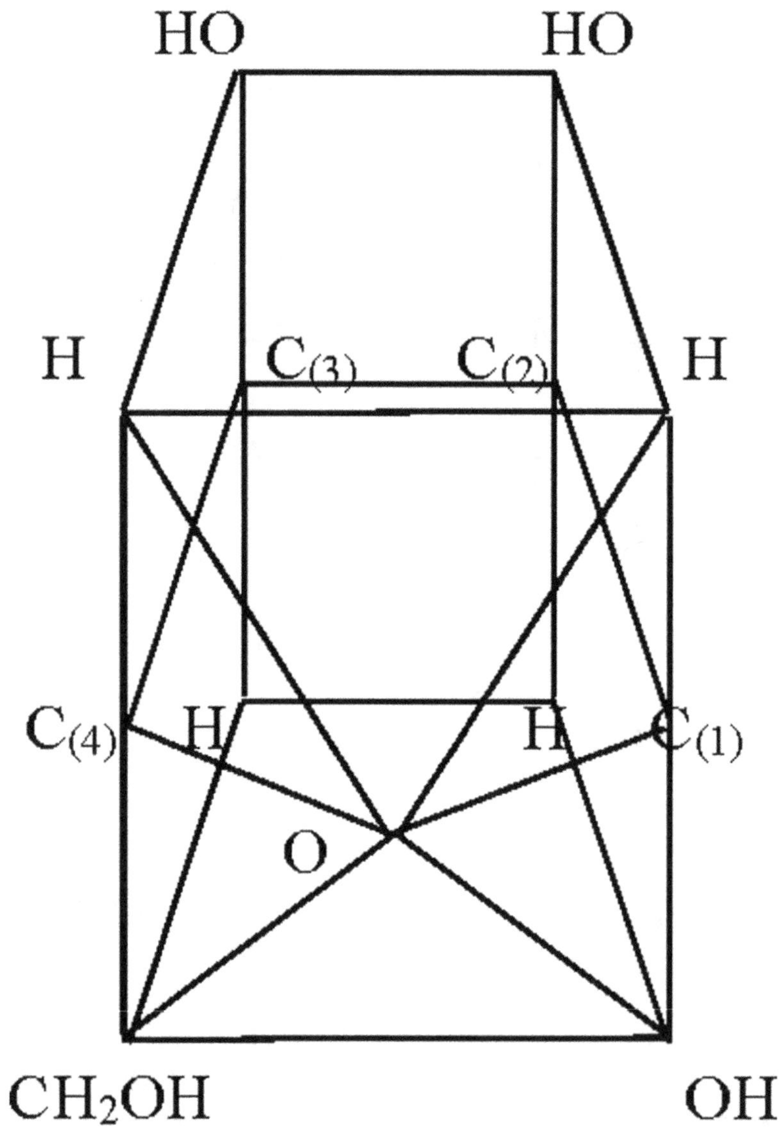

Figure 22. The three – dimensional model of the sugar molecule D - ribose A

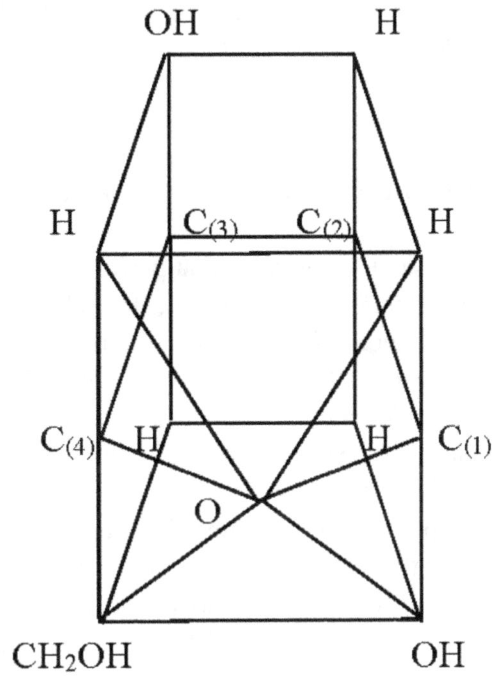

Figure 23. The three – dimensional model of the sugar molecule D - ribose B

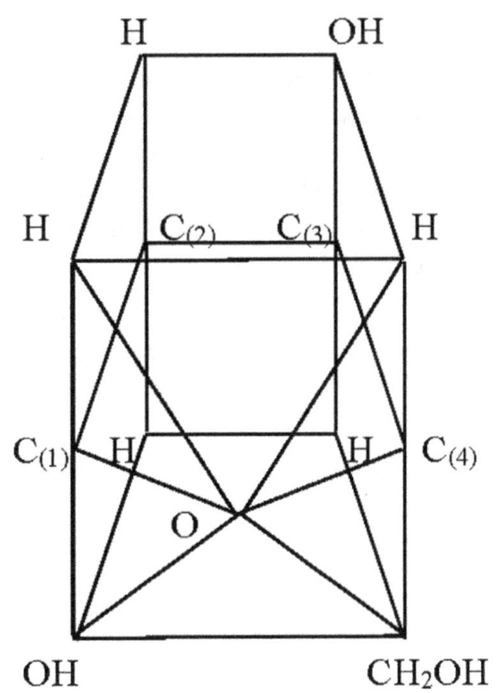

The arrangement of atoms in the sugar molecule in Figures 20 – 23 is determined by their arrangement in the previously obtained Figures 11 – 19. These options for the arrangement of atoms and functional groups in three - dimensional models of five - carbon sugar molecules will be taken into account when analyzing the possibilities of sequential connection of five - carbon sugar molecules and phosphoric acid residues in the formation of a chain of nucleic acids.

The Dimensions and Spatial Images of Protein Molecules

Every living organism contains thousands of proteins that perform various functions. They are all made up of chains of amino acids. Studying their structure is an extremely important task. Back in the middle of the last century, the diffraction patterns of crystals of amino acids and simple di- and tri- peptides were studied (Pauling, Corey & Branson, 1951; Pauling & Corey, 1950, 1953; Corey & Pauling, 1953; Dickerson & Geis, 1969). The characteristic lengths of valence bonds and the values of bond angles were determined, and the existence of repeating structural units was established. Based on the results of these studies, three-dimensional models of protein structures in various conformations were built. They have found widespread use up to the present time (Metzler, 1980; Lehninger, 1982; Koolman & Roehm, 2013). However, it has recently been found that amino acid molecules are of the highest dimension (Zhizhin, 2016, 2018). In this paper, the spatial models of the conformations of polymeric protein molecules are considered taking into account the highest dimension of protein molecules using the metric data obtained in the previous period. It is shown that the simplifying assumptions made in the works of Pauling joint authors contradict the results of the analysis of the spatial structures of protein molecules. Rejection of these simplifying assumptions leads to fundamentally different structures of the conformations of protein molecules.

Dimensions of Protein Molecules

Amino acids are chemical compounds that simultaneously include a carboxyl group COO^- and an amine group H_3N^+. These groups are covalently linked to a carbon atom (α - carbon), to which a hydrogen atom H and a side chain R are also covalently attached (Koolman & Roehm, 2013). Amino acids have two enantiomorphic forms L and D, which are Fischer planar views shown in Figure 24. From Figure 24 it can be seen that the two enantiomorphic forms are related by mirror image relative to the line passing through the amine and carboxyl groups.

Figure 24. Enantiomorphic forms of the amino acid molecule
a) L - amino acid, b) D - amino acid

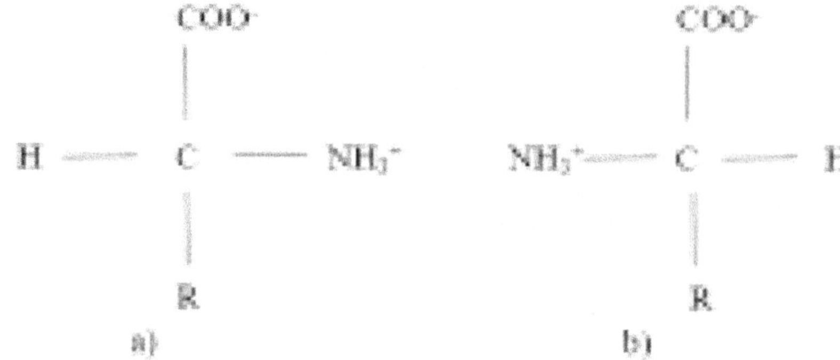

The side chains in Figure 24 have 20 variants in classic amino acids (Koolman & Roehm, 2013). They determine the properties of the corresponding amino acids. Fisher's projections do not reflect the spatial structure of amino acids, although it is clear that these are spatial objects. Spatial images of enantiomorphic amino acids (Zhizhin, 2016) are presented in Figure 25.

Figure 25. Spatial images of amino acids
a) L - amino acid, b) D - amino acid

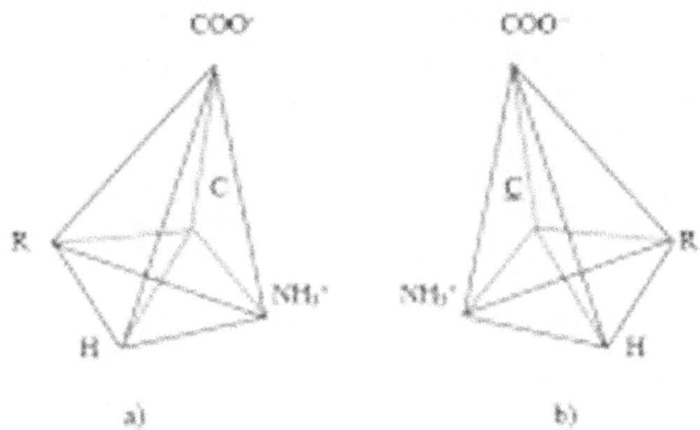

In Figure 25, covalent bonds are indicated in red. The rest of the segments define the spatial shape of the molecule, they are marked in black.

From Figure 25 it follows that the amino acid molecule is a tetrahedron with a center. The dimension of such a polyhedron can be determined by the Euler-Poincaré equation (3.3) (Poincaré, 1895)

From Figure 25 it follows that in any of the two enantiomorphic forms of the amino acid molecule, the number of vertices is 5, the number of edges is 10, the number of two-dimensional faces is 10, the number of three-dimensional faces is 5. In this case $f_0=5, f_1=10, f_2=10, f_3=5$.
Substituting the obtained values f_i, $i=0\div 3$ into equation (3.3), we obtain,

$$5 - 10 + 10 - 5 = 0.$$

i.e. Euler-Poincaré's equation for a tetrahedron with a center holds for $n = 4$. This proves that a tetrahedron with a center (amino acid molecule) has dimension 4.

Linear Polypeptide Chain Structure

The study of diffraction patterns of crystals of amino acids (Pauling, Corey & Branson, 1951; Pauling & Corey, 1950, 1953; Corey & Pauling, 1953) made it possible to establish some values of bond bonds and bond angles in an amino acid. However, the interpretation of the spatial arrangement of atoms in an amino acid requires additional instructions. An example of the possibility of ambiguous interpretation of diffraction patterns can be found in numerous works on the description of quasicrystals. In particular, the apparent icosahedrical symmetry of the diffraction patterns of quasicrystals was proposed to be explained (Pauling, 1987) by multiple twinning of cubic crystals. Pauling based his arguments on the radial arrangement of the spots in the diffractograms. However, high-resolution photomicrographs refuted this assumption (Gratias & Cahn, 1986). The statement about the absence of translational symmetry in quasicrystals (Shechtman, et al., 1984) was constructed using the representations of the three-dimensional geometry of Euclid. Later this statement was refuted (Zhizhin, 2013 a, b, 2014; Zhizhin & Diudea, 2016). It was proved that translational symmetry in the diffractograms of quasicrystals appears if the diffractograms are considered as projections of structures from the space of higher dimensions. When studying diffraction patterns of amino acid crystals in the works of Pauling joint authors the values of a number of parameters were not determined, for example, bond lengths R – C depending on the species R, bond angles between carbon atoms and nitrogen atoms, between hydrogen atoms and R, etc. To compensate for these disadvantages, it was proposed to use a simplifying assumption about the arrangement of a group of atoms of a peptide bond and centers of tetrahedrons in a plane (Zhizhin, 2020 e) (Figure 26)

Figure 26.

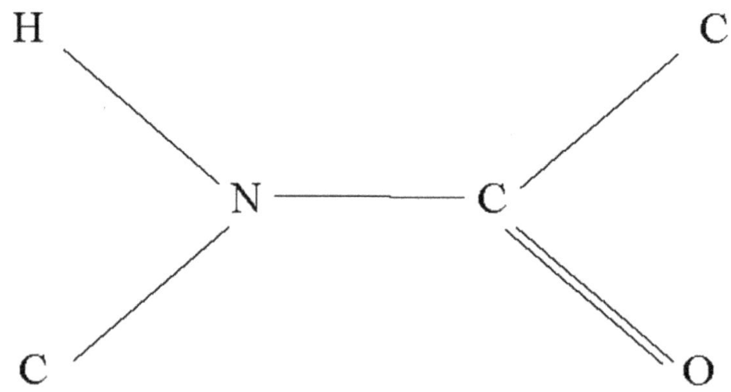

Under this assumption, knowledge of the indicated quantities was not required. It should be noted that in subsequent publications by other authors (Metzler, 1980; Lehninger, 1982; Koolman & Roehm, 2013) with a reference to Pauling's work, it was argued that the arrangement of these atoms in the plane was proved in Pauling's works. But this is not so, to be convinced of this it is enough to refer to these works. This unproven statement became the basis for the construction of three-dimensional models of protein molecules of various conformations in Pauling's works. Let us consider this issue in detail when constructing spatial structures of a linear polypeptide chain. In this case, the atoms of amino acid molecules must be located on a set of parallel lines. Then the polypeptide chain of two molecules of L- amino acids, taking into account the four-dimensionality of amino acid molecules, has the form shown in Figure 27.

Figure 27. Linear polypeptide chain of L – molecules amino acids

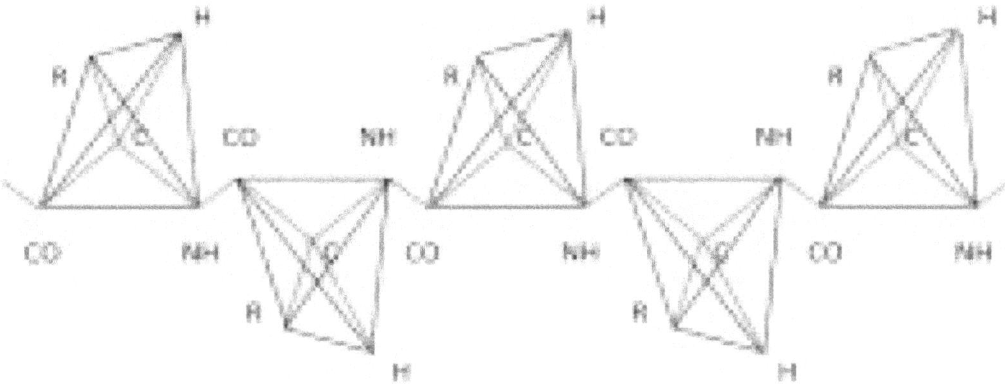

In the case of a D - amino acid, such a polypeptide chain is shown in Figure 28.

Figure 28 Linear polypeptide chain of D – molecules amino acids

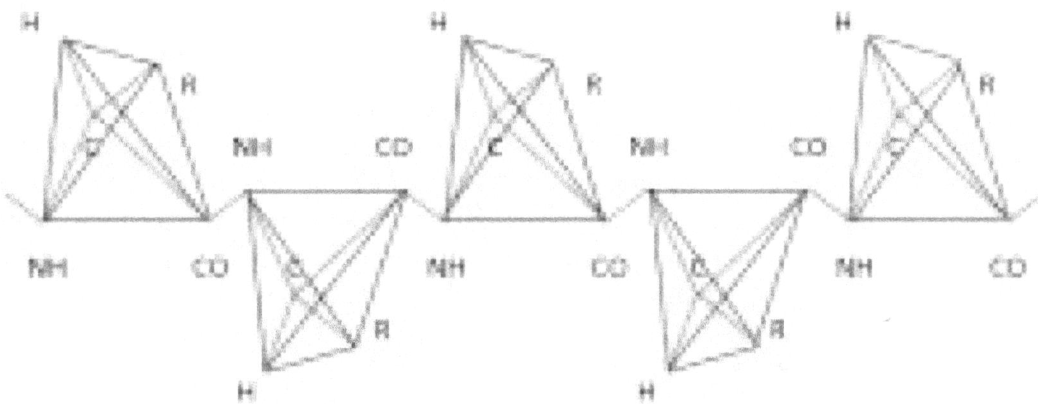

From Figure 27 and Figure 28 it can be seen that in order for the corresponding vertices of tetrahedrons with the center (i.e., specific atoms or functional groups) could lie on a system of parallel lines (i.e., form a linear polypeptide chain), it is necessary that the amino acid molecules in the chain alternate with their mirror image relative to the edge in the tetrahedron connecting the bond centers of the polypeptide chain (i.e., relative to the segment CO – NH). The chains in Figure 27 and Figure 28 differ by the mirror reflection of the molecules relative to the perpendicular to the segment CO – NH (L- and D - molecules).

In Figures 27 and 28, you can see the translational symmetry of the circuits. Moreover, the elementary translational element in this case is a group of two linked amino acid molecules, one of which is a mirror image of another amino acid molecule relative to the segment. From these two linked amino acid molecules, you can create a convex polytope by connecting edges of each vertex of any of the molecules with all other vertices in the group. This is how a simplex polytope of dimension 9 is formed, since the dimension of a simplex polytope is one less than the number of vertices in a simplex (Zhizhin, 2019 a), and the number of vertices in two tetrahedra with a center is 10. Therefore, a polypeptide linear amino acid chain has translation symmetry, in which an elementary translation element is a simplex polytope of dimension 9. In Figures 27 and 28, as well as in Fischer's projections (Figure 24) and in spatial images of molecules (Figure 25), in accordance with the definition of functional dimension, functional groups R, CO, NH were used. If we open the images of functional groups CO, NH and accept Pauling's assumption about their arrangement in the plane, then we will have a sequence of their alternation as shown in Figure 29 in the case of D-molecules of amino acids.

Figure 29. The location of the edges of peptide bonds in the chain of D-molecules of amino acids

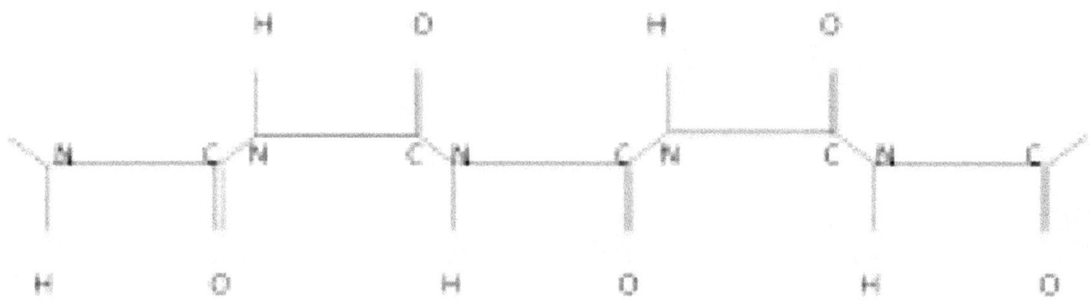

However, it is essential (see Figure 29) that the α-carbon atoms do not lie in the same plane with the peptide bond atoms. From a comparison of Figures 27 – 29 it follows that α-carbon atoms lie either above the plane in which the atoms of the peptide bond are located, or below this plane, and can never lie on this plane. Therefore, Pauling's idea that α-carbon atoms lie in the same plane with functional groups CO, NH is erroneous. Therefore, Pauling's approximation cannot be used both in the construction of linear polypeptide chains and in the construction of more complex conformations of amino acid molecules.

Spiral Polypeptide Chains

In the previous section, it was proved that the centers of tetrahedrons cannot lie on a flat sheet in which the peptide bond is located. Therefore, Pauling's idea of the connection of such flat leaves at the centers of tetrahedrons to form a helical peptide chain, as Pauling did, cannot be used. There is a relatively simple way of forming a helical polypeptide chain based on the concept of the four-dimensionality of amino acid molecules. This method was proposed in 2016 (Zhizhin, 2016). Suppose another amino acid molecule (not in a mirror image) is attached to an amino acid molecule (tetrahedron with a center) by a peptide bond C – N (Figure 30) so that in the general case between the valence bond C – C of the first molecule and the peptide bond C – N there is a certain angle less than 180 degrees (since there is no reason to assert that one of these connections is an exact continuation of the other).

Figure 30. Spiral polypeptide chains

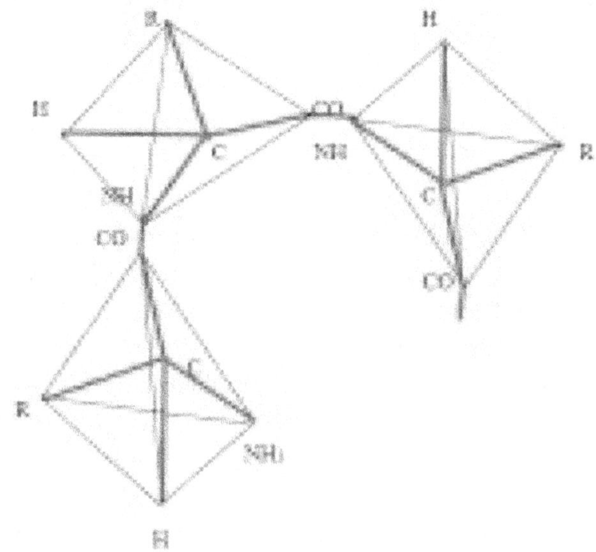

Therefore, the second amino acid molecule is rotated by some angle relative to the first amino acid molecule. At the same time, while maintaining the shape of the second amino acid molecule, the second peptide bond, with the help of which the third amino acid molecule can be attached, turns out to be significantly (almost 90 degrees) rotated with respect to the first peptide bond. The third peptide bond is thus almost antiparallel to the first peptide bond. There is a rotation of the attached amino acid molecules relative to some center O (Zhizhin, 2016, 2018). If we designate the length of the peptide bond "*a*", the length of the edge of the tetrahedron between the atoms C and N "*b*", then in the projection onto the plane perpendicular to the axis of rotation, we get a polygon (Figure 31), which corresponds to the particular with the period of rotation of the molecules equal to 4.

Figure 31. Rotation of peptide bonds

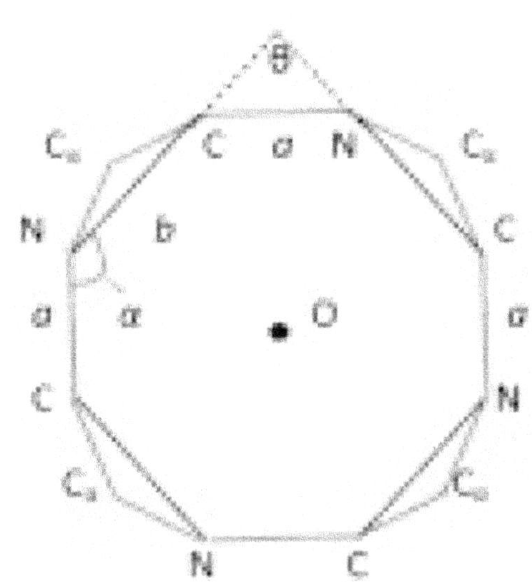

If we continue the nearest sides of the polygon *b* until they intersect with each other, we find the angle between these extensions. This is the angle θ=180–2(180–α), α is the angle between sides a and b. From here we find the connection between the period n and the angle α

$$n\theta = 360, \alpha = \frac{1}{2}(\frac{360}{n} + 180).$$

If the period $n = 4$ so $\alpha = 135^0$. Pauling had a minimum period of 3.6, but it should be noted that the rotation period of an amino acid molecule can only be an integer. If we now assume that the lines connecting the centers of the tetrahedrons do not lie in the plane of rotation, but have a certain angle with respect to this plane, then we get a helicoid of rotation of the amino acid molecule (i.e., α - helix, since α -carbon atoms are located at the centers of the tetrahedrons). It can be shown that if we use the values of the valence bond lengths and the bond angle determined in Pauling's works, then to achieve the distance between the turns of 0.54 nm in the spiral (adopted in Pauling's works), the required value of the angle between the line connecting the centers of the tetrahedra and the plane of rotation, is equal to 1.5 degrees. It should be noted that when constructing a helix, a fundamentally different method of connecting amino acid molecules is used in comparison with a linear peptide chain. This method is more natural since it does not require the use of mirror-opposite amino acid molecules. Between adjacent turns of the spiral of amino acid molecules there is a stabilizing hydrogen bond between the *H* atoms of the peptide bond and the *O* atoms of the peptide bond with a double bond.

β - Structures of the Amino Acids

One of the important principles of protein structure formation is the formation of as many hydrogen bonds between groups as possible. This is especially pronounced in the so-called β - structures in which elongated chains of amino acid molecules are hydrogen bonded to each other to form a continuous layer. In these structures, the hydrogen bond plays a decisive role, and when calculating the dimensions of the compounds, it must be taken into account on an equal footing with the covalent bond. The repeating element in these β - structures is a polyhedron composed of two amino acid molecules. Figure 32 shows one of two variants of such polyhedrons.

Figure 32. Polyhedron composed of two amino acid molecules

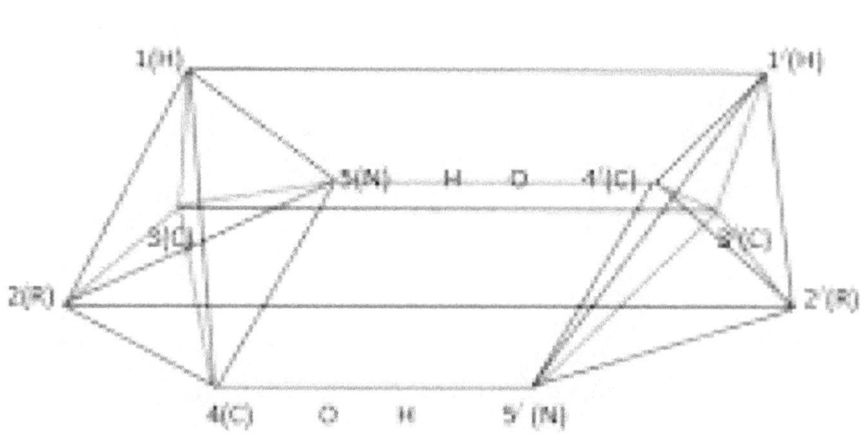

The ribs of the CN in Figure 32 are in an antiparallel position. It should be noted that the H and O atoms are not the tops of the figure. Therefore, the polyhedron in Figure 32 has 10 vertices (f_0=10), 24 edges (f_1=24), 29 two-dimensional faces (f_2=29), 20 three-dimensional faces (f_3=20), 7 four-dimensional faces (f_4=7). Four-dimensional faces are faces 12345, 1′2′3′4′5′, like tetrahedrons with a center. In addition, tetrahedrons are four-dimensional faces located on four triangular faces of tetrahedra with centers 13451′3′4′5′, 13251′3′2′5′, 12341′2′ 3′4′, 32543′2′5′4′. These figures are topologically equivalent to the products of tetrahedrons and a one-dimensional segment (Zhizhin, 2019 a). The difference between these figures and the products of tetrahedra and a one-dimensional segment is only that in this case the one-dimensional segment as a factor in these figures discretely changes its length. But this does not change the number of elements of different dimensions included in the figure. Therefore, as well as the product of a tetrahedron and a one-dimensional segment, these figures include 8 vertices (f_0=8), 16 edges each (f_1=16), 14 flat faces (f_2=14), 6 three-dimensional faces (f_3=6). Substituting these numbers into the Euler-Poincaré equation (3.3), we obtain

$$8 - 16 + 14 - 6 = 0.$$

This proves that for all the listed figures with vertices 3 and 3′ the Euler-Poincaré equation is fulfilled for n = 4, i.e. all of them have dimension 4. Another four-dimensional figure is formed by two tetrahedrons 1254 and 1′2′5′4′ connected by parallel lines. They are also topologically equivalent to the product of a tetrahedron and a one-dimensional segment. Thus, the total number of four-dimensional figures included in Figure 8 is 10.7. Substituting the values obtained for Figure 11.32 into the Euler Poincaré equation (3.3), we find

$$10 – 24 + 29 – 20 + 7 = 2.$$

This proves that in this case the Euler-Poincaré equation is fulfilled for n = 5, i.e. Figure 32 has dimension 5.

In a continuous layer of amino acid molecules, the polyhedron in Figure 32 alternates with another polyhedron formed by two tetrahedrons with center at other positions (Figure 33).

Figure 33. Second polyhedron formed by two tetrahedrons with center

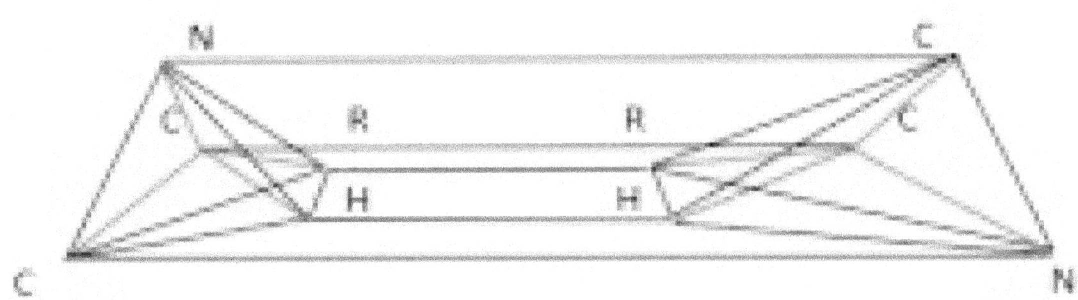

It can be seen that Figure 33 is also topologically equivalent to the product of a tetrahedron with center on a one-dimensional segment and therefore has dimension 5. The polyhedrons in Figure 32 and Figure 33 cannot directly adjoin each other. Between them are other polyhedron formed from two three-dimensional pyramids attached to each other along flat faces. One of these topologically equivalent polyhedrons is shown in Figure 34.

Figure 34. Polyhedron of two three-dimension pyramids

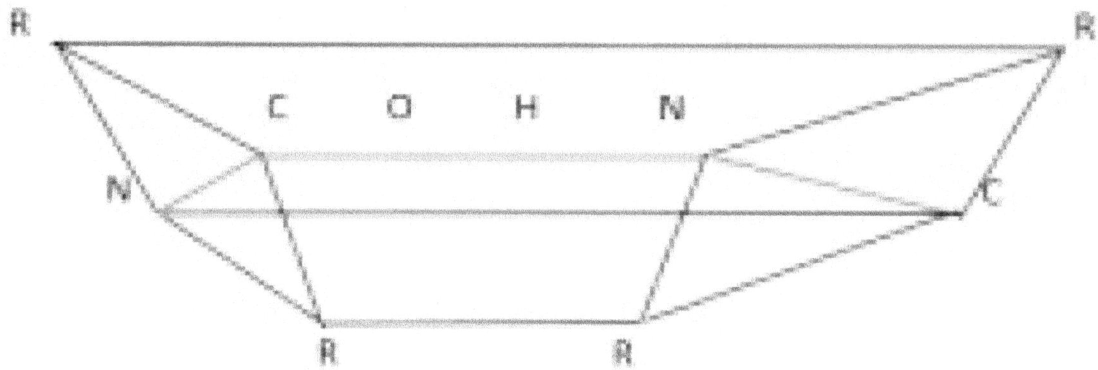

You can see that it has 8 vertices (f_0=8), 14 edges (f_1=14), 9 two-dimensional faces (f_2=9), 3 three-dimensional faces (f_3=3). Substituting these numbers into the Euler - Poincaré equation (3.3), we find

$$8 - 14 + 9 - 3 = 0.$$

Therefore, Figure 34 satisfies the Euler - Poincaré equations for n = 4, i.e. it has dimension 4. A topologically equivalent Figure 35 also has dimension 4.

Figure 35. Polyhedron of two three-dimension pyramids equivalent Figure 34

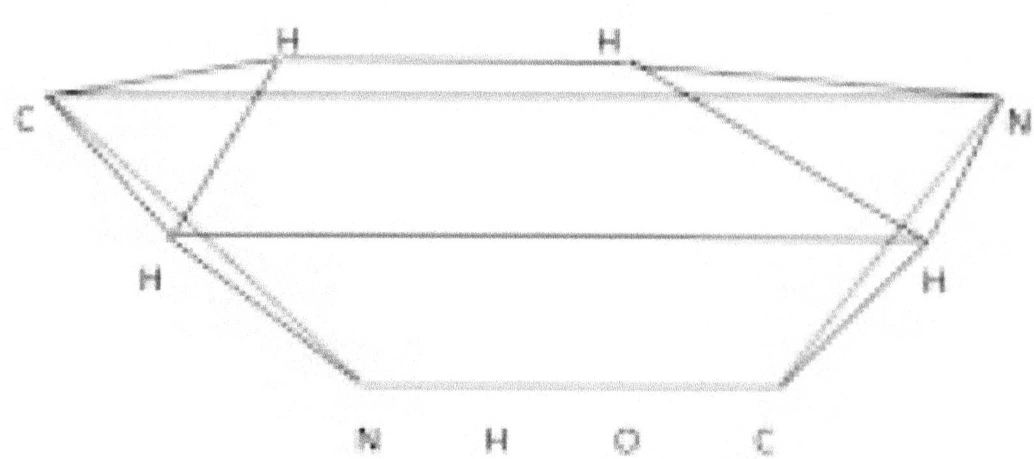

A part of the common layer composed of periodically repeating polyhedrons 8, 9, 10, 11 is shown in Figure 36.

Figure 36. Layer composed of periodically repeating polyhedrons 8,9,10, 11.

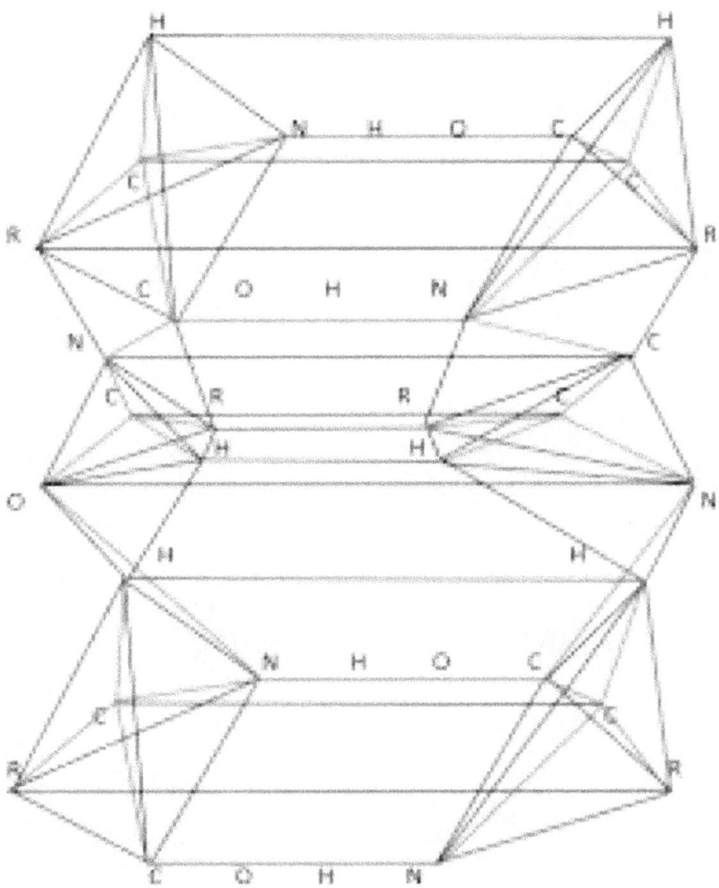

Figure 36 shows that the β - structure has translational symmetry. An elementary element of translational symmetry is the sequence of polyhedrons shown in Figures 32 – 35. If we denote the polyhedron in Figure 32 by symbol 5 (1), the polyhedron in Figure 33 by symbol 5 (2), and the polyhedron in Figure 34 by symbol 4 (1), the polyhedron in Figure 35 by the symbol 4 (2), then the translational element of the β - structure in Figure 36 is the polyhedron 5 (1)-4 (1)-5 (2)-4 (2). This polyhedron contains 24 vertices. If we connect each vertex of this polyhedron with all other vertices by edges, we get a single convex simplex polyhedron of dimension 23. Thus, the translational element of the β - structure in Figure 36 is a simplex of dimension 23.

In the literature, the β - structures traditionally described are called antiparallel and are depicted as alternating flat two-dimensional stripes (sheets).

However, it is clear that the name anti-parallel in this case does not reflect much of the essence of the structure in Figure 36. More precisely, this structure could be called non-parallel. In addition, and this is essential, the β - structures are fundamentally not flat and represent periodically repeating 5-dimensional and 4-dimensional regions, and each of these regions has two topologically equivalent images.

There are β - structures that include parallel 4-dimensional amino acid molecules with two different orientations. In these structures, alternation of 5-dimensional and 4-dimensional regions (polyhedrons) also occurs. In continuous form, it is presented in Figure 37. Figure 37 shows that the β - structure has translational symmetry too.

Figure 37. Parallel β - structure

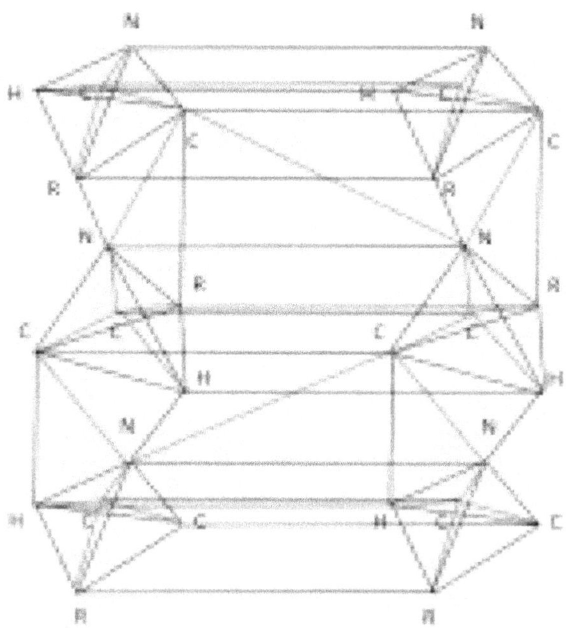

For this structure, you can conditionally save the name of the parallel structure.

Similar this the β - structure in Figure 36, it is also possible to define the sequence of polyhedrons of dimensions 5 and 4, which make up the translational element of the β - structure in Figure 37. Given the difference in the form of polyhedrons in Figures 36 and 37, the sequence of polyhedrons that make up the translational element of the structure in Figure 37 can be written as 5 (3)-4 (3)-5 (4)-4 (4). The particular type of polyhedrons of dimensions 5 and 4 is easy to determine from Figure 37. The translational element of β - structure in Figure 37 also includes 24 vertices and, therefore, is a simplex polyhedron of dimension 23.

Most proteins in a living cell are characterized by a more complex arrangement of amino acid molecules associated with a significant influence of hydrogen bonds and the formation of quaternary structures based on secondary and tertiary structures (i.e., secondary structures with the addition of unstructured fragments). The dimension of such structures (hemoglobin, collagen, etc.) should be yet higher.

CONCLUSION

It has been proven that the molecules of living organisms have the highest dimension (more than three). This is confirmed by the data of calculating the dimensions of molecules of various carbohydrates, proteins, and other biomolecules. Water molecules, which are known to average 80% of the mass of any living organism, also have the highest dimension. Since it is believed that biomolecules were formed in the process of evolution from the LUKA protocell more than 4.5 billion years ago in layers of water on the surface of minerals, apparently the higher dimension of biomolecules is a consequence of the higher dimension of water molecules. The highest dimension of water molecules is a consequence of the electronic structure of the atoms of the water molecule and the processes of hydration around the ions of mineral substances in water and on the surface of minerals. In this case, it should be taken into account that the elementary cells of minerals also have a higher dimension and water during hydration changes its structure complementary to the vicinity of mineral ions.

Three - dimensional geometric metric models of the pyranose molecule and chains of these molecules, constructed on the basis of simplification of complete higher - dimensional models of such molecules, are studied in detail. It is shown that a simplified three - dimensional model of the pyranose molecule consists of two parts of a straight pentagonal prism and a pyramid with pentagonal and hexagonal bases connected by a carbon cycle. It was found that due to the difference in the geometric three - dimensional figures that form the sugar molecule, the spiral chains of these molecules are characterized by a seemingly paradoxical property: rotation of the figures in two perpendicular planes in opposite directions.

It is shown that the widespread quasi - plane Pauling models do not reflect and even contradict the spatial structure of polypeptide amino acid molecules, taking into account their highest dimension. This applies to all types of polypeptide chain conformation. In this regard, new models of linear polypeptide chains were construct and it was found that these chains have translational symmetry in the space of higher dimensions. The elementary element of the translational symmetry of these chains is the 9-dimensional polyhedron, which consists of two mirror-opposite amino acid molecules. Helical conformations of polypeptide chains are also composed of higher-dimensional elements. The rotation of monomeric amino acid units in a helical conformation is determined only by the alternation of active centers in an asymmetric amino acid molecule. The study of the amino acid β - structures showed that they also have translational symmetry in the space of the highest dimension. Moreover, both the so-called "parallel" and "antiparallel" β - structures have elements of translation symmetry with dimension 23, consisting of two topologically equivalent polyhedrons of dimension 5 and two topologically equivalent polyhedrons of dimension 4. The resulting polyhedrons in high dimension, as translation elements of the construction of spaces protein molecules, should be further compared with the general theory of normal partitioning of n-dimensional spaces and corresponding high-dimensional parallelohedrons (Zhizhin, 2021).

Based on the study of the polytope of dimension 12, corresponding to the D-ribose molecule, a simplified three-dimensional model of this molecule was constructed. It will be used in the next chapter to analyze the processes of heredity.

REFERENCES

Corey, R. B., & Pauling, L. (1953). Fundamental dimensions of polypeptid chains. *Proc R Soc Lord B Biol Sci*, *141*(902), 10-20.

Dickerson, R. E., & Geis, I. (1969). *The Structure and Action of Proteins*. Benjamin.

Eisenberg, D., & Kautsman, V. (1975). *Structure and properties of water*. Gidrometyoizdat.

Ershov, Y. A. (1989). *Mechanisms of toxic action of inorganic compounds*. Medicine.

Gratias, D., & Cahn, J. W. (1986). Periodic and qusiperiodic crystals. *Scripta Metallurgica, 20*(9), 1193–1197. doi:10.1016/0036-9748(86)90030-X

Grünbaum, B. (1967). *Convex Polytopes*. Springer.

Ivanov, K. P. (1993). Fundamentals of the body's energy. Biological oxidation and oxygen supply. *Science*.

Klassen, V. I. (1973). *Water and magnet*. Science.

Koolman, J., & Roehm, K. (2013). *Color Atlas of Biochemistry*. Thieme.

Kostromina, N. A. (1990). *Chemistry of coordination compounds*. Chemistry.

Lalalyants, I. (1990). "Closure" in Paris. *Science and Life, 1*, 86–89.

Lehninger, A. L. (1982). *Principes of Biochemistry*. Worth Publishers, Inc.

Lyashchenko, A. K., & Dunyashev, V. S. (2003). Complementary organization of the water structure. *Journal of Structural Chemistry, 44*(5), 909–915. doi:10.1023/B:JORY.0000029822.94836.67

Martynova, O. I., Gusev, B. T., & Leont'ev, E. A. (1969). To the question of the mechanism of the influence of the magnetic field on aqueous solutions of salts. *Успехи физических наук, 98*(1), 195–199. doi:10.3367/UFNr.0098.196905n.0195

Metzler, D. E. (1977). *Biochemistry. The Chemical Reactions of Living Cells*. Academic Press.

Miller, E. V. (1969). On the spontaneous decrease in the density of water condensed from steam. *Reports of the USSR Academy of Sciences, 184*(1), 136–141.

Pauling, L. (1987). So-called Icosahedral Decagonal Quasicrystals are Twins of an 820-atom Cubic Cristal. *Physical Review Letters, 58*(4), 365–368. doi:10.1103/PhysRevLett.58.365 PMID:10034915

Pauling, L., & Corey, R. B. (1950). Two Hydrogen-bonded Spiral Configurations of the Polypeptide Chain. *Journal of the American Chemical Society, 72*(11), 5349. doi:10.1021/ja01167a545

Pauling, L., & Corey, R. B. (1953). Compound Helical Configurations of the Polypeptide Chain:Structure of Proteins of the α - keratin Type. *Nature, 171*(4341), 59–61. doi:10.1038/171059a0 PMID:13025480

Pauling, L., Corey, R. B., & Branson, H. R. (1951). The Structure of Proteins: Two Hydrogen-bonded Helical Configurations of the Polypeptide Chain. *Proceedings of the National Academy of Sciences of the United States of America, 37*(4), 205–211. doi:10.1073/pnas.37.4.205 PMID:14816373

Poincare, A. (1895). Analysis situs. *J. de e` Ecole Polytechnique, 1.*

Qian, G., Brugger, J., Skinner, W. M., Chen, G., & Pring, A. (2010). An experimental study of the mechanism of the replacement of magnetite by pyrite up to 300 ^{0} C. *Geochimica et Cosmochimica Acta, 74*(19), 5610–5630. doi:10.1016/j.gca.2010.06.035

Rahman, A., & Stillinger, F. H. (1973). Hydrogen-Boden patterns in liquid water. *Journal of the American Chemical Society*, *95*(24), 7943–7948. doi:10.1021/ja00805a003

Rodnikova, M. N. (1993). Features of solvents with a spatial network of H-bonds. *Journal of Physical Chemistry*, *67*(2), 275–280.

Russel, M. G., & … . (2005). On hydrothermal convection systems and the emergence of life. *Economic Geology and the Bulletin of the Society of Economic Geologists*, *100*, 419–438.

Samoilov, O. Y. (1957). *Structure of aqueous solutions and hydration of ions*. Science.

Semenchenko, V. N. (1932). *What are solutions*. Science.

Shechtman, D., Blech, I., Gratias, D., & Cahn, J. W. (1984). Metallic phase with longerange orientational order and no translational symmetry. *Physical Review Letters*, *53*(20), 1951–1953. doi:10.1103/PhysRevLett.53.1951

Simakov, S. (2018). Nano - and micro - sized diamond genesis in nature: An overview. *Geoscience Frontiers*, *9*(6), 1849–1858. doi:10.1016/j.gsf.2017.10.006

Simakov, S. (2020). Sicilian serpentinite xenoliths containing abiotic organics with nanodiamond clusters as key model for prebiotic processes. *Geoscience Frontiers*. Advance online publication. doi:10.1016/j.gsf.2020.04.008

Tatarsky, F. (1989). Struggle for "strange" water. *Chemistry and Life*, *12*, 57–59.

Zhizhin, G. V. (2013 a). Relations for the number of faces of different dimensions in the tower of n - dimensional convex polytopes. In Materials 9th All-Russian Scientific School "Mathematical research in the natural sciences." Geological Institute KSC RAS.

Zhizhin, G. V. (2013 b). Images of convex regular and semiregular n-dimensional polytopes. In Materials 9th All-Russian Scientific School "Mathematical research in the natural sciences." Geological Institute KSC RAS.

Zhizhin, G. V. (2014). *World – 4D*. Polytechnic Service.

Zhizhin, G. V. (2016). The structures, topological and functional dimension of biomolecules. *International Journal of Chemoinformatics and Chemical Engineering*, *5*(2), 44–58. doi:10.4018/IJCCE.2016070104

Zhizhin, G. V. (2018). *Chemical Compound Structures and the Higher Dimension of Molecules: Emerging Research and Opportunities*. IGI Global. doi:10.4018/978-1-5225-4108-0

Zhizhin, G. V. (2019a). *The Geometry of Higher-Dimensional Polytopes*. IGI Global. doi:10.4018/978-1-5225-6968-8

Zhizhin, G. V. (2019b). *Attractors and higher dimensions in population and molecular biology: emerging research and opportunities*. IGI Global. doi:10.4018/978-1-5225-9651-6

Zhizhin, G. V. (2020a). The Dimension and Life. *Acta Scientific Biotechnology*, *1*(11), 35–40.

Zhizhin, G. V. (2020b). Spirals of Polysaccharides Simultaneously Rotating in Opposite Directions. *Acta Scientific Biotechnology*, *1*(7), 13–16.

Zhizhin, G.V. (2020c). Higher Dimensions in Linear, Helicoidal and Folder Structures of Protein Molecules. *Acta Scientific Biotechnology, 1*(12), 2-9.

Zhizhin, G. V. (2021). *Normal partitions and hierarchical fillings of n-dimensional spaces*. IGI Global. doi:10.4018/978-1-7998-6768-5

Zhizhin, G. V., & Diudea, M. V. (2016). Space of Nanoworld. In M. V. Putz & C. M. Marius (Eds.), *Sustainable nanosystems, development, properties, and applications* (pp. 221–245). IGI Global.

Zundel, G. (1972). *Hydrotation and intermolecular interactions*. Mir.

KEY TERMS AND DEFINITIONS

Functional (Topological) Dimension of a Molecule: The dimension of a convex polytope, as a model of a molecule, at the vertices of which not only individual atoms but also functional groups of the molecule can be located.

Geometrical Image of a Chemical Compound: The geometrical image of a chemical compound (molecule) is a convex polytope, at the vertices of which atoms (or functional groups) are located. The edges of the polytope connecting the vertices correspond to the chemical bonds of the compound. The part of edges only carry a geometric function. They are necessary to give the molecule the image of a convex geometric figure. The dimension of the polytope is determined by the Euler-Poincare equation.

Hybridization of Electronic Orbitals: This is the interaction of the electronic orbitals of atoms entering the molecule, leading under certain conditions to the formation of higher dimensionality of molecules.

N–Cross-Polytope: The convex polytope of dimension n in which opposite related of centrum edges not have connection of edge.

N–Simplex: The convex polytope of dimension n in which each vertex is joined by edges with all remain vertices of polytope.

Tetrahedral Coordination of Electron Pairs: The location of the electronic pairs of the outer and the pre-outer electron layer at the vertices of the tetrahedron.

Transitional Elements: Chemical elements in which electrons fill *d*- and *f*-orbitals of an atom (*d*- and *f*-elements).

Chapter 11
Polytopes of the Highest Dimension in the Theory of Heredity

ABSTRACT

On the basis of images of higher dimension of nucleic acids, the properties of the interaction of nucleic acids with each other have been studied. It was found that the region of interaction of nucleic acids is a cross-polytope of dimension 13 (the polytope of genetical information). Its structure indicates that the interaction of nucleic acids has a hidden order, according to which the number of standard amino acids exactly corresponds to the number of possible locations of nitrogenous bases in the polytope of hereditary information. This leads to the absence of the need to introduce the concept of the genetic code. It has been shown that DNA methylation leads to a significant increase in DNA informational activity, which makes it possible to resist a viral infection.

INTRODUCTION

In the forties of the twentieth century, primarily on the basis of the cytological and biochemical observations of Casperson (1941) and Brachet (1941, 1942), the idea began to take shape that DNA localized in the nuclei of cells, in their chromosomes, is closely connected with the apparatus of heredity, and RNA - an essential component of the cellular cytoplasm, responsible for protein synthesis. Deoxyribonucleic acid (DNA), as a chemical substance, it was isolated by Johann Friedrich Micher in 1869 from the remains of cells contained in the pus. He singled out a substance that includes nitrogen and phosphorus. When Misher determined that this substance has acid properties, the substance was called nucleic acid (Dahm, 2005). Gradually it was proved that it was DNA, and not proteins, as previously thought, and which is the carrier of genetic information. One of the first decisive proofs was the experiments of Oswald Avery, Colin MacLeod and McLean McCarthy (1944) on the transformation of bacteria.

The structure of the double helix DNA it was proposed by Francis Crick and James Watson in 1953 on the base of the X-ray structural data obtained by Maurice Wilkins and Rosalind Franklin and the

DOI: 10.4018/978-1-7998-8374-6.ch011

"Chargaff rules" according to which in each DNA molecule the strict relationships connecting the quantity of nitrogenous bases of different (Watson, & Crick, 1953a, b). For outstanding contributions to this discovery, Francis Crick, James Watson and Maurice Wilkins were awarded the 1962 Nobel Prize in Physiology or Medicine. Deoxyribonucleic acid (DNA) is a biopolymer, the monomer of which is the nucleotide (Albert, et al., 2002; Butler, 2005). Each nucleotide consists of a phosphoric acid residue attached to sugar deoxyribose, to which one of the four nitrogen bases is attached also. The bases that make up the nucleotides are divided into two groups: purines (adenine [A] and guanine [G]) and pyrimidines (cytosine [C] and thymine [T]) are formed by combined five - and six - membered heterocycles.

They managed to show that the DNA isolated from the pneumococci corresponds to the so-called transformation (the acquisition of pathogenic properties by a harmless culture as result of the addition of dead pathogenic bacteria to it). The experiment of American scientists Alfred Hershey and Martha Chase (Hershey - Chase experiment, 1952) with radioactively labeled proteins and bacteriophage DNA showed that only the phage nucleic acid is transmitted to the infected cell, and the new generation of phage contains the same proteins and nucleic acid, as the initial phage (Hershey & Chase, 1952). Deciphering the structure of DNA (1953) has become one of the turning points in the history of biology.

Nucleotides are long polynucleotide chains covalently linked. These chains in the overwhelming majority of cases (except for some viruses possessing single - stranded DNA genomes) are combined pairwise by means of hydrogen bonds into a secondary structure, called the double helix (Watson & Crick, 1953 a, b; Berg, Tymoczko & Stryer, 2002). Each a base on one of the chains is connected to one definite base on the second chain. This specific binding is called complementary. Purines are complementary to pyrimidines (that is, they are capable of forming hydrogen bonds with them): adenine forms bonds only with thymine, and cytosine - with guanine. In a double helix, chains are also linked by hydrophobic interactions and stacking, which do not depend on the DNA base sequence (Ponnuswamy & Gromiha, 1994). Complementarity of the double helix means that the information contained in one chain is also contained in another chain. Different base pairs form a different number of hydrogen bonds. In the future, the existence of nucleic acids differing in the length of the period and shape with rotation of the spiral both to the right and to the left was experimentally established (Ha, et al., 2005; Cantor, & Schimmel, 1980).

Watson and Crick postulated the spiral form of the DNA molecule, but they did not discuss the reasons for the formation of such a DNA molecule. Until now there have been no works explaining the existence of a spiral in the DNA molecule.

In the book (Zhizhin, 2018 a), the molecules of practically all the elements of the periodic system were studied in detail and it was shown that many of them, including magnesium and calcium, form compounds of higher dimension (see also Zhizhin, 2015 a, b; Zhizhin, & Diudea, 2016; Zhizhin, Khalaj, & Diudea, 2016). In the works (Zhizhin, 2016 a, b), the structures of biomolecules, which also form compounds of higher dimension, are also studied. When analyzing the metric space of nucleic acid structures, one should use the representations of their constituents in the form of polytopes of higher dimension. This primarily applies to the phosphoric acid residues and the *D* - ribose molecules, since nitrogenous bases have a dimensionality of 2 and do not have a significant effect on the geometric structure of nucleic acids.

The determination of the geometric forms of chemical compounds according to the Euler-Poincaré (Poincare, 1895) relation translates classical stereochemistry in three-dimensional space into sulfur chemistry in a space of higher dimension. Here new discoveries are possible in the laws of chemical compounds. Consideration of the geometry of molecules using spaces of higher dimension makes it possible to explain the facts observed in reality, which previously had no clear explanations.

The concept of the geometric correspondence of the structure of substances in the formation of a chemical compound is essential for the form of the compound. However, the complementarity of DNA spirals when connecting spirals can`t be the reason for the formation of these spirals. The geometric shape of the constituent parts of the spiral itself must be important here. They are phosphoric acid residues and D - ribose molecules. This Chapter is devoted to the investigation of this problem. In this case it is necessary to take into account the dimension of these parts, which, as shown in (Zhizhin, 2018 a), is higher than three. It should be noted that there is no contradiction in the representation of the structure of DNA as a spiral in a three - dimensional space with the highest dimension of its components. The point is that according to the geometry of Riemann (Riemann, 1854), a space of higher dimension has a boundary. Therefore, a space of a smaller dimension can be located outside this space.

In addition, for the convenience of images, a simplified three - dimensional model of a sugar molecule was built on the basis of a full twelve - dimensional model of a sugar molecule (Zhizhin, 2019 c, 2021 d). This model can be used in further constructions, not forgetting the dimension of its complete model. This model was used to construct images of single nucleic acid helices. When passing to the analysis of nucleic acids consisting of two linked helices, it is necessary to take into account the emerging anti-parallelism of geometric elements in linked spirals (Zhizhin, 2018 b, 2019 c). In this Chapter, continuing the research begun in the works (Zhizhin, 2018 b, 2019 a), it will be shown that sugar molecules located in linked nucleic acid helices form a polytope of dimension 13 of the type of cross - polytope in which there are exactly 12 coordinate planes. In these coordinate planes outlined by rectangles with antiparallel edges, exactly 12 compounds of nitrogen bases currently known can be located (Spirin, 2019). The image of the polytope and its coordinate planes is obtained, its structure is determined. This polytope can be called the polytope of hereditary information, since it transmits hereditary information from one spiral to another using a sequential alternation of nitrogenous bases.

It has been shown that the polytopic of hereditary information is characterized by a powerful stream of incidents between geometric elements of different dimensions (Zhizhin, 2019 a, b), providing an extremely intense exchange of information between the components of nucleic acids. The evidence of the existence of such an exchange of information can be found in the recently discovered inheritance of changes not related to the modification of sequences in DNA, i. e. with methylation - the binding of a methyl group CH_3 to the nitrogenous base of DNA (Mancuso, 2017; Lindquist, et al., 2016).

The geometric features of nucleic acids during methylation of their nitrogenous bases, which is typical for viral nucleic acids in infectious diseases, were studied. It was found that methylation of viral nucleic acids leads to the transition of their nitrogenous bases into the space of higher dimension and sharply increases (more than 2000 times) the intensity of information processes in the field of interaction of nucleic acids (in the polytopic of hereditary information). The activity of methylated viral nucleic acids can be countered by activation, i. e. a fairly high degree of methylation, nucleic acids of the organism (Zhizhin, 2021 b).

THE POLYTOPE OF HEREDITARY INFORMATION

Let us consider in detail the formation of a polytope of two sugar molecules with antiparallel edges. Here, as in the case of the tetrahedron, to form a polytope with antiparallel edges from two sugar molecules, you must have one sugar molecule on one helix to turn this helix together with the sugar molecule and move the sugar molecule along this reversed helix in opposite direction to the original helix direction.

When the sugar molecule rotates 180 degrees to the right while moving, then the original sugar molecule and the sugar molecule on the reversed helix are two polytopes with antiparallel edges. Both of these sugar molecules in a simplified form are represented in Figure 1.

Figure 1. The two sugar molecules in a simplified form with antiparallel edges.

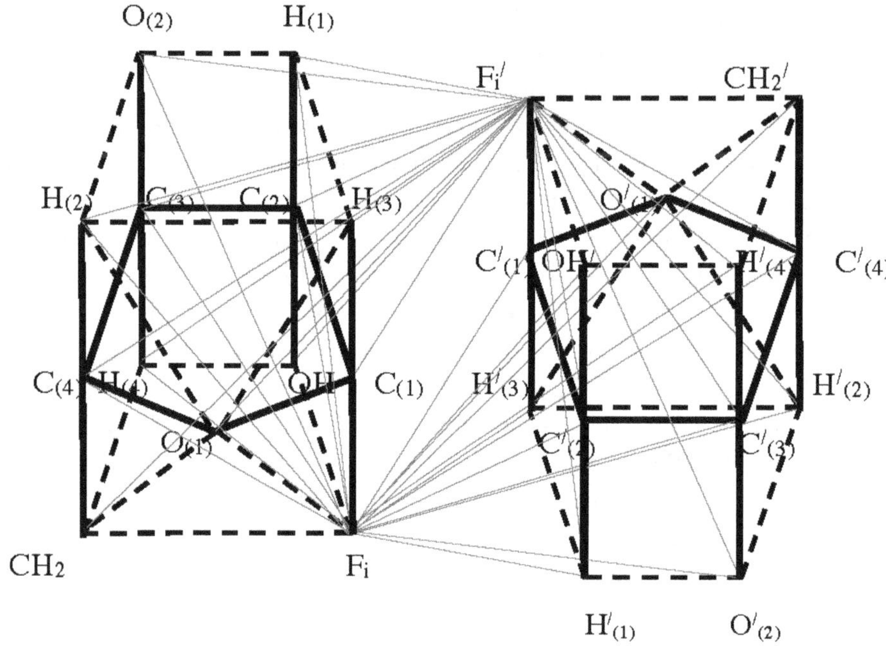

When the five - carbon sugar molecule is rotated to the left by 180 degrees, the nitrogenous bases are on opposite sides of both molecules, so that for their connection it is necessary to cross the entire set of atoms of two molecules. This is un existent, therefore this option is not considered. When the sugar molecule is rotated 180 degrees to the left, it is possible to connect the sugar molecules through nitrogenous bases only between two different chains of nucleic acids or in the case of turning the chain itself in the opposite direction.

In full, the sugar molecules have a dimension of 12, in the corresponding polytope each vertex must have an edge connection with all the other vertices. Knowledge of this now one need. For the formation of a polytope of dimension 13, it is necessary to connect each vertex of one polytope with the vertices of another polytope so that there are no vertex connections with the same letters. All connecting edges break into pairs of antiparallel edges. At the same time, a set of coordinate a two - dimensional planes emanates from the center of the formed polytope as from the origin of coordinates. Their number is equal to the number of combinations from 13 to 2, i. e. 48 coordinate planes. In order to clarify the possible geometrical circumstances of the connection of helices in double - helix nucleic acid molecules with nitrogenous bases, we are primarily interested in the coordinate planes containing these nitrogenous bases $F_i, F_i^{/}$. There are exactly 12 such coordinate planes in the obtained polytope of dimension 13. They are depicted in Figure 1 by blue solid lines and are indicated below by the vertices of the polytopes contained in them:

$$F_i^/ H_{(1)} \ F_i^/ O_{(2)} \ F_i^/ H_{(2)} \ F_i^/ C_{(3)} \ F_i^/ C_{(2)} \ F_i^/ H_{(3)}$$
$$H_{(1)}^/ F_i \text{'} \ O_{(2)}^/ F_i \text{'} \ H_{(2)}^/ F_i \text{'} \ C_{(3)}^/ F_i \text{'} \ C_{(2)}^/ F_i \text{'} \ H_{(3)}^/ F_i \text{'}$$

$$F_i^/ C_{(4)} \ F_i^/ H_{(4)} \ \ F_i^/ OH \ \ F_i^/ C_{(1)} \ F_i^/ O_{(1)} \ \ F_i^/ CH_2$$
$$C_{(4)}^/ F_i \text{'} \ H_{(4)}^/ F_i \text{'} (OH)^/ F_i \text{'} \ C_{(1)}^/ F_i \text{'} \ O_{(1)}^/ F_i \text{'} (CH_2)^/ F_i \cdot$$

Other edges of the polytope of dimension 13 are not shown in Figure 1, so as not to ignite the figure. In the center of each parallelogram, indicated by its four vertices, is the origin of coordinates and the corresponding pair of coordinate axes (they are not shown). To identify the different hydrogen and oxygen atoms at vertices of polytope, they indicated by numbers in brackets at the lower indices.

It is surprising that the number of coordinate planes containing vertices is exactly equal to the number of possible compounds of nitrogenous bases 12 (Spirin, 2019)

$$A:U,U:A,G:C,C:G,G:U,G:A,U:U,U:C,A\cdot A,A\cdot C,U:G,A:G. \tag{11.1}$$

Since each coordinate plane designated by the vertices of the parallelograms has a specific atomic environment, it can be assumed that each of the 12 possible compounds of nitrogenous bases is located on one particular coordinate plane out of 12 possible. This solves the question of the possible orientation of the bond of nitrogenous flat bases in nucleic acids using ideas about the high dimensionality of the constituent nucleic acids. It is also surprising that in order to create 13 – cross - polytopes, providing the connection with nitrogenous bases $F_i, F_i^/$, nature specially created double - stranded nucleic acids with oppositely directed spirals. This is realized in DNA and RNA when creating regions with inverted helices.

In a variety of nucleic acid molecules, the issue of chain interaction is important. In ribosomes, RNA interacts with each other due to bivalent metal ions (mainly magnesium ions). Positively charged magnesium ions attract negative charges of phosphoric acid residues, ensuring the stability of the ribosomes. In double - stranded nucleic acids, the helices are connected to each other by means of nitrogenous bases complementarily interacting with each other by a hydrogen bond. However, the magnesium ions and nitrogen bases in nucleic acids could not be specifically located. It has been established that magnesium ions and flat nitrogenous bases can be located inside special polytopes of higher dimension. Here knowledge is needed of the higher dimension of phosphoric acid residues and sugar molecules. Such polytopes are polytopes with antiparallel edges, i. e. cross - polytopes of higher dimension. Binding agents are located on the free coordinate planes of these polytopes in the vicinity the center of the polytope.

In this case, the two - dimensional coordinate plane on the boundary of the polytope should contain the objects to be joined. In the case of magnesium ions, there are four specific coordinate planes inside the 5 – cross - polytope, in which an ion can accommodate, combining negative charges. In the case of nitrogenous bases, the existence of 12 coordinate planes inside a cross - polytope of dimension 13, in which flat nitrogenous bases can be located, connecting the helix of nucleic acids, was discovered. Exactly as much as there are options for combining nitrogenous bases. It is given, that each coordinate plane of these 12 planes has a specific environment of atoms. It should be assumed that only one of the 12 possible compounds of nitrogenous bases is placed in each of these planes. It is surprising that the existence of higher - dimensional polytopes with anti - parallel edges is possible only in the case of the opposite direction of interacting helices, and this is exactly what nature provides in the double - helix DNA and in the RNA segments with self - inversion of the helix in the opposite direction.

To build the polytopic of hereditary information, we need (Zhizhin, 2019 a) four – dimensional polytope with anti – parallel edges to supplement with edges that create a closed and convex figure. In this case, it is not necessary to connect the vertices symmetrically relative to the center of figure with the edges. Each vertex will be connected by an edge to the other vertices. In this case, the polytope will be a cross -polytope and its dimension is equal to half the number of vertices. Indeed, according to (Zhizhin, 2019 b), there is a relation between f_i the number of dimension elements i in the cross - polytope and the dimension d of the cross - polytope itself:

$$f_i(d) = 2^{1+i} C_d^{d-1-i}. \tag{11.2}$$

At the vertex $i = 0$, therefore, according to (11.2)

$$f_0 = 2C_d^{d-1} = 2d, d = f_0 / 2 = 13.$$

Thus, the polytope of the hereditary information has dimension 13. When portraying this polytope, we will use the technique that is used in portraying cross-polytopes. We distribute all 26 vertices on the circle so that the vertices opposite in Figure 1 remain opposite and there is no edge between them (Figure 2).

Figure 2. The polytope hereditary information

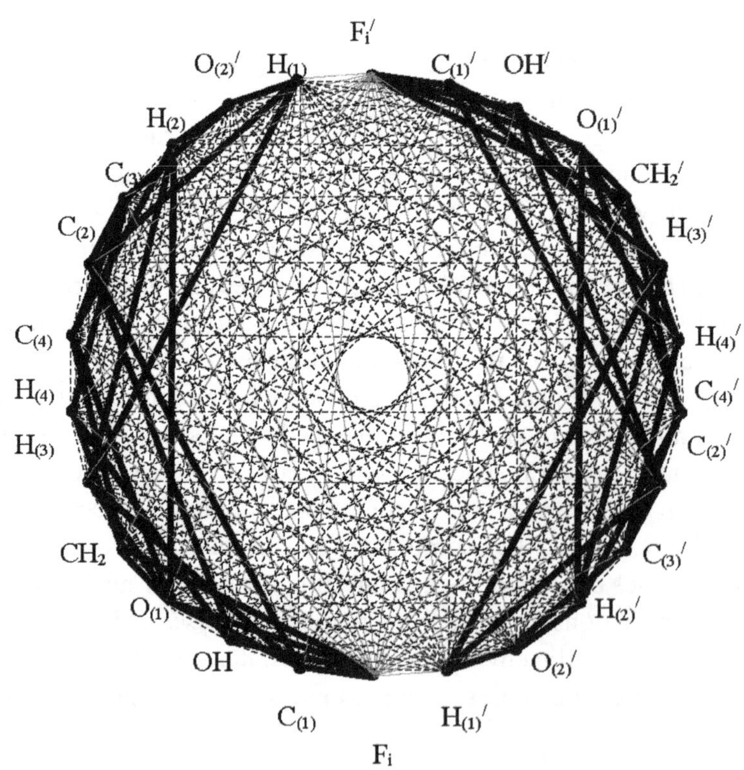

The edges corresponding to the chemical bonds are indicated in this figure by thick black solid lines. Antiparallel edges highlighting coordinate planes with vertices corresponding to nitrogenous bases are indicated by bluey lines. The remaining edges are indicated by thin dash-dotted black lines. We emphasize that there is a chemical bond between the nitrogenous bases, but it is not covalent. Therefore, this connection is not indicated by an edge.

When the polytope of hereditary information moves along the common axis of the double helix of nucleic acids in accordance with the sequence of bases in DNA, the connection between the vertices $F_i^{/}, F_i^{//}$ is carried out by one of the twelve base pairs, which occupies one of the 12 coordinate planes of the polytope.

By relation (11.2), one can determine the number of elements of different dimensions in the polytope of hereditary information.

The number of edges is $f_1(13) = 2^2 C_{13}^{11} = 312$,
the number of triangles is $f_2(13) = 2^3 C_{13}^{10} = 2288$,
the number of tetrahedrons is $f_3(13) = 2^4 C_{13}^{9} = 10560$,
the number of fourth-dimensional faces (simplexes) is $f_4(13) = 2^5 C_{13}^{8} = 41184$,
the number of fives-dimensional faces (simplexes) is $f_5(13) = 2^6 C_{13}^{7} = 109824$,
the number of six-dimensional faces (simplexes) is $f_6(13) = 2^7 C_{13}^{6} = 219648$,
the number of seven-dimensional faces (simplexes) is $f_7(13) = 2^8 C_{13}^{5} = 164736$,
the number of eight – dimensional faces (simplexes) is $f_8(13) = 2^9 C_{13}^{4} = 183040$,
the number of nine-dimensional faces (simplexes) is $f_9(13) = 2^{10} C_{13}^{3} = 292864$,
the number of ten-dimensional faces (simplexes) is $f_{10}(13) = 2^{11} C_{13}^{2} = 159744$,
the number of eleven-dimensional faces (simplexes) is $f_{11}(13) = 2^{12} C_{13}^{1} = 26624$,
the number of twelve-dimensional faces (simplexes) is $f_{12}(13) = 2^{13} = 4096$.

The obtained numbers determine the structure of the polytope of hereditary information.

THE LAW OF CONSERVATION OF INCIDENTS IN POLYTOPE OF HEREDITARY INFORMATION

The monograph (Zhizhin, 2019 c) introduced the concept of the incidence coefficients of elements of lower dimension with respect to elements of the higher dimension and elements of higher dimension with respect to elements of the lower dimension. The first characterizes the number of elements of a certain higher dimension to which the given element of a lower dimension belongs. The second characterizes the number of elements of a given lower dimension that are included in a particular element of a higher dimension. Here we must remember that the vertices of geometric elements of various dimensions are atoms, molecules or functional groups. Therefore, the incidence of geometric elements to a friend means contacted between particles of the matter, including living matter. The contact between particles of matter can be interpreted as the transfer of information on material structures, including biological structures. We introduce the notation: $k_{d_j^i d_u}$ is the number of elements of dimension *u*, which include an element of dimension j ($u > j$) with number i. Thus, $k_{d_j^i d_u}$ is the incidence factor of element i with dimension j relative to elements with dimension u. We introduce the notation also: $k_{d_j d_u^i}$ is the number of elements of dimension j, which included in element i with a dimension u ($u > j$). Thus, $k_{d_j d_u^i}$ is the incidence factor of element i with dimension u relatively to elements with dimension j.

The smallest dimension of the cross - polytope is 4. From (11.2) it follows that in this polytope there are 8 vertices, 24 edges, 32 two - dimensional faces, 16 three - dimensional faces. The factors of incidents (from smaller dimension to larger) are:

$$k_{d_0^i d_1} = 6, i = 1 \div 8; k_{d_0^i d_2} = 12, i = 1 \div 8; k_{d_0^i d_3} = 8, i = 1 \div 8; k_{d_0^i d_4} = 1, i = 1 \div 8; k_{d_1^i d_2} = 4, i = 1 \div 24;$$
$$k_{d_1^i d_3} = 4, i = 1 \div 24; k_{d_1^i d_4} = 1, i = 1 \div 24; k_{d_2^i d_3} = 2, i = 1 \div 32; k_{d_2^i d_4} = 1, i = 1 \div 32; k_{d_3^i d_4} = 1, i = 1 \div 16.$$

Sum up the incidence coefficients for all vertices, edges, two - dimension faces and three - dimension faces of the 4 - cross – polytope:

$$\sum_{i=1}^{8} k_{d_0^i d_1} + \sum_{i=1}^{8} k_{d_0^i d_2} + \sum_{i=1}^{8} k_{d_0^i d_3} + \sum_{i=1}^{8} k_{d_0^i d_4} + \sum_{i=1}^{24} k_{d_1^i d_2} + \sum_{i=1}^{24} k_{d_1^i d_3} + \sum_{i=1}^{24} k_{d_1^i d_4} + \sum_{i=1}^{32} k_{d_2^i d_3} + \sum_{i=1}^{16} k_{d_3^i d_4} + \sum_{i=1}^{32} k_{d_2^i d_4} = 544. \quad (11.3)$$

The factors of incidents (from larger dimension to smaller) are

$$k_{d_0 d_1^i} = 2, i = 1 \div 24; k_{d_0 d_2^i} = 3, i = 1 \div 32; k_{d_0 d_3^i} = 4, i = 1 \div 16; k_{d_0 d_4} = 8; k_{d_1 d_2^i} = 3, i = 1 \div 32;$$
$$k_{d_1 d_3^i} = 6, i = 1 \div 16; k_{d_1 d_4} = 24; k_{d_2 d_3^i} = 4, \mathrm{i} = 1 \div 8; k_{d_2 d_4} = 32; k_{d_3 d_4} = 16.$$

Sum up the incidence coefficients for all elements of the 4 – cross - polytope with dimension larger of zero:

$$\sum_{i=1}^{24} k_{d_0 d_1^i} + \sum_{i=1}^{32} k_{d_0 d_2^i} + \sum_{i=1}^{32} k_{d_1 d_2^i} + \sum_{i=1}^{16} k_{d_0 d_3^i} + \sum_{i=1}^{16} k_{d_1 d_3^i} + \sum_{i=1}^{16} k_{d_2 d_3^i} + k_{d_0 d_4} + k_{d_1 d_4} + k_{d_2 d_4} + k_{d_3 d_4} = 544. \quad (11.4)$$

Comparing (11.4) and (11.3) you can see that the sum of incidents in a 4 - cross - polytope from elements with a lower dimension to elements with a higher dimension is equal to the sum of incidents from elements with a higher dimension to elements with a lower dimension. Thus, the sum of incidents retains its value when changes the direction of the relationship between the elements (the law of conservation of incidents, (Zhizhin, 2019 d)).

Theorem 11.1. In any cross - polytope of dimension n, the sum of all incidents of elements of a lower dimension with respect to all elements of a higher dimension is equal to the sum of all incidents of elements of a higher dimension with respect to all elements of a lower dimension and equals the sum of the series:

$$f_0(n)3^{n-1} + f_1(n)3^{n-2} + f_2(n)3^{n-3} + ... + f_{n-1}(n)3^0, f_d(n) = C_n^{n-1-d} 2^{1+d}, d = 0 \div (n-1).$$

Proof. According to equation (11.2), each cross - polytope of dimension n has $2n = f_0(n)$ vertices. The peculiarity of a cross - polytope is that each of its vertices has an opposite vertex, with which it is not connected by an edge. Moreover, there is one edge between this vertex and all other vertices. We subtract from the total number of vertices two vertices (the selected vertex and its opposite) 2n - 2. This is the possible number of edges emanating from the selected vertex. Thus, the incidence coefficient of the edges of any vertex is $k_{d_0^i d_1} = 2(n-1) = 2\,\mathrm{C}_{n-1}^1, i = 1 \div f_0(n)$. A cross - polytope of any dimension can be depicted as a projection on a two - dimensional plane (Zhizhin, 2019 a). In this image, all its vertices are located on a circle, with the selected vertex and its opposite vertex located symmetrically relative to the center of the circle. A mentally drawn line through these two vertices halves the circle (Figure 2). Therefore, the number of variants the location of the edges from the selected vertex to the other vertices in one of the halves of the circle is n - 1, i. e. the number of combinations of n - 1 vertices one by one. Since there are two halves of a circle, then for the total number of vertex selection options for edge formation, this number of combinations should be multiplied by 2. This is the meaning of the expression for the incidence rate of any vertex in the n - cross - polytope with respect to the edge. To further prove the theorem 11.1 and clarify the nature of the formation of the coefficients of incidence in the n - cross - polytope, we will use this technique. We arrange the vertices of the 4 - cross - polytope in two lines, so that the vertices unlinked by an edge form vertical pairs: each vertex in the top line is not connected by an edge to the vertex in the bottom line located strictly under this vertex in the top line:

$$\begin{matrix} d_0^1 & d_0^2 & d_0^3 & d_0^4 \\ d_0^5 & d_0^6 & d_0^7 & d_0^8 \end{matrix}. \qquad (11.5)$$

We choose some vertex in the top line (11.5), for example d_0^1 . It cannot be connected by an edge with a vertex d_0^5 in the bottom line, but it can be connected by an edge with other vertices of the bottom line. The number of such options is 3, i. e. with n = 4 is $3 = C_{n-1}^1 = C_3^1$. However, a vertex d_0^1 may be connected by an edge with any of the remaining three vertices in the top line (11.5). Therefore, the incidence coefficient of a vertex with respect to an edge in 4 - cross - polytope $k_{d_0^i d_1} = 2\,\mathrm{C}_{n-1}^1 = 6, i = 1 \div 8$. When considering the belonging of a vertex d_0^1 to two - dimensional faces, it is necessary to determine the number of options for the participation of two vertices in the top line (11.5) (without a vertex d_0^1) and in the bottom line (11.5) (without a vertex d_0^5). This will be the number $2C_3^1$. In addition, as the vertices of the triangle, there may be vertices located in the upper and lower lines (11.4) in a cross - section way. There are 6 such variants in (11.5), i. e. more $2C_3^1$. Therefore, the incidence coefficient of a vertex with respect to a two - dimensional element in 4 - cross - polytope is $k_{d_0^i d_2} = 2^2\,\mathrm{C}_{n-1}^2 = 12, i = 1 \div 8$. Multiplying this number by the number of vertices in 4 - cross - polytope you can get the total number of incidents of vertices to two - dimensional elements in 4 - cross - polytope equal to 96. This number coincides with the corresponding number, defined earlier in Figure 2. Combination d_0^1 with three the different vertices from (11.5) can get the incidence coefficient of vertex to the three - dimension body (tetrahedron) for condition absent in combination opposite vertices. Can show that this combination is

8, $k_{d_0^i d_3} = 2^3 C_3^3 = 8, i = 1 \div 8$. Multiplying this number by the number of vertices in 4 - cross - polytope you can get the total number of incidents of vertices to three - dimensional elements in 4 - cross - polytope equal to 64. This number coincides with the corresponding number, defined earlier in Figure 2. Combination two vertices, for example d_0^1 and d_0^2 (the edge $d_0^1\ d_0^2$) with two the different vertices from (11.5) you can get the incidence coefficient of edge to three - dimension body (tetrahedron) for condition absent in combination opposite vertices. Can show that this combination is 4, $k_{d_1^i d_3} = 2^2 C_2^2 = 4, i = 1 \div 8$. Multiplying this number by the number of edges in 4 - cross - polytope you can get the total number of incidents of edges to three - dimensional elements in 4 - cross - polytope equal to 96. This number coincides with the corresponding number, defined earlier in Figure 2. Combination two vertices (the edge $d_0^1\ d_0^2$) with one the vertex from (11.5) can get the incidence coefficient of edge to triangle for condition absent in combination opposite vertices. Can show that this combination is 4, $k_{d_1^i d_2} = 2C_2^1 = 4, i = 1 \div 8$. Multiplying this number by the number of edges in 4 - cross - polytope you can get the total number of incidents of edges to triangle in 4 - cross - polytope equal to 24. This number coincides with the corresponding number, defined earlier in Figure 2. In general case can write of $2n$ vertices of the n - cross - polytope in two lines

$$\begin{matrix} d_0^1 & d_0^2 & d_0^3 & \ldots & d_0^n \\ d_0^5 & d_0^6 & d_0^7 & \ldots & d_0^{2n} \end{matrix}, \tag{11.6}$$

where vertex in the top line is not connected by an edge to the vertex in the bottom line located strictly under this vertex in the top line. The incidence coefficient of a vertex with respect to an edge in n - cross - polytope is $k_{d_0^i d_1} = 2\mathrm{C}_{n-1}^1, i = 1 \div f_0(n)$. The incidence coefficient of a vertex with respect to a two - dimensional element in n - cross - polytope is $k_{d_0^i d_2} = 2^2\ \mathrm{C}_{n-1}^2, i = 1 \div f_0(n)$. The incidence coefficient of a vertex with respect to a three - dimensional edge in n - cross - polytope is $k_{d_0^i d_3} = 2^3\ \mathrm{C}_{n-1}^3, i = 1 \div f_0(n)$. Go on can to say the incidence coefficient of a vertex with respect to a $(n-1)$ - dimensional element in n - cross - polytope $k_{d_0^i d_{n-1}} = 2^{n-1}\ \mathrm{C}_{n-1}^{n-1}, i = 1 \div f_0(n)$. Obviously, that the incidence coefficient of a vertex with respect to n - cross - polytope is $k_{d_0^i d_n} = 1, i = 1 \div f_0(n)$.

Multiplying of the incidence coefficients of a vertex with respect to different dimension elements in n - cross - polytope and sum the product you can get the common express for the number of incident vertices to elements of different dimension in a n - cross – polytope:

$$f_0(2^0 + 2^1 C_{n-1}^1 + 2^2 C_{n-1}^2 + \ldots + 2^{n-1} C_{n-1}^{n-1}) = f_0(n) \sum_{i=0}^{n-1} 2^i C_{n-1}^i = f_0(n) 3^{n-1}.$$

In this way you can get the common express for the number of incident edges to elements of different dimension more one in n - cross - polytope

$$f_1(2^0 + 2^1 C_{n-2}^1 + 2^2 C_{n-2}^2 + ... + 2^{n-2} C_{n-2}^{n-2}) = f_1(n) \sum_{i=0}^{n-2} 2^i C_{n-2}^i = f_1(n) 3^{n-2},$$

and go on.

In result, one can get the common express for the sum of all incidents of elements of a lower dimension with respect to all elements of a higher dimension in a n - cross – polytope:

$$f_0(n)3^{n-1} + f_1(n)3^{n-2} + ... + f_{n-1}3^0 = \sum_{i=0}^{n-1} f_i(n)3^{n-1-i}, f_i(n) = C_n^{n-1-i} 2^{1+i}. \tag{11.7}$$

Let $k_{d_h d_j^i}$ be the number of elements of dimension h in a n - cross - polytope belonging to some one element of dimension j ($h<j$), that is d_j^i . Obviously, this number is equal to $f_h(j)_s$ for simplex, and $i=1 \div f_j(n)_{cr}$ for n - cross - polytope. So, elements of a cross - polytope are simplexes the product $f_h(j)_s f_j(n)_{cr}$ corresponds to the number of elements of dimension h belonging to all elements of dimension j for a simplex. This product is equal to $f_d(n) = C_{n+1}^{d+1}$

$$C_{j+1}^{h+1} 2^{1+j} C_n^{n-1-j} = 2^{1+j} \frac{(j+1)!}{(h+1)!(j-h)!} \frac{n!}{(n-1-j)!(1+j)!}.$$

Let us compare this number with the number of elements of dimension j, which have elements of dimension h in the n - cross – polytope:

$$k_{d_h^i d_j} f_h(n)_{cr} = C_n^{n-1-h} 2^{1+h} C_{n-h-1}^{j-h} 2^{j-h} = 2^{1+j} \frac{n!}{(n-1-h)!(1+h)!} \frac{(n-h-1)!}{(n-j-1)!(j-h)!}.$$

Obviously, these numbers are equal to each other. This proves that the number of elements of dimension h in a n - cross - polytope belonging to all dimensions of j ($j>h$) in a n - cross - polytope is equal to the number of elements of dimension j, which have elements of dimension h as elements. Thus, the total number of incidents of elements of a smaller dimension with respect to elements of a higher dimension is equal to the total number of incidents of elements of a higher dimension with respect to elements of a smaller dimension. The total number define expression (11.7). Q.E.D.

The polytope of hereditary information is a cross-polytope of dimension 13. Substituting the value $n = 13$ and the values of $f_i(n)$ calculated in the previous section in expression (11.7), we find the total value of the incident flow in the polytope of hereditary information

$$\sum_{i=0}^{12} f_i(13)3^{12-i} = 1.78 \cdot 10^9 .$$

A significant value of the total incidence stream in the polytope of hereditary information indicates an intensive exchange of information between elements of the polytope of hereditary information. For example, this value is 400 times larger than the incident flux in a simplex of dimension 13 (Zhizhin, 2019 b). This may explain for the recently discovered epigenetic principle of the transmission of hereditary information without changing the sequence of genes in DNA and RNA molecules.

HIDDEN NUCLEIC ACID BOND ORDER

It was established that a polytope of hereditary information exists between two nucleic acids in each region of their bond with nitrogen bases (Figure 2). The dimension of this polytope is 13, its type is a cross-polytope. This polytope has 12 flat coordinate planes, i. e. exactly as many different pairs of connected nitrogenous bases (11.1). The polytope of hereditary information is the result of combining two sugar molecules located in two nucleic acids in the antiparallel position. In this segment, examining the geometry of the polytopic of hereditary information, it was found that out of 12 coordinate planes of 13 - cross - polytope pairs of flat nitrogenous bases can be located in only five coordinate planes (Zhizhin, 2020, 2021 c).

Investigation of the Possible Location of Canonical Pairs of Nitrogenous Bases in the Polytope of Hereditary Information

A simplified image of two sugar molecules with antiparallel ribs is shown in Figure 1. The blue color in Figure 1 indicates 12 coordinate planes passing through the vertices of the polytope, including through the vertices between which flat nitrogenous bases could be located in the coordinate planes. From Figure 1 it follows, however, that not all 12 coordinate planes can have flat nitrogenous bases. They could be located only between the vertices of the polytope belonging to different sugar molecules closest to each other, i.e. between the planes $OH - H_{(3)}$ - $H_{(1)}$ - F_i and $OH^{/} - H_{(3)}{}^{/} - H_{(1)}{}^{/} - F_i^{/}$. Between the distant peaks located on the opposite planes of sugar molecules, the location of flat nitrogenous bases is difficult, due to the obstacles encountered in the form of sugar molecule atoms and covalent chemical bonds. Therefore, out of the 12 possible coordinate planes, only five coordinate planes designated by the vertices of the polytope:

$$\begin{matrix} F_i^{/}H_{(1)} & F_i^{/}C_{(2)} & F_i^{/}H_{(3)} & F_i^{/}C_{(1)} & F_i^{/}OH \\ H_{(1)}^{/}F\text{'} & C_{(2)}^{/}F_i\text{'} & H_{(3)}^{/}F_i\text{'} & C_{(1)}^{/}F_i\text{'} & (OH)^{/}F_{ii} \end{matrix}. \tag{11.8}$$

If among 12 pairs (11.1) canonical (Watson & Crick, 1953 a, b) base pairs for linked DNA molecules are distinguished, then there are only 4 of these pairs:

$$A:U, G:C, C:G, U:A. \tag{11.9}$$

If RNA molecules are linked, then in (11.9), base U must be replaced with base T.

Considering that there are 4 canonical base pairs, and there are 5 planes in which each pair can be located, the total number of base pairs location options is 20, i. e. exactly as many different amino acids exist. Thus, each variant of the arrangement of base pairs corresponds to a certain amino acid connected in the ribosome to transport RNA. As is known (Spirin, 2019), the attachment of an amino acid to transport RNA carries out a specific enzyme that recognizes the amino acid and its corresponding transport RNA when they are combined. In this case, transport RNA is associated with messenger RNA, and between these RNAs at the junction point there again exists a polytopic of hereditary information. Thus, information on the state of messenger RNA is transmitted by transport RNA, on which the given amino acid joins. As we see in this process, there is no need to use the concept of a genetic code to describe it. The latter was introduced as an attempt to explain the difference in the number of canonical base pairs from the number of amino acids. However, this concept together with the concept of the codon does not explain anything. Taking into account the dimension of the polytope of genetic information allows us to determine the internal (hidden) order in the location of the bases. The situation is similar to the situation with the discovery of quasicrystals. X- ray diffraction patterns of some intermetallic compounds revealed a lack of translational symmetry in them (Shechtman, et al., 1984). This caused a stormy reaction of the scientific community. Such intermetallic compounds were called quasicrystals. However, in works of author (Zhizhin, 2014, 2018) it was shown that translational symmetry immediately appears if we consider diffractograms as projections onto a two-dimensional plane of a structure of higher dimension (a hidden order of diffractograms is revealed).

Currently, there are no results of special experiments to determine a unique correspondence between the location of this pair of bases in the polytopic of hereditary information and a specific amino acid. However, one can qualitatively confirm the existence of five different arrangements of base pairs using experimental data on the correspondence of codons (nucleotide triplets) to amino acids (Spirin, 2019). If we assume that the average nucleotide in the codon plays the main influence in this correspondence, then five amino acids correspond to triplets with an average nucleotide *U*: *Phe*, *Leu*, *Ile*, *Met*, *Val*. Since each amino acid corresponds to one of the positions of nucleotide *U*, it turns out that these directions exist 5. A similar situation is repeated with triplets with an average nucleotide *C*. They correspond to 5 amino acids: *Ser*, *Pro*, *Thr*, *Ala*, *Cus*. For triplets with an average nucleotide *G* there correspond the amino acids: *Ser*, *Cys*, *Trp*, *Arg*, *Gly*. Triplets with an average nucleotide A correspond to 7 amino acids: *Tyr*, *His*, *Glu*, *Asn*, *Lys*, *Gln*, *Asp*. Some violations of the strict unambiguous equality of 5 amino acid numbers to each nucleotide can be explained by the influence of lateral nucleotides in triplets and noncanonical parasitic bases.

It should be noted that taking into account the highest dimension of the polytopic of hereditary information is also necessary in the analysis of pandemics in which viral RNA penetrates the cell and forces the ribosomes to produce proteins necessary for viruses to reproduce.

METHYLATED POLYTOPE OF HEREDITARY INFORMATION

Recently, the question of counteracting viral infections, their spread and development has become acute. Various stages of the development viruses from capsids to return to capsids are studied (Agol, et al., 1990; Novikova, et al., 2002; Novikova, 2007; Spirin, 1990; Wurm, et al., 2001; Wulan, et al., 2015; Lundberg, et al., 2013; Crotty, et al., 2002). Various drugs are being developed to inhibit various stages of viral population development (Gonzalez, et al., 2008; Dong, et al., 2020; De Clerey, 2004; Crotty et al., 2002; Caly et al., 2012; Razonable, 2011). However, we are forced to state that viruses have amazing properties of attacking a living organism, penetrating it, suppressing the protective properties of the organism, using the life-supporting functions of the organism to our advantage, reproducing inside the organism, often leading ultimately to the death of the organism. The drugs developed against viruses can often only temporarily extinguish the infection, and viruses are modified and, in a new form, attack living organisms again. What allows viruses to often be the winner in the fight against living organisms? It should be remembered here that viruses are equal ancient players in the development of life. It is assumed that they preceded the emergence of universal LUCA (Last Universal Common Ancestor) protocells, since structural and functional similarities were found between archaeal viruses and some DNA - containing eukaryotic viruses (Novikova, 2007). This leads to the idea that it was viruses that contributed to the creation of life on Earth.

During their existence (about 5 billion years), viruses have developed complex systems of control, information transfer, recognition of dangers and ways to overcome them. Viruses participated in the creation of DNA and are themselves carriers of genetic information. Communication of information is always an important issue in a complex self-regulating system.

Following the model of Watson and Crick (Watson and Crick, 1953 a, b), the transfer of genetic information in a double - stranded DNA molecule occurs as a result of the complementary nitrogenous bases of the two helices of the molecule. At the same time, the genetic information is incorporated not only in the sequence of arrangement along the spirals of various nitrogenous bases (nucleotide), but also in the geometric characteristics of complementary compounds: distances between atoms in compounds (centers of glycosidic bonds, locations of glycosidic centers, angles between glycosidic bonds) (Spirin, 2019).

It was recently discovered (Zhizhin, 2019 a, b) that the vicinity of nitrogenous bases at the place of their connection in DNA is a polytope of dimension 13 (polytope of hereditary information), on the coordinate two - dimensional planes of which nitrogenous bases can be located. In this case, the geometric characteristic of the complementary compound of nitrogenous bases is the geometric characteristic of the structure of the hereditary information polytope. The main parameter here is incidence (Zhizhin, 2019 c), i.e. belonging of one element of some dimension of the polytope to other elements of another dimension. The structure of the polytope will include a listing of all such incidents. The law of conservation of incidents was established (Zhizhin, 2019 d), according to which the integral sum of incidents from elements of low dimension to elements of high dimension is equal to the integral sum of incidents from elements of high dimension to elements of low dimension. An analytical expression for calculating the integral sum of incidents over the polytope is obtained. The integral sum of incidents over the polytope is interpreted as the flow of information between the elements of the polytope. It is shown that an increase in the dimension of the polytope leads to a sharp increase in the value of the information flow between the elements of the polytope.

A distinctive feature of the DNA of a number of viruses is the presence of methylated nitrogenous bases in them, i. e. a methyl group is attached to nitrogenous bases. For example, in bacterial DNA methylated bases are $5'$ - methylcytosine, $6'$ - methylaminopurine (Novikova, 2007; Agol, et al., 1990). The origin of such bases is the enzymatic methylation of the already synthesized DNA strand. This process is carried out by virus - specific methylases, which use the methyl groups of S - adenosylmethionine of the host cell as a donor (Novikova, 2007).

In this section, the effect of methylation of nitrogenous bases on an increase in the intensity of transmission of hereditary information is theoretically investigated. This is due to an increase in the dimension of the hereditary information polytope during its methylation. It was shown that the flow of information during methylation of the hereditary information polytope increases 2000 times as compared with the flow of information in the absence of methylation. This may explain the high degree of organization of viruses when they attack living organisms.

It should be noted that the experimentally intensifying effect of methylation on the transmission of hereditary information is currently observed in plants (Hawkes, et al, 2016; Lindquist, et al., 2016; Mancuso, 2017).

NUCLIEC ACIDS METHYLATION

The concept of nucleic acids as chains containing only four types of nucleotides is somewhat simplified. In DNA, as follows from various sources, there is a certain number of methylated nitrogen bases, i. e. bases containing a methyl group – CH_3. For example, the Fisher formula of cytosine $C_4H_5N_3O$ has the form (**Figure** 3).

Figure 3. Fisher's formula for cytosine

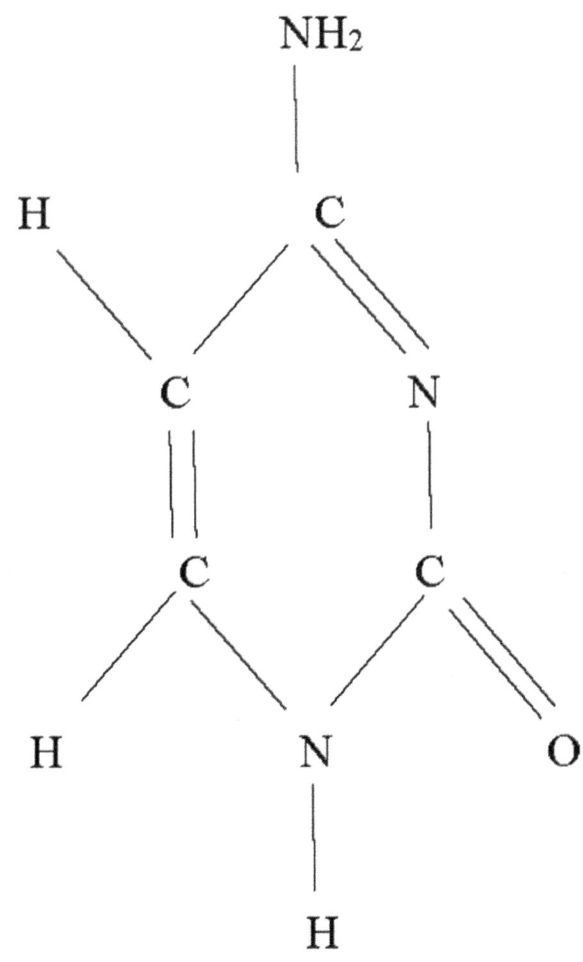

Fisher's formula for methylated cytosine (5 - methylcytosine) is shown in **Figure** 4.

Figure 4. Fisher Formula for Methylated Cytosine

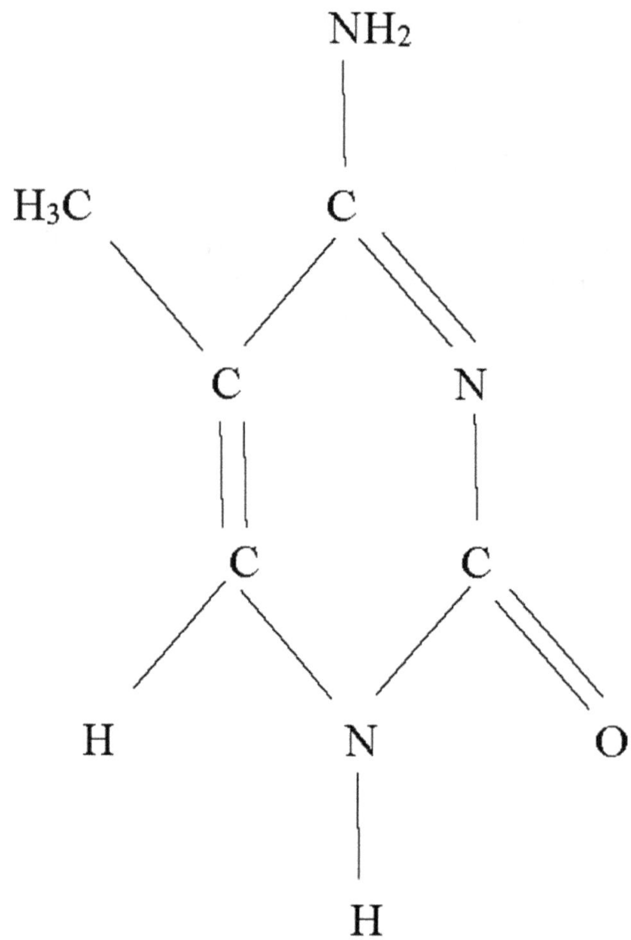

The nitrogenous base adenine (aminopurin) $C_5H_5N_5$ has the Fisher formula (**Figure** 5)

*Figure 5. Fisher Adenine Formula Methylated adenine (6 - methyladenine) has the Fisher formula (**Figure** 6)*

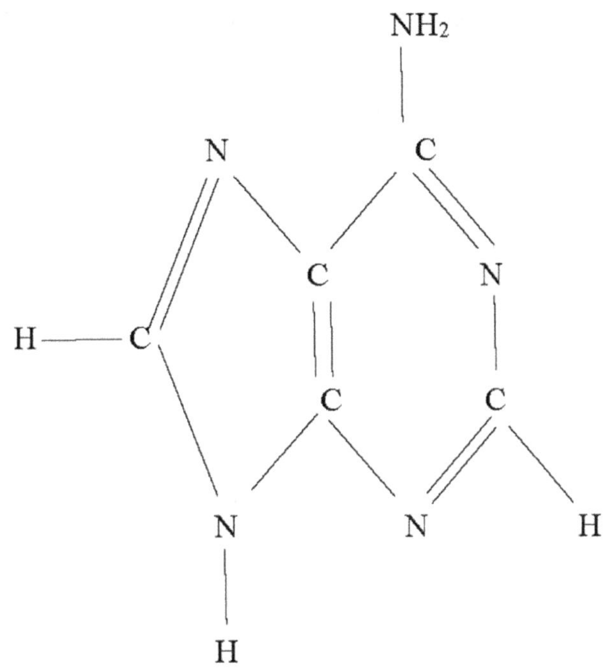

Figure 6. Fisher Formula for Methylated Adenine

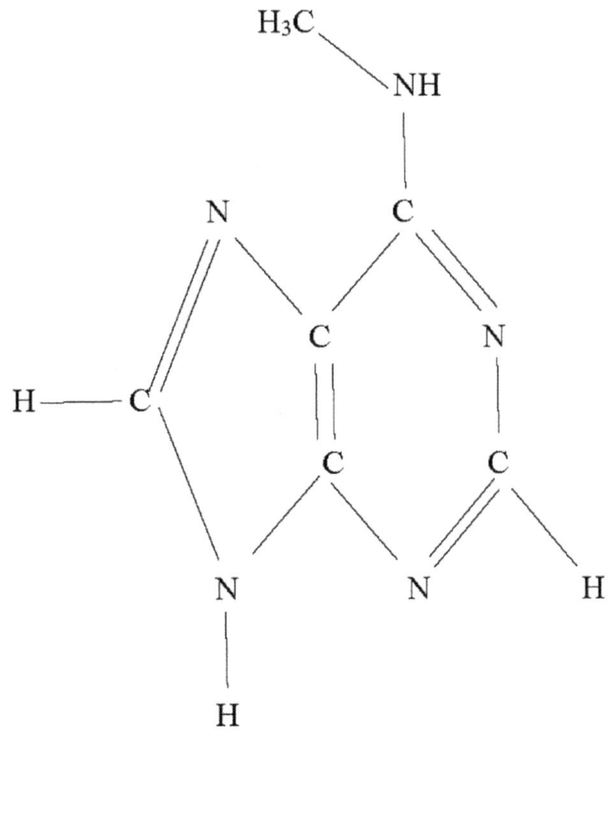

Methylation occurs after polynucleotide synthesis. It is used by organisms in its fight against viruses. However, viruses also use methylation to counteract the defenses of organisms. For example, viruses in nucleic acids contain oxymethylcytosine instead of cytosine (**Figure** 7)

Figure 7. Fisher's Formula of Oxymethylcytosine

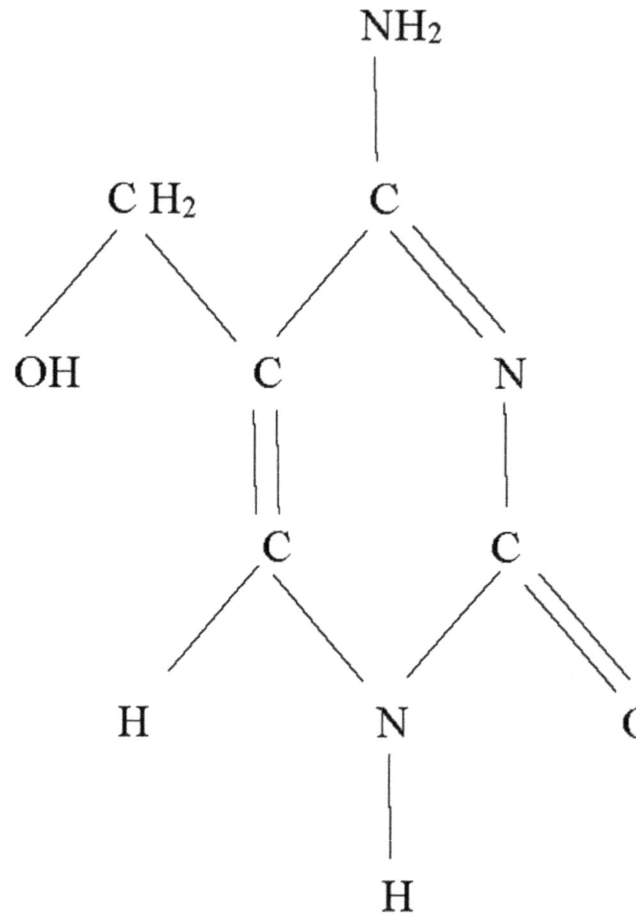

Here, a hydroxyl group is included in the methyl groups. Emerging hydroxyl groups may carry one or two glucose residues.

One of the bacteriophages that infect *Bacillus subtitis* contains 5 - hydroxymethyluracil instead of uracil, and 5 - dioxipentyluracil instead of thymine (**Figure** 8)

Figure 8. Fisher's Formula 5 - Dioxipentyllуracil

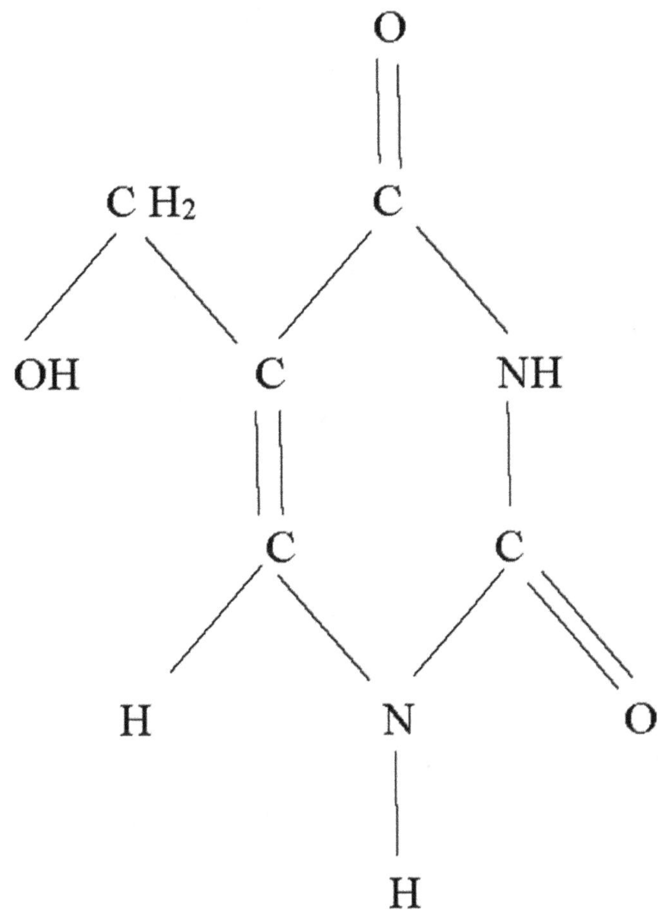

Instead of guanine (**Figure** 9) as a result of methylation, O - methylguanine is formed (**Figure** 10)

Figure 9. Fisher`s Formula Guanine

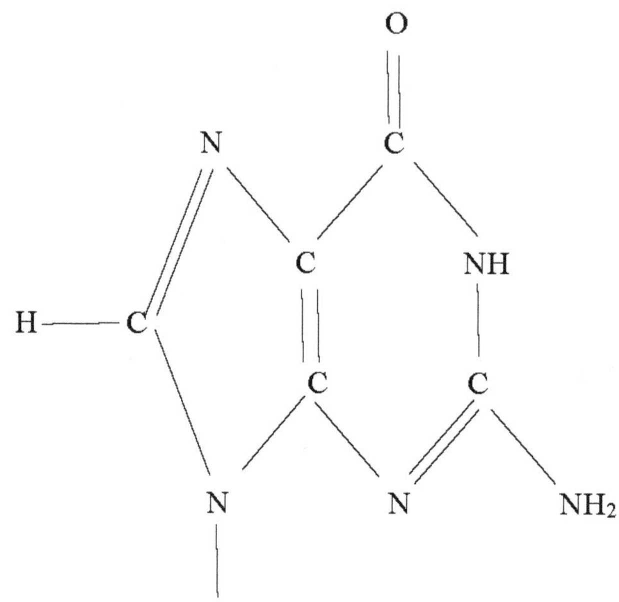

Figure 10. Fisher's Formula O^6 - Methylguanine.

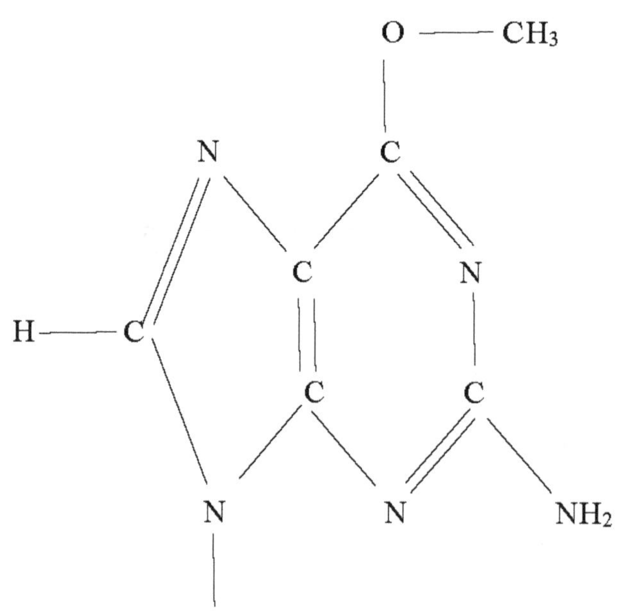

Thus, all classic nitrogenous bases as a result of methylation can take a different form. Moreover, it was established that these changes are very significant for the opposing cells of organisms and viruses.

In eukaryotic DNA, about 5% of the total cytosine content is 5′ - methylcytosine. The degree of DNA methylation changes during development, and also depends on the age of the body and hormonal influences. DNA in vertebrates is especially strongly methylated. Methylation can accompany DNA replication; it is inherited and transmitted to offspring. Modified DNA can affect the local structure of the chromosome. Methylation promotes the transition to DNA Z - conformation, controls the activity of genes (Spirin, 1990). However, the mechanism of action of methylation has not yet been disclosed.

Thus, all nucleotides of the classical type can, under appropriate conditions, be replaced by nature itself with modified nucleotides with methylated nitrogen bases. In this case, the complementarity of the base compound is preserved, and the symmetry of nucleic acids is preserved. What is the secret to the need for such modifications of nitrogen bases? Try to answer this question. It was proved that many biomolecules have a higher dimension (Zhizhin, 2016 a, b; 2018 a, 2019 e). In particular, it was proved that the connection of a carbon atom with four other atoms or functional groups has a dimension of 4. A methyl group attached to nitrogen bases is a special case of such a compound. Consequently, methylation changes the dimension of nitrogenous bases. Without a methyl group, the dimension of nitrogen bases is 2, with the addition of a methyl group, the dimension of nitrogen bases becomes 4 or more. Find out what this leads to.

When nucleic acids are labeled, their nitrogenous bases cease to be flat, since a tetrahedron with a center is formed in the vicinity of one of the carbon atoms of these bases, i. e. four - dimensional polytope. In this case, due to the antiparallel nature of molecular structures formed by antiparallel nucleic acid helices, it should be assumed that such a four-dimensional polytope appears in both linked nitrogen bases (Figure 11).

Figure 11. The formation of four - dimensional polytopes in nitrogen bases during nucleation of nucleic acids.

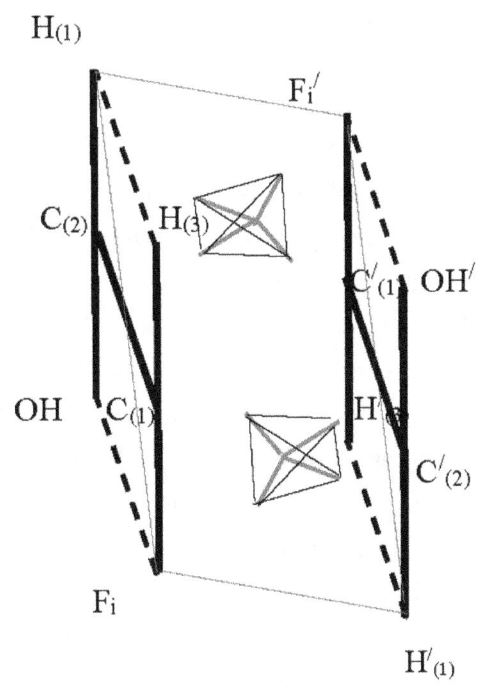

In Figure 11, the coordinate plane $\frac{F_i^/ H_{(1)}}{H_{(1)}^/ F_i}$ is selected as an example. In Figure 11, the edges of the tetrahedra with the center, marked in red, correspond to chemical covalent bonds. Thin black ribs are needed to create a closed convex body. The general construction of the polytope of hereditary information synchronizes the type of cross - polytope, but the number of vertices in it is 10 more than that of the non -methylated polytope of hereditary information. The complementarity of the nitrogenous base compound under these conditions is preserved, the symmetry of nucleic acids is preserved. This is facilitated by the mechanisms laid down by nature in nucleic acids that support complementarity and break the chains of nucleic acids when such conditions are violated (Spirin, 1990). According to (Zhizhin, 2019 c), there is a relation between f_i the number of dimension elements i in the cross - polytope and the dimension d of the cross - polytope itself (11.1).

Using formula (11.1) from the number of vertices at which the dimension $i = 0$, we can calculate the dimension of the methylated polytope of hereditary information

$$f_0 = 2C_d^{d-1} = 2d = (26+10) = 36, d = f_0 / 2 = 18. \qquad (11.10)$$

Thus, the methylated polytope of the hereditary information has dimension 18.

Consequently, the dimension of the methylated polytope of hereditary information increased by 5 units compared with the dimension of the unmethylated polytope of hereditary information (13). By the ratio (11.10), it is possible to determine the number of elements of different dimensions and in the methylated polytope of hereditary information.

The number of edges is $f_1(18) = 2^2 C_{18}^{16} = 612$,
the number of triangles is $f_2(18) = 2^3 C_{18}^{15} = 6528$,
the number of tetrahedrons is $f_3(18) = 2^4 C_{18}^{14} = 48960$,
the number of fourth - dimensional faces (simplexes) is $f_4(18) = 2^5 C_{18}^{13} = 274176$,
the number of fives - dimensional faces (simplexes) is $f_5(18) = 2^6 C_{18}^{12} = 1188096$,
the number of six - dimensional faces (simplexes) is $f_6(18) = 2^7 C_{18}^{11} = 4073472$,
the number of seven - dimensional faces (simplexes) is $f_7(18) = 2^8 C_{18}^{10} = 11202048$,
the number of eight – dimensional faces (simplexes) is $f_8(18) = 2^9 C_{18}^{9} = 24893440$,
the number of nine - dimensional faces (simplexes) is $f_9(18) = 2^{10} C_{18}^{10} = 44808192$,
the number of ten - dimensional faces (simplexes) is $f_{10}(18) = 2^{11} C_{18}^{11} = 65175552$,
the number of eleven - dimensional faces (simplexes) is $f_{11}(18) = 2^{12} C_{18}^{12} = 76038144$,
the number of twelve - dimensional faces (simplexes) is $f_{12}(18) = 2^{13} C_{18}^{13} = 70189056$,
the number of thirteenth - dimensional faces (simplexes) is $f_{13}(18) = 2^{14} C_{18}^{14} = 50135040$,
the number of fourteenth - dimensional faces (simplexes) is $f_{14}(18) = 2^{15} C_{18}^{15} = 26738688$,
the number of fifteenth - dimensional faces (simplexes) is $f_{15}(18) = 2^{16} C_{18}^{16} = 10027008$,
the number of sixteenth - dimensional faces (simplexes) is $f_{16}(18) = 2^{17} C_{18}^{17} = 2359296$,
the number of eighteenth - dimensional faces (simplexes) is $f_{17}(18) = 2^{18} C_{18}^{18} = 262144$.

The obtained numbers determine the structure of the methylated polytope of hereditary information (Zhizhin, 2021 b). To prove the existence of a methylated polytope of hereditary information, we use the Euler-Poincaré equation (3.3) (Poincaré, 1895)
Substitute the obtained values f_i in the Euler - Poincaré equation (3.3)

36 – 612 + 6528 – 48960 + 274176 – 1188096 + 4073472 – 11202048 + 24893440 – 44808192 + 65175552 – 76038144 + 70189056 – 50135040 + 26738688 – 10027008 + 2359296 – 262144 = 0.

Equality to zero of the familiar variable proves that the methylated polytope of hereditary information exists, its dimension is even and equal to 18.

The Law of Conservation of Incidents in the Methylated Polytope of Hereditary Information

The monograph (Zhizhin, 2019 c) introduced the concept of the incidence coefficients of elements of lower dimension with respect to elements of the higher dimension and elements of higher dimension with respect to elements of the lower dimension. The first characterizes the number of elements of a certain higher dimension to which the given element of a lower dimension belongs. The second characterizes the number of elements of a given lower dimension that are included in a particular element of a higher dimension. Here we must remember that the vertices of geometric elements of various dimensions are atoms, molecules or functional groups. Therefore, the incidence of geometric elements to a friend means contacted between particles of the matter, including living matter. The contact between particles of matter can be interpreted as the transfer of information on material structures, including biological structures. We introduce the notation: $k_{e_j^n e_u}$ is the number of elements of dimension *u*, which include an element of dimension *j* ($u > j$) with number *n*. Thus, $k_{e_j^n e_u}$ is the incidence factor of element *n* with dimension *j* relative to elements with dimension *u*. We introduce the notation also: $k_{e_j e_u^n}$ is the number of elements of dimension *j*, which included in element *n* with a dimension *u* ($u > j$). Thus, $k_{e_j e_u^n}$ is the incidence factor of element *n* with dimension *u* relatively to elements with dimension *j*.

In work (Zhizhin, 2019 c) was proved state Theorem 11.1. According to this theorem in any cross - polytope of dimension *d*, the sum of all incidents of elements of a lower dimension with respect to all elements of a higher dimension is equal to the sum of all incidents of elements of a higher dimension with respect to all elements of a lower dimension and equals the sum of the series:

$$f_0(d)3^{d-1} + f_1(d)3^{d-2} + f_2(d)3^{d-3} + ... + f_{d-1}(d)3^0 = \sum_{i=0}^{d-1} f_i(d)3^{d-1-i},$$
$$f_i(d) = C_d^{d-1-i} 2^{1+i}, i = 0 \div (d-1). \quad (11.11)$$

Substituting in the formula (11.11) the values of $f_i(d)$ for $d = 18$ we find that the sum of the incidents (information flow) for the methylated polytope of hereditary information takes the value

$$\sum_{i=0}^{d-1} f_i(d)3^{d-1-i} = 3.8143 \cdot 10^{12}. \tag{11.12}$$

Recall that the value of the information flow for the unmethylated polytope of hereditary information (see Zhizhin, 2019 b) was 1.78•10^9. Thus, the methylation of nucleic acids led to an increase in the flow of information between elements of the polytope of hereditary information by more than 2000 times.

CONCLUSION

The currently known forms of the helical structure of DNA molecules *A*, *B*, *Z*, and DNA molecules in the form of a closed ring were experimentally found in the works of Watson, J.D., Crick, F.H.C., Ha, S.C., and other scientists.

The representations of the sugar molecule and the residue of phosphoric acid in the form of polytopes of higher dimension are used. Based on these ideas and their simplified three - dimensional images, a three-dimensional image of nucleic acids is constructed. The geometry of the neighborhood of the compound of two nucleic acid helices with nitrogen bases has been investigated in detail. It is proved that this neighborhood is a polytope with anti-parallel edges of dimension 13 (13 – cross - polytope). This polytope is called of the polytope of hereditary information. The geometry of the polytope of hereditary information is investigated. It is shown that in the flat coordinate planes of the polytope of hereditary information there are located flat complementary hydrogen compounds of the nitrogenous bases of two nucleic acid helices. It turned out that the number of these coordinate planes (12) is exactly as many as there are various options for hydrogen compounds of nitrogenous bases. An image of the polytope of hereditary information with a specific indication of its coordinate planes is constructed. It is shown that not all 12 coordinate planes can have flat nitrogenous bases. Only five coordinate planes can have nitrogenous bases. Considering that there are 4 canonical base pairs, and there are 5 planes in which each pair can be located, the total number of base pairs location options is 20. This leads to a coordination of the number of nucleotides and the number of amino acids without the need for a genetic code.

Impact investigated of nucleic acid methylation, i. e. the connection of methyl groups - CH_3 to the nitrogenous bases of nucleic acids, on the geometric characteristics of nucleic acids. Such methylation is characteristic of viral nucleic acids when infected with viruses of organisms. It was shown that methylation of nitrogenous bases leads to an increase in their dimension, creating a higher dimension of the bases. This, in accordance with the recently established law for the conservation of incidents (Zhizhin, 2019 d), sharply increases the intensity of information transfer between elements of different dimension of nucleic acids. At the same time, the intensity of transmission of hereditary information in the polytopic of hereditary information (a local geometric region of the highest dimension of nucleic acid interaction) increases significantly. It has been established that the intensity of information transfer in the methylated polytope of hereditary information increases by more than 2000 times in comparison with the unmethylated polytope of hereditary information. It should be assumed that it is this property that activates viral nucleic acids when they infect organisms, weakening organisms, including due to the use of methyl groups of organism cells by viruses.

REFERENCES

Albert, B. (2002). *Molecular Biology of the Cell* (4th ed.). Garland Science.

Berg, J., Tymoczko, J., & Stryer, L. (2002). *Biochemistry*. W.H. Freeman & Co Ltd.

Brachet, J. (1941-1942). La detection histochimique et la microdosage des acides pentosenucleiques. *Enzymologia*, *10*, 87–96.

Butler, J. M. (2005). *Forensic DNA Typing*. Gardners Books.

Caly, L., Wagstaff, K. M., & Jans, D. A. (2012). Nuclear trafficking of proteins from RNA viruses: Potenial targetfor anti-viral? *Antiviral Research*, *95*(3), 202–206. doi:10.1016/j.antiviral.2012.06.008 PMID:22750233

Cantor, C. R., & Schimmel, P. R. (1980). *Biophysical Chemistry. Part I. The Conformation of Biological Macromolecules*. W.H. Freeman & Co Ltd.

Crotty, S., Cameron, C., & Audino, R. (2002). Rubaviriu's antiviral mechanism of action: Lethal mutagenesis? *Journal of Molecular Medicine*, *80*(2), 86–95. doi:10.100700109-001-0308-0 PMID:11907645

Dahm, R. (2005). Friedrich Miescher and the discovery of DNA. *Developmental Biology*, *278*(2), 274–288. doi:10.1016/j.ydbio.2004.11.028 PMID:15680349

De Clerey, E. (2004). Antiviral drugs in current clinical use. *Journal of Clinical Virology*, *30*, 2. PMID:15125867

Dong, L., Hu, S., & Gao, J. (2020). Discovering drugs to treat coronavirus disease 2019(COVID -19). *Drug Discoveries & Therapeutics*, *14*(1), 58–60. doi:10.5582/ddt.2020.01012 PMID:32147628

Gonzalez Canga, A., Sahagún Prieto, A. M., Diez Liébana, M. J., Fernández Martínez, N., Sierra Vega, M., & García Vieitez, J. J. (2008). The pharmacokinetics and interactions of ivermectin in humans-a mini-review. *The AAPS Journal*, *10*(1), 42–46. doi:10.120812248-007-9000-9 PMID:18446504

Ha, S. C., Lowenhaupt, K., Rich, A., Kim, Y.-G., & Kim, K. K. (2005). Cristal structure of a junction between B – DNA and Z – DNA revels two extruded bases. *Nature*, *437*(7062), 1183–1186. doi:10.1038/nature04088 PMID:16237447

Hawkes, E. J., Hennelly, S. P., Novikova, I. V., Irwin, J. A., Dean, C., & Sanbonmatsu, K. Y. (2016). COOLAIR antisense RNAs from evolutionarily conserved elaborate secondary structures. *Cell Reports*, *16*(12), 3087–3096. doi:10.1016/j.celrep.2016.08.045 PMID:27653675

Hershey, A., & Chase, M. (1952). Independent functions of viral protein and nucleic acid in growth of bacteriophage. *The Journal of General Physiology*, *36*(1), 39–56. doi:10.1085/jgp.36.1.39 PMID:12981234

Lindquist. (2016). Luminidependens (LD) is an Arabidopsis protein with prion behavior. (2016). *Proceedings of the National Academy of Sciences of the United States of America*, *113*(21), 6065–6070. doi:10.1073/pnas.1604478113 PMID:27114519

Lundberg, L., Pinkham, C., Baer, A., Amaya, M., Narayanan, A., Wagstaff, K. M., Jans, D. A., & Kehn-Hall, K. (2013). Nuclear import and export inhibitors alter capsid protein distribution in mammalian cells and reduce Venezuelan Equine Encephalitis Virus replication. *Antiviral Research*, *100*(3), 662–672. doi:10.1016/j.antiviral.2013.10.004 PMID:24161512

Mancuso, S. (2017). *Plant Revolution. Le piante hanno già inventato il nostro futuro*. GUIUTI EDITORE.

Novikova, N.A. (2002). *Virology*. Lower Novgorod State University named N.I. Lobachevsky.

Novikova, N. A. (2007). *Storage and sale of virus genetic information*. Lower Novgorod State University named N.I. Lobachevsky.

Poincare, A. (1895). Analysis situs. *J de e'Ecole Polytechnique.*, *1*, 1–121.

Ponnuswamy, P., & Gromiha, M. (1994). On the conformational stability of oligonucleotide duplexes and tRNA molecules. *Journal of Theoretical Biology*, *169*(4), 419–432. doi:10.1006/jtbi.1994.1163 PMID:7526075

Rasonable, R. R. (2011). Antiviral Drugs for Viruses Otber Than Humanimmunodeficiency virus. *Mayo Clinic Proceedings*, *86*(10), 1009–1026. doi:10.4065/mcp.2011.0309 PMID:21964179

Riemann, G. F. B. (1854). On the Hypotheses Which Lie at the Foundations of Geometry. Gottingen Observatory.

Shechtman, D., Blech, I., Gratias, D., & Cahn, J. W. (1984). Metallic phase with longer range orientational order and no translational symmetry. *Physical Review Letters*, *53*(20), 1951–1953. doi:10.1103/PhysRevLett.53.1951

Spirin, A. S. (Ed.). (1990). *Molecular biology. The structure and biosynthesis of nucleic acids*. High School.

Spirin, A. S., & Gavrilova, L. P. (1971). *The Ribosome*. Science.

Watson, J. D., & Crick, F. H. C. (1953a). Molecular structure of nucleic acids. *Nature*, *171*(4356), 738–740. doi:10.1038/171737a0 PMID:13054693

Watson, J. D., & Crick, F. H. C. (1953b). Genetical implications of the structure of deoxyribose nucleic acid. *Nature*, *171*(4361), 964–967. doi:10.1038/171964b0 PMID:13063483

Wulan, W. N., Heydet, D., Walker, E. J., Gahan, M. E., & Ghildyal, R. (2015). Nucleocytoplasmic transport of nucleocapsid proteins of enveloped RNA viruses. *Frontiers in Microbiology*, *6*, 553. doi:10.3389/fmicb.2015.00553 PMID:26082769

Wurm, T., Chen, H., Hodgson, T., Britton, P., Brooks, G., & Hiscox, J. A. (2001). Localization to the nucleolus is a common feature of coronavirus nucleoproteins, and the protein may disrupt host cell division. *Journal of Virology*, *75*(19), 9345–9356. doi:10.1128/JVI.75.19.9345-9356.2001 PMID:11533198

Zhizhin, G. V. (2015a). The dimensions of supramolecular compounds. *Biosphere*, *7*(2), 149–154.

Zhizhin, G. V. (2015b). Dimensions of compounds in supramolecular chemistry. *Int. J. Chem. Modeling*, *7*(2), 215–238.

Zhizhin, G. V. (2016a). From multidimensional phase spaces to multidimensionality of space in Nature. *Biosphere*, *8*(3), 258–267.

Zhizhin, G. V. (2016b). The structures, topological and functional dimension of biomolecules. *J Chemoinformatics and Chemical Engineering*, *5*(2), 44–58. doi:10.4018/IJCCE.2016070104

Zhizhin, G. V. (2018a). *Chemical Compound Structures and the Higher Dimension of Molecules: emerging Research and Opportunities*. IGI Global. doi:10.4018/978-1-5225-4108-0

Zhizhin, G. V. (2018b). The Metric Structure of nucleic Acids and the higher Dimension of Their Constituents. *International Journal of Chemoinformatics and Chemical Engineering*, *7*(2), 1–15. doi:10.4018/IJCCE.2018070101

Zhizhin, G. V. (2019a). The Polytope of Hereditary Information the Structure, Location, Signification. *Biochemistry and Modern Applications*, *2*(1), 56–62. doi:10.33805/2638-7735.124

Zhizhin, G. V. (2019b). The Structure of the Polytope of Hereditary Information. *International Journal of Chemoinformatics and Chemical Engineering*, *8*(2), 7–22. doi:10.4018/IJCCE.2019070102

Zhizhin, G. V. (2019c). *The Geometry of Higher-Dimensional Polytopes*. IGI Global. doi:10.4018/978-1-5225-6968-8

Zhizhin, G. V. (2019d). The Law of Conservation of Incidents of the Space of Nanoworld. *International Journal of Chemoinformatics and Chemical Engineering*, *8*(1), 25–46. doi:10.4018/IJCCE.2019010103

Zhizhin, G. V. (2019e). *Attractors and Higher Dimensions in Population and Molecular Biology: Emerging Research and Opportunities*. IGI Global. doi:10.4018/978-1-5225-9651-6

Zhizhin, G. V. (2020). Hidden Nucleic Acid Bond Order? *Acta Scientific Biotechnology*, *1*(4), 34–36.

Zhizhin, G. V. (2021a). The Structure of DNA Taking into Account the Higher Dimension of Its Components. In *Encyclopedia of Information Science and Technology* (5th ed., pp. 730–747). Ed. Mehdi Khosrow- Pour. doi:10.4018/978-1-7998-3479-3.ch050

Zhizhin, G. V. (2021b). Methylated Polytope of Hereditary Information. *Int J of Applied Research on Public Health Management*, *6*(2), 49–60. doi:10.4018/IJARPHM.2021070104

Zhizhin, G. V. (2021c). Interaction of Nucleic Acids: Hidden Order of Interaction. *Int J of Applied Research in Bioinformatics*, *11*(2), 1–8. doi:10.4018/IJARB.2021070101

Zhizhin, G. V. (2021d). Spatial Models of Sugars and Their Components. *Int J of Applied Research in Bioinformatics*, *11*(2), 9–22. doi:10.4018/IJARB.2021070102

Zhizhin, G. V., & Diudea, M. V. (2016). Space of nanoworld. In M. V. Putz & C. M. Marius (Eds.), *Sustainable nanosystems, development, properties, and applications* (pp. 221–245). IGI Global.

Zhizhin, G. V., Khalaj, Z., & Diudea, M. V. (2016). Geometrical and topology dimensions of the diamond. In A. R. Ashrafi & M. V. Diudea (Eds.), *Distance, symmetry and topology in carbon nanomaterials* (pp. 167–188). Springer. doi:10.1007/978-3-319-31584-3_12

KEY TERMS AND DEFINITIONS

Dimension of the Space: The member of independent parameters needed to describe the change in position of an object in the space.

Incidence Coefficients of Elements of Lower Dimension With Respect to Elements of Higher Dimension: The number of elements of a certain higher dimension to which the given element of a lower dimension belongs.

N–Cross-Polytope: The convex polytope of dimension n in which opposite related of centrum edges not have connection of edge.

Polytope: The polyhedron in the space of the higher dimension.

Afterword

The study showed that in various natural science disciplines, when describing chemical, physical, and biological systems, it is necessary to use polytopes of the highest dimension. A distinctive feature of the classes of polytopes of the highest dimension is the different degree of their deviation from the correct metric forms. The most common polytopes in chemical, physical, and biological systems are n-simplex and n-cross-polytopes. As the complexity of chemical compounds increases in the transition from fairly simple compounds of inorganic chemistry to clusters of chemical compounds and biomolecules, the structure of polytopes becomes more complex, and their dimensions increase. It can be argued with a sufficient degree of confidence that the highest dimension of polytopes corresponding to chemical compounds is an essential fundamental property of matter, from the moment of the emergence of life on Earth to the present time. The highest dimension reflects the hidden order of Nature from quasicrystals to the interaction of nucleic acids that determine the processes of heredity. Ignoring the highest dimension of chemical compounds means refusal of attempts to understand the patterns of complex processes in nature.

About the Author

Gennadiy Vladimirovich Zhizhin was evacuated from Leningrad during the blockade. He has been head of the Russian Federation Branch of the European Society of Mathematical Chemistry, an academician of the Russian Academy of Natural Science, and a corresponding member of the Russian Ecological Academy. He has also been a senior fellow at the Instiute of Silicate Chemistry, Russian Academy of Sciences and a professor at North-Western State Technical University.

Index

5-Cross-Prism 183
5-Simplex-Prism 183
6-Complex-Polytopic Prismahedrons 183
6-Complex-Tetrahedral Prismahedrons 183
6-Simplex-Polytopic Prismahedron 183

C

convex regularly faceted polyhedrons 20-21, 63, 185, 220, 253
Cube-Polytopic Prismahedron 183

D

Dimension of the Space 71, 90, 122, 145, 362
Divided Electron Pair 292
Dodecahedral Prism 154, 183
Dodeca-Polytopic Prismahedron 183
Duality in Polytopes 253

F

Functional (Topological) Dimension of a Molecule 334

G

geometrical image of a chemical compound 334
gonohedron 2-3, 19, 22-45, 47, 52-53, 59-60, 63, 217, 220-221, 247, 249, 251, 253
graph theory 3, 18-19

H

Hierarchical and Translation Filling Spaces 183
Homo-Element Metal Chains 292
Hybridization of Electronic Orbitals 334

I

Icosahedral Prism 154, 183
Icosa-Polytopic Prismahedron 183
Incidence Coefficients of Elements of Higher Dimension With Respect to Elements of Lower Dimension 90, 122
Incidence Coefficients of Elements of Lower Dimension With Respect to Elements of Higher Dimension 90, 122, 145, 362
Incidence in polytopes 253

L

Linear Metal Chains 272, 278, 292

M

Mechanical Operations on a Polyhedron 19, 63, 253
metal chains 272, 275, 278-279, 292

N

N*3-Angular Prismahedron 183, 220, 253
Nanocluster 122
N–Cross-Polytope 90, 122, 145, 334, 362
N–Cube 90, 122, 145
N–Simplex 122, 145, 292, 334

O

Octahedral Prism 153, 183
octahedral prismahedron 165, 183
Octahedral Prismahedrons 183
Operation on maps 19

P

P-Elements 292
poly-incident polytopes 240, 245, 253
polytope 1, 3, 16-18, 59, 64-70, 73-85, 87-88, 90-93, 95-96, 100-128, 130-137, 139-143, 145-148, 150-161, 163-178, 180-181, 183, 185, 187-191,

193-195, 197-200, 203-206, 208-211, 213-218, 220-222, 224-230, 232-243, 245, 251-254, 257-259, 262, 264-265, 267-268, 274-275, 277-279, 281-283, 285-286, 292, 295, 302-304, 313-314, 323-324, 331, 334-335, 337-348, 355-358, 361-362
polytopic prismahedron 166-170, 176-178, 180, 183, 220, 253, 257
Polytypichedron (Max) 220, 253
polytypichedrons 184-185, 218, 220, 245, 251, 253
polytypictops 181, 221, 245, 251-253

R

regular polyhedron 3-5, 15, 19, 52, 59, 63, 65-66, 111, 220, 247, 249-250, 253

S

semiregular polyhedrons 1, 3-5, 10, 12, 14-15, 17-20, 47-48, 51, 57, 59, 61-63, 121, 219-220, 252-253

T

Tetrahedral Coordination of Electron Pairs 292, 334
tetrahedral prism 151-152, 178-179, 183, 220, 253, 283, 285, 287-289
Tetrahedral Prismahedron 163-164, 183, 220, 253
Transitional Elements 289, 334
triangular prismahedron 160, 183, 220, 232, 234, 253

U

Undivided Electron Pair 292

www.ingramcontent.com/pod-product-compliance
Lightning Source LLC
LaVergne TN
LVHW062121300926
849161LV00010B/2515
9781799883746